developmental nutrition

▼

NORMAN KRETCHMER, M.D., PH.D.

PROFESSOR EMERITUS, UNIVERSITY OF CALIFORNIA

MICHAEL ZIMMERMANN, M.S., M.D.

SWISS FEDERAL INSTITUTE OF TECHNOLOGY

ALLYN AND BACON

BOSTON • LONDON • TORONTO • SYDNEY • TOKYO • SINGAPORE

Senior Series Editor: Suzy Spivey
Vice President, Editor and Chief, Social Sciences: Sean W. Wakely
Marketing Manager: Quinn Perkson
Production Administrator: Christopher Rawlings
Editorial-Production Services: Thomas E. Dorsaneo
Text Designer and Text Art: Seventeenth Street Studios
Composition: Seventeenth Street Studios
Cover Administrator: Suzanne Harbison
Composition and Prepress Buyer: Linda Cox
Manufacturing Buyer: Megan Cochran

Library of Congress Cataloging-in-Publication Data

Kretchmer, Norman, 1923-1995
    Developmental nutrition / Norman Kretchmer, Michael Zimmermann.
       p.   cm.
    Includes bibliographical references and index.
    ISBN 0-13-303744-4
    1. Nutrition. 2. Developmental biology. I. Zimmermann, Michael, M.D.
II. Title
QP141.K688 1996   96-14114
612.3'9—dc20   CIP

Printed in the United States of America
10 9 8 7 65 4 3 2 1 01 00 99 98 97

IN MEMORIAM

Norman Kretchmer, M.D., Ph.D. (1923–1995)

Biochemist and physician, pioneer in developmental biology and nutrition, spirited supporter of nutrition education and research and warmhearted friend.

CONTENTS

3

4

LACTATION AND HUMAN MILK   170

5

6

INFANCY AND NUTRITION   305

7

CHILDHOOD AND NUTRITION   385

8

ADOLESCENCE AND NUTRITION 457

9

ADULTHOOD AND NUTRITION   523

Among the most exciting events in human biology are the dramatic changes that take place during growth, development and aging. Although nutrition plays a fundamental and dynamic role in this process, nutritional science has traditionally focused on the adult and neglected the very young and very old. However, increasing awareness of the importance of nutrition during early life, together with growing appreciation for the nutritional health of our aging population, has led to a shift in emphasis. It is now recognized that although all ages share certain fundamental nutritional needs, dietary requirements—both qualitative and quantitative— during different stages of the lifecycle vary enormously. Not only are we beginning to identify what constitutes a healthy diet at different stages of the lifecycle, but also to understand how nutrients act on the cellular, tissue, and whole organism level to influence human growth, development and aging. *Developmental Nutrition* addresses this dynamic new focus in nutritional science.

The field of developmental nutrition draws on a broad range of disciplines; it requires an understanding of human physiology, developmental biology, behavioral and food science, and public health. This book is written by two physicians who specialize in nutrition and teach nutrition to undergraduates and graduates in the Unites States and Europe. Their individual backgrounds are complementary: Dr. Norman Kretchmer specializes in nutrition for pregnancy and pediatrics, Dr. Michael Zimmermann in the field of nutrition and preventive health for adults and older adults. This unique perspective allows them to provide in-depth insight into how diet and nutrition affect growth, development, health maintenance and disease prevention through the entire lifespan.

FEATURES

Developmental Nutrition presents the basic principles of developmental nutrition in clear, understandable terms, and covers both basic science and practical applications. It uses a clearly-organized approach that has been classroom-tested at the University of California at Berkeley, Stanford University and the Swiss Federal Institute of Technology in Zürich. Each chapter introduces the essential concepts of growth and development for that life stage, discusses the unique dietary needs of the stage, and considers the impact of optimal, marginal and deficient nutrient intake on health at that age and later in life. Moreover, for each stage of the life cycle, *Developmental Nutrition* examines how psychosocial environment and cultural influences interact to influence dietary patterns and practices.

Nutrition is a dynamic science, characterized by vigorous debate and new discovery. This book encourages the student to examine and actively consider controversial and unresolved issues in the field. Its goal is to provide an integrated understanding of underlying principles of developmental nutrition that will help the student make informed decisions as the field evolves. Because nutritional science is expanding so rapidly, the authors have made every effort to keep the book up-to-date. To do this, the most recent developments and references have been added, even during production of the text.

Developmental Nutrition repeatedly reminds the student of the continuity of the life cycle: the underlying concept that, beginning in utero, the health of each succeeding stage in the cycle is built on the nutritional foundation established in earlier stages, and that dietary choices early in life can have repercussions many years later. For example, the importance of adequate calcium intake in achieving peak bone mass is introduced early in the book, and is reiterated in the adult and older adult chapters in the discussion of mineral loss and osteoporosis. Similarly, the common thread of fat and cholesterol intake and vascular disease is touched on in the chapters on infancy, childhood and adulthood.

This book emphasizes human diversity. Integrated throughout the book are discussions of gender differences in nutritional needs, including the unique nutritional needs of women through the life cycle. Ethnic differences in health concerns and dietary practices are highlighted. Students are presented with a 'global' view of topics in developmental nutrition, with discussions of nutritional concerns in both the industrial and the economically developing regions of the world throughout the book.

Developmental Nutrition is written for students who have had introductory courses in biology and nutrition, but provides clear explanations of many basic concepts in nutrition and development to make the subject approachable to students from many fields. The text is useful to students and health-care practitioners from a broad range of backgrounds, including nutrition, developmental biology, dietetics, nursing, medicine, and public health.

ORGANIZATION

The first chapter provides an in-depth review of the basic concepts of developmental biology and human nutrition. It discusses the role of nutrition in growth and development and provides an overview of the major developmental stages of the human lifecycle. It reviews basic nutritional terms and concepts, including a discussion of the nutrients, energy metabolism, nutritional assessment, dietary guidelines, and the RDAs.

Chapters 2 and 3 focus on planning for pregnancy and pregnancy. Chapter 2 begins with discussions of the influence of nutrition on fertility and the importance of nutrition before conception. The unique nutritional environment of pregnancy, with separate discussions of the mother, the fetus and the placenta are presented. Chapter 3 examines the profound impact of maternal nutrition on growth in utero, and the dietary and environmental factors that can influence pregnancy outcome. It emphasizes the central role of nutrition in prenatal care, and the impact of the WIC program on maternal and infant health in the U.S.

Chapters 4 and 5 discuss lactation and breastfeeding. Chapter 4 examines the physiology of lactation and the unique composition of human milk. Maternal nutritional needs and the impact of dietary quality on breast milk composition are presented. The process of breastfeeding and its influence on maternal and child health worldwide are discussed in Chapter 5. Chapter 6 focuses on nutrition in infancy, the development of the digestive tract, infant formulas, and the relative merits of breast vs. bottle feeding. Chapters 7 and 8 examine childhood and adolescence, including nutritional needs for growth, eating patterns and behavior, and the importance of the psychosocial environment of the teen years. An in-depth discussion of the eating disorders are presented in Chapter 8.

Chapter 9 looks at the impact of the modern 'affluent ' diet on the health of our adult population, and our rapidly expanding knowledge of the myriad connections between nutrition and chronic disease. Finally, Chapter 10 discusses the influence of nutrition on the physiological changes of aging (and vice versa) and examines the links between diet and the common age-associated degenerative diseases. It also looks at the sociocultural context of aging and its affect on diet patterns and nutritional programs for older adults.

LEARNING AIDS

The chapters open with a short introduction to 'set the stage', engage the student, and provide a perspective for the information to follow.

The chapter headings and subheadings are carefully chosen and set in special type to enhance the organization of the material.

Special emphasis was placed on including a large number of figures and illustrations, many of which were developed specifically for this text. This generous art program will enhance understanding of key concepts throughout the book.

Bold-face and italics are used throughout the book to identify important terms and concepts. These terms are either defined in context or provided as marginal definitions to aid in understanding.

Brief chapter summaries highlight and review the key material presented in each chapter.

References. The text is extensively and carefully referenced, citing essential and up-to-date research, reviews and books available in the field.

ACKNOWLEDGEMENTS

First, we would like to thank Allyn & Bacon and the many people there who made this book possible, including Suzy Spivey and Christopher Rawlings. We are also grateful to Tom Dorsaneo and his staff at Publishing Consultant & Production Services for superb editorial assistance.

A special thanks to Raleigh Wilson, for his constant support and enthusiasm in developing and guiding this book to completion, and without whom this project would not have been possible.

We would also like to thank our many students, fellows, and colleagues who have taught us so much about teaching and writing. In addition, we would like to acknowledge and thank the following reviewers for their constructive suggestions and comments throughout the development of this manuscript:

Leta P. Aljadir	University of Delaware
Laura E. Nagy	University of Guelph
Ronald E. Puhl	Bloomsburg Unviersity
Joanne Slavin	The University of Minnestoa
Beth Stewart	University of Arizona
Adrienne White	University of Maine
Carol T. Windham	Utah State University
John Worobey	Rutgers University

Finally, and most of all, we would like to thank our families for their support, enthusiasm, patience and encouragement.

Michael Zimmermann

Norman Kretchmer

NORMAN KRETCHMER, M.D., PH.D., received his BS degree in animal physiology from Cornell University and his MS and PhD degrees in physiological chemistry from the University of Minnesota. He obtained his MD at the State University of New York in Brooklyn. During his early academic years he studied the development of amino acid metabolism during fetal and perinatal life. From 1959 to 1974, he was professor and chair of the Department of Pediatrics at Stanford University Medical School where he developed the interdepartmental program in Human Biology. He then served for seven years as the Director of the National Institute of Child Health and Human Development at the National Institutes of Health. Since 1981, Dr. Kretchmer was on the faculty of the University of California as professor in the Department of Nutritional Sciences at Berkeley and professor in the Departments of Pediatrics and Obstetrics at the School of Medicine in San Francisco. He also served as the director of the Koret Center for Human Nutrition at San Francisco General Hospital, where he focused on the nutritional problems of immigrant children and on intrauterine growth. He was a member of the Institute of Medicine of the US National Academy of Sciences and past president of the American Pediatric Society. He published more than 200 papers and edited 16 books, and was Editor-in Chief of the American Journal of Clinical Nutrition from 1991-1995.

MICHAEL ZIMMERMANN, M.S., M.D., received his BA degree in Biological Sciences with Honors from the State University of New York at Binghamton and his MD from Vanderbilt University School of Medicine in 1988. He obtained his MS in Nutritional Sciences at The University of California at Berkeley. He was a National Institutes of Health Research Fellow in the Department of Nutritional Sciences at the University of California at Berkeley and a Clinical Fellow in Diabetes and Metabolism in the Department of Pediatrics at the University of California at San Francisco. He is a member of the Alpha Omega Alpha Medical Honor Society and the Swiss Diabetes Society. He has published a number of papers in the fields of nutrition and diabetes. He is currently a Senior Research Scientist in the Laboratory for Human Nutrition and directs the postgraduate course in Human Nutrition at the Swiss Federal Institute of Technology in Zürich, Switzerland.

nutrition and human development

▼

NUTRITION DEFINED

Nutrition is a broad interdisciplinary subject—a discipline that draws from anthropology, biochemistry, molecular biology, dietetics, geography, and other areas. Studies in nutrition, consequently, encompass basic and applied science as well as social problems with global impact.

Because of this broadness and complexity, human nutrition is not easily defined. Essentially *nutrition* is the study of the interactions between the human organism and its food. *Opportunities in the Nutrition and Food Sciences*, a book developed by the Food and Nutrition Board of the Institute of Medicine (NAS 1994), emphasizes the diversity of the field:

> More than forty years ago, the first issue of *The American Journal of Clinical Nutrition* defined nutrition as "the corner stone of preventive

medicine, the hand maiden of curative medicine and the responsibility of every physician." Almost two decades later, Alfred Harper suggested that nutrition be called an "integrating science" or perhaps an "applied science," since it is concerned with solving practical problems. Harper added, "There is no such thing as a nutritionist; there are nutritionists…We are chemists, biochemists, physiologists, pathologists, microbiologists, physicians, dentists, economists, dietitians, sociologists, animal scientists, food technologists, toxicologists, and many others." Admittedly, he added, "this makes it much more difficult to decide where we are going because we are in different places going in different directions."

Figure 1.1 illustrates the scope of nutrition and food sciences and divides the field into four groups:

1 Disciplines that look within cells or *in vitro* environments

2. Disciplines that look at specific organs—either human or animal

3. Disciplines that look at entire populations

4. Disciplines that look at food supply

Source: Adapted from Institute of Medicine, Committee on Opportunities in the Nutrition and Food Sciences, Opportunities in the Nutrition and Food Sciences, (Washington, D.C., National Academy of Sciences, 1994).

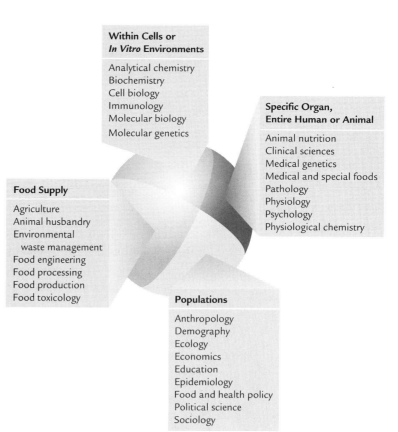

FIGURE 1.1

The scope and coverage of nutrition and food sciences.

THE ROLE OF NUTRITION IN GROWTH AND DEVELOPMENT

Among the more exciting events in human biology are the dramatic changes that take place during human growth and development. Nutrition plays a fundamental and dynamic role in this process, as many developmental changes are controlled by the adequacy of nutrient intake. Although individuals of all ages share certain fundamental nutritional needs, requirements during different stages of the life cycle vary greatly.

Traditionally, much of applied biology and medicine has focused on the adult (generally the adult male) and neglected the very young and very old; but recently there has been a shift in emphasis. Awareness of the importance of nutritional health during early life has increased and become a major public health concern. The health of our rapidly aging population has become a matter of governmental policy worldwide. Clearly, there is a need to broaden our scope within the discipline of nutrition.

Although often used interchangeably, it is important to distinguish between the terms *growth* and *development*. Growth refers to a measurable increase in size. Height and weight are the standard measures of growth in humans, but growth can also be measured in metabolic studies, in which body retention of substances—such as calcium and nitrogen—are determined. Development describes an increase in complexity of function and is characterized by **differentiation** and **maturation.**

CELLULAR GROWTH AND DIFFERENTIATION

differentiation: the act or process of acquiring characteristics or functions different from the original; implies increasing specialization of function

maturation: the stage or process of attaining maximal biological, intellectual, and emotional development

germinal cell: a cell in a primitive stage of development from which other cells are grown

In order to understand development as applied to the entire organism, it is important to appreciate the underlying changes that characterize cell growth and differentiation.

Tissue growth can result from (See Figure 1.2):

1. cell division (mitosis) and an increase in cell numbers, or *hyperplastic* growth;

2. increasing cell size, or *hypertrophic* growth;

3. a combination of both hyperplastic and hypertrophic growth.

Most tissues grow and develop using all of these mechanisms. In human tissues, cells exhibit three general patterns of growth (Goss 1986):

1. In regenerating tissues, cells are continually dying and being replaced by new cells. For example, mature skin cells are highly differentiated but incapable of dividing; they have a determined lifespan. Dead cells are steadily shed from the skin surface and new cells, produced by mitosis of **germinal cells** deep in the skin, take

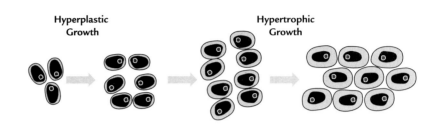

Hyperplastic
Growth

Hypertrophic
Growth

FIGURE 1.2
Phases of cell growth

bone marrow: the soft
tissue that fills the
cavities of bones

hypertrophy: increase in
volume of a tissue or
organ produced entirely
by enlargement of
existing cells

genotype: the funda-
mental hereditary
constitution or assort-
ment of genes of an
individual

**deoxyribonucleic acid
(DNA)**: a nucleic acid
containing deoxyribose
as the sugar component;
found in the nucleus
usually loosely bound to
protein; the autorepro-
ducing component of
chromosomes and the
repository of hereditary
characteristics

chromosome: a structure
in the cell nucleus con-
taining a linear thread of
DNA, which transmits
genetic information

**messenger ribonucleic
acid (mRNA)**:
informational or tem-
plate RNA that carries
the message for protein
synthesis from DNA to
the cytoplasmic areas

their place. A similar pattern exists for red blood cells, with the germinal cells residing in the **bone marrow**.

2. Nerve and muscle cells display a different pattern of development. They form early in life and are meant to last for a lifetime. These tissues are nonregenerating—their cells are highly specialized and can only divide early in life. In the brain, nerve cells stop dividing very early—the neurons are formed by about the 20th week in utero. Although the brain continues to increase in size, growth occurs through hypertrophy of formed cells and development of supporting cells (neurons cannot be replaced if lost). Similarly, the number of muscle fibers is fixed early in development. Weight lift-ing later in life can **hypertrophy** muscle cells but not increase the number of fibers.

3. A third pattern of cell growth is exhibited by the kidney and many secretory tissues throughout the body. These renewing tissues do not continually replace themselves. However, in special situations—in response to injury or increased demand—they can undergo mitosis and proliferate.

HEREDITY, ENVIRONMENT, AND NUTRITION

Human growth and development is a result of the continuous and com-plex interaction between heredity and environment. Each individual inherits a unique set of genes from his or her mother and father (their **genotype**), a genetic endowment consisting of about two meters of **deoxyribonucleic acid (DNA)** carefully divided up and packed into 46 **chromosomes.** Each gene is a linear portion of the DNA that codes for the structure of a specific protein, and most cell nuclei contain over 100,000 working genes.

DNA, through the formation of **messenger ribonucleic acid (mRNA),** directs the assembly of structural and functional proteins from amino acid precursors. Production of proteins enables cells to grow, divide, and specialize. For example, as cells destined to become red blood cells differ-entiate, they contain an abundance of messenger RNA that codes for the production of the protein hemoglobin. Other proteins function as hormones or growth factors, leaving the cells in which they are produced to interact with other developing cells. Growth and development of tissues and organs reflect division and differentiation at the cellular level.

Through the fundamental sequence DNA -> RNA -> protein, genes elaborate a master plan that codes for the development of the **phenotype** of the person (see Figure 1.3). Genes—often an array of genes acting together—contain information on hair color, stature, susceptibility to disease, and multiple other characteristics that determine phenotype. However, the final expression of phenotype is strongly influenced by the environment.

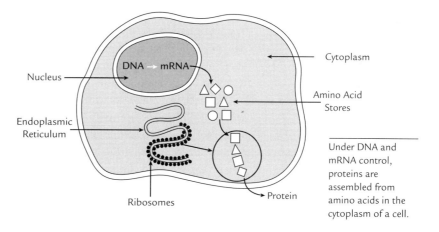

FIGURE 1.3

Assembly of proteins under DNA and mRNA control

A striking example of the interaction between nature and nurture—and the importance of nutrition in early development—is the differences in size of monozygotic and dizygotic twins at birth. Monozygotic (identical) twins carry identical genotypes and develop from the same fertilized ovum, whereas dizygotic twins are formed from separate ova and have distinct genotypes.

On genetic grounds alone, greater similarity in size would be expected in the monozygotic twins. However, on average, monozygotic twin pairs show greater differences in size at birth than dizygotic twins (Wilson 1976). Why? Differences are likely due to unequal distribution of nutrients and oxygen in utero–monozygotic twins share the same placenta and compete for blood flow and nutrient supply while dizygotic twins each have their own placenta.

Nutrition is a major environmental influence on phenotype, affecting everything from clearly visible characteristics, such as height and weight, to the more subtle attributes of blood pressure and concentration of blood cholesterol.

phenotype: the outward, visible expression of the hereditary constitution of an individual

Critical Periods in Development

Although the inherent drive to grow and develop is strong in early life, environmental insults such as undernutrition or disease can modify the growth process: whether complete recovery and resumption of normal growth occurs afterward depends on the nature, severity, and duration of the insult. See Figure 1.4. Timing is also important. In human development, generally the earlier the insult, the greater potential for serious harm.

In development, the term *critical period* is used to describe the time of greatest vulnerability to a specific action or insult. Critical periods are typically times of rapid cell division. If cell division is limited during a critical period of organ growth, the final cell number in the organ can be irreversibly reduced and the organ permanently damaged. The embryonic period of human development is a particularly critical time for adequate nutrition. An insufficient supply of nutrients in utero can produce severe and irreversible deficits in growth.

Recovery is usually possible if nutritional deprivation occurs at a less crucial time in development. Temporary undernutrition during childhood can slow growth markedly, but if normal food intake resumes, a period of accelerated growth occurs until former levels of height and weight are achieved. This phase of rapid growth occurring after a period of growth impairment is referred to as *catch-up growth*.

Vitamin A: A Paradigm of the Influence of Nutrition on Heredity

The interplay between nutrition and heredity in development occurs at many levels, from the molecular level to the organism level. The complexity of this interaction is illustrated by the role vitamin A plays in growth, development, repair, and maintenance throughout the lifespan.

On the molecular level, vitamin A can directly affect gene expression. In tissues throughout the body, vitamin A is taken up by cells, combines with specific binding proteins, and passes into cell nuclei. See Figure 1.5. In the nucleus, vitamin A interacts with DNA, regulating gene transcription (the production of mRNA from DNA) (Oong 1994).

By influencing gene expression, vitamin A promotes the differentiation of **epithelial cells** from simpler to more specialized forms. Distinct types of epithelial cells are found in specific locations throughout the body because they serve distinct functions. This epithelial 'specialization' is modulated by vitamin A (De Luca 1994), and insufficient intake of vitamin A results in a reversion of the differentiated cells to a less complex phenotype.

epithelial cells: the cells that cover the internal and external surfaces of the body, including the lining of blood vessels and other small cavities

A shows normal development. In B an adverse influence (such as nutritional deprivation), felt late, temporarily impairs development but is followed by full recovery. In C an adverse influence felt during the critical period impairs development and full recovery does not occur.

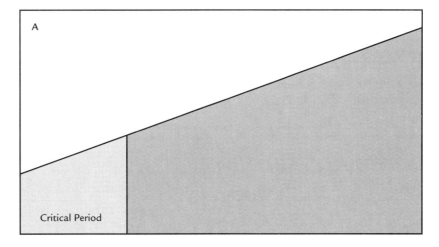

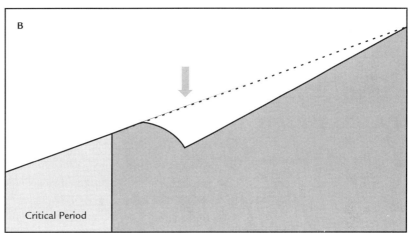

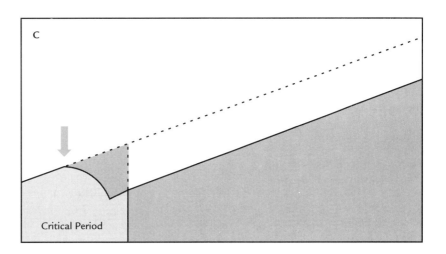

FIGURE 1.4
Critical periods in development

(1) vitamin A circulating in the blood bound to retinol binding protein;
(2) binding of retinol to specific cell receptors;
(3) entry of retinol into the cell and its conversion to retinol metabolites;
(4) binding to nuclear receptors; and
(5) interaction with DNA to regulate transcription.

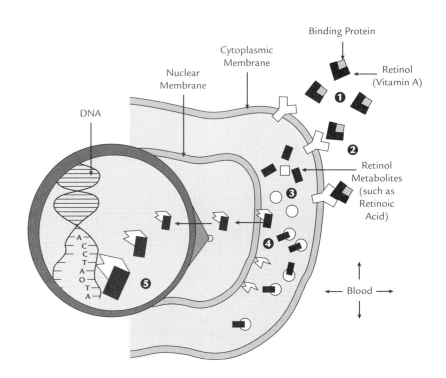

FIGURE 1.5

Vitamin A and gene expression

embryogenesis: the phase of prenatal development in which the characteristic organ systems and basic configuration of the embryonic body is established

morphogenesis: differentiation of cells and tissues in the early embryo that establishes the basic form and structure of various organs and parts of the body

metabolic pathway: sequence of chemical reactions leading from one compound to another

The effects of vitamin A on development at the whole organism level begin early in **embryogenesis,** when vitamin A plays a fundamental **morphogenetic** role in the organization of the embryo (Maden 1994). Vitamin A is one of the signals that direct specific portions of the embryo to develop into limbs, torso, and nervous system.

Later in childhood, vitamin A continues to influence growth and development profoundly, often in less direct ways. Inadequate dietary vitamin A impairs functioning of the immune system and sharply increases the risk of serious infections, such as measles and diarrheal diseases (Sommer 1994). Consequently, in impoverished regions of the world where vitamin A intake is poor, many children suffer from chronic or recurrent illness and grow poorly—entering adulthood without having achieved their genetic potential for size.

Finally, in the later stages of life, vitamin A is a vital component of cellular mechanisms of maintenance and repair. Adequate vitamin A status may reduce the risk of certain cancers, including those of the mouth and breast (Willett 1994).

Nutrition and Inborn Errors of Metabolism

Inborn errors of metabolism are inherited diseases caused by impairment of normal **metabolic pathways.** These disorders have gained a new prominence over the past 30 years because of improvements in diagnosis and a better understanding of the metabolic pathways involved. Nutrition is increasingly important in the management of many of the inborn errors of metabolism.

A number of these disorders result from mutations in specific genes. The mutation leads to the deletion or impairment of the protein coded by the affected gene, which is usually an **enzyme.** Loss of a critical enzyme cripples a metabolic pathway and leads to the production of a toxic by-product or the deficiency of an important metabolite. Depending on the metabolic pathway(s) affected, the gene defect produces a characteristic disorder or disease.

Many of these diseases have been shown to be amenable to **nutritional** therapy. Therapy can involve supplying the missing metabolite, stimulating the pathway with a vitamin, or adjusting the diet to restrict production of the toxic product. Table 1.1 describes several prominent inborn errors of metabolism.

An example of an inborn error of metabolism is phenylketonuria (PKU)—discussed in detail in Chapter 5. In PKU, a genetic defect impairs production of phenylalanine hydroxylase, the enzyme that converts the amino acid phenylalanine to tyrosine. Phenylalanine and toxic metabolites (phenylketones) build up and irreversibly damage the developing nervous system. However, if the diagnosis is made in the newborn, institution of a diet low in phenylalanine reduces production of the toxic phenylketones and prevents damage (Matalon 1991).

Many diseases of later life have roots in heredity. Diseases such as diabetes, heart disease, and cancer have a strong genetic component. However, the genetic contribution is more complex (perhaps involving many different genes), making it difficult to identify a responsible metabolic step or pathway. Although we have not yet identified the exact genetic causes, risk of these diseases can be reduced and symptoms ameliorated with appropriate diet.

enzyme: a protein that acts as a catalyst, increasing the rate at which a reaction occurs

OVERVIEW OF HUMAN DEVELOPMENT

LIFE CYCLE VERSUS LIFESPAN

Normal human development is distinguished by an orderly series of progressive and irreversible changes. Although there is variation in the rate of human development, all individuals begin as a fertilized egg, develop through intrauterine life to birth, and move through childhood to reach sexual maturity at puberty. The changes from conception through puberty constitute the human *life cycle,* for at puberty the individual begins producing eggs or sperm that will form the next generation.

Disease Name	Defect or Deficiency	Result of Defect or Deficiency	Main Effects	Therapy
Galactosemia	Galactose-1-phosphate uridyl transferase	· Failure to convert galactose-1-phosphate to glucose-1-phosphate	· Failure to thrive · Jaundice, enlarged liver · Renal dysfunction · Cataracts · Hemolytic anemia · Seizures · Coma, death	· Eliminate galactose from the diet (feed soybean-based formulas containing sucrose or glucose, without lactose)
Phenylketonuria (PKU)	Phenylalanine hydroxylase	· Failure to convert phenylalanine to tyrosine	· Skin rash and light pigmentation · Delayed development · Neurological defects	· Restrict protein sources; supplement with a protein containing formula with low phenylalanine content
Von Gierke's disease (glycogen storage disease type 1)	Glucose-6 phosphatase	· Incomplete gluconeogenesis · Incomplete glycogenolysis · High concentrations of serum uric acid, cholesterol, triglycerides, and lactate	· Hypoglycemia · Enlarged liver · Growth failure (due to low insulin production) · Adiposity · Hemorrhagic tendency	· Provide glucose between meals and during the night
Hartnup disease	Disruption of amino acid absorption, transport, and excretion (tryptophan, alanine, asparagine, glutamine, histidine, isoleucine, leucine, phenylalanine, serine, theonine, tryptophan, tyrosine, and valine)	· Low amino acid concentrations · Excessive amino acid excretion · Minimal biosynthesis of nicotinic acid from tryptophan	· Muscular incoordination due to brain disease · Pellegra and associated mental disturbances	· Megadoses of nicotinic acid

TABLE 1.1
Selected inborn errors of metabolism

TABLE 1.1 *Continued*

Disease Name	Defect or Deficiency	Result of Defect or Deficiency	Main Effects	Therapy
Cobalamin-responsive methylmalonic acidemia	Methylmalonyl CoA mutase, Methyltetra-hydrofolatehomo-cysteine methyltransferase	· Failure to convert hydroxycobalamin to its active coenzymes	· Protein intolerance · Failure to thrive ketoacidosis · Neurological abnormalities · Death (in severe, untreated cases)	· Pharmacological doses of cobalamin
Homocystinuria	Cystathionine synthetase	· Disruption of sulfur amino acid (cysteine, cystine, methionine) metabolism; blocks transsulfuration of methionine to cysteine · High methionine concentrations and related compounds · High homocysteine concentrations and related compounds · Low cystathionine and cystine concentrations	· Dislocation of eye lenses · Mental retardation · Skeletal abnormalities	· Restrict methionine · High cystine diet · Pharmacological doses of pyridoxine (cystathionine synthetase requires pyridoxal phosphate as a coenzyme)
Propionic acidemia	Propionyl-CoA carboxylase (an enzyme required in branched-chain amino acid metabolism)	· High concentrations of ammonia · High concentrations of glycine · High concentrations of propionic acid · High concentrations of organic acids · Ketonuria	· Vomiting; lethargy · Ketoacidosis · Seizures · Osteoporosis · Developmental retardation	· Restrict total protein or restrict isoleucine, methionine, threonine, and valine · Megadoses of biotin (in conjunction with other therapy)

Although the life cycle is biologically complete at puberty, human development is functional as well as morphological: a more expansive view of human development includes the entire *lifespan*. During the later stages of life, although physical growth ceases and degenerative changes appear, individuals continue to develop and mature psychologically and socially. Optimal nutrition, vital for growth and development in early life, continues to be important in supporting maintenance and repair throughout the lifespan.

MAJOR DEVELOPMENTAL STAGES

The stages of human development used in this book are listed in the following table.

Developmental Stage	Approximate Age
Prenatal	
Embryonic	0 to 8 weeks
Fetal	8 weeks to birth
Infancy	
Newborn	First month after birth
Early infancy	1 month to 1 year
Later infancy	Second year
Childhood	
Early (preschool)	2 to 5 years
Late (school age)	5 to 10 years
Adolescence	
Females	10 to 18 years
Males	12 to 20 years
Adult	18 or 20 to 65 years
Older Adult	Greater than 65 years

These stages, along with the special conditions of pregnancy, lactation, and breast feeding in the adult female, provide a logical framework for a discussion of nutrition and development.

The Prenatal Period

Development begins at **fertilization,** and the human organism begins as a single cell—the fertilized ovum. During the first few days after conception, the rapidly dividing ovum is self-sufficient, living off food stored in its **yolk sac.** As the ovum divides, cell differentiation occurs. Some cells are destined to form the embryo while others eventually develop into the **placenta.**

About a week after fertilization the ovum has developed into a blastocyst, a hollow ball of dozens of rapidly dividing cells. The blastocyst buries itself in the **endometrium** of the uterus. As the placenta forms, the embryo

fertilization: in human reproduction, the process by which the male's sperm unites with the female's ovum; by this event, the genes of the male and female are combined to determine the sex and other biological traits of the new individual; also called conception

yolk sac: a membranous sac of nutritive material in the ovum for nutrition of the embryo

placenta: an organ characteristic of true mammals during pregnancy, joining the mother and fetus, and providing endocrine secretion and metabolic interchange of soluble bloodborne substances

endometrium: the epithelium that lines the inner wall of the uterus

becomes completely dependent on nutrition derived from the maternal bloodstream. Although growth is modest during the embryonic period, rapid differentiation establishes all of the major organ systems. The fetal period is characterized by further development and extraordinary growth.

The fetus is entirely formed from basic nutrients supplied by the mother. Simple substrates—glucose, amino acids, vitamins, and minerals derived from the maternal diet and maternal stores—are transferred in a continuous stream across the placenta to the fetus. In nine months the fetus grows from a single cell to a fully formed infant weighing 3 kilograms.

Infancy

At birth, the newborn infant must begin to process and assimilate food: digestion and absorption of complex substances must occur before nutrients are available for energy production and growth. Dependence on external sources of nutrition triggers the lifelong search for food—a driving force in the life of all organisms.

Digestion and absorption are fundamental biological processes essential in all stages of the life cycle, but the ability to digest certain substances varies with age (Lebenthal 1983). For example, starch is poorly digested by an infant because of the lack of pancreatic amylase, but infants can readily digest and absorb lactose (the main sugar in milk) because activity of lactase in the intestine is high. As the child grows, the situation reverses itself—the ability to digest starch increases while the digestion of lactose often decreases.

Childhood and Adolescence

growth spurt: the period of accelerated growth during adolescence

puberty: the period during which the secondary sexual characteristics begin to develop and the capability of sexual reproduction is attained

sexual maturity: the state of complete development of the sexual organs; attainment of the capability for reproduction

Compared to most other animals, human growth and development after birth proceeds relatively slowly—and is distinguished by a prolonged period of dependence during infancy and early childhood. During childhood, physical growth is slow but steady. Psychological and social development is brisk, and there is increasing coordination of functions as intellect and personality develop.

A distinguishing feature of human development—one shared only with other primates—is a period of accelerated growth in early adolescence. (Most other animals pursue a steady pattern of decelerating growth with increasing age.) Adolescence is characterized by the **growth spurt, puberty** and the attainment of **sexual maturity.**

Although some animals retain the potential for growth throughout life, humans achieve maximum stature after about two decades of growth (at about 21 years for males and 17 for females). Muscle mass usually continues to increase into the mid-20s, and skeletal mass into the early 30s.

Adult and Older Adulthood

Although physiological maturity is reached in the second or third decade, psychosocial development continues through adulthood. Nutrient needs for physical growth cease, but adequate nutrition continues to be important to support maintenance and repair. Optimal nutrition enables individuals to tolerate the psychological and physiological stresses of family, the workplace, and the many other responsibilities of adult life.

Senescence, the state of later maturity, is characterized by a gradual loss of cells from tissues and functional decline in tissues throughout the body. Aging is inherent and genetically determined, but the rate of the aging process is strongly influenced by the environment and diet. Many chronic degenerative diseases associated with advancing age can be prevented or favorably influenced by diet (National Research Council 1989).

NUTRITION BASICS

A *nutrient* is a substance obtained from the diet that is used for energy, growth, and maintenance of life. There is a continual flow of nutrients into and out of the human body; molecules entering from our diet remain only a short time before being excreted. Although our outward appearance remains the same from day to day, the constituents of our bodies are in a constant state of turnover. Substances turn over at varying rates: the average molecule of calcium generally stays in the body for several months, the average nitrogen for only a few weeks. To support this dynamic state, the body must be continually replenished with adequate energy and nutrients from the diet.

There are six major classes of nutrients, as shown in Table 1.2. Four of them— carbohydrates, protein, fat, and vitamins—are **organic** compounds. Organic compounds contain carbon, typically as **carbon-carbon** or **carbon-hydrogen bonds.** The other two classes of nutrients, water and minerals are inorganic compounds—they contain no carbon molecules.

Carbohydrates, fats, and proteins are termed energy-yielding nutrients because they can be metabolized to produce energy (discussed later in this chapter). They also serve important structural roles in the body.

The body can manufacture many nutrients, but others, termed *essential nutrients*, must be obtained preformed from the diet. Essential nutrients cannot be **synthesized** by the body or cannot be synthesized in adequate amounts to meet physiological needs. Nine of the amino acids, two fatty acids, water, and all the vitamins and minerals are essential nutrients.

organic: denoting chemical substances containing carbon

carbon-carbon and carbon-hydrogen bonds: covalent (shared electron) bonds joining neighboring carbon atoms or carbons and hydrogens

synthesis: creation of a compound by union of elements or simpler compounds

	Organic Compounds	
	Not essential	**Essential**
1. Carbohydrates	monosaccharides disaccharides complex carbohydrates	
2. Fats	fats and oils sterols phospholipids	linoleic acid linolenic acid
3. Proteins (amino acids)	alanine arginine asparagine aspartic acid cysteine glutamic acid glycine proline serine tyrosine	histidine isoleucine leucine lysine methionine phenylalanine tryptophan valine threonine

4. Vitamins		Fat-soluble
		  vitamin A
		  vitamin D
		  vitamin E
		  vitamin K
		Water-soluble
		  ascorbic acid
		  thiamin
		  riboflavin
		  niacin
		  biotin
		  pantothenic acid
		  folic acid
		  vitamin B6
		  vitamin B12

Inorganic Compounds (All Essential)

| 5. Minerals | Major minerals
  calcium
  magnesium
  phosphorus
  sulfur
  sodium
  chloride
  potassium | Trace minerals
  iron
  zinc
  iodine
  copper
  manganese
  fluoride
  chromium
  selenium
  molybdenum |
| 6. Water | | |

TABLE 1.2

The six major classes of nutrients

WATER

Water is the major constituent of the body, making up nearly 70% of the weight of a newborn baby and about 60% of an adult's body weight. Simple and abundant, it is essential to the life of the human organism. It is the milieu in which metabolic reactions take place, and it is the medium in which substances move between cells and throughout the body. The amount of water consumed each day dwarfs the intake of other nutrients: an adult needs about 2 liters, or 2000 grams, each day—compared to about 50 grams of protein and only 0.06 grams of vitamin C (National Research Council 1989).

ENERGY

The body requires a constant supply of energy for growth, for physical activity, to digest and utilize nutrients, and to maintain the **basal metabolic rate.** The body obtains energy by breaking the chemical bonds of carbohydrates, fats, and protein.

Energy Recovery

The metabolism of glucose illustrates the fundamentals of energy recovery by cells. When a molecule of glucose is formed, energy is required to synthesize the molecule, and this energy is stored as potential energy in the bonds that link the six carbon atoms.

In the cytoplasm of the cell, a molecule of glucose moves through a series of biochemical steps (termed *glycolysis*) and is broken into two three-carbon fragments. As the chemical bonds of glucose are broken, energy is transferred and stored as a high-energy bond in adenosine triphosphate (ATP). During glycolysis there is a net yield of two molecules of ATP.

The two three-carbon fragments (pyruvate) produced from glucose are then converted to acetate and enter the central furnace of energy metabolism, the *tricarboxylic acid (TCA) cycle.* Most of the energy-yielding nutrients consumed in food are ultimately metabolized in the TCA cycle.

Metabolizing carbohydrates, fats, and proteins for energy involves many separate biochemical pathways. However, these pathways all converge on the central hub of energy metabolism: the simple two-carbon fragment of acetate. Acetate is the fuel broken down in the TCA cycle to produce carbon dioxide and energy. The energy, like that obtained from glycolysis, ends up stored for future use as high-energy bonds in ATP. The complete metabolism of a molecule of glucose produces 38 molecules of ATP.

These two pathways—glycolysis and the TCA cycle—are the fundamental pathways of molecular energetics; a simplified scheme of these reactions is shown in Figure 1.6. The energy recovered in the high-energy

basal metabolic rate: the energy required by the body to maintain life-sustaining activities, such as breathing and body temperature; measured for a specific amount of time

bonds of ATP can be subsequently used by the body to support a variety of activities, such as synthesis of protein, maintenance of body temperature, and physical activity.

Energy released from the energy-yielding nutrients can be measured in calories. Calories are measures of heat energy—1000 calories (a *kilocalorie,* or kcal) is the amount of heat required to raise the temperature of 1 kilogram of water 1°C.

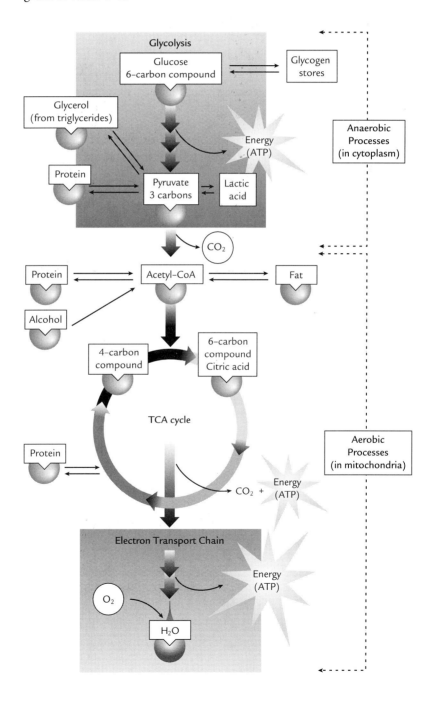

FIGURE 1.6
Energy recovery: Glycolysis and the TCA cycle

The energy content of a food depends on its quantity of fat, carbohydrate, and protein. When completely metabolized, 1 g of fat yields 9 kcal; 1 g of carbohydrate and protein yield 4 kcal. The relative importance of the energy-yielding nutrients at different stages of the life cycle varies: infants derive around 60% of their dietary energy as fat (Fomon and Bell 1993), while the average adult diet in the United States provides about 30 to 35% of calories as fat (National Research Council 1989).

ENERGY-YIELDING NUTRIENTS

Although protein, fats, and carbohydrates can be metabolized to provide energy, they also have important structural and functional roles in the body.

Carbohydrates

These substances, as their name implies, are composed of carbon and hydrogen. Carbohydrates supply the greatest percentage of calories and bulk in an average diet, yet they make up less than one percent of total body weight. The main function of dietary carbohydrate is to supply energy, and carbohydrate that is not metabolized for energy is stored as glycogen (a readily available form of carbohydrate stored mainly in muscle) or converted to fat. Figure 1.7 shows the structure of some simple and complex carbohydrates.

The **monosaccharides** glucose, galactose, and fructose are the basic carbohydrates. These sugars contain a backbone of six carbons and combine to form **disaccharides** such as sucrose in honey, maltose in grains, and lactose in milk, and starches. Starches are complex carbohydrates found in plants. They contain long chains of glucose molecules and are important dietary constituents.

Carbohydrates have important structural roles in the body. A small percentage of dietary carbohydrate is combined with proteins to form the nucleotides and nucleoproteins that are found inside every cell. Carbohydrates also help form connective tissue and are combined with fat in cell membranes.

Fiber is a general term used to describe a number of complex carbohydrates containing bonds that cannot be digested by human digestive enzymes. Most dietary fibers are nonstarch polysaccharides such as cellulose, pectins, gums, and mucilages derived from plant sources. Although they are resistant to human digestion, bacteria in the large intestine can metabolize some of these fibers so that they can be absorbed. However, they are not important sources of energy or nutrients in most human diets.

monosaccharide: a simple sugar; a carbohydrate that cannot be broken down to simpler substances by hydrolysis

disaccharide: a class of sugars each molecule of which yields two molecules of monosaccharide on hydrolysis

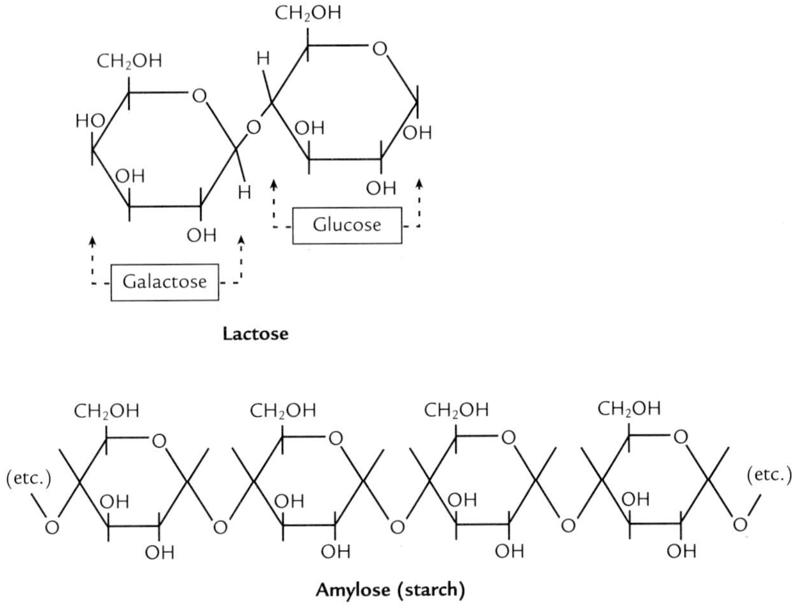

FIGURE 1.7
Structure of simple and complex carbohydrates

Lipids

The lipids are a large family of compounds that include three major subsets: the triglycerides (fats and oils), the sterols (such as cholesterol), and the phospholipids (an example is lecithin).

Fatty acids are linear chains of carbon atoms surrounded by hydrogen; one end of the carbon chain terminates in an acid group (-COOH). Fatty acids can be saturated or unsaturated depending on the number of hydrogens attached to the carbon chain. In *saturated fats,* the carbon molecules are fully loaded with hydrogens. If hydrogens are missing, the carbons form double bonds with each other. Double bonds between carbons are characteristic of unsaturated fats. *Polyunsaturated fats,* such as the essential fatty acids linoleic and linolenic acid shown in Figure 1.8, contain more than two carbon-carbon double bonds.

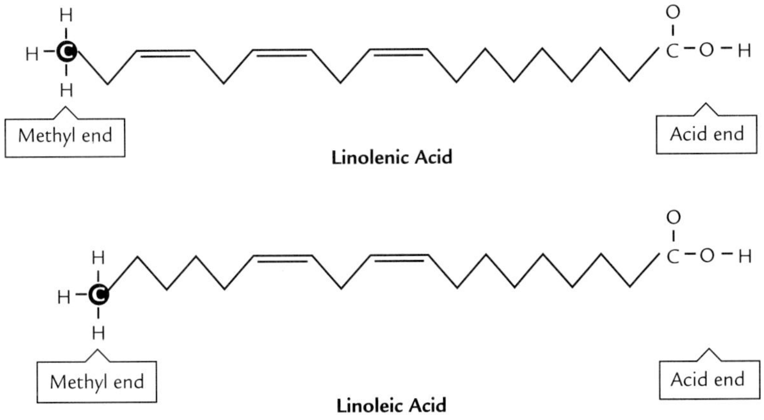

FIGURE 1.8
The essential fatty acids

In addition to their prominent role as sources of concentrated energy, lipids have important functional and structural roles. Adipose cells support, insulate, and cushion tissues, and certain fatty acids are precursors to prostaglandins and other important physiological compounds. *Prostaglandins* are a family of regulatory compounds with diverse roles in the body, including blood vessel contraction, immune function, and nervous transmission.

The phospholipids and sterols are significant constituents of cell membranes, and cholesterol is a precursor for the synthesis of hormones, vitamin D, and the bile acids.

Amino Acids and Proteins

All *amino acids* have the same basic structure containing a central carbon atom surrounded by four groups: a hydrogen atom, an amine group ($-NH_2$), an acid group ($-COOH$), and a distinctive side group. The side group, which can vary from a single hydrogen atom to a complex ringed structure, distinguishes the 21 different amino acids, as shown in Figure 1.9.

In adults, 12 of the amino acids can be synthesized, and 9 are essential. Early in infancy, several amino acids that are nonessential in later life are conditionally essential. For example, cysteine and arginine are conditionally essential for the newborn—the synthetic pathways that produce these amino acids are immature and do not synthesize adequate amounts (Rassin 1991).

Proteins are formed by amino acids linked together by peptide bonds, as shown in Figure 1.10. They vary greatly in size—ranging from a dozen to hundreds of amino acids in length. Proteins, found in all cells and body fluids, are the major structural and functional components of the body.

All of the body's enzymes are proteins. Also, many hormones and growth factors are proteins or protein derivatives. Antibodies and clotting factors are proteins, and proteins transport many substances through the bloodstream.

Cell division and tissue growth require synthesis of new proteins, and needs are particularly high during periods of rapid growth, for example, during pregnancy, infancy, and early adolescence. Because certain amino acids are essential, protein needs for growth and maintenance are both quantitative and qualitative. For each stage of life, an adequate protein intake is one that contains sufficient nitrogen and enough of the essential amino acids to meet maintenance needs and support normal growth.

In order for a protein to be absorbed it must be broken down to amino acids by proteolytic enzymes in the digestive tract. Once absorbed, amino acids can meet several different fates in the body. Many are resynthesized into structural and functional proteins. If protein intake exceeds requirements for maintenance and growth, the surplus amino acids in food are

1 Amino acids with aliphatic side chains, which consist of hydrogen and carbon atoms (hydrocarbons):

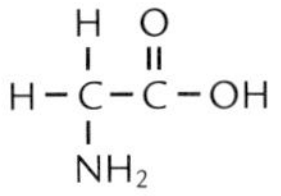

Glycine (Gly)

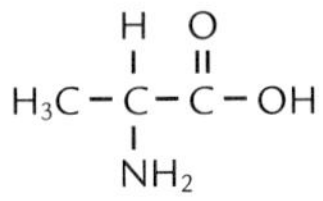

Alanine (Ala)

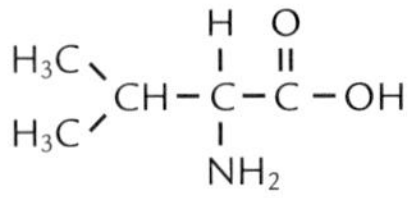

Valine (Val)

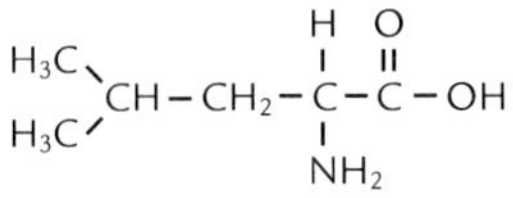

Leucine (Leu)

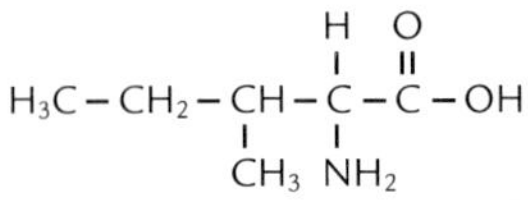

Isoleucine

2 Amino acids with hydroxyl (OH) side chains:

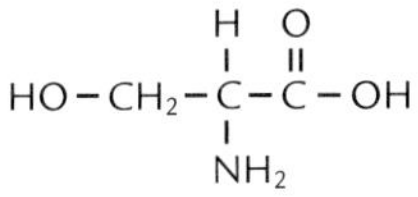

Serine (Ser)

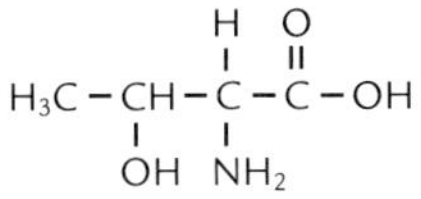

Threonine (Thr)

3 Amino acids with side chains containing acidic groups or their amides, which contain the group NH_2:

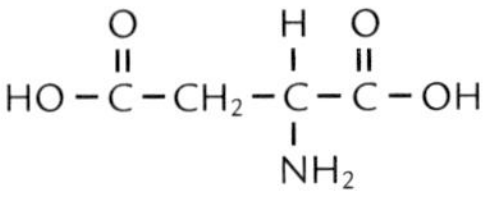

Aspartic acid (Asp)

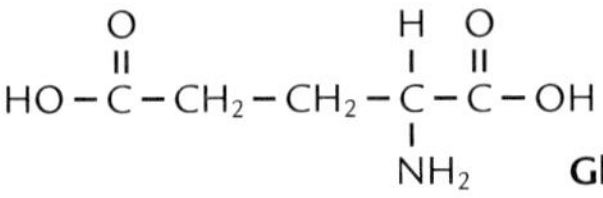

Glutamic acid (Glu)

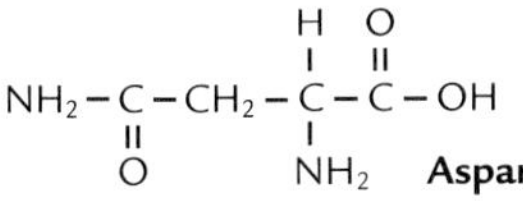

Asparagine (Asn)

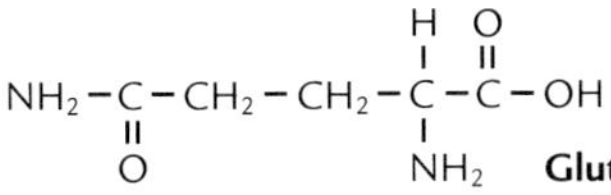

Glutamine (Gln)

4 Amino acids with basic side chains:

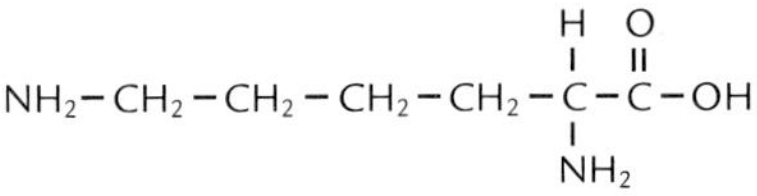

Lysine (Lys)

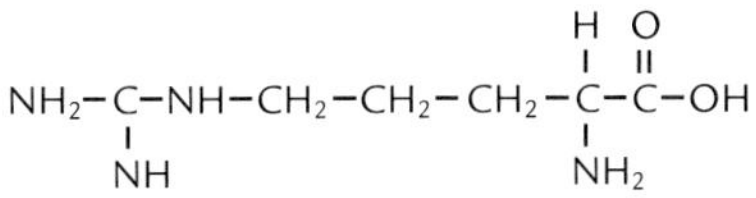

Arginine (Arg)

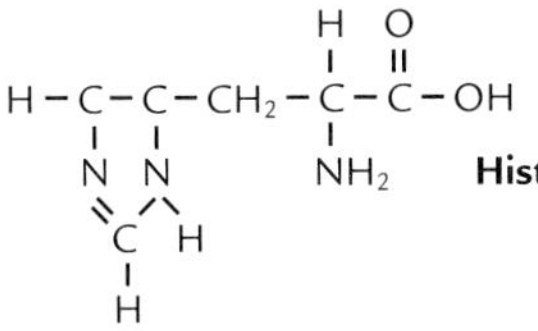

Histidine (His)

5 Amino acids with aromatic side chains which are characterized by the presence of a ring structure:

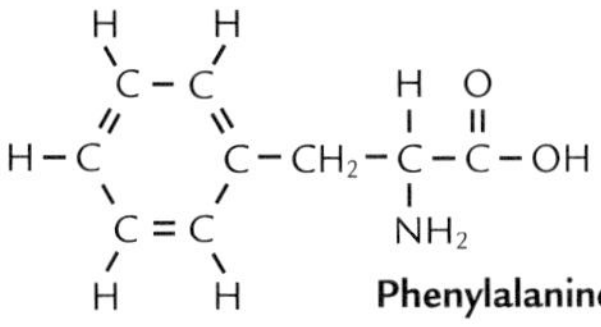

Phenylalanine (Phe)

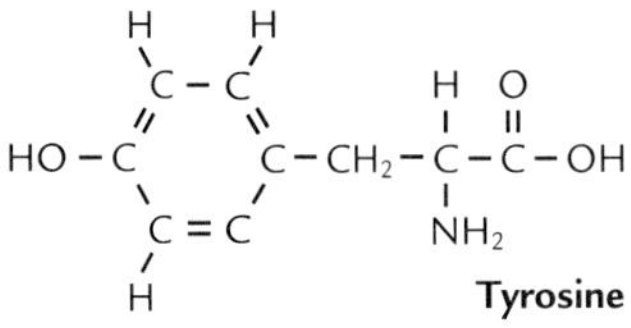

Tyrosine (Tyr)

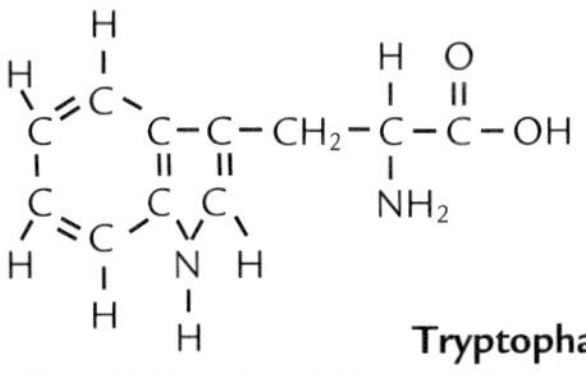

Tryptophan (Trp)

6 Amino acids with side chains containing sulphur atoms:

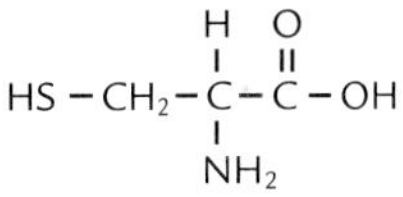

Cystine (Cys)

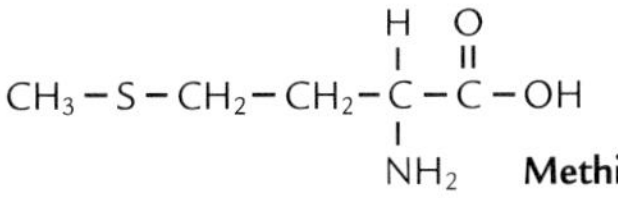

Methionine (Met)

7 Imino acid:

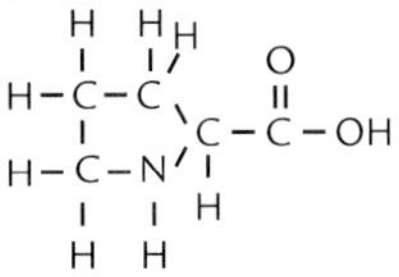

Proline (Pro)

FIGURE 1.9

The amino acids

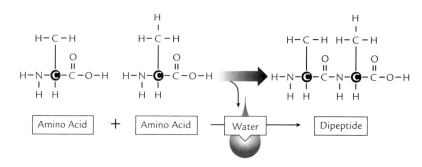

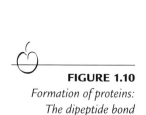

FIGURE 1.10
*Formation of proteins:
The dipeptide bond*

deaminated, the nitrogen excreted, and the carbon skeletons converted to glucose or fat.

Endogenous amino acids (derived from body proteins) can also be converted to glucose or fat and used for energy if supplies of carbohydrates and lipids are limited. For these reasons protein is considered an energy-yielding nutrient, although under normal conditions this is not its primary role.

MICRONUTRIENTS

The term *micronutrient* is reserved for the vitamins and minerals, substances required in minute amounts in the diet. Micronutrient requirements are in the microgram to milligram range, as opposed to the macronutrients (lipid, carbohydrate, and protein) for which requirements are measured in grams.

VITAMINS

A *vitamin* is an organic micronutrient essential to human life. Although vitamins do not yield energy when metabolized, many of the vitamins (particularly the B vitamins) participate in energy metabolism as **coenzymes.**

The vitamins are divided into those that are water soluble and those that are fat soluble. The water-soluble vitamins are the B vitamins and vitamin C; the fat-soluble ones are vitamins A, D, E, and K. Table 1.3 lists the water-soluble vitamins and their functions and dietary sources; Table 1.4 lists the fat-soluble vitamins.

Solubility affects the way the body handles vitamins. Fat-soluble vitamins are readily stored (mostly in the liver and fatty tissues) and only slowly excreted. In contrast, the body does not store water-soluble vitamins in appreciable amounts, and excess intakes are rapidly excreted by the kidneys.

Because of these differences, deficiencies of the water-soluble vitamins occur more readily if short-term intake is inadequate. Because the body contains stores of the fat-soluble vitamins, short periods of low dietary

deamination: the removal of an amine group (-NH₂) from an amino acid

coenzyme: a substance that enhances or is necessary for the action of an enzyme

intake are better tolerated. Also, because they are avidly stored and slowly excreted, fat-soluble vitamins are more likely to cause toxicity when consumed in excess amounts.

MINERALS

The essential minerals are inorganic micronutrients. Based on amounts in the body, they are often divided into two groups: the *major minerals* are those present in amounts greater than 5 g, the *trace minerals* in amounts less than 5 g. Although all of the essential minerals are critical to life, body content varies greatly. For example, the average adult body contains over 1000 g of calcium, but only about 0.025 g of iodide.

The major minerals are calcium, sodium, potassium, phosphorus, magnesium, and sulfur. The trace minerals currently recognized as essential are iron, zinc, iodide, copper manganese, fluoride, chromium, selenium, and molybdenum. Several additional trace minerals (such as nickel and silicon) are essential in animals and are currently under study as potentially essential minerals in humans.

Although they compose less than 5 percent of body weight, the minerals have a large number of important physiological and structural functions. Their major functions and their significant sources in the diet are shown in Table 1.5.

ASSESSING NUTRITIONAL HEALTH

A dietitian, physician, or health-care worker uses nutritional assessment methods to evaluate nutritional status based on four components:

1. History
2. Physical examination
3. Anthropometric measures
4. Laboratory tests

Taken together, data collected by these four methods can be integrated and interpreted to provide an appraisal of nutritional health.

HISTORICAL INFORMATION

Usually the first step in the nutritional assessment is a thorough questioning about medical history, socioeconomic situation, use of drugs or alcohol, and diet. A detailed history can bring to light a disease or chronic condition that interferes with appetite, digestion, or excretion of nutrients.

Name	Characteristics	Biochemical Action	Effects of Deficiency	Effects of Excess	Sources
· *Thiamin:* Vitamin B$_1$	· Water-soluble; fat-insoluble; stable in slightly acid solution; labile to heat, alkali, sulfites	· Component of thiamin pyrophosphate carboxylases, which act in various oxidative decarboxylations, including that of pyruvic acid	· Beriberi—fatigue, irritability, anorexia, constipation, headache, insomnia, cardiac failure, edema	· None from oral intake	· Liver, meat, especially pork, milk, whole grain or enriched cereals, wheat germ, legumes, nuts
· *Riboflavin:* Vitamin B$_2$	· Soluble in water, sensitive to light light and alkali; stable to heat, oxidation, acid	· Constituent of flavoprotein enzymes, important in hydrogen transfer reactions, amino acid, fatty acid, and carbohydrate metabolism and cellular respiration	· Photophobia, blurred vision, burning and itching of eyes; poor growth; skin problems	· Not harmful	· Milk, cheese, liver and other organs, meat, eggs, fish, green leafy vegetables, whole or enriched grains
· *Niacin:* Nicotinamide; nicotinic acid	· Water-soluble; stable to acid, alkali, light, heat, oxidation	· Cofactor in a number of dehydrogenase systems	· Pellagra: dermatitis, diarrhea, dementia	· Nicotinic acid (not the amide) is vasodilator; skin flushing and itching, may induce liver damage	· Meat, fish, poultry, liver, whole grain and enriched cereals, green vegetables, peanuts
· *Folacin:*	· Slightly soluble in water; labile to heat, light, acid	· Formation and metabolism of one-carbon units; participates in synthesis of purines, pyrimidines, nucleoproteins, and methyl groups	· Megaloblastic anemia	· Unknown	· Liver, green vegetables, nuts, cereals, cheese

TABLE 1.3
*Water-soluble vitamins:
their functions and
dietary sources*

TABLE 1.3 *Continued*

Name	Characteristics	Biochemical Action	Effects of Deficiency	Effects of Excess	Sources
· Vitamin B$_6$: 3 active forms: pyridoxine, pyridoxal, pyridoxamine	· Water-soluble; destroyed by ultraviolet light and by heat	· Constituent of coenzymes for decarboxylation, transamination, transsulfuration, fatty acid metabolism	· Irritability, convulsions, anemia; nerve disorders	· Nerve damage	· Meat, liver, kidney, whole grains, peanuts, soybeans
· Cobalamin: Vitamin B$_{12}$	· Slightly soluble in water; destroyed by light · Intrinsic factor of the stomach required for absorption	· Transfer of one-carbon units; essential for maturation of red blood cells in bone marrow; metabolism of nervous tissue	· Anemia; nervous system disorder	· Unknown	· Muscle and organ meats, fish, eggs, milk, cheese
· Biotin	· Soluble in water	· Coenzyme of acetyl coenzyme A carboxylase; involved in CO_2 transfer	· Skin disorders	· None known	· Yeast, animal products
· Pantothenic acid	· Soluble in water	· Part of coenzyme A	· Vomiting, insomnia, fatigue	· Diarrhea, water retention	· Widespread in foods
· Vitamin C: Ascorbic acid	· Water-soluble; easily oxidized, destroyed by heat, light, oxidative enzymes	· Integrity and maintenance of intercellular material in all tissues; facilitates absorption of iron and conversion of folic acid to folinic acid; metabolism of tyrosine and phenylalanine	· Scurvy and poor wound healing	· Oxalates in urine	· Citrus fruits, tomatoes, berries, cantaloupe, cabbage, green vegetables

Name	Characteristics	Biochemical Action	Effects of Deficiency	Effects of Excess	Sources
· *Vitamin A:* Retinol · *Provitamin A-* The plant pigments, alpha-. beta-, and gamma-carotenes and cryptotoxanthin	· Fat-soluble heat-stable; destroyed by oxidation; bile necessary for absorption; stored in liver, protected by vitamin E	· Component of retinal pigments, for vision in dim light; bone and tooth develop-ment; formation and maturation of epithelia	· Night blindness, xerophthalmia, conjunctivitis, keratomalacia leading to blindness; faulty bone formation; defective tooth enamel; keratini-zation of mucous membranes and skin; retarded growth	· Anorexia, slow growth, drying and cracking of skin, enlargement of liver and spleen, swelling and pain of long bones, bone fragility, increased intracranial pressure · Excessive carotene intake may pro-duce carotenemia	· Liver, fish-liver oils, whole milk, milk fat products, egg yolk, fortified margarines · Carotenoids from plants—green vegetables, yellow fruits and vegetables
· *Vitamin D:* Ergocalciferol (plant or yeast origin) Cholecalciferol (animal origin)	· Fat-soluble; stable to heat, acid, and oxidation; bile necessary for absorption	· Regulates absorption and deposition of calcium and phosphorus	· Rickets, poor growth, osteomalacia	· Nausea, diarrhea, weight loss, polyuria, nocturia, calcification of soft tissues, including heart, kidney, blood vessels	· Vitamin D— fortified milk and margarine, fish-liver oils, exposure to sunlight or other ultraviolet sources
· *Vitamin E:* Group of related chemical compounds— tocopherols— with similar biologic activities	· Fat-soluble; unstable to ultraviolet light; readily oxidized · Antioxidant; bile necessary for absorption	· Minimizes oxidation of carotene, vitamin A, and lipids	· Requirements related to poly-unsaturated fat intake; red blood cell disorders in premature infants; loss of neural integrity	· Unknown	· Germ oils of various seeds, green leafy vegetables, nuts, legumes

TABLE 1.4
*Fat-soluble vitamins:
Their functions and
dietary sources*

TABLE 1.4 *Continued*

Name	Characteristics	Biochemical Action	Effects of Deficiency	Effects of Excess	Sources
· *Vitamin K:* Group of naph-thoquinones with similar biologic activities; K_1 is phyto-quinone	· Natural compounds are fat-soluble; water-soluble products have been developed (menadione); · Stable to heat; labile to oxidizing agents, strong acids, light; bile salts necessary for intestinal absorption of fat-soluble forms	· Prothrombin formation, coagulation factors II, VII, IX, X and osteocalcin are K-dependent	· Hemorrhagic manifestations; impairment of bone metabolism	· Not established	· Green leafy vegetables, liver · Widely distributed

Questioning may uncover regular use of a medicinal or recreational drug that influences nutritional needs. Lifestyle factors, such as significant alcohol or tobacco use, may increase the risk for nutrient deficiency. A discussion of socioeconomic conditions may reveal underlying financial or social barriers to nutritional health.

A diet history is an important tool in nutritional assessment. Carefully done, it provides an estimate of energy and nutrient intake. In taking a diet history, the assessor collects and analyzes data on what foods an individual eats. There are several ways to do this. A person can be questioned about what foods he or she typically eats and how much. Or, the individual can be asked to remember everything eaten or drunk in the previous 24 hours (termed a *24-hour recall*). Usually, several 24-hour recalls are collected on nonconsecutive days, and the information is combined to estimate typical daily intakes.

Another type of diet history is the *food diary*. Data on food intake can be collected by having the individual record in a diary all the foods eaten during a defined period, for example, a week. For a food diary to be accurate, it is crucial that the person eat a representative diet during the chosen period and carefully record portions. Records are then analyzed using established tables of food composition, and they can be compared to recommended intakes for age, size, gender, and activity.

Mineral	Function and Metabolism	Effects of Deficiency	Effects of Excess	Sources
Calcium	· Structure of bone and teeth, muscle contraction, nerve irritability, coagulation of blood, cardiac action · Absorbed from upper small intestine; aided by vitamin D, lactose, acid reaction; hindered by excesses of dietary oxalic acid, phytic acid, fat, fiber, phosphate	· Poor mineralization of bones and teeth; osteomalacia; osteoporosis; tetany; rickets; impairment of growth	· Unknown (dietary) · Heart block and kidney stones (parenteral)	· Milk, cheese, green leafy vegetables, canned salmon, clams, oysters
Chloride	· Osmotic pressure; acid-base balance; HCl in gastric juice · Readily absorbed	· Hypochloremic alkalosis may occur with prolonged vomiting or excessive sweating; muscle cramps; poor growth	· Unknown	· Table salt, meat, milk, eggs
Chromium	· Glycemic regulation and insulin metabolism	· Impaired glucose metabolism	· None known	· Yeast, liver, seafood, meat, vegetables
Cobalt	· Component of B_{12} (cobalamin) molecule	· None known	· May produce heart weakness	· Widely distributed in animal products
Copper	· Essential for production of red blood cells; catalyst in hemoglobin formation; absorption of iron · Associated with activities of tyrosinase, catalase, cytochrome C oxidase, lysyl oxidase; · Transported in plasma bound to plasma proteins and in ceruloplasmin; highest concentration in liver and central nervous system; excreted mainly via the intestinal wall and bile	· Anemia, osteoporosis, depigmentation · Increased serum cholesterol	· Liver damage	· Liver, oysters, meats, meats, fish, whole grains, nuts, legumes
Fluoride	· Tooth and bone structure · Excreted in urine and sweat; deposited in bones	· Tendency to dental caries	· Bone deformation, mottling of teeth	· Water, sea foods, plant and animal foods (dependent upon content in soil and water)

TABLE 1.5

The minerals: Their functions and dietary sources

TABLE 1.5 *Continued*

Mineral	Function and Metabolism	Effects of Deficiency	Effects of Excess	Sources
Iodine	· Constituent of thyroid hormones · Readily absorbed from intestine; selectively concentrated in the thyroid gland; excretion mainly in urine	· Simple goiter, endemic cretinism	· May cause goiter	· Iodized salt, sea food, water
Iron	· Structure of hemoglobin and myoglobin for O_2 and CO_2 transport; oxidative enzymes: cytochrome C and catalase · Absorption aided by gastric juice and ascorbic acid, hindered by fiber, phytic acid · Transported in plasma bound to transferrin; stored in liver, spleen, bone marrow, kidney as ferritin and hemosiderin; carefully conserved and reused; minimal losses in urine and sweat; about 90% of intake excreted in the stool	· Anemia, weakness, impaired immune functions, cognitive impairments	· Acute: shock, death · Chronic: liver damage, cardiac failure	· Liver, meat, egg yolk, green vegetables, whole grains, legumes, nuts
Magnesium	· Structure of bones and teeth; activation of enzymes in carbohydrate metabolism; muscle and nerve irritability · Absorption from small intestine varies with intake; urinary excretion	· Hypokalemia, hypocalcemia, neurologic disturbances	· None (dietary); toxicity from intravenous medication	· Cereals, legumes, nuts meat, milk
Manganese	· Enzyme activation, especially superoxide dismutase; normal bone structure, carbohydrate metabolism · Poor absorption from intestine; transported in plasma; excretion mainly via the intestine in bile	· Abnormal bone and cartilage	· None (dietary); toxicity from chronic inhalation	· Legumes, nuts, whole grain cereals, green leafy vegetables

TABLE 1.5 *Continued*

Mineral	Function and Metabolism	Effects of Deficiency	Effects of Excess	Sources
Molybdenum	· Component of enzymes: xanthine oxidase for conversion to uric acid and mobilization of ferritin iron in liver · Readily absorbed from intestine; excreted chiefly in urine, some in bile	· Disorder of nitrogen metabolism	· Impairs copper metabolism	· Legumes, grains, dark green leafy vegetables, animal organs
Phosphorus	· Constituent of bones and teeth; structure of nucleus and cytoplasm of all cells; acid-base balance; key position in energy transformations and transmission of nerve impulses; metabolism of carbohydrate, protein, and fat · Absorbed as free phosphates from intestine; vitamin D and para-thormone implicated in intestinal absorption and kidney retention; excreted in urine and feces	· Muscle weakness; demineralization of bone	· Reduced absorption of other minerals	· Milk, milk products, egg yolk, legumes, nuts, whole grains
Potassium	· Muscle contraction; nerve impulse conduction; intracellular osmotic pressure and fluid balance; heart rhythm · Primarily intracellular; excretion mainly in urine—some in sweat and feces	· Muscle weakness, anorexia, nausea, abdominal distention, nervous irritability, drowsiness, confusion	· Heart block; muscle weakness	· All foods
Selenium	· Cofactor for glutathione peroxidase; thyroid hormone activation	· Muscle diseases; heart damage	· Nausea and vomiting, loss of hair and nails	· Vegetables, meats, seafood
Sodium	· Osmotic pressure; acid-base balance; water balance, muscle and nerve irritability · Readily absorbed from intestine; excreted chiefly in urine	· Nausea; diarrhea, muscle cramps, dehydration	· Edema if inadequate excretion	· Table salt, milk, eggs, sodium compounds as baking soda and powder, glutamate seasonings and preservatives

TABLE 1.5 *Continued*

Mineral	Function and Metabolism	Effects of Deficiency	Effects of Excess	Sources
Sulfur	· Constituent of cellular protein, connective tissues, cartilage, heparin; insulin; metabolism of nerve tissue; detoxification mechanisms; tissue metabolism as SH group in coenzyme A, cystathionine, and glutathione	· Growth failure from protein deficiency may be due in part to deficiency of S-containing amino acids	· Unknown	· Protein foods
Zinc	· Constituent of several enzymes: carbonic anhydrase (in erythro-cytes) essential for CO_2 exchange; carboxypeptidase of intestine for hydrolysis of protein, dehydrogenase of liver, and many others	· Dwarfism, iron deficiency anemia, hyperpigmentation and hypogonadism, depression of immunocompetence, poor wound healing, dermatitis	· Copper deficiency, depressed immune function, nausea and vomiting	· Meat, grain, nuts, cheese

PHYSICAL EXAMINATION

Thorough inspection of an individual by a skilled observer can provide valuable clues to nutritional status. A careful physical exam includes scrutiny of the hair, eyes, tongue, skin, and posture along with listening to heart rhythms and breathing patterns.

Abnormal findings can suggest possible nutrient imbalances or deficiencies. Table 1.6 lists some of the findings on a physical exam that may reflect nutritional problems. Often clues from the physical exam need to be confirmed by laboratory testing.

ANTHROPOMETRIC MEASURES

Anthropometric (*anthropos*-person; *metric*-measuring) measurements are often part of the physical exam. Measurements of stature, weight, limb circumferences, and skinfold measurements can be compared to standards or previous measurements for an individual (shown in Table 1.7). Calculation of *percent ideal body weight* (%IBW) or *body mass index* (BMI) can provide useful indications of nutritional status.

Deviation from standards for age and gender, or changing patterns in a series of measurements, may provide useful information on growth, weight loss or gain, and changes in body composition. Table 1.8 shows standards for %IBW and BMI.

TABLE 1.5 *Continued*

Body System	Normal Findings	Malnutrition Findings	What the Findings May Reflect
Skin	Smooth, good color	Dry, rough, sores, lack of fat under skin, "sandpaper" feel	Protein-energy malnutrition (PEM), vitamin A, the B vitamins, and vitamin C status
Hair	Shiny, soft, firm in the scalp	Brittle, dry, dull, loose	PEM
Eyes	Bright, pink membranes, vision adjusts quickly in changing light	Dull, pale membranes, redness, adjust slowly to darkness	Iron, vitamin A, the B vitamins, and zinc status
Teeth and gums	Teeth not decayed or missing, gums firm and pink	Decayed, missing teeth, gums spongy and swollen, bleed easily	Mineral and vitamin C status
Tongue	Rough, bumpy, and red	Purplish, smooth, sore	B vitamin status
Thyroid gland	Not enlarged	Swollen at lower front of neck	Iodine status
Fingernails	Smooth, nailbeds pink	Spoon-shaped, brittle, ridged, pale nailbed	Iron, mineral status
Muscle	Muscles firm and with good tone	Loss of muscle mass	PEM
Skeleton	Bone development and shape appropriate for age	Bony swellings on ends of bones, ribs and skull, bowed legs	Vitamin D status
Overall growth	Normal for age	Stunted	PEM, vitamin and mineral status

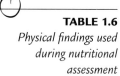

TABLE 1.6
Physical findings used during nutritional assessment

Type of Measurement	What It Reflects
Height-weight	Overnutrition and undernutrition; growth in children
Percent ideal body weight, recent weight change	Overnutrition and undernutrition
Head circumference	Brain growth and development in infants and children under 2 years
Midarm circumference	Muscle mass and subcutaneous fat
Fatfold	Subcutaneous and total body fat
Midarm muscle circumference	Muscle mass (i.e., protein status)

TABLE 1.7
Anthropometric measurements used during nutritional assessment

Percentage of Ideal or Usual Body Weight	Status
>120	Obese
110–120	Overweight
80–90	Mildly undernourished
70–79	Moderately undernourished
<70	Severely undernourished

Body Mass Index (kg/m2)		Status
Men	Women	
<20.7	<19.1	Underweight
20.7–27.8	19.1–27.3	Acceptable weight
≥27.8	≥27.3	Overweight
≥31.1	≥32.3	Severely overweight

TABLE 1.8
Body weight and BMI as indicators of nutritional status

LABORATORY TESTS

Appropriately selected laboratory tests are an important component of a nutritional assessment (see Table 1.9). Samples of body fluids or tissues can be analyzed in the laboratory and compared to normal values. Urine and blood samples are commonly used because they are easy to collect and provide an inside look at body physiology and function. Laboratory tests can uncover signs of nutritional deficiency or imbalance early, in many cases before symptoms appear on the physical exam.

Nutrient	Assessment Tests
Protein	Urinary creatinine excretion, serum albumin, serum transferrin, retinol-binding protein, total lymphocyte count, nitrogen balance
Vitamins	
Vitamin A	Serum retinol, serum carotene
Thiamin	Erythrocyte (red blood cell) transketolase activity, urinary thiamin
Riboflavin	Erythhrocyte glutathione reductase activity, urinary riboflavin
Vitamin B_6	Urinary xanthurenic acid excretion after tryptophan load test, urinary vitamin B_6, erythrocyte transaminase activity
Niacin	Urinary metabolites: NMN (N-methyl-nicotinamide) or 2-pyridone or both expressed as a ratio
Folate	Folate in the blood, erythrocyte folate, urinary formimino-glutamic acid (FIGLU)
Vitamin B_{12}	Serum vitamin B_{12} erythrocyte vitamin B_{12}
Biotin	Serum biotin, urinary biotin
Vitamin C	Serum or plasma vitamin C, vitamin C in white blood cells
Vitamin D	Serum 25-hydroxyvitamin D
Vitamin E	Serum tocopherol, erythrocyte hemolysis
Vitamin K	Blood clotting time (prothrombin time)
Minerals	
Potassium	Serum potassium
Magnesium	Serum magnesium
Iron	Hemoglobin, hematocrit, serum ferritin, total iron-binding capacity (TIBC), transferrin saturation, serum ferritin, mean corpuscular volume (MCV), serum iron
Iodine	Serum protein-bound iodine, radioiodine uptake, urinary iodine excretion
Zinc	Plasma zinc, hair zinc
Copper	Serum copper, ceruloplasmin
Calcium	Urinary calcium
Sodium	Serum sodium
Selenium	Plasma selenium, glutathione peroxidase activity in blood

TABLE 1.9

Biochemical tests useful in nutritional assessment

SURVEYS OF POPULATIONS

To evaluate the nutritional status of a population, researchers use one or more of the assessment techniques described in the previous sections, but apply them to large numbers of people. A *food consumption survey* measures the types and amounts of foods eaten by a population, using individual diet histories that are combined and analyzed. Results are usually compared to standards such as the RDAs (discussed in the next section).

An example of a large food consumption survey is the periodic U.S. Nationwide Food Consumption Survey (NFCS). The latest NFCS, done in 1987 and 1988, compiled three-day diet histories on 8000 people and provided valuable information on food consumption patterns and nutrient intakes of the U.S. population (Washington D.C.: Public Health Service 1989).

A broader type of population survey is a *nutrition status survey.* These surveys evaluate nutritional status with both diet histories and direct examination of individuals, using anthropometric measurements and laboratory testing.

In the United States, the National Center for Health Statistics, Centers for Disease Control and Prevention conducts the National Health and Nutrition Examination Survey (NHANES). The most recent, NHANES III (1988–1994), provides a broad range of data on health and nutrition in the United States (Wotecki 1990), and comparisons can be made with previous information gathered in NHANES II and I (from the early 1980s and 1970s).

RECOMMENDED NUTRIENT INTAKES AND DIETARY GUIDELINES

RECOMMENDED DIETARY ALLOWANCES (RDAS)

In the United States, the Food and Nutrition Board of the National Academy of Sciences (NAS) periodically reviews new scientific evidence and publishes the RDAs. The RDAs are defined as "the levels of intake of essential nutrients that, on the basis of scientific knowledge, are judged to be adequate to meet the known nutrient needs of practically all healthy persons" (NAS 1989).

The RDAs are established by estimating the average physiological requirements for an absorbed nutrient, by sex and age. This estimate is adjusted to compensate for variation in the requirements among individuals and the **bioavailability** of the nutrient in the food supply. The RDAs are not simply minimal requirements to avoid deficiency; they are fairly generous allowances that actually exceed the requirements of most individuals. *Note*: An exception is energy. For energy, only the mean requirement is

bioavailability: the degree to which a nutrient or other substance becomes available to an organism or tissue after ingestion or administration

offered for each age group, to avoid the intake of excess energy by the average individual (NAS, 1989).

The RDAs are typically used for five purposes:

1. Evaluating the adequacy of food supplies and planning food supplies for populations

2. Interpreting food consumption surveys

3. Establishing standards for food assistance programs

4. Designing nutrition education programs

5. Developing new products in industry

Although the RDAs are most appropriate for populations and not individuals, comparing long-term individual intakes to the RDAs allows an estimate about the probable risk of deficiency for that individual (NAS 1989).

DIETARY GUIDELINES

The NAS has published recommendations meant to complement the RDAs in planning diets to support health. These guidelines are contained in the report: *Diet and Health: Implications for Reducing Chronic Disease Risk* (National Research Council 1989). Unlike the RDAs, which focus on individual nutrients, the *Diet and Health* guidelines give advice on foods individuals should include, limit, or avoid to reduce the risks of chronic disease. The *Diet and Health* guidelines are shown in Table 1.10. Table 1.11 lists an additional set of recommendations, the *1995 Dietary Guidelines for Americans*, which are published by the U.S. Department of Agriculture and the U.S. Department of Health and Human Services (USDA 1995).

NUTRITION OBJECTIVES OF HEALTHY PEOPLE 2000

Healthy People 2000: National Health Promotion and Disease Prevention Objectives is an ambitious program of the Department of Health and Human Services that began in 1990 (U.S. Public Health Service, 1991). This initiative sets national goals for the year 2000 intended to reduce preventable death and disability, enhance the quality of life, and reduce disparities in the health status of various population groups within the U.S.

There are 21 objectives that focus on nutrition. They are aimed at decreasing the prevalence of obesity and iron-deficiency anemia, reducing coronary heart disease death rates, and reversing the rise in cancer mortality. In addition, they target specific dietary changes, such as decreasing consumption of total fat, saturated fat, and sodium, as well as increasing consumption of grains, fruits and vegetables, and calcium-rich foods. These wide-ranging objectives are shown in Table 1.12.

- Reduce total *fat* intake to 30 percent or less of kcalories. Reduce saturated fatty acid intake to less than 10 percent of kcalories and the intake of cholesterol to less than 300 milligrams daily.[a]

- Increase intake of starches and other *complex carbohydrates*.[b]

- Maintain *protein* intake at moderate levels.[c]

- Balance food intake and physical activity to maintain appropriate *body weight*.

- For those who drink *alcoholic beverages*, limit consumption to the equivalent of less than 1 ounce of pure alcohol in a single day.[d] Pregnant women should avoid alcoholic beverages.

- Limit total daily intake of *salt* (sodium chloride) to 6 grams or less.[e]

- Maintain adequate *calcium* intake.

- Avoid taking dietary *supplements* in excess of the RDA in any one day.

- Maintain an optimal intake of *fluoride*, particularly during the years of primary and secondary tooth formation and growth.

Note: italics added to highlight the areas of concern.

a   The intake of fat and cholesterol can be reduced by substituting fish, poultry without skin, lean meats, and lowfat or nonfat dairy products for fatty meats and whole-milk products; by choosing more vegetables, fruits, cereals, and legumes and by limiting oils, fats, egg yolks, and fried and other fatty foods.

b   Every day eat five or more servings of a combination of vegetables and fruits, especially green and yellow vegetables and citrus fruits; and six or more daily servings of a combination of breads, cereals, and legumes.

c   Meet at least the RDA for protein; do not exceed twice the RDA.

d   The committee does not recommend alcohol consumption. One ounce of pure alcohol is the equivalent of two cans of beer, two small glasses of wine, or two average cocktails.

e   Limit the use of salt in cooking; avoid adding it to food at the table. Consume salty, highly processed salty, salt-preserved, and salt-pickled foods sparingly. Six grams of salt are the equivalent to 2.4 grams of sodium.

Source: Adapted from the National Academy of Sciences report. *Diet and Health: Implications for Reducing Chronic Disease Risk* (Washington D.C.: National Academy Press, 1989).

TABLE 1.10
National Academy of Sciences' Diet and Health Recommendations

- Eat a variety of foods

- Balance the food you eat with physical activity—maintain or improve your weight

- Choose a diet with plenty of grain products, vegetables, and fruits

- Choose a diet low in fat, saturated fat, and cholesterol

- Choose a diet moderate in sugars

- Choose a diet moderate in salt and sodium

- If you drink alcoholic beverages, do so in moderation

Note: *Dietary Guidelines for Americans* is a government document developed by the U.S. Department of Agriculture and the U.S. Department of Health and Human Services. The basis for these guidelines derives from *The Surgeon General's Report on Nutrition and Health* DHHS (PHS) publication no. 88-50211 (Washington D.C.: Government Printing Office, 1988); *Diet and Health: Implications for Reducing Chronic Disease Risk* (Washington D.C.: National Academy Press, 1989) and the *Recommended Dietary Allowances,* 10th ed. (Washington D.C.: National Academy Press, 1989).

TABLE 1.11
1995 Dietary Guidelines for Americans

	Objective	Baseline	1991	1992	Target 2000
1.1	Coronary heart disease deaths (age adjusted per 100,000)	135	118		100
	African American (age adjusted per 100,000)	168	158		115
1.2	Cancer deaths (age adjusted per 100,000)	134	135	133	130
1.3	Overweight prevalence				
	Adults, 20–74 yr.	26%	34%		20%
	Males	24%	32%		
	Females	27%	35%		
	Adolescents, 12–19 yr.	15%			15%
	Low-income females, 20–74 yr.	37%			25%
	African-American females, 20–74 yr.	44%	48%		30%
	Hispanic females, 20–74 yr.	27%			25%
	Mexican-American	9%	47%		
	Cuban	34%			
	Puerto Rican	37%			
	American Indians/Alaska Natives, ≥20 yr.	29–75%	40%	36%	30%
	People with disabilities, ≥20 yr.	36%	36%	37%	25%
	Females with high blood pressure, 20–74 yr.	50%			41%
	Males with high blood pressure, 20–74 yr.	39%			35%
2.4	Growth retardation among low-income children, ≤ 5 yr.	11%	9%	8%	10%
	African American, <1 yr.	15%	15%	15%	10%
	Hispanic, <1 yr.	13%	8%	8%	10%
	Hispanic, 1 yr.	16%	11%	9%	10%
	Asian/Pacific Islander, 1 yr.	14%	13%	12%	10%
	Asian/Pacific Islander, 2–4 yr.	16%	12%	11%	10%
2.5	Dietary fat intake among people ≥ 2 yr. (% of calories)				
	From total fat	36%	34%		30%
	From saturated fat	13%	12%		10%
	People, 20–74 yr. (% of calories)				
	From total fat	36%	34%		
	From saturated fat	13%	12%		
	Females, ≥20 yr. (% of calories)				
	From total fat	36%	34%		
	From saturated fat	13%	12%		
	Males, ≥20 yr. (% of calories)				
	From total fat	37%	34%		
	From saturated fat	13%	12%		

TABLE 1.12

Healthy People 2000:
Nutrition objectives

TABLE 1.12 *Continued*

	Objective	Baseline	1991	1992	Target 2000
2.6	Daily intake of vegetables, fruits, and grain products (no. of servings)				
	Vegetables and fruits				
	Adults (≥20 yr.)	4.0			5.0
	Males				
	20–39 yr.	4.1			
	40–59 yr.	4.3			
	≥60 yr.	4.4			
	Females				
	20–39 yr.	3.4			
	40–59 yr.	4.0			
	≥60 yr.	3.9			
	19–50 yr.	2.5			
	Grain products				
	Adults (≥20 yr.)				6.0
	Females, 19–60 yr.	3.0			
2.7	Weight loss practices among overweight people, ≥12 yr.				50%
	Overweight females, ≥18 yr.	30%	22%		
	Overweight males, ≥18 yr.	25%	19%		
2.8	Foods rich in calcium				
	3 or more servings daily				
	People, 12–24 yr.	15%			50%
	Males, 19–24 yr.	14%	14%		
	Females, 19–24 yr.	7%	7%		
	Pregnant and lactating females	24%	16%		50%
	2 or more servings daily				
	People, ≥25 yr.	19%			50%
	Males, 25–50 yr.	23%	23%		
	Females, 25–50 yr.	15%	16%		
2.9	Salt and sodium intake				
	Prepare foods without adding salt	54%	43%		65%
	Adults who avoid using salt at table	68%			80%
	Adults who regularly purchase foods lower in sodium	20%			40%
2.10	Iron deficiency				
	Children, 1–4 yr.				3%
	Children, 1–2 yr.	9%			3%
	Children, 3–4 yr.	4%			3%

TABLE 1.12 *Continued*

	Objective	Baseline	1991	1992	Target 2000
2.10	Iron deficiency				
	Females of childbearing age (20–44 yr.)	5%			3%
	Iron deficiency prevalence				
	Low-income children, 1–2 yr.	21%			10%
	Low-income children, 3–4 yr.	10%			5%
	Low-income females, 20–44 yr.	8%			4%
	Anemia prevalence				
	Alaska native children, 1–5 yr.	22–28%			10%
	African-American, low-income pregnant females, 15–44 yr. (3rd trimester)	41%	42%	43%	20%
2.11	Breastfeeding				
	During early postpartum period	54%	53%	54%	75%
	Low-income mothers	32%	33%	35%	75%
	African-American mothers	25%	26%	28%	75%
	Hispanic mothers	51%	52%	52%	75%
	American Indian/Alaska native mothers	47%	48%	53%	75%
	At age 5–6 mo.	21%	18%	19%	50%
	Low-income mothers	9%	9%	9%	50%
	African American mothers	8%	7%	9%	50%
	Hispanic mothers	16%	16%	17%	50%
	American Indian/Alaska native mothers	28%	22%	24%	50%
2.12	Baby bottle tooth decay				
	Parents and caregivers who use preventative feeding practices	55%			76%
	Parent and caregivers with less than high school education	36%			65%
	American Indian/Alaska Native parents and caregivers	74%			65%
2.13	Use of food labels	74%	76%	74%	85%
2.14	Informative nutrition labeling				
	Processed/packaged foods	60%	66%		100%
	Fresh produce			77%	40%
	Fresh seafood			75%	40%
	Fresh meat/poultry				40%
	Take-out foods				40%
2.15	Availability of reduced-fat processed foods	2,500	5,618		5,000
2.16	Low-fat, low-calorie restaurant food choices	70%	75%		90%
2.17	Nutritious school and child care food services				90%

TABLE 1.12 *Continued*

	Objective	Baseline	1991	1992	Target 2000
2.18	Home-delivered meals for older adults	7%			80%
2.19	Nutrition education in schools	60%			75%
2.20	Worksite nutrition/weight management programs				
	Nutrition education	17%		31%	50%
	Weight control	15%		24%	50%
2.21	Nutrition assessment, counseling, and referral clinicians	40–50%			75%
	Percent of clinicians routinely providing service to 81–100% of patients				
	Inquiry about diet/nutrition				
	Pediatricians	53%			
	Nurses	46%			
	Obstetricians/gynecologists	15%			
	Internists	36%			
	Family physicians	19%			
	Formulation of a diet/nutrition plan				
	Pediatricians	31%			
	Nurses	31%			
	Obstetricians/gynecologists	19%			
	Internists	33%			
	Family physicians	24%			

Note: There are 21 nutrition objectives. This table lists the objectives, shows the baseline data and targets for each objective, and shows the current progress toward the targets as reported in the 1994 *Nutrition Progress Review*. Adapted from C.J. Lewis et al., "Healthy People 2000: Report on the 1994 Nutrition Progress Review," *Nutrition Today* 29 (1994):13.

REFERENCES

De Luca, L.M. et al.,"Vitamin A in epithelial differentiation and skin carcinogenesis," *Nutr Rev* 52 (1994):45–53.

Federation of the American Societies for Experimental Biology, Life Sciences Research Office, "Nutrition Monitoring in the United States—An Update Report on Nutrition Monitoring," prepared for the U.S. Dept. of Agriculture and the U.S. Dept. of Health and Human Services, DHHS (PHS) publication no. 89-1255 (Washington D.C.: Public Health Service, 1989).

Fomon, S.J. and Bell, E.F., "Energy" in *Nutrition of Normal Infants* (St. Louis: Mosby-Year Book, 1993).

Goss, R.J., "Modes of Growth and Regeneration: Mechanisms, Regulation, Distribution," in *Human Growth*, 2d ed., eds. F. Falkner and J.M. Tanner (New York: Plenum Press, 1986), 1–26.

Institute of Medicine, Committee on Opportunities in the Nutrition and Food Sciences, Opportunities in the Nutrition and Food Sciences, (Washington D.C., National Academy of Sciences, 1994).

Lebenthal, E., "Impact of Development of the Gastrointestinal Tract on Infant Feeding," *J Pediatr* 102 (1983):1.

Maden, M., "Vitamin A in Embryonic Development," *Nutr Rev* 52 (1994):3–13.

Matalon, R., and Michals, K., "Phenylketonuria: Screening, Treatment and Maternal PKU," *Clin Biochem* 24 (1991):337.

National Academy of Science, Food and Nutrition Board of the National Research Council, *Recommended Dietary Allowances*, 10th ed. (Washington D.C.: National Academy Press, 1989).

National Research Council, "Diet and Health: Implications for Reducing Chronic Disease Risk", report of the Committee on Diet and Health, Food and Nutrition Board (Washington D.C.: National Academy Press, 1989).

Ong, D.E., "Cellular Transport and Metabolism of Vitamin A: Roles of Cellular Retinoid-Binding Proteins," *Nutr Rev* 52 (1994):24–32.

Rassin, D.K., "Amino Acid and Protein Metabolism in the Premature and Term Infant," in *Neonatal Nutrition and Metabolism*, ed. W.W. Hay (St Louis: Mosby-Year Book, 1991), 110–122.

Sommer, A., "Vitamin A: Its Effect on Childhood Sight and Life," *Nutr Rev* 52 (1994):60–67.

Tanner, J.M., "Fetus into Man: Physical Growth from Conception to Maturity" (Cambridge: Harvard University Press, 1990), 165–78.

U.S. Public Health Service, Healthy People 2000: National Health Promotion and Disease Prevention Objectives, PHS publication no. 91-50212 (Washington D.C.: U.S. Dept. of Health and Human Services, 1991).

USDA, "Report of the Dietary Guidelines Advisory Committee for Americans" (Washington D.C.: USDA 1995).

Willet, W.C., and Hunter D.J. "Vitamin A and Cancers of the Breast, Large Bowel and Prostate: Epidemiologic Evidence," *Nutr Rev* 52 (1994):53–60.

Wilson, R.S., "Concordance in Physical Growth for Monozygotic and Dizygotic Twins," *Ann Human Biol* 3 (1976):1–10.

Wotecki ,C.E., et al., "Selection of Nutrition Status Indicators for Field Surveys: The NHANES III design," *J Nutr* 120 (1990):1440–45.

pregnancy and nutrition

▼

INFANT MORTALITY, LOW BIRTHWEIGHT, AND NUTRITION DURING PREGNANCY

Despite enormous health care expenditures and wealth, the U.S. infant mortality rate is much higher than that of most other industrialized countries (Wegman 1994). Defined as the number of deaths in children under one year of age for every thousand births, the infant mortality rate of the United States is twice that of Japan and nearly 50% higher than those of Sweden, Finland, and Switzerland (see Table 2.1). A look beneath the surface of the statistics reveals why infant deaths are more common in the United States.

Although often used to compare quality of health care among countries, infant mortality rates reflect a complicated mix of socioeconomic, demographic, and general health care factors, and measure much more than simple differences in health care.

Poverty increases infant mortality sharply and reduces access to health care and preventive services. The number of children growing up in poverty in the United States is markedly higher than in many developed countries (Wise 1990). Compared to other industrialized countries, social services in the United States for pregnant women and children are limited, and fewer women—particularly among the poor—obtain routine **prenatal care** (Liu et al. 1992). Teenage pregnancy rates are high in the United States, as shown in Table 2.2, and infant deaths are twice as likely to occur in teen pregnancies (Alexander and Guyer 1993).

These social conditions are surrogates for many other risk factors: smoking, alcohol and illicit drug use, infection, limited education, and poor nutrition during pregnancy. All of these factors influence the major determinant of infant mortality in the United States—low infant weight at birth. Babies weighing less than 5.5 pounds— a standard definition of low birthweight—account for more than two thirds of all infant deaths in the

prenatal care: care of the pregnant woman before delivery of the infant

Country	Infant Mortality Rate
Japan	4.5
Singapore	5.0
Sweden	5.4
Finland	5.6
Norway	5.9
Canada	6.1
Germany	6.2
Netherlands	6.3
Hong Kong	6.4
Switzerland	6.4
France	6.5
Denmark	6.5
United Kingdom	6.6
Ireland	6.6
Australia	7.0
New Zealand	7.3
Spain	7.4
Austria	7.5
Belgium	7.5
Italy	8.3
Greece	8.4
United States	8.5

Adapted from M.E. Wegman, "Annual Summary of Vital Statistics", *Pediatrics* 94, no.6 (1994):792–803.

TABLE 2.1

Infant mortality rates in selected countries

Country	Younger Than 20 (%)	Older Than 40 (%)
United States	12.7	0.8
United Kingdom	8.7	1.1
West Germany	3.5	1.3
Norway	4.2	1.3
France	3.2	1.2
Netherlands	2.3	1.0
Canada	6.1	0.7
Sweden	3.2	1.7
Japan	1.3	0.6

Adapted from K. Liu et al., "International Infant Mortality Rankings", *Health Care Fin Rev* 13, no.4 (1992): 105–18.

TABLE 2.2

Percentage of mothers younger than 20 and older than 40 in industrialized countries

United States (Institute of Medicine, 1985), largely explaining the high infant mortality in this country. Compare the rates of low birthweight in industrialized countries in Table 2.3 (2500 grams is approximately 5.5 pounds).

Proper nutrition and adequate maternal weight gain during pregnancy have a substantial impact on birthweight (Institute of Medicine, 1990). Ensuring adequate nutrition during pregnancy is a major public health priority in the United States, for which we spend over $2.5 billion each year.

Although many factors affect pregnancy outcome and birthweight, none is more important than nutrition. Moreover, the quality of nutrition during pregnancy has reverberations long after birth: health during childhood and later life is built to a large extent on a nutritional foundation established in utero.

Country	Less Than 1500 g (%)	Less Than 2500 g (%)
United States	1.2	6.7
Canada	0.9	5.8
Japan	0.4	5.4
Norway	0.6	4.1
Sweden	0.7	4.4
West Germany	0.2	5.7

Adapted from K. Liu "International Infant Mortality Rankings", *Health Care Fin Rev* 13, no. 4 (1992): 105–18.

TABLE 2.3

Percentage of births less than 1500 g and less than 2500g in industrialized countries

NUTRITION AND FERTILITY

Nutrition has a broad and profound influence on reproduction. Optimal nutrition during pregnancy and lactation is vital to maternal health, and it plays a pivotal role in pregnancy outcome and the future health and development of the infant. But the importance of nutrition in reproductive health begins long before conception. Women have special nutritional needs throughout their childbearing years, and fertility, the biological ability to conceive and maintain a pregnancy, can be strongly influenced by nutritional status.

hormone: a chemical messenger produced by cells and transported by the blood to other cells and organs on which it has a regulatory effect

estrogen: a generic term for substances that exert biological effects characteristic of estrogenic hormones; estrogenic hormones are formed by the ovaries and placenta (small amounts are also formed by the testes and adrenal gland) and stimulate the secondary sex characteristics and the menstrual cycle

progesterone: a steroid sex hormone produced by the ovary that plays a major role in the menstrual cycle; it fosters growth of the placenta and breasts during pregnancy

menses: female menstrual bleeding normally occurring at approximately 4-week intervals, having its source from the uterine epithelium

ovum: the female sex cell produced by the ovary

ovulation: release of an ovum from the ovary

THE FEMALE SEXUAL CYCLE

Throughout life, the brain produces small amounts of two **hormones,** follicle-stimulating hormone (FSH) and luteinizing hormone (LH). These hormones, secreted by the pituitary gland into the blood, send signals to the sex organs. LH and FSH control the production of the sex hormones **estrogen** and **progesterone** by the ovaries. In childhood, the brain produces very small amounts of FSH and LH, but at puberty, production increases markedly. This stimulates a sudden upsurge in the production of the sex hormones by the ovaries, and the appearance of the sex hormones in turn triggers the changes of puberty.

At puberty, the female becomes sexually mature and fertile. Onset of reproductive maturity is signaled by the appearance of the **menses** and the beginning of the rhythmic female sexual cycle (menstrual cycle). The monthly female sexual cycle is characteristic of the female reproductive years, between puberty (at about ages 10 to 14) and the menopause (at about age 45 to 55).

The female sexual cycle has two purposes: the monthly production of a mature **ovum** (ovulation), and the preparation of the uterus to receive and support the ovum if fertilization occurs. The cycle (lasting about 28 days) is characterized by three distinct phases: the follicular phase, the luteal phase, and the menstruation phase, as shown in Figure 2.1.

- *The follicular phase.* During the first half of the cycle, under the influence of LH and FSH, the ovum develops and matures, nourished in the ovarian follicle. At the same time, estrogen produced by the ovary stimulates growth and proliferation of the epithelium of the uterus (the endometrium).

Midway through the cycle, at about day 14, **ovulation** occurs. A surge in LH and FSH production stimulates the follicle to swell and rupture, sending the mature ovum into the fallopian tube.

- *The luteal phase.* After ovulation, the follicle evolves into the corpus luteum, which begins to secrete large amounts of progesterone. During

the postovulatory, or luteal phase, progesterone stimulates further proliferation of the endometrium, preparing it to receive the fertilized ovum.

- *The menstruation phase.* At about day 23 of the cycle, if fertilization
  has not occurred, levels of estrogen and progesterone abruptly fall,
  and the endometrium is sloughed and expelled by contractions of the
  muscle in the wall of the uterus. The menstruation phase typically
  lasts about five days, after which the cycle repeats itself.

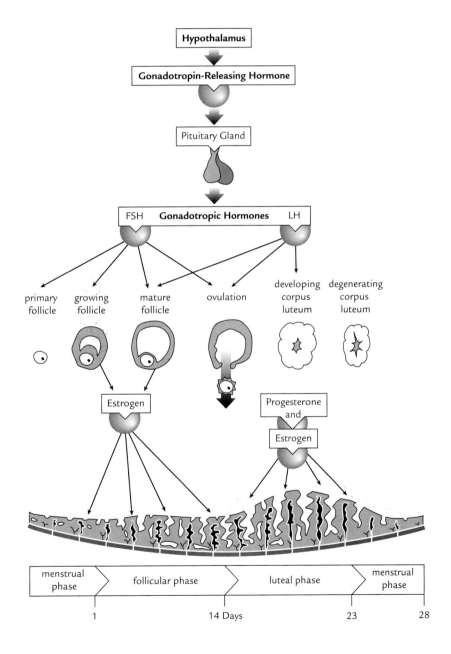

FIGURE 2.1

The female sexual cycle

Nutrition and Menarche

The attainment of female sexual maturity at puberty is modulated by a large increase in circulating estrogen. Many scientists think the trigger that initiates the surge in estrogen and the onset of puberty is attainment of a critical body weight and body fat percentage. Menarche in females is associated with achieving a weight of about 47 kg (104 lb), or about 17–22% body fat (Frisch and McArthur 1974).

In the industrialized world since the early 1800s, menarche has been occurring in females at younger ages. This decline in the age of menarche (and in males, the age of reproductive maturation) is generally attributed to better nutrition, as improvements in diet and higher caloric and protein intake allow children to reach a critical weight or body fat at younger ages. Young women today become fertile on average about two years earlier than women did 50 years ago (Tanner 1990).

Influence of Low Body Fat on the Female Sexual Cycle

How does body fat influence the menstrual cycle and fertility? Along with the ovaries, **adipose cells** play important roles in the production and metabolism of estrogens. Undernourished women with low fat stores produce less estrogens and tend to metabolize estrogens to inactive forms (Garner 1984). In addition, undernutrition impairs the secretion of LH and FSH by the pituitary, so the signal to the ovaries to produce estrogen is lost. Therefore, women with low body fat are hypoestrogenic, a condition that impairs ovulation and produces **amenorrhea** and **infertility.**

Weight loss of about 10 to 15% of normal weight for height delays menarche in the prepubertal female and causes amenorrhea in the mature female (Luke and Johnson 1993).

Whereas normal, nonathletic women have body composition with about 26% fat, competitive runners often have body fat percentages as low as 8 to 12%. Female endurance athletes and dancers are at high risk of developing amenorrhea and delayed menarche (Warren et al. 1986). See Figure 2.2. For example, over a third of competitive long-distance runners aged 12 to 45 have irregular menstrual periods or amenorrhea, and the more intense the training the greater the occurrence (Luke and Johnson 1993). Amenorrhea is also very common among women with the eating disorder anorexia nervosa (discussed in Chapter 8). Abnormally low weight and low body fat are common causes of infertility in industrialized countries. Fortunately, in almost all women, a return to a more normal body composition, usually about 22% fat, will produce a return of the menses and fertility.

adipose cell: a cell which stores fat; body fat consists of masses of adipose cells

amenorrhea: absence or cessation of the menses

infertility: the inability to conceive and produce viable offspring

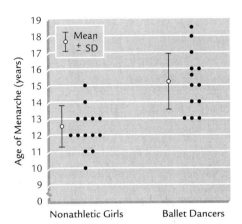

(Source: Adapted from Warren et al., 1986)

FIGURE 2.2

Age of merarche: nonathletic females vs. ballet dancers

Appetite and the Menstrual Cycle

The rhythmic rise and fall of hormones during the sexual cycle influences energy needs and appetite. During the luteal phase, the basal metabolic rate rises and, in most women, appetite increases. Caloric intake tends to be greater in the ten days prior to menstruation than in the ten days afterward (Barr and Janelle 1995).

In a recent study, Martini and others found that women consumed significantly more energy, protein, carbohydrate, and fat during the luteal phase of their cycles. Because overall food intake increased, micronutrient intakes were generally higher during the luteal phase as well (Martini et al. 1994).

Hormonal changes in the days preceding menstruation also cause sodium and water retention, and often, modest weight gain. In some women, these physiological changes become severe and may be symptoms of the **premenstrual syndrome (PMS).** PMS is characterized by the cyclic appearance of a variety of physical and emotional symptoms just prior to the menses, and the links between PMS, appetite, and nutrition are discussed in Chapter 9.

UNDERNUTRITION AND FERTILITY: THE 1944-45 DUTCH FAMINE

premenstrual syndrome (PMS): the regular monthly experience of distressful physical and/or emotional symptoms usually occurring several days before the menses; often characterized by nervousness, depression, fluid retention, and weight gain

Acute malnutrition experienced by populations during famine causes a dramatic reduction in fertility. Late in World War II, food supplies were severely restricted in parts of Holland. During the Dutch "hunger winter" of 1944 to 1945, mean daily calorie intake fell below 1000 kcal/day and protein intake to 30–40 g/day.

The famine occurred in a previously well-nourished population and lasted for about six months. Fifty percent of the women who experienced the famine developed amenorrhea, and the fertility rate fell to 53% of what was expected (Stein et al. 1975). Reinstitution of the food supply resulted in a rapid recovery in fertility.

OBESITY AND FERTILITY

Too much body fat can also produce amenorrhea and infertility. Because adipose cells produce estrogens, obese women have larger amounts of circulating estrogens than normal-weight women. Extreme obesity, possibly due to high levels of estrogens, disrupts the normal signaling between the pituitary and ovary, and the impairment of LH and FSH secretion produces anovulation and amenorrhea. Eight to ten percent of very obese women—defined as more than 75% above ideal body weight—lose their cycle or have abnormal menstrual cycles (Shoupe 1991).

Bates and Whitworth studied the effects of diet and weight loss in 18 obese, anovulatory women (average weight was greater than 40% of ideal weight) who were infertile. Thirteen women lost about 15% of their body weight, and in 11 of them, ovulation returned and 10 conceived (Bates and Whitworth 1982).

MALE FERTILITY AND NUTRITION

Severe undernutrition in males also impairs fertility. Chronic undernutrition associated with weight loss in prepubertal males will delay the onset of puberty and sexual maturation. Starvation in postpubertal men causes loss of libido, and a decrease in sperm motility. When weight loss approaches 25% of normal weight for height, sperm production ceases. Sperm number and motility return to normal as weight is regained. Chronic zinc deficiency has also been linked to pubertal delay and impaired fertility in men (Prasad, 1982).

NUTRITION BEFORE CONCEPTION

NUTRIENT NEEDS OF WOMEN DURING CHILDBEARING YEARS

This section briefly reviews the subject of women's nutrient needs during childbearing years. It is discussed in detail in Chapter 9. During the female sexual cycle, repeated blood loss in the menses increases a woman's requirements for nutrients involved in **hematopoiesis.** Because iron losses average about 1 mg/day during menstruation, the iron needs of adult women are 50% higher than those for men (the RDA for women

hematopoiesis: the formation and development of the blood cells

during the reproductive years is 15 mg, compared with 10 mg for adult men) (NAS 1989). Adequate folate status is also important to replace lost blood cells, as folate plays a central role in synthesis of nucleic acid during cell division in the bone marrow.

Independent of the menstrual cycle, a third nutrient of particular importance to women is calcium. During the reproductive years, with the demands of pregnancy and lactation and later during menopause, women lose more calcium from their bones than men. As a consequence, many more women than men suffer from osteoporosis during their later years.

In addition to these increased physiological needs, the diet and life-style patterns of many women may further jeopardize their nutritional health. Societies that prize thinness and leanness pressure young women to pursue chronic, low-calorie dieting. In the United States, national surveys have found that over half of women of childbearing age regularly diet to lose weight (Bendich 1993). Most low-calorie diets, whether self-selected or from popular dieting programs, are nutritionally inadequate. Nutrients often missing from low-energy diets include the ones that women need most: iron, calcium, zinc, and folate. Table 2.4 shows the percentage of RDAs of certain vitamins in five weight loss diets.

About 20% of women between 15 and 45 drink alcoholic beverages in moderate or heavy amounts. Nearly 30% of women smoke cigarettes, and 10 to 15% use oral contraceptives (Bendich 1993). These substances increase the requirements for certain nutrients. For example, women taking oral contraceptives often have greater needs for folate (discussed in Chapter 9), and heavy alcohol use increases iron losses.

Certain dietary patterns can also increase the risk of nutrient deficiencies. Women on vegetarian diets tend to have lower intakes of iron and zinc than those who are nonvegetarians (Institute of Medicine, *Nutrition during Pregnancy* 1990). Many women, particularly Blacks and women of Asian descent, are intolerant to lactose. Because most calcium-rich foods

TABLE 2.4

Vitamin content of five popular weight loss diets as a percentage of the RDAs

Vitamin	Set Point (% RDA)	Fit for Life (% RDA)	Immune Power (% RDA)	Family Circle (% RDA)	Eat to Succeed (% RDA)
Vitamin D	33	0	0	24	76
Vitamin E	30	83	18	99	117
Vitamin B_6	75	151	103	124	135
Folacin	74	130	65	79	81
Vitamin B_{12}	51	6	42	33	82

Adapted from A. Bendich, "Lifestyle and Environmental Factors", *Ann NY Acad of Sci* 678 (1993) 255-65.

are high in lactose, lactose intolerance sharply increases the risk of inadequate calcium consumption.

All of these factors are potential obstacles to optimal nutrition for women during their reproductive years. Adequate nutritional status before conception and in the first few weeks of pregnancy can be of vital importance to a successful pregnancy.

Over 40% of all pregnancies in the United States are unplanned, and optimal nutrition is critical in the periconceptional period, often before a woman knows she is pregnant. Beginning to plan for pregnancy when pregnancy is confirmed, usually well into the first trimester, may be too late. Throughout their reproductive years, women should strive to consume a high-quality diet and maintain normal body weight. But how well do diets of women in the United States prepare them for the challenges of pregnancy?

NUTRITIONAL SURVEYS OF WOMEN OF REPRODUCTIVE AGE

For folate and several other micronutrients thought to be important in the periconceptional period, nationwide surveys of women of childbearing age have found that most women consume amounts well below recommended levels. Current guidelines recommend an intake of 0.4 mg/day of folate by all women of childbearing age capable of becoming pregnant, but fewer than 10% achieve an intake near 0.4 mg, and about a quarter of women of childbearing age have intakes of only about 0.1 mg/day (U.S. Human Nutrition Information Service 1987).

In preparation for pregnancy, adequate consumption of zinc, calcium, and iron is also important, but 50 to 60% of women consume less than 70% of the RDA for zinc; the average intake of calcium is only about two thirds of the RDA; and only 10% of women achieve the RDA for iron (Block and Abrams 1993). Table 2.5 shows mean nutrient intakes of women between the ages of 19 and 50.

Approximately one in five women of reproductive age have depleted iron stores (Institute of Medicine, *Nutrition during Pregnancy* 1990). Figure 2.3 breaks down the percentage of iron deficiency by different groups. Women with incomes less than 130% of the poverty level are at particularly high risk for nutrient deficiencies (see Figure 2.4). More than half of women at this income level are between 15 and 24 years old, and they consume less than two thirds of the RDA for vitamins A, C, and B_6, and calcium, iron, and zinc (Block and Abrams 1993).

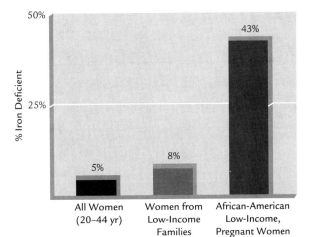

(Source: NHANES II and III, National Health and Nutrition Examination Surveys)

FIGURE 2.3

Iron deficiency in women in the United States

Nutrient	Daily Intake	Percent of RDA (nonpregnant levels)
Energy	**1528 kcal**	**69**
Protein	60.7 gm	121
Vitamin A	828 µg RE	104
Vitamin E	7.1 mg	89
Vitamin C	77 mg	128
Thiamin	1.05 mg	95
Riboflavin	1.34 mg	103
Niacin	16.1 mg	107
Vitamin B$_6$	**1.16 mg**	**73**
Folic acid	189 µg	105[a]
Vitamin B$_{12}$	4.85 µg	243
Calcium	**614 mg**	**77**
Phosphorus	966 mg	121
Magnesium	**207 mg**	**74**
Iron	**10.1 mg**	**67**
Zinc	**8.6 mg**	**72**
Copper[b]	1.0 mg	67
Sodium[c]	2368 mg	474
Potassium[c]	2066 mg	103

Bolded items are ≤ 80% of the RDA or ESADDI.

a < 50% of currently recommended levels of 400 µg (U.S. Public Health Service, 1992)

b Based on estimated safe and adequate daily dietary intake.

c Based on estimated minimum requirement.

Adapted from USDA, Human Nutrition Information Service, Nutrition Monitoring Division, *Nationwide Food Consumption Survey, Continuing Survey of Food Intakes by Individuals,* NFCS, CSFII Report No. 85-4 (Washington D.C.: U.S. Govt. Printing Office, 1987) 46–49.

TABLE 2.5

Mean nutrient intakes; Continuing Survey of Food Intakes by Individuals (CSFII), women aged 19–50

(Source: Adapted from Block and Abrams, 1993.)

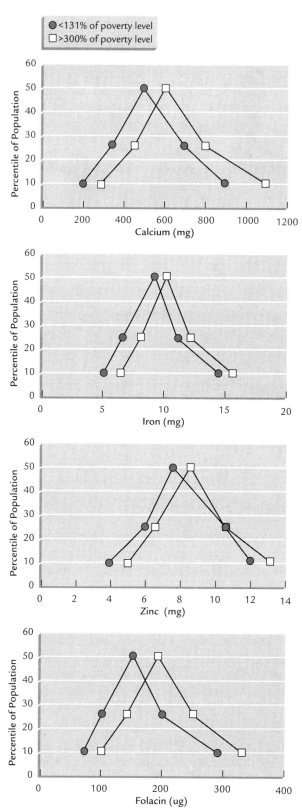

FIGURE 2.4

Distribution of nutrient intake based on income level, women aged 19–50: folate, zinc, calcium, and iron

IMPORTANT NUTRIENTS FOR WOMEN CONSIDERING PREGNANCY

Iron

Women should try to achieve ample iron reserves prior to conception. Pregnancy sharply increases the need for iron. Total iron requirements during gestation average approximately 1 g—nearly half of an adult woman's total body iron pool of 2.2 g. The daily need for iron during pregnancy is over 4 mg—considerably greater than the typical daily iron absorption of 1.3 mg in nonpregnant women. Maintaining adequate iron status during pregnancy helps prevent **iron-deficiency anemia** which can be harmful to both the mother and the fetus (discussed in more detail in the next chapter).

Folate

Folate is a nutrient of vital importance for women considering pregnancy. Poor folate status during the periconceptional period appears to increase the incidence of certain birth defects, termed *neural tube defects* (NTDs). Neural tube defects is a collective term used to describe various **congenital abnormalities** of the fetal spine and nervous system.

The most common NTD is spina bifida, a condition in which the spine does not completely develop and close around the spinal cord. Babies born with spina bifida have a permanently damaged spinal cord. They often cannot walk, their bladder and bowel functions are abnormal, and they have many other lifelong problems. Other NTDs are fatal to the fetus and result in miscarriages and stillbirths. NTDs are not uncommon: nearly half a million infants worldwide are born with NTDs each year. In the United States, NTDs affect 2 to 3 percent of pregnancies (Centers for Disease Control 1991).

NTDs are thought to occur because of interactions between genetics and environmental factors. Genetic factors are important because a woman who has an NTD-affected pregnancy has a much greater risk of having a second occurrence in a subsequent pregnancy (Centers for Disease Control 1992). Environment during the periconceptional period also plays a role, and scientists have focused on the importance of adequate micronutrients during this period.

Folic acid is the nutrient most strongly implicated in studies of NTDs. Folate has a central role in the metabolism of several amino acids and is vital for the synthesis of nucleic acids. Folate deficiency impairs cell growth and replication, and inadequate folate in early pregnancy may produce abnormalities in fetal development. Research in animals

iron-deficiency anemia: a reduction below normal in the number or volume of the red blood cells, or the quantity of hemoglobin in the blood, due to insufficient iron

congenital abnormality: present at and existing from the time of birth; may be hereditary or due to an influence during fetal growth

indicates impaired folate status can cause NTDs, and use of the anticonvulsant valproic acid during human pregnancy blocks the metabolism of folic acid to its active form, which increases risk of NTDs 20-fold (Trotz et al. 1985).

In the last 15 years, several large epidemiological studies around the world have provided evidence that folate supplementation during the periconceptional period protects against NTDs (Centers for Disease Control 1991). In a recent British study, 1000 women who had already had a NTD-affected pregnancy and who were planning a new pregnancy were randomly divided into four groups (MRC Vitamin Study Research Group 1991). The first group received 4 mg of folic acid, the second a multivitamin plus 4 mg of folic acid, the third a multivitamin without folic acid, and the fourth group neither multivitamins nor folic acid.

The results were dramatic. Use of multivitamins without folic acid had no protective effect. Folic acid, either alone or in conjunction with the multivitamin, was associated with a 71% reduction in the recurrence of NTDs. Because of this substantial protective effect, the study was halted earlier than planned so all the women could receive the benefits of supplementation with folate. The U.S. Centers for Disease Control (CDC) subsequently recommended that women who have had a previous NTD-affected pregnancy receive 4 mg of folic acid beginning at least four weeks before conception and through the first trimester of pregnancy.

A recent Hungarian study examined the efficacy of a multivitamin supplement containing 0.8 mg of folic acid in reducing the incidence of first occurrences of congenital malformations, including NTDs. Over 7000 women planning a pregnancy were randomly assigned to receive either the multivitamin containing folic acid or a trace element supplement. The number of the women who became pregnant was 4750. The women who received the folic acid-containing multivitamin had over a third fewer congenital malformations. There were no NTDs in the folic-acid/multivitamin group, while six NTDs occurred in the group treated with the trace element supplement (Czeizel and Dudas 1992).

Several other studies of NTD occurrence and folic acid intake in the periconceptional period support these findings (Centers for Disease Control 1992). Based on these data, in September 1992 the U.S. Public Health Service (USPHS) recommended that all women of childbearing age who are capable of becoming pregnant consume 0.4 mg of folic acid per day to reduce the risk of NTDs.

The recommendation by the USPHS also identified several questions that remain unanswered. These include the precise intake of folate necessary to reduce the risk of NTDs, and the potential role of other micronutrients in the prevention of NTD. Also, it is unclear whether simple dietary deficiency of folate or a possible defect in folate metabolism is responsible for the increased risk of NTDs. In spite of these uncertainties, US government agencies estimate a 50% reduction in NTD births could occur if all women consumed 0.4 mg of folic acid daily throughout their childbearing years.

To meet this recommendation will require significant changes in dietary intake, as national surveys have shown about 50% of women consume less than 0.2 mg of folate on any given day (US Human Nutrition Service 1987), and only about ten percent of women of childbearing age achieve intakes near 0.4 mg. Adequate folate intake can be achieved by diet, but the diet must be carefully chosen and include ample vegetables and fruits.

Scientists are currently debating the most appropriate method for increasing intake of folic acid among women of childbearing age. One approach being discussed is fortification of grain products with folate, as has been done with vitamin D in milk and iodine in table salt. Tables 2.6 and 2.7 list a sample day's menu providing an adequate supply of folate and typical sources of folate, respectively.

Food, Amount	Folic Acid (µg)
Breakfast	
Fortified cereal, 1 oz.	100
Milk, 1 cup	12
Banana, 1 med.	33
Orange juice, 1 cup	62
Lunch	
Whole wheat bread, 2 slices	26
Peanut butter, 1 Tbs	12
Carrot, 1 med.	26
Milk, 1 cup	12
Dinner	
Chicken, 1 breast	2
Broccoli. ½ cup	52
Potato, baked	19
Orange, 1	60
Milk, 1 cup	12
	428

TABLE 2.6

Sample day's menu providing 400+ micrograms folate

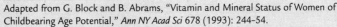

Adapted from G. Block and B. Abrams, "Vitamin and Mineral Status of Women of Childbearing Age Potential," *Ann NY Acad Sci* 678 (1993): 244–54.

Food	Folate per Usual Serving (µg)	Folate per 100 g (µg)
Liver	383	426
Superfortified cereals	242	991
Cold cereals (not bran or super fortified)	112	275
Pinto, navy, and other dried beans (cooked)	84	100
Asparagus	82	101
Spinach	70	128
Instant breakfast, diet bars, supplements	65	112
Bran and granola cereals	58	236
Broccoli	53	65
Avocados	49	62
Okra	49	84
Brussels sprouts	47	60
Orange juice	43	41
Artichokes	43	46

TABLE 2.7
Food sources high in folate

Adapted from Subar et al., "Folate Intake and Food Sources in the U.S. Population" *Am J Clin Nutr* 50 (1989): 508.

IMPORTANCE OF PREPREGNANCY WEIGHT

Women of childbearing age should strive to maintain a normal body weight, as prepregnancy maternal weight is a major determinant of fetal growth and pregnancy outcome. Table 2.8 shows normal weight ranges for specific height.

Underweight

Women who are lighter than ideal body weight before pregnancy tend to deliver infants that are smaller than those of heavier women, even if they gain as much weight during pregnancy. Women who enter pregnancy underweight have a higher risk of delivering preterm, and of delivering a low-birth-weight infant (Institute of Medicine 1990). In addition, maternal complications, including anemia, are more common in women who are underweight before conception.

Overweight

Women who are overweight also are at higher risk for a poor pregnancy outcome. In women who are more than 130% of their ideal body weight, serious maternal complications, such as diabetes and hypertension

| | Women | | |
Height	Small Frame	Medium Frame	Large Frame
4′10″	102–111	109–121	118–131
4′11″	103–113	111–123	120–134
5′	104–115	113–126	122–137
5′1″	106–118	115–129	125–140
5′2″	108–121	118–132	128–143
5′3″	111–124	121–135	131–147
5′4″	114–127	124–138	134–151
5′5″	117–130	127–141	137–155
5′6″	120–133	130–144	140–159
5′7″	123–136	133–147	143–163
5′8″	126–139	136–150	146–167
5′9″	129–142	139–153	149–170
5′10″	132–148	142–156	152–173
5′11″	135–148	145–159	155–176
6′	138–151	148–162	158–179

Adapted from height and weight tables issued by the Metropolitan Life Insurance Company. In indoor clothing weighing 3 lbs; shoes with 1 ″ heels.

TABLE 2.8
1983 Metropolitan height and weight table for women with small, medium, and large frames

during pregnancy, are more common. (These topics will be discussed in more detail in the next chapter.) However, weight loss, even in the very obese woman, should never be attempted during pregnancy (discussed later in this chapter).

NUTRITIONAL PLANNING FOR PREGNANCY

Although attention is often focused on nutrition during pregnancy, one of the best ways to achieve a healthy pregnancy outcome is to actively plan for pregnancy and enter pregnancy in good nutritional health.

Many factors influence a woman's nutritional needs and affect her ability to achieve adequate nutrition before conception. Risks associated with poor pregnancy outcome and low birthweight can often be recognized before a woman becomes pregnant. The U.S. Institute of Medicine has identified risk factors in the preconceptional period, shown in Table 2.9, and many of them have an impact on a woman's nutritional needs. Proper preconceptional care includes counseling on avoidance of smoking, alcohol, and substance abuse, and treatment of medical problems. Several maternal illnesses—including diabetes, hypertension, and phenylketonuria—if adequately controlled before pregnancy, present a less serious risk to the health of the mother and fetus during pregnancy.

- Certain chronic illnesses such as diabetes and hypertension

- Smoking

- Moderate to heavy alcohol use and substance abuse

- Inadequate weight for height and poor nutritional status

- Susceptibility to rubella and other infectious agents

- Age (under 17 and over 34)

- Likelihood of a very short interval between pregnancies

- High parity

Adapted from: Institute of Medicine Committee to Study the Prevention of Low Birth-weight, *Prevention of Low Birthweight* (Washington, D.C.: National Academy Press, 1985)

TABLE 2.9
Preconceptional risk factors for poor pregnancy outcome

Vitamin and Mineral Supplements

Although women who are preparing for pregnancy should try to build ample nutrient reserves prior to conception, excessive use of vitamin and mineral supplements can be harmful and should be avoided. High doses of vitamin A, for example, are a known **teratogen** (discussed in Chapter 3). Women considering pregnancy should avoid vitamin A supplements, particularly at doses greater than 4000 IU daily. The best way to ensure adequate nutritional status before pregnancy is to consume a balanced diet and follow a healthy lifestyle.

Achieving Normal Weight

Because both high and low pregnancy weight are associated with a number of unfavorable pregnancy outcomes, women considering pregnancy should try to attain normal body weight. A woman who is underweight should gradually raise her energy intake by consuming a balanced diet and eating more nutrient-dense foods. Overweight women who are planning a pregnancy should avoid crash diets, as we have seen that they are usually nutritionally inadequate and can produce deficiences of essential nutrients. Regular exercise can be helpful in achieveing weight goals.

Dietary Guidelines for Women Considering Pregnancy

teratogen: a substance capable of causing abnormal fetal development

The U.S. National Academy of Sciences Committee on Nutrition provides broad dietary guidelines for women during the reproductive years and special recommendations for women considering pregnancy (Institute of Medicine 1992). These important guidelines are shown in Table 2.10.

Recommendations for All Women
Based on Dietary Guidelines for Americans

- Eat a variety of foods.

- Choose a diet low in fat, saturated fat, and cholesterol. This reduces the risk of chronic disease and may help you manage your weight. Keeping fat intake under control also helps make room for foods that are rich in vitamins and minerals. Aim to have 2 servings daily of low-fat meat, fish, or poultry or of legumes. Also aim to have 2 to 3 servings of low-fat, calcium-rich milk products such as low-fat milk, cheese, or yogurt. One cup of milk or yogurt is an example of one serving.

- Choose a diet with plenty of vegetables, fruits, juices, and grain products. Choose whole-grain products at least part of the time. Aim to have 2 or more servings of fruit or juice, 3 or more servings of vegetables, and 6 to 11 servings of grains each day. One slice of bread and ½ cup of rice, fruit, or vegetables are examples of one serving.

- Use sweets, sugars, and soft drinks only in moderation.

- Use salt and salty foods only in moderation. This helps prevent or control hypertension. However, restricted salt intake is not believed to be useful for the prevention or control of hypertension during pregnancy.

Special Recommendations
Before Pregnancy

- Maintain a healthy weight. Regular physical activity will help.

- If you need to gain or lose weight, do so gradually (no more than 1 to 2 lb/week).

- If you are trying to become pregnant and you ordinarily drink alcoholic beverages, omit alcohol or cut way back on the amount you drink.

- Avoid the use of any kind of medication or supplement unless it is prescribed or approved by a physician.

- Avoid cigarettes, smokeless tobacco, alcoholic beverages, and illegal drugs.

Adapted from Institute of Medicine, Subcomittee for a Clinical Application Guide, *Nutrition During Pregnancy and Lactation: An Implementation Guide* (Washington D.C.: National Academy Press, 1992).

TABLE 2.10

Dietary guidelines for women of reproductive age and those considering pregnancy

FIGURE 2.5
Fertilization and implantation: development of the blastocyst

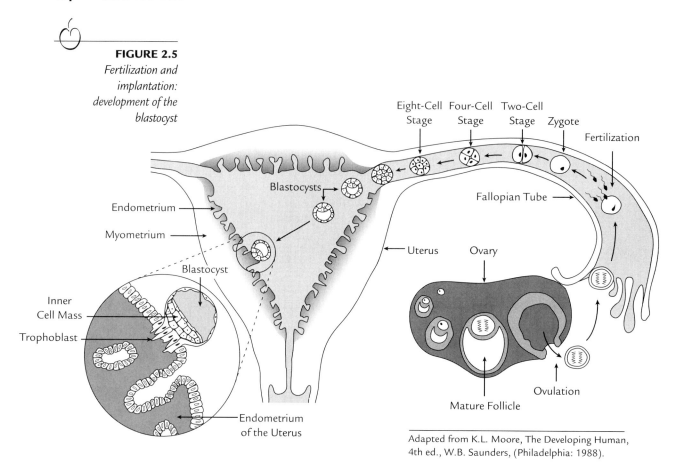

Adapted from K.L. Moore, The Developing Human, 4th ed., W.B. Saunders, (Philadelphia: 1988).

FERTILIZATION AND IMPLANTATION

blastocyst: the spherical structure produced by cleavage of a fertilized ovum, consisting of cells surrounding a fluid-like cavity

fetus: the developing baby in the uterus; specifically the unborn offspring in the postembryonic period which in humans is from eight weeks after fertilization until birth

A pregnancy begins with fertilization. Although 200 to 300 million sperm are deposited in the vagina, only around 100 succeed in moving through the uterus and fallopian tubes and getting close to the ovulated egg. The genetic material of the mother and father join when a single sperm penetrates the egg, usually in the fallopian tube near the ovary.

The fertilized egg begins dividing as it moves toward the uterus, as shown in Figure 2.5, and implantation occurs about six days after fertilization. At the time of implantation, the tiny **blastocyst** is composed of dozens of rapidly dividing cells, and is about 0.1 mm in size (Moore 1990). The developing blastocyst buries itself in the endometrial cells lining the wall of the uterus. It contains two distinct groups of cells: an inner embryonic cluster that will develop into the **fetus,** and an outer shell composed of **trophoblastic cells.** The trophoblasts will join with cells from the uterine endometrium to form the placenta. By about 14 days after fertilization, the developing embryo and placenta have enlarged to form a tiny protrusion visible on the uterine wall.

THE PLACENTA

The developing fetus is completely dependent on nutrients and oxygen coming from the mother's bloodstream. Nutrients in maternal blood must move across the placenta to reach the fetus, but the placenta is much more than a simple filter. It is a dynamic organ that selectively transfers some substances, metabolizes others, and provides excretory and respiratory functions for the fetus. The placenta also serves as an important endocrine organ during pregnancy. It produces at least ten major hormones that are secreted into the maternal circulation, and it profoundly affects maternal metabolism.

DEVELOPMENT AND STRUCTURE OF THE PLACENTA

After implantation, the tissues and blood vessels of the uterus directly surrounding the blastocyst break down to form lacunae—small spaces filled with maternal blood. As shown in Figure 2.6, maternal blood circulates through the lacunae, as fresh blood enters through uterine arteries and exits through veins.

Meanwhile, the trophoblast develops fingerlike projections (called villi) that grow into the newly formed lacunae. These villi contain rich beds of fetal capillaries. As the villi develop, they form delicately branched structures resembling tree roots that float in the lacunar spaces. Although the fetal and maternal tissues intermingle in the developing placenta, fetal and maternal blood does not mix. The fetal and maternal blood are always separated by a thin placental membrane only a few cells thick.

The multiple, branched villi provide a large surface area (over 13 square meters) for efficient exchange of gases, nutrients, and waste products between the mother and fetus. The weight, blood flow, and villous surface of the placenta increase progressively during gestation, to supply the increasing nutrient needs of the growing fetus. A normal, mature placenta weighs between 1 and 2 pounds at term.

trophoblastic cells: the cell layer covering the blastocyst, which attaches the fertilized ovum to the uterine wall and becomes the placenta and the membranes that protect the developing organism

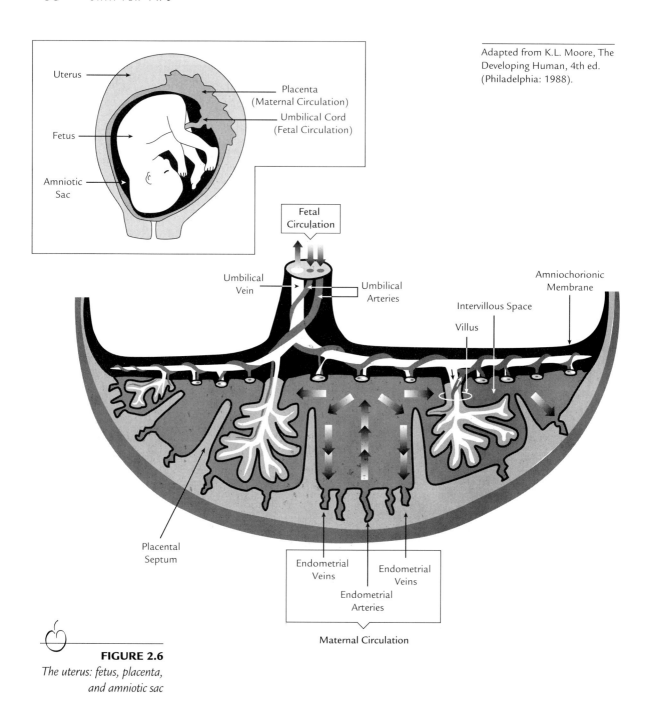

Adapted from K.L. Moore, The Developing Human, 4th ed. (Philadelphia: 1988).

Uterus

Placenta (Maternal Circulation)

Umbilical Cord (Fetal Circulation)

Fetus

Amniotic Sac

Fetal Circulation

Umbilical Vein

Umbilical Arteries

Amniochorionic Membrane

Intervillous Space

Villus

Placental Septum

Endometrial Veins

Endometrial Veins

Endometrial Arteries

Maternal Circulation

FIGURE 2.6

The uterus: fetus, placenta, and amniotic sac

MATERNAL–FETAL NUTRIENT TRANSFER

During pregnancy a substantial amount of blood is directed to the placenta, and uterine blood flow at term represents about 7% of the total cardiac output—or about 30 liters of blood each hour. This blood flow is a major determinant of nutrient transfer to the fetus: a small placenta with limited blood flow can transfer fewer nutrients and impair fetal growth. Maternal malnutrition during pregnancy can reduce placental size and, consequently, fetal growth (Gonzalez-Cossio and Delgado 1991).

Nutrient transfer across the thin placental membrane is a complex and carefully regulated process. Transfer occurs via four major mechanisms: **simple diffusion,** facilitated diffusion, **active transport,** and **endocytosis** (see Figure 2.7). In addition, placental metabolism influences the transfer of certain substances from mother to fetus. For example, a portion of the glucose and amino acids taken up by the placenta are used for placental growth and metabolism, and only the remainder is transported into the fetal circulation.

Transfer of Carbohydrates

Glucose crosses the placenta via facilitated diffusion. The capacity of the placenta to transport glucose is enormous, with transport mechanisms reaching their limit only at very high levels of maternal blood glucose (about 450 mg/dl—over four times normal levels) (Oakley et al. 1986).

Transfer of Lipids

Although both cholesterol and free fatty acids (FFA) can cross the placenta via passive diffusion, cholesterol is readily transferred, while the transplacental movement of FFA is slow. Triglycerides and phospholipids do not cross the placenta, but are taken up by the placenta, metabolized, and some of the FFA subsequently released into the fetal blood (Coleman 1986).

Transfer of Amino Acids and Protein

Most neutral amino acids are actively transported across the placental membrane, and they are present at higher concentrations in the fetal than in the maternal blood (Cetin et al. 1988). Nonneutral amino acids cross the placental membrane more slowly and are found in lower concentrations in fetal blood, indicating that transfer is probably via simple or facilitated diffusion.

Most intact proteins do not cross the placental membrane because of their large size, so the fetus must synthesize almost all of its protein from

simple diffusion: the random movement of molecules toward a uniform distribution throughout the available volume

active transport: the passage of substances across a membrane by an energy-consuming process

endocytosis: the process through which substances are taken into the cell by the invagination of the cell membrane

FIGURE 2.7

Placental transfer mechanisms and substances transported by each mechanism

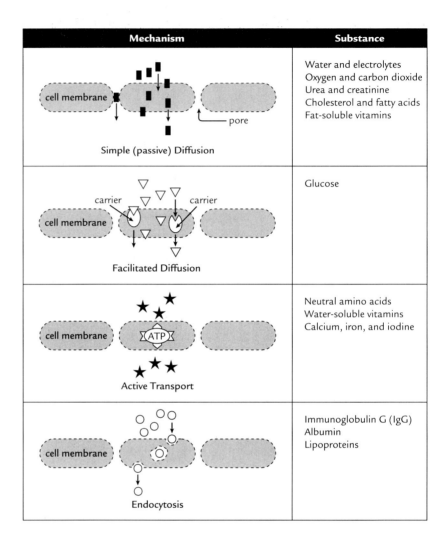

Mechanism	Substance
Simple (passive) Diffusion	Water and electrolytes Oxygen and carbon dioxide Urea and creatinine Cholesterol and fatty acids Fat-soluble vitamins
Facilitated Diffusion	Glucose
Active Transport	Neutral amino acids Water-soluble vitamins Calcium, iron, and iodine
Endocytosis	Immunoglobulin G (IgG) Albumin Lipoproteins

albumin: in humans, a blood protein formed mainly in the liver; it constitutes over half of all protein in the plasma

fibrinogen: a blood protein that is converted to fibrin during blood clotting

IgG: immunoglobulin G, a protein synthesized by lymphocytes that is capable of acting as an antibody in the blood and tissues

individual amino acids. However, a few proteins are selectively transported through the placenta, including **albumin** and **fibrinogen** (Rosso 1990). The maternal immune protein **IgG** is also rapidly and efficiently transported. It provides the fetus with significant protection from infections.

Transfer of Vitamins

Lipid-soluble vitamins move through the placenta via simple or facilitated diffusion, as concentrations in fetal blood are generally lower than in maternal blood. There are specific binding proteins for vitamins A and D in the placenta, and the placenta is able to synthesize both 25-OH vitamin D and 1,25-(OH) vitamin D (Whitsett et al. 1981).

In contrast, water-soluble vitamins move across the placenta via active transport, and most are present in higher concentrations in fetal than in maternal blood. For example, ascorbic acid moves across via a carrier-mediated, energy dependent mechanism, and fetal plasma levels of vitamin C are 50% higher than those of the mother (Institute of Medicine 1990). There are specific placental receptors for transcobalamin II (Friedman et al. 1977), a carrier protein of vitamin B_{12}, and vitamin B12 is actively transported through the placenta to the fetus.

Transfer of Minerals

The placenta has specific receptors for maternal transferrin, and iron is removed from transferrin by the placenta and transported to the fetal circulation for incorporation into fetal hemoglobin (Brown et al. 1983). Calcium needed for the formation of the fetal skeleton is bound to a specific placental calcium-binding protein(Lester 1986), and it is actively transported across the placenta against a concentration gradient.

There is little information about placental transfer of the other minerals and trace elements. The concentration of some minerals, such as zinc, is higher in fetal than maternal blood, while certain minerals, including copper, are found at lower concentrations in fetal blood (Rosso 1990).

In general, nutrient transfer by the placenta accelerates during the last weeks of pregnancy, building up fetal stores and increasing fetal plasma concentrations in preparation for birth. For example, calcium transfer reaches a peak in the last month, when over 300 mg are moved across to the fetus each day. Infants born prematurely miss this critical late period in utero, and plasma concentrations of vitamins E, B_6, B_{12}, and folate are lower in **preterm infants** compared to term newborns. Also, stores of iron are lower in preterm newborns, making them more prone to iron-deficiency anemia during infancy.

MATERNAL–FETAL RESPIRATORY AND EXCRETORY EXCHANGE

Oxygen diffuses across the placental membrane from maternal red blood cells to fetal red cells. The fetus requires a constant supply of oxygen for normal growth, and the mother's nutritional status can have a significant impact on oxygen delivery to the fetus.

Insufficient iron and folate during pregnancy are common causes of maternal anemia (discussed in the next chapter), which reduces the oxygen-carrying capacity of the blood and, if severe, can impair growth of the fetus. Anemia during pregnancy also increases the risk of preterm birth and other complications (Murphy et al. 1986).

Substances also move from the fetus to the mother through the placenta. During pregnancy, maternal respiratory rate increases and lowers the amount of CO_2 in the maternal blood, allowing carbon dioxide to move readily (by passive diffusion) from fetal blood (where concentrations are

preterm infant: an infant born after less than 37 weeks gestation

higher) to maternal blood, where it is excreted through the mother's lungs. Metabolic wastes of the fetus, such as urea and creatinine, move through the placenta into the maternal bloodstream and are subsequently excreted by the maternal kidney and liver.

THE ENDOCRINE FUNCTION OF THE PLACENTA

The placenta is a major endocrine organ during pregnancy and secretes steroid and peptide hormones that have profound effects on both mother and fetus. Placental progesterone and estrogen are essential for fetal development, and they help maintain uterine blood flow to the growing fetus. Hormones secreted by the placenta modify maternal physiology and have wide-ranging effects on maternal nutrition and metabolism (discussed in more detail later in this chapter). These alterations allow the mother to support the developing fetus and maintain her own health during pregnancy.

EFFECTS OF NUTRITION ON PLACENTAL GROWTH

Proper growth and function of the placenta is vital to the health of the fetus, and placental development is dependent on adequate maternal nutrition. Compared to well-nourished mothers, poorly nourished mothers have smaller placentas with fewer cells and reduced blood flow (Rosso 1983). A smaller placenta is less able to supply the developing fetus with necessary nutrients and oxygen (and can limit fetal growth).

A study of poorly nourished Guatemalan women found that an inadequate diet during pregnancy was associated with both lower placental weight and lower infant birthweight (Lechtig 1975). Fetal growth impairment in mothers who are poorly nourished may be due to both inadequate placental size and blood flow, as well as low levels of nutrients available in the maternal blood.

THE FETUS

After fertilization, blastocyst development, and implantation, prenatal life can be divided into two major periods: the embryonic phase and the fetal phase. The embryonic phase consists of the first eight weeks of gestation, and the fetal phase is the period from the beginning of the third month until term. Figure 2.8 illustrates embryonic and fetal development.

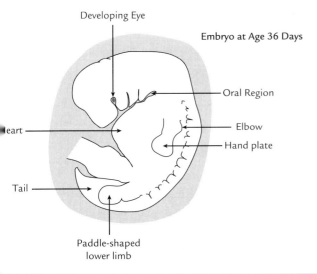

Developing Eye

Embryo at Age 36 Days

Oral Region

Elbow

Hand plate

Heart

Tail

Paddle-shaped
lower limb

Adapted from K.L. Moore,
The Developing Human, 4th ed.
(Philadelphia: 1988).

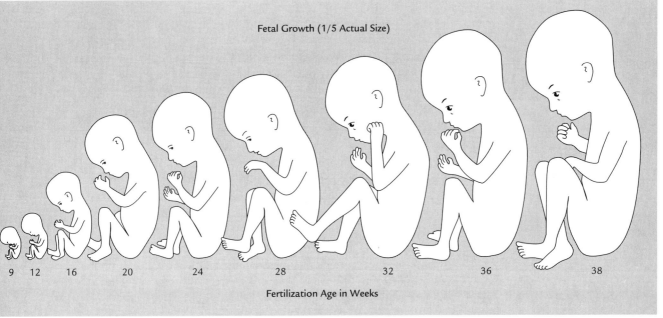

Fetal Growth (1/5 Actual Size)

| 9 | 12 | 16 | 20 | 24 | 28 | 32 | 36 | 38 |

Fertilization Age in Weeks

FIGURE 2.8

*Overview of embryonic and
fetal development*

EMBRYONIC PHASE

The first few weeks after conception are a hazardous time for the fertilized ovum: 10% fail to implant, and over 50% of those that do implant and form embryos spontaneously abort—most without the mother recognizing conception (Tanner 1990). Many embryos fail because of developmental abnormalities in the embryo itself, or in the surrounding nutritive tissues.

When the embryo is about three weeks old, organs and body parts begin to form. As mentioned in Chapter 1, this critical period of the embryonic phase is termed *morphogenesis*. The first step is the differentiation of the embryonic cells into three germinal layers. The outer layer (the ectoderm) forms the brain, nervous system, and skin. The middle layer (the mesoderm) gives rise to the cardiovascular, urinary, and musculoskeletal systems. The innermost layer (the endoderm) forms the digestive and respiratory systems and most of the major glands. By nine to ten weeks, although the fetus weighs only about 6 g, all of the major organ systems are present, the fetal heart begins to beat, and the fetus can move.

During the embryonic phase, growth is unique in that it results from cell division and a rapid increase in the number of new cells (hyperplastic growth). Later growth in utero (after about 25 weeks), and during childhood and adolescence, is due mainly to an increase in size of existing cells (hypertrophic growth), many of which formed early in prenatal life (Tanner 1990).

Optimal, balanced nutrition is important during the first trimester because the embryo is extremely vulnerable to changes in nutrient supply during morphogenesis. Absence or excess of nutrients during this critical period may impair development and produce a permanent growth anomaly or congenital malformation.

In animals, high doses of vitamin A, or deficiencies of zinc, folate, and vitamin B_6 during the first trimester, increase the incidence of malformations in offspring (Apgar 1992), (Kirksey and Wasynczuk 1993). Comparable effects may occur in humans, as will be discussed later in this chapter.

FETAL PHASE

During the fetal phase, growth velocity is higher than at any other time in life. The major organs and tissues, established in the embryonic phase, mature and acquire their basic adult characteristics. The weight of the developing fetus increases from 6 g to over 3000 g during the second and third trimesters—a remarkable 500-fold increase (see Figure 2.9). In contrast to the embryonic phase, during which nutrient deficiencies may produce developmental malformations, nutrient deficiencies later in pregnancy are more likely to affect the overall growth of the fetus (Tanner 1990).

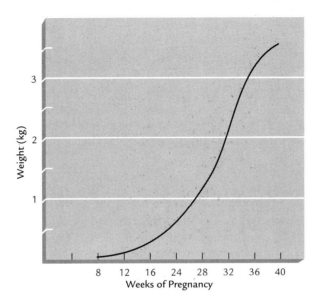

FIGURE 2.9
Changes in fetal body weight during gestation

Body Composition during Fetal Growth

Substantial changes in fetal body composition take place during its growth and development. The overall trend is a progressive decrease in the proportion of body water and an increase in protein, fat, and minerals (Widdowsen 1981). See Table 2.11.

During the first 25–30 weeks of gestation, almost all tissue formed is lean tissue, but during the last trimester the fetus stores considerable amounts of fat. In healthy term infants, body fat is approximately 15%.

Gestational Age (weeks)	Body Weight (g)	Per 100 g Body Weight			
		Water (g)	Protein (g)	Lipid (g)	Other (g)
24	690	88.6	8.8	0.1	2.5
26	880	86.8	9.2	1.5	2.5
28	1,160	84.6	9.6	3.3	2.4
30	1,480	82.6	10.1	4.9	2.4
32	1,830	80.7	10.6	6.3	2.4
34	2,230	79.0	11.0	7.5	2.5
36	2,690	77.3	11.4	8.7	2.6
38	3,160	75.6	11.8	9.9	2.7
40	3,450	74.0	12.0	11.2	2.8

Adapted from P. Rosso, *Nutrition and Metabolism in Pregnancy*. (Oxford: Oxford University Press, 1990).

TABLE 2.11
Changes in fetal body composition during gestation

Fetal fat deposition during late pregnancy provides the infant with reserves of energy that can be used, if needed, in the first few months after birth.

Similarly, increasing amounts of calcium, magnesium, and iron are deposited in the fetus in the final trimester. Calcium concentration in the fetus increases as the skeleton forms, from 3 g/kg lean tissue at 13 weeks gestation to 10 g/kg at term (Rosso 1990). The fetus also requires substantial iron in the last trimester to support growth of red blood cells and muscle, drawing heavily on the resources of the mother in the final weeks of pregnancy.

Disturbances in Fetal Development

During a normal, well-nourished pregnancy, optimal fetal growth produces a newborn with the full capability for further physical and mental development. Two major goals of pregnancy concern gestational period and birthweight:

1. A gestational period of at least 37 weeks is desirable. The average length of pregnancy is 40 weeks, and between 37 and 42 weeks is considered normal. Babies born after 37 weeks are called term infants. Those born earlier are called preterm infants.

   In preterm infants, fetal tissues (particularly the lungs) have not developed fully and are not completely ready for life outside the womb. Physiological immaturity increases the risk of respiratory problems, abnormal bleeding, infection, and death.

2. A normal birthweight is greater than 2500 g (5 lb, 8 oz). Babies weighing less than 2500 grams at birth are termed low-birth-weight (LBW) infants. LBW infants are mainly of two types: either small because they are preterm (but appropriately sized for their age) or term (or near-term) infants but abnormally small. LBW infants born after 37 weeks are called small for gestational age (SGA). These two types are shown in Figure 2.10. The baby marked A is born a month early (preterm), weighs 2500 g, and is of normal weight for her age. The baby B is full term, weighs 2500 g, and is small for gestational age.

   Birthweight is a strong predictor of infant health. Compared to normal-weight infants, LBW infants have much higher rates of morbidity and mortality. Complications in the **neonatal period** include low blood sugar, low blood calcium, and breathing difficulties. Because of these problems, LBW infants are 40 times more likely to die in the first few weeks of life, compared with normal-weight infants, making LBW a major cause of infant mortality in the United States (Institute of Medicine 1985).

   The distinction between low birthweight due to preterm birth and low birthweight due to being small for gestational age is an important one. Most preterm infants who are otherwise unimpaired (and who survive

neonatal period: the first four weeks after birth

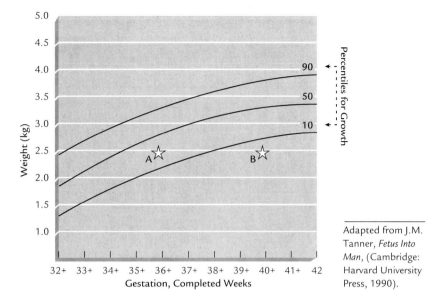

Adapted from J.M. Tanner, *Fetus Into Man*, (Cambridge: Harvard University Press, 1990).

FIGURE 2.10

Two LBW infants: one preterm and one SGA

the neonatal period) catch up in growth and development and have few—if any—long-term problems.

In contrast, SGA infants often don't catch up—even with adequate nutrition after birth—and most will be shorter than average for the rest of their lives (Nilsen 1984). Many will show long-term impairments in intellect and mental development (Perez-Escamilla and Pollitt 1992). SGA infants also tend to have more chronic health problems in later life. Thus, poor nutrition in utero may have profound effects that cannot be reversed after birth.

The short and long-term consequences of low birthweight are major public health problems, both in the developing world and in the United States. In the United States, studies have identified young age (less than 20 years), race, and lower socioeconomic status as major risk factors for having LBW infants (Institute of Medicine 1985). In the developing world, the overwhelming cause of LBW infant is maternal malnutrition. Table 2.12 gives a further breakdown of risk factors for giving birth to LBW infants. All of these issues will be discussed in detail in the next chapter.

Low Birthweight and Race in the United States

There are large racial differences in rates of low birthweight in the United States. The incidence among Blacks, American Indians, and Asians is higher than the rate among Whites (Wegman 1994). Rates of low birthweight for Blacks are over twice that of Whites, and it appears that socioeconomic factors such as poverty, lack of access to quality prenatal and perinatal care, teenage pregnancy, smoking and alcohol use, and poor nutrition all contribute to these differences (Institute of Medicine 1985).

Demographic	Prepregnancy
Race (Black)	Low weight for height
Low socioeconomic status	Short stature
	Chronic medical illness
Behavioral	Poor nutrition
Low educational status	Previous low-birth-weight infant
Smoking	Parity (none or more than five)
No care or inadequate prenatal care	
Poor weight gain during pregnancy	**Pregnancy**
Alcohol abuse	Multiple gestation
Illicit and prescription drugs	Poor nutrition
Short interpregnancy interval	Anemia
(less than 6 months)	Fetal disease
Age (less than 16 or over 35)	Preeclampsia and hypertension
Unmarried	Infections
Stress (physical and psychological)	Placental problems

Many of these variables are risk factors for both IUGR and preterm delivery.
Adapted from Institute of Medicine, Committee to Study the Prevention of Low Birthweight, *Prevention of Low Birthweight* (Washington D.C.: National Academy Press, 1985).

TABLE 2.12
*Overview of risk factors
for low birthweight*

THE MOTHER

ANATOMICAL AND PHYSIOLOGICAL CHANGES

Implantation and the rapid growth of the placenta and fetus induce dramatic changes in maternal physiology and anatomy. These changes, modulated primarily by placental hormones, represent adjustments to support fetal growth while continuing to satisfy maternal needs and prepare for **lactation.** Maternal adaptations improve utilization of nutrients, through increased absorption, reduced excretion, and alterations in metabolism.

The Cardiovascular System

During pregnancy the mother must transport large amounts of nutrients and oxygen to the fetus, and excrete more metabolic waste. To accomplish this, marked functional changes occur in the cardiovascular system.

Cardiac output increases 30 to 50% over nonpregnant values, due to increases in both heart rate and volume of blood pumped with each contraction (Robson 1989). Much of this increase flows to the placenta and the maternal kidneys.

The mother produces large amounts of new blood, and plasma volume increases 45%. Synthesis of plasma proteins—mainly albumin—is sharply

lactation: the secretion of milk by the breasts; the period after birth during which a child is breast-fed

cardiac output: the volume of blood pumped by the heart

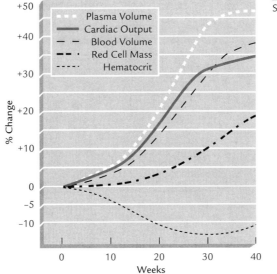

Source: Rosso, 1990

FIGURE 2.11

Changes in plasma volume, blood volume, and red cell mass during pregnancy

erythropoiesis: the formation of red blood cells

physiological anemia of pregnancy: the normal dilution of the number of red blood cells in the blood caused by the increase in plasma volume during pregnany

renin-aldosterone system: a hormonal system important in regulating blood pressure and blood volume. Renin is an enzyme secreted by the kidneys. It catalyzes the production of several compunds that constrict blood vessels and stimulate thirst. One of these compounds is aldosterone, a hormone secreted by the adrenal gland which signals the kidneys to reduce excretion of sodium and water.

increased. As a result, the mother needs more dietary iron, and other micronutrients (such as folate and zinc) involved in **erythropoiesis.**

Early expansion of plasma volume is followed by a more gradual increase in the number of red cells, but the increase in red cell mass is proportionately smaller than the expansion of the plasma volume. Therefore, hemodilution produces a **physiological anemia of pregnancy.** Although the actual number of red cells in the blood increases, there are fewer red blood cells per unit volume in the expanded plasma compartment. The hematocrit falls from normal values of about 35% percent to near 30%. Hemoglobin also declines to values as low as 10–11 g/100 ml. See Figure 2.11.

Hemodilution is a term used to describe the dilution of normal concentrations of nutrients, vitamins, and minerals by the expansion of the volume of maternal plasma during pregnancy. Although the total amount of a substance in the blood may stay the same or actually increase, per-volume values will fall. These observations may be misinterpreted as evidence of deficiency. During pregnancy, healthy blood concentrations of substances in healthy blood can be significantly lower than values for nonpregnant women.

The Kidney

The kidney expands the plasma and extracellular volumes by increasing sodium and water retention. The activity of the **renin-aldosterone system** increases 2 to 20 times in pregnancy (Seely and Moore 1994), adding over 7 liters of water and 17 grams of sodium to the maternal system.

Blood flow to the kidneys increases 50 to 85% during pregnancy (Rosso 1990). This allows the mother to excrete increasing amounts of fetal and maternal waste products, but also increases urinary losses of glucose, amino acids, and other nutrients. The rise in renal blood flow increases filtration of these nutrients to levels that exceed the tubular capacity for nutrient reabsorption, producing losses in the urine.

Glucose losses begin during the first month of gestation and increase progressively until term (Davidson and Dunlop 1980). Near term, pregnant women excrete up to ten times more glucose in their urine than nonpregnant women (Lind and Hytten 1972). Substantial amounts of amino acids are also lost, as urinary amino acid excretion increases 200 to 700% over nonpregnancy levels, with average daily losses of approximately 2 g per day (Rosso 1990).

Vitamins and minerals also spill over into the urine during pregnancy. For example, urinary excretion of folate doubles during the first trimester and increases progressively until term. Although average daily losses of folate are usually small (10–15mg), in some women losses are substantially higher (over 50 mg/day) (Landon and Hytten 1971). These losses may adversely affect folate status during pregnancy if dietary intake is marginal.

The Gastrointestinal System

Pregnancy alters taste acuity: the ability to taste salty foods diminishes, while discrimination of sweet tastes remains unchanged (Brown and Toma 1986). Pregnancy may produce marked food preferences and aversions in many women (discussed later in this chapter).

Increased circulating progesterone causes smooth muscle cells to relax throughout the gastrointestinal tract, and reduced tone in the muscular sphincter at the end of the esophagus allows stomach acid to slip back up into the esophagus—causing irritation and discomfort (heartburn) (Baron et al. 1993). As activity of smooth muscle slows, stomach emptying is delayed, and mothers may feel full after small meals.

Progesterone slows the smooth muscle contractions that normally propel material through the intestines, and slower transit through the colon allows more water resorption from the stool and contributes to constipation—a common complaint during pregnancy. (Constipation and other digestive problems during pregnancy—heartburn, nausea, and vomiting—are discussed in more detail in the next chapter.) Gallstones occur more often in pregnancy: the gallbladder empties more slowly and the bile contains more cholesterol, making stone formation more likely (Braverman et al. 1980).

Pregnancy alters the absorption of certain nutrients. Absorption efficiency of iron, calcium, and vitamin B_{12} is increased (Rosso 1990), but glucose tends to be absorbed more slowly into the bloodstream after a meal because the stomach empties more slowly.

ALTERATIONS IN MATERNAL METABOLISM

Balancing Maternal and Fetal Needs

As we have seen, during pregnancy dramatic changes in maternal metabolism alter the way nutrients are absorbed, metabolized, stored, and excreted. A coordinated series of physiological adjustments enables the mother to supply the developing fetus with ample energy and nutrients, and at the same time, to provide for her own requirements. Metabolic adaptations during pregnancy must balance certain fetal and maternal needs:

- The fetus must be provided with energy and substrates for growth and development in utero, as well as adequate stores of energy and certain nutrients needed for the transition to **extrauterine life.**

- The mother requires energy and substrates to meet the increased physiological demands of pregnancy, to synthesize new tissues, and to accumulate energy stores to support late pregnancy, labor, and lactation.

The fetus needs a continuous supply of energy, protein, and nutrients for growth and metabolism. Throughout pregnancy, the placenta steadily pulls glucose and amino acids (primarily alanine) from the maternal bloodstream and transfers them to the fetus. Glucose provides about 80% of the energy needs of the fetus, with the remaining needs being supplied by the breakdown of amino acids (Rosso 1990). A primary aim of maternal metabolism is to ensure that ample amounts of these nutrients are circulating in the maternal blood and available for placental transport. After meals, absorbed glucose, amino acids, and gluconeogenic precursors (such as alanine and glycerol) are diverted for use by the fetus; between meals, these nutrients must be drawn from maternal stores.

Because most glucose and amino acids are conserved for fetal use, maternal needs must be met by an alternate energy substrate. As pregnancy progresses, the mother becomes increasingly dependent on fatty acids for energy, and maternal blood levels of free fatty acids and triglycerides rise (Leturga et al. 1987). Fatty acids are not transferred to the fetus and remain available for maternal use.

Anabolic and Catabolic Phases of Maternal Metabolism

Overall, pregnancy is an intensely anabolic state: appetite and food intake increase, physical activity decreases, and large amounts of new fat and protein are deposited in the mother and fetus. While the growing fetus and placenta are anabolic throughout pregnancy, the mother shifts from an anabolic state during the first half of pregnancy to a catabolic state in the later half (Norton et al. 1994).

extrauterine life: life outside the uterus

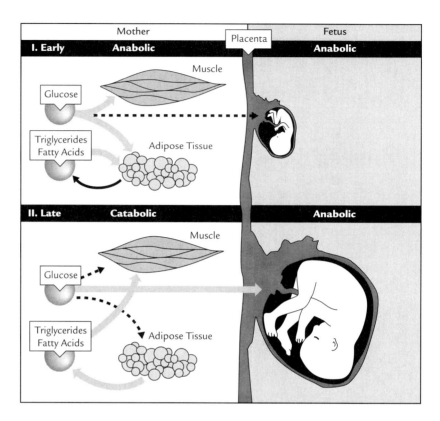

FIGURE 2.12

Anabolic and catabolic phases of pregnancy

In the anabolic stage, preparation for the demands of later pregnancy and lactation take place. Energy from the diet is directed toward building stores of maternal fat, and protein and nutrients are used to build new blood cells and other maternal tissues. Weight gain is mostly in new maternal tissues and the placenta. During the second half of pregnancy, maternal metabolism becomes more catabolic, as stored fat is broken down to supply energy, and glucose and protein are directed to the rapidly growing fetus. Weight gain is primarily in the fetus and placenta. See Figure 2.12.

Metabolic changes spread energy and protein costs over the entire nine months of pregnancy. In early pregnancy, energy is conserved by the mother, followed by later redirection of energy (glucose) to the fetus. The mother uses protein economically, conserving amino acids for fetal use.

Early pregnancy (*the anabolic phase*) has the following characteristics:

- Extra dietary carbohydrate is readily stored as glycogen or converted to fat and stored in maternal adipose tissue. Fetal needs for glucose are small during the first half of gestation.

- Dietary fatty acids are rapidly synthesized into triglycerides and stored. Lipolysis is slowed, conserving maternal fat stores.

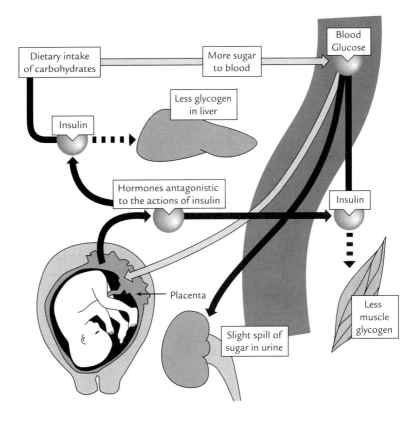

FIGURE 2.13
*Carbohydrate metabolism
after a meal in late
pregnancy*

- Maternal protein synthesis is increased, and protein is deposited into maternal tissues (primarily new red blood cells and placenta).

Placental hormones play a central role in resetting maternal metabolism during pregnancy. During pregnancy, insulin secretion after meals increases sharply, and high levels of insulin mediate many of the anabolic changes of early pregnancy. As pregnancy progresses, levels of estrogen, progesterone, and human placental lactogen (hPL) rise (Norton et al. 1994). All of these hormones oppose the anabolic effects of insulin, hPL being the most potent insulin antagonist. In later pregnancy, the anabolic effects of insulin are blunted, allowing maternal fat, glycogen, and protein to be broken down for use by the rapidly growing fetus and placenta.

Later pregnancy (*the catabolic phase*) has the following characteristics:

- As shown in Figure 2.13, carbohydrate metabolism after a meal funnels dietary glucose to the fetus. Because the actions of insulin are blunted by circulating hormones, less glucose is taken up by the liver and muscles. Glucose levels in maternal blood are elevated and glucose circulates for a longer time. Most of it is passed to the fetus and some is spilled in the urine. Between meals, as the placenta avidly transfers glucose to the fetus, maternal blood glucose levels are 10–20% lower, compared to nonpregnant levels (Norton et al. 1994).

- Maternal fat stores are broken down and increasingly used by maternal tissues for fuel, conserving glucose. As fatty acids are oxidized for energy, ketogenesis increases, and levels of ketone bodies in maternal blood rise, providing an alternative energy substrate to glucose. Cholesterol levels also rise in pregnancy, due to increased synthesis and decreased breakdown.

- Between meals, maternal protein is catabolized to provide additional amino acids for the fetus. Amino acids absorbed after meals are directed toward the fetus, and new protein synthesis occurs mainly in the developing placenta and fetus. Maternal blood levels of amino acids are lower than in nonpregnant women.

The Importance of Regular Meals and Snacks

Later pregnancy has been described as a state of accelerated starvation for the mother. Many of the adaptations in maternal metabolism between meals resemble those that occur in nonpregnant individuals who fast for long periods. Starvation is also characterized by lower fasting blood glucose and amino acid levels, reduced **glucose tolerance,** and increased levels of free fatty acids, triglycerides, and ketones.

Pregnant women more rapidly develop **fasting hypoglycemia** and **ketosis,** particularly after an overnight fast or if meals are skipped during the day. Some women experience feelings of lightheadedness or faintness if they miss meals. After 16 to 18 hours of fasting (skipping breakfast), glucose available for transfer to the fetus falls sharply (Metzger et al. 1982). Also, **ketones** readily cross the placenta, and high levels may adversely affect fetal development. Clearly, women should consume regular meals and snacks and avoid long periods of fasting while pregnant.

Changes in Micronutrient Metabolism

Pregnancy also affects the metabolism of micronutrients. High estrogen levels increase liver production of many plasma proteins involved in micronutrient transport. For example, plasma levels of **transferrin** and **ceruloplasmin** are elevated during pregnancy, increasing the capacity of the blood to transport iron and copper. Intestinal absorption of calcium and iron increases, and there is greater turnover of many nutrients, including vitamin C and vitamin B6 (Institute of Medicine 1990). Overall, maternal plasma levels of many vitamins and minerals show a gradual, modest decline during gestation. This reflects both hemodilution in the expanded plasma volume and greater fetal and maternal demands.

glucose tolerance: the ability to maintain a normal range of blood glucose concentrations in response to intake of dietary carbohydrate

fasting hypoglycemia: an abnormally low level of glucose in the blood between meals when no glucose is being absorbed into the blood from the digestive tract

ketosis: the condition characterized by enhanced production and accumulation of ketones from the breakdown of fats in the liver; occurs in starvation and in diabetes

ketones: the compounds acetone, acetoacetic acid, and betahydroxybutyrate that are produced during fat metabolism in the liver

transferrin: a blood protein that binds and transports iron

ceruloplasmin: a blood protein important in copper transport

WEIGHT GAIN DURING PREGNANCY

Components of Weight Gain

Of the total weight gain during pregnancy, about a third is due to growth of the products of conception, with the fetus accounting for 25% of the total gain, the placenta 5%, and the **amniotic fluid** about 6%. Maternal tissue and fluid accretion account for the remaining two thirds. See Figure 2.14.

The composition of the weight gain is about 62% water, 8% protein, and 30% fat (Hytten and Leitch 1971). Nearly two thirds of the accumulated protein is in the fetus and placenta, while 90% of the fat goes into maternal fat stores (see Figure 2.15). The average mother accumulates about 2 kg of new fat during a healthy pregnancy. Fat is used to meet the needs of lactation and as energy reserves in late pregnancy.

Table 2.13 gives a complete breakdown of the components of weight gain during pregnancy. During the first trimester, the small weight gain is almost all new maternal tissue; in the second, weight gain is still mostly maternal tissue and only small amounts of fetal tissue; in the third trimester, it is mostly new fetal tissues. Maternal tissue accumulation sets the stage for the rapid development of the fetus in the later half of pregnancy.

amniotic fluid: the fluid in the amniotic sac that surrounds the fetus during gestation

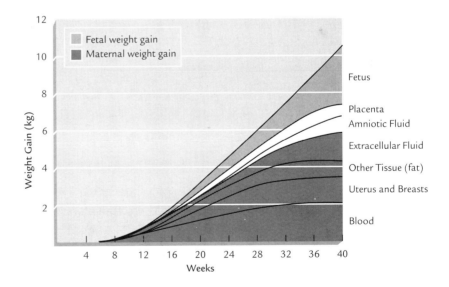

FIGURE 2.14

Components of weight gain during pregnancy

From Institute of Medicine, Subcommittee on Nutritional Status and Weight Gain during Pregnancy, *Nutrition during Pregnancy* (Washington D.C. National Academy Press, 1990).

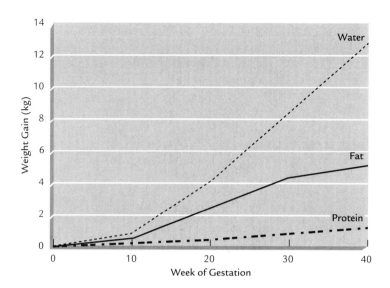

FIGURE 2.15

Composition of weight gained in pregnancy

	Weight gain (g) at			
	10 Weeks	**20 Weeks**	**30 Weeks**	**40 Weeks**
Maternal:				
Maternal stores (fat)	310	2,050	3,480	3,345
Interstitial fluid	0	30	80	1,680
Blood	100	600	1,300	1,250
Uterus	140	320	600	970
Mammary gland	45	180	360	405
	595	3,180	5,820	7,650
Fetal:				
Fetus	5	300	1,500	3,400
Amniotic fluid	30	350	750	800
Placenta	20	170	430	650
	55	820	2,680	4,850
Total gain	650	4,000	8,500	2,500

TABLE 2.13

Components of weight gain during pregnancy

Adapted from P. Rosso, Nutrition and Metabolism in Pregnancy (Oxford: Oxford Universty Press, 1990).

Weight-for-Height Category	Recommended Total Gain	
	kg	lb
Low (BMI < 19.8)	12.5–18	28–40
Normal (BMI 19.8–26.0)	11.5–16	25–35
High* (BMI > 26.0–29.0)	7–11.5	15–25

Young adolescents and black women should strive for gains at the upper end of the recommended range. Short women (<157 cm, or 62 in.) should strive for gains at the lower end of the range.

*The recommended target weight gain for obese women (BMI >29.0) is at least 6.0 kg (15 lb).

From Institute of Medicine, Subcommittee on Nutritional Status and Weight Gain during Pregnancy, *Nutrition During Pregnancy* (Washington D.C.: National Academy Press, 1990).

TABLE 2.14

Recommended weight gain ranges for pregnant women by prepregnancy BMI

Recommended Patterns for Weight Gain

Normal, steady weight gain is characteristic of a pregnancy that is progressing well. Although average weight gain during pregnancy is 10.5–12.5 kg, it can vary widely in well-nourished women with normal, healthy pregnancies. The 15th and 85th percentiles of weight gain are about 16 and 40 lb, respectively, for women in the United States who deliver babies of optimal weight (3–4 kg) at term (Institute of Medicine 1990).

In 1990 the U.S. Academy of Sciences, after an extensive review, produced new guidelines (shown in Table 2.14) for weight gain in pregnancy. Weight gain within the recommended ranges is associated with improved maternal and fetal outcomes, and—in particular—desirable newborn birthweight. The emphasis on optimal birthweight reflects its importance for infant and child health. The ranges of recommended weight gain are based on prepregnancy weight for height, which is a major determinant of fetal growth. Women who are lighter before pregnancy tend to deliver smaller infants than heavier women—even with the same gestational weight gain (Institute of Medicine 1990). Because of this, desirable weight gains for thinner women are higher than those for heavier women. Weight gain during pregnancy should be routinely measured and recorded on a standard prenatal weight gain chart.

Detrimental Effects of Low or High Weight Gain

Maternal weight gain that is too low may have adverse effects on both the short- and long-term health of the infant. Gestational weight gain, particularly during the last two trimesters, is an important determinant of fetal growth. Low gestational weight gain (less than 7 kg, or 16 lb) is associated with an increased risk of giving birth to a growth-retarded infant (Institute of Medicine 1990).

In industrialized countries such as the United States, excess weight gain during pregnancy is more common than inadequate gain (Institute of Medicine 1990). Large gestational weight gains increase the risk of delivering a high-birth-weight infant, and babies that are too big may be difficult to deliver. High birthweight increases the risk of complications during labor and delivery—including prolonged labor, birth trauma, neonatal asphyxia, and increased infant and maternal mortality. Figure 2.16 shows the relationship between weight gain and infant mortality. High-birth-weight infants also tend to be heavier throughout childhood and have an increased risk of obesity in later life (Binkin et al. 1988).

Large weight gains during pregnancy may also increase the risk of retaining extra weight **postpartum.** In the United States, the average woman who gains about 13 kg during pregnancy loses about 5 kg during delivery. Another 2–3 kg will be lost during the first week, mostly through **diuresis** of body water. During lactation, further weight loss occurs, as body fat is mobilized to supply the needs of the breast-feeding infant. An average woman ends up retaining only about 1 kg of weight after lactation is complete.

However, women who gain larger amounts of weight during pregnancy are more likely to retain more gestational weight postpartum. As shown in Table 2.15, a recent study compared postpartum weights in two groups of well-nourished women: one group had normal gestational weight gains (12.7 kg), the other had high weight gains (19.3 kg). At nine months postpartum, the women who had higher weight gains still retained an average of 6 kg of weight—substantially more than the group with normal gestational weight gains (Parham et al. 1990). In women who are well nourished prior to pregnancy, large gestational weight gains (particularly if a woman has several pregnancies) may contribute to the subsequent development of obesity and its detrimental effects on health.

postpartum: after birth

diuresis: increased secretion of urine

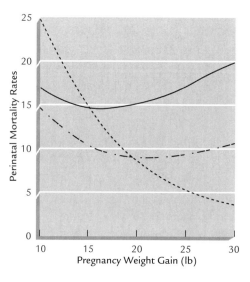

FIGURE 2.16

Relationship between prepregnancy weight, weight gain in pregnancy and perinatal mortality

The curves represent 3 categories of prepregnancy weight (as percentages of mean weight-for-height values from Metropolitan Life Insurance tables).

Source: Naeye, R.L., *Am J Obstet Gynecol* 135, (1979): 3.

Maternal Weight Gain During Pregnancy	12.7 kg	19.3 kg
Months postpartum	**Residual Weight**	
1–3	3.9 kg	8.9 kg
3–6	4.9 kg	8.6 kg
6–9	3.2 kg	6.1 kg

Residual weight = postpartum weight – prepregnancy weight. All values are means.

Adapted from E.S. Parham, M.F. Astrom, and S.H. King, "The Association of Pregnancy Weight Gain with the Mother's Postpartum Weight", *J Am Diet Assoc* 90 (1990): 550-54.

TABLE 2.15
Residual weight gain one to nine months postpartum in relation to weight gain during pregnancy

Recommendations for gestational weight gain have shifted markedly in the past 75 years. Before 1940, unrestricted weight gain was encouraged, but large weight gains were later found to increase complications in delivery and have other adverse effects. During the 1940s and 1950s, based on the erroneous belief that small weight gains would reduce maternal problems during pregnancy, it was popular to limit weight gain to 4–6 kg. Such small weight gains were found to have equally undesirable consequences.

The 1990 guidelines from the U.S. Institute of Medicine (shown in Table 2.14), represent the latest consensus. They are based on a more thorough understanding of the impact of weight gain on pregnancy outcome.

Determinants of Weight Gain

Prepregnancy weight, age, **parity,** physical activity, ethnic origin, socioeconomic status, and substance abuse all influence gestational weight gain. A complete list of risk factors is shown in Table 2.16.

parity: the number of completed pregnancies a woman has experienced

- High prepregnancy weight for height
- Short stature
- Black, Hispanic, or Southeast Asian background
- Very young age (<15 years)
- Multiparicity
- Unmarried
- Low income
- Cigarette smoking
- Substance abuse

Adapted from Institute of Medicine, Subcommittee on Nutritional Status and Weight Gain during Pregnancy, *Nutrition During Pregnancy* (Washington D.C.: National Academy Press, 1990).

TABLE 2.16
Maternal risk factors for low weight gain during pregnancy

- A major determinant of gestational weight gain is prepregnancy weight for height. Women who are overweight (BMI greater than 26.0) gain less weight during pregnancy than do thinner women.

- Maternal age does not greatly influence weight gain, with the exception of very young adolescents (less than two years after menarche). Very young mothers gain less weight than older pregnant women.

- Parity influences weight gain: primiparous women in all age groups gain about 1 kg more than multiparous women, and the risk of low weight gain is significantly higher among multiparous women.

- Ethnic origin influences weight gain in pregnancy. Black, Hispanic, and Southeast Asian women tend to gain less weight than their White counterparts.

- Socioeconomic factors affect weight gain. Unmarried mothers with low incomes are at higher risk for low weight gain.

- Cigarette smoking and substance abuse are associated with lower gestational weight gain (discussed in detail in the next chapter).

NUTRITIONAL NEEDS OF THE MOTHER AND FETUS IN PREGNANCY

ENERGY

The fetus needs energy for growth, metabolism, and to a limited extent, physical activity. Current estimates of energy requirements for the near-term fetus are 50–95 kcal/kg/day (see Figure 2.17), or about 175–350 kcal/day in a 3.5 kg fetus (Rosso 1990). Approximately a third of the energy is for growth, and the remainder is required for metabolism.

The mother needs energy to support growth of new maternal tissue (including red blood cells, mammary tissue, and fat), and growth of the placenta. Also, the increased activity of the maternal cardiovascular, respiratory, and excretory systems during pregnancy increases metabolic needs. Table 2.17 summarizes the energy needs of mother and fetus. In most women, the increased demands for energy are partially offset by a decline in physical activity, particularly in the third trimester (Prentice et al. 1994).

For a healthy term pregnancy, in which a 63 kg mother gains 12.5 kg and delivers a 2.5 kg infant, the total energy cost of gestation is estimated to be 65–80,000 kcal. Table 2.18 shows the actual energy cost computed in one study. By dividing this number by the duration of an average pregnancy (250 days after the first month), the World Health Organization has calculated the daily energy allowance for pregnant women to be an additional 300 kcal/day (WHO 1985).

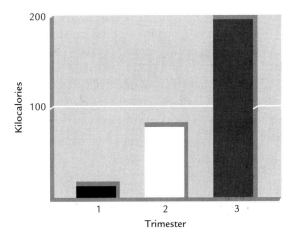

FIGURE 2.17

Approximate daily energy needs of the fetus for each trimester

Fetus
    Growth
    Metabolism

Mother
    Increased basal metabolism
    Fat stores
    Growth of uterus and mammary tissue
    Increased blood volume
    Change in activity

TABLE 2.17

Determinants of energy needs during pregnancy

Significant weight gain does not occur until after the first ten weeks of gestation. Therefore, the RDA for pregnancy provides an additional 300 kcal/day only during the second and third trimesters (NAS 1989). This recommendation applies to well-nourished women, and if a woman begins pregnancy with depleted nutritional reserves, she will need to increase energy intake at the beginning of pregnancy.

	Amount Protein and Fat Deposited (g)	Energy Cost (kcal)
Tissue deposition		
Gain in fat stores	2,000	22,000
Fat deposition	480	5,280
Protein deposition	925	6,475
Cumulative increase in basal metabolism	—	34,505
Total		68,260

Adapted from J.M.A. van Raaij, "Energy Requirements of Pregnancy in the Netherlands", *Lancet* 2 (1987):953–55.

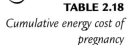

TABLE 2.18

Cumulative energy cost of pregnancy

Although a pregnant woman is "eating for two", she does not need to double her food intake during pregnancy. For example, to obtain an extra 300 kcal, an additional two cups of milk and an extra apple or banana each day would suffice. For most pregnant women, diet quality is more important than quantity. Weight gain within the recommended ranges is a good indication that energy intake is appropriate. Table 2.19 compares the RDAs for pregnant and nonpregnant women.

CARBOHYDRATE

The fetus near term requires more than 40 grams of glucose each day. Glucose is the most important fetal energy source, with about 80% of the daily energy requirement of the fetus supplied by oxidation of glucose. Glucose is also needed for synthesis of lipid, glycogen, and structural polysaccharides.

Maternal glucose concentrations and placental blood flow determine the availability of glucose to the fetus. As maternal blood glucose rises, more glucose is transported to the fetus, and fetal insulin secretion is stimulated. Women with sustained hyperglycemia give birth to larger babies who have higher insulin levels (Silverman et al. 1993) (discussed with gestational diabetes in Chapter 3).

| | Age | | |
Nutrient	19–24	25–50	Pregnancy
Energy (kcal)	2200	2200	+ 300*
Protein (g)	46	50	60
Vitamin A (µg RE)	800	800	800
Vitamin D (µg)	10	5	10
Vitamin E (mg α-TE)	8	8	10
Vitamin C (mg)	60	60	70
Folate (µg)	180	180	400
Niacin (mg NE)	15	15	17
Riboflavin (mg)	1.3	1.3	1.6
Thiamin (mg)	1.1	1.1	1.5
Vitamin B_6 (mg)	1.6	1.6	2.2
Vitamin B_{12} (µg)	2.0	2.0	2.2
Calcium (mg)	1200	800	1200
Phosphorus (mg)	1200	800	1200
Iodine (µg)	150	150	175
Iron (mg)	15	15	30
Magnesium (mg)	280	280	320
Zinc (mg)	12	12	15
Selenium (µg)	55	55	65

*Second and third trimesters.

TABLE 2.19
RDAs for females: nonpregnant vs. pregnant

From National Academy of Sciences, Food and Nutrition Board, National Research Council, *Recommended Dietary Allowances*, 10th ed. (Washington D.C.: National Academy Press, 1989).

LIPIDS

The fetus derives fat from both maternal sources via the placenta and *de novo* synthesis in fetal tissues. Fetal liver and adipose tissue can synthesize fatty acids from glucose, lactate, and amino acids. In the fetus, minimal amounts of fatty acids are oxidized for fuel, as most are stored as triglycerides in adipose tissue. Lipid accumulation by the fetus is most rapid after the 35th week, when storage averages 5.3 mg of fat/day (Rosso 1990).

In humans, linoleic and linolenic acid are essential fatty acids and must be supplied by the diet. Several additional long-chain fatty acids appear to be essential to the fetus. The fatty acids **EPA** and **DHA** play important structural roles in the developing fetus, particularly in the eye and brain (Crawford 1993). Adults are able to synthesize EPA and DHA, but because the necessary metabolic pathways have not fully developed, the fetus appears to have only limited ability to form these fats. Like certain of the amino acids that are essential only during early fetal life, these fatty acids need to be supplied to the fetus by the mother.

AMINO ACIDS AND PROTEIN

EPA: eicosapentaenoic acid, a polyunsaturated fatty acid containing 20 carbons and 5 double bonds; synthesized from linolenic acid

DHA: docosahexaenoic acid, a polyunsaturated fatty acid containing 22 carbons and 6 double bonds; synthesized from EPA

essential amino acid: amino acids that cannot be synthesized in the human body in amounts sufficient to meet physiological needs

nonessential amino acid: amino acids that can be synthesized in the human body in amounts sufficient to meet physiological needs

Total fetal protein requirements during gestation have been estimated to be 350–450 g (Institute of Medicine 1990). The fetal requirements for protein is not linear: early fetal needs are minimal, but by the third trimester, they become significant. Fetal protein requirements near term have been estimated to be about 2 g/kg/day. The fetus may obtain a small percentage (about 5%) of its amino acid requirement from swallowing protein contained in the surrounding amniotic fluid, but most must come from the maternal diet. Table 2.20 shows amino acid requirements in a full-term fetus.

During the first trimester, all the **amino acids** are **essential** for the fetus—unlike the mature liver, the fetal liver cannot synthesize the **nonessential amino acids.** After about week 20, the fetus generally requires the same essential amino acids as an adult. However, several amino acids that are nonessential for adults (such as arginine and cystine) are not synthesized in adequate quantities by the fetus even late in gestation (Rassin 1991). Requirements for these amino acids must be supplied at least partially from maternal sources.

During pregnancy, the maternal diet must supply ample protein. Protein is required for synthesis and maintenance of fetal, placental, and maternal tissues (see Table 2.21). About 925 g of total protein is deposited during a pregnancy in which maternal weight gain is 12.5 kg and the weight of the newborn infant is 3.3 kg (NAS 1989). Mean daily increments increase progressively: protein deposition is 0.6, 1.8, 4.8, and 6.1 g/day in the successive quarters of pregnancy. To calculate dietary protein needs, these values must be adjusted for the efficiency at which dietary protein is converted to new maternal and fetal tissue. In women eating mixed diets containing at least a third of protein from high-quality animal foods, the efficiency of conversion is about 70%. In women on mostly vegetarian diets, efficiency is decreased, and more protein should to be consumed.

Considering these factors and normal biological variability, 10 g/day of additional dietary protein will cover the needs of most pregnant women. The RDA and WHO recommendations are therefore set at an extra 10 g/day (NAS 1989). Since 300 kcal/day is the estimated increased energy need for pregnancy, the proportion of the additional daily energy need derived from protein would be about 13%.

TABLE 2.20

Amino acid requirements in a full-term fetus

Requirements	g/day
Amino acid consumed in oxidative metabolism	5.2
Amino acid retained in tissue growth	2.5
Total	7.7

Adapted from P. Rosso, *Nutrition and Metabolism in Pregnancy* (Oxford: Oxford University Press, 1990).

TABLE 2.21

Estimate of protein components of weight gain in a normal full-term pregnancy

Component	Weight (g)	Protein (g)
Fetus	3,400	440
Placenta	650	100
Amniotic fluid	800	3
Uterus	970	166
Blood	1,250	81
Extracellular fluid	1,680	135
Total	8,750	925

Adapted from Institute of Medicine, Subcommittee on Nutritional Status and Weight Gain During Pregnancy, *Nutrition During Pregnancy* (Washington D.C.: National Academy Press, 1990).

FAT-SOLUBLE VITAMINS

Vitamin A

Although vitamin A is vital for cellular differentiation and normal fetal development, total vitamin A requirements of the fetus are small. Only small amounts are stored in the fetal liver (less than 25 mg/g) (Wallingford and Underwood 1986), and fetal utilization of vitamin A in the third trimester is only 200 mg/day, or a total of about 18 mg (NAS 1989).

Since the total body pool of vitamin A in the average adult woman in the United States is greater than 200 mg, over the course of pregnancy only about 9% of maternal stores are needed by the growing fetus. Therefore, well-nourished women do not need to increase vitamin A intake during gestation (NAS 1989). Pregnant women should maintain intake near the adult RDA of 800 retinol equivalents (RE). Large doses of vitamin A are potentially dangerous teratogens.

Vitamin D

The fetus obtains most of its vitamin D from the mother as 25-OH vitamin D, and it is stored in fetal muscle and liver. Increased concentrations of 25-OH vitamin D in maternal serum are associated with higher fetal concentrations, and synthesis of $1,25\text{-}(OH)_2$ vitamin D by both the placenta and fetal kidney contribute to fetal vitamin D status (Mimouri and Tsang 1994).

Maternal blood levels of $1,25\text{-}(OH)_2$ vitamin D rise during pregnancy (see Figure 2.18). At 38 weeks gestation, plasma concentrations are twice those in nonpregnant women, reflecting both maternal and placental synthesis (Mimouri et al. 1989). Increased $1,25\text{-}(OH)_2$ vitamin D levels stimulate intestinal absorption of dietary calcium during pregnancy.

The precise requirement for vitamin D during pregnancy is not known, but it is estimated to be 10 mg/day. Therefore, the RDA calls for an additional 5 mg/day for women 25 years or older (the RDA for nonpregnant women younger than 25 is already set at 10 mg) (NAS 1989). Pregnant women should obtain regular exposure to sunlight and consume ample vitamin D-fortified milk.

Researchers have found ethnic and regional differences in vitamin D status during pregnancy. White mothers and their infants tend to have

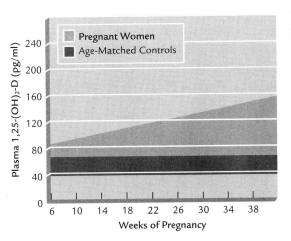

FIGURE 2.18
Plasma 1,25(OH)$_2$ vitamin D in pregnant and nonpregnant women

Source: Kumar, R. et al. *J Clin Invest* 63 (1979) : 342.

higher levels of vitamin D than their Black counterparts. This may be because vitamin D synthesis in the skin of Blacks tends to be slower than in Whites. In several regions of the world, pregnant women are clothed so that exposure to the sun is limited. For example, Saudi and Bedouin women typically have very low levels of 25-OH vitamin D during pregnancy (Institute of Medicine 1990).

Vitamin E

The fetus accumulates vitamin E during the third trimester, particularly during the last eight to ten weeks of pregnancy when fat accumulation is high. The full-term neonate therefore has larger stores than preterm infants, and preterm infants are routinely given vitamin E supplements to protect them from hemolytic anemia due to vitamin E deficiency (discussed in Chapter 6). To allow for growth of the fetus, the RDA for vitamin E during pregnancy is an additional 2 mg/day (NAS 1989).

Vitamin K

Concentrations of vitamin K and vitamin K-dependent **clotting factors** are low in the fetus, and the newborn infant is functionally vitamin K deficient. Supplemental vitamin K is recommended for all newborns (AAP 1988) (discussed in Chapter 6). Placental transport of vitamin K from mother to fetus is minimal, and supplementing maternal diets with the vitamin does not significantly affect fetal levels. There is little information about gestational requirements for vitamin K. Since usual adult diets of nonpregnant women in the United States typically exceed the RDA for vitamin K, no additional intake is currently recommended during pregnancy (NAS 1989).

WATER-SOLUBLE VITAMINS

Vitamin C

Vitamin C is actively transported by the placenta, and the fetus has a relatively large pool of the vitamin (per unit body weight) compared with the mother (Salmenpera 1984). Although there is little data on fetal requirements for vitamin C, if fetal needs per unit of body weight are comparable to adults, the increment in the fetal requirement for the vitamin near term would be 3–4 mg/day. To offset losses from the maternal pool to the fetus and to provide adequate vitamin C to the mother for gestational needs, the RDA calls for an additional 10 mg/day during pregnancy (NAS 1989).

clotting factor: any of the various blood components involved in the coagulation process

Thiamin

Measurements of maternal thiamin status, including urinary thiamin excretion and activity of thiamin-dependent enzymes such as **red cell transketolase,** indicate that the requirement for thiamin increases early in pregnancy and remains elevated throughout gestation (NAS 1989). To accommodate fetal and maternal growth and increased maternal caloric intake, it is estimated that an additional 0.4 mg/day is required during pregnancy.

Niacin and Riboflavin

During pregnancy, high estrogen levels stimulate increased conversion of tryptophan to niacin (Wolf 1971). This enhancement of the biosynthesis of niacin from tryptophan appears to be an adaptive response of pregnancy because it increases the availability of niacin to provide for increased gestational needs. Because of increased energy requirements during gestation, an additional 2 mg/day of niacin and 0.3 mg/day of riboflavin are recommended during pregnancy (NAS 1989).

Vitamin B_6

The growing fetus lacks the ability to **phosphorylate** sufficient **pyridoxal** for its needs, due to developmental immaturity of the metabolic pathway. The fetus is therefore dependent on a steady maternal supply of pyridoxal phosphate (PLP), which is actively transported across the placenta. Concentrations in fetal blood are two to five times higher than those in maternal blood (Rosso 1990).

Maternal plasma PLP declines substantially between the fourth and eighth months of gestation, when fetal growth is most rapid (Schuster et al. 1984). There are also significant decreases in vitamin B_6-dependent enzyme activity in pregnancy. The efficiency of absorption of vitamin B_6, its metabolism to PLP, and rates of excretion are unchanged.

Because vitamin B_6 is important in amino acid metabolism, the additional protein intake during pregnancy should be accompanied by extra vitamin B_6. Active placental transport of PLP to the fetus also increases requirements for the vitamin. Therefore, it is recommended that pregnant women consume an additional 0.6 mg/day (NAS 1989).

Folic Acid

The ability to absorb dietary folate does not change during pregnancy, but folate excretion in the urine is more than twice that of nonpregnant

red cell transketolase: an enzyme of the transferase class involved in carbohydrate metabolism in red blood cells; its activity is dependent on adequate thiamin status

phosphorylate: to introduce a phosphate group into an organic molecule

pyridoxal: one of the forms of vitamin B_6; as pyridoxal phosphate it is a major coenzyme involved in amino acid metabolism

women (Landon and Hytten 1971). There is a progressive fall in the concentration of folate in serum and red blood cells through gestation; near-term values are 20–30% lower than those in nonpregnant women (see Figure 2.19). Also, formiminotransferase activity (a folate-dependent enzyme) decreases, and **formiminoglutamic acid (FIGLU)** excretion is increased, suggesting impaired folate status (O'Connor 1994).

Many pregnant women whose intake of folate is marginal show signs of deficiency, particularly in the second and third trimesters, when folate requirements increase rapidly (Institute of Medicine 1990). The large expansion in red cell mass requires ample folic acid, and many pregnant women who are folate deficient develop signs of anemia. In order to maintain maternal stores and support rapidly growing fetal and placental tissues, the RDA for pregnant women is 400 mg/day of folate, a greater than 100% increase over the RDA for nonpregnant women (NAS 1989). As discussed earlier, folate deficiency in the periconceptional period may increase risk of birth defects.

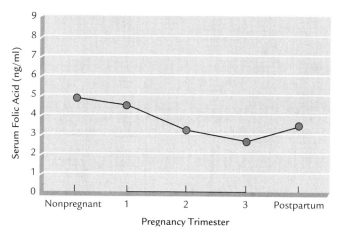

Source: Adapted from Rosso, 1990

FIGURE 2.19

Changes in folic acid levels in serum during pregnancy

formiminoglutamic acid (FIGLU): a compound that is an intermediate in the degradation of histidine to glutamate; it may be excreted in the urine in folate or vitamin B_{12} deficiency

Vitamin B_{12}

The placenta actively transports vitamin B_{12} to the fetus, and serum levels of newborn infants are twice that of their mothers (Giugliani et al. 1985). Estimates of fetal requirements for vitamin B_{12} are 0.1–0.2 mg/day (Food and Agriculture Organization 1988). Although in well-nourished women, maternal stores are adequate to meet these increased needs a 10% increase in intake during pregnancy is recommended, to 2.2 mg/day (NAS 1989).

Biotin and Pantothenic Acid

There is no recommended increase in intake of biotin or pantothenic acid during pregnancy, as present levels of consumption of these vitamins in adult diets are thought to be adequate to cover gestational needs (NAS 1989).

| Vitamin | Recommended Dietary Allowance | | Rationale for Increased Allowance for Pregnancy |
	Nonpregnant Women	Pregnant Women	
Vitamin C	60 mg	70 mg	To provide for fetal needs; at term, fetal plasma levels are 50% higher than maternal levels
Thiamin	1.1 mg	1.5 mg	To accommodate maternal and fetal growth and increased energy allowance during pregnancy
Riboflavin	1.3 mg	1.6 mg	To provide for increased maternal and fetal growth
Niacin (NE)	15 mg	17 mg	Based on energy increase of 300 kcal/day for pregnancy
Vitamin B_6	1.6 mg	2.2 mg	Based partially on the additional protein allowance of 10 g/day for pregnancy
Folate	190 μg	400 μg	Based on 50% food folate absorption; to build or maintain maternal folate stores and to provide for increased folate turnover in rapidly growing tissue
Vitamin B_{12}	2.0 μg	2.2 μg	Fetal needs (0.1–0.2 μg/day) based on analyses of stillborn fetuses; metabolic needs of pregnancy estimated at 0.2 μg/day

From National Academy of Sciences, Food and Nutrition Board, National Research Council, *Recommended Dietary Allowances*, 10th ed. (Washington D.C.: National Academy Press, 1989).

TABLE 2.22
Comparison of RDAs for water-soluble vitamins in pregnant and nonpregnant women

MINERALS

Calcium

The placenta actively transports calcium to the fetus, and concentrations of serum calcium in the fetus are substantially higher than those in the mother. The rate of calcium transfer to the fetus at 20 weeks gestation has been estimated to be 50 mg/day, and transfer increases markedly to approximately 330 mg/day near term.

Nearly 30 g of calcium are transferred to the fetus during the second half of pregnancy, most of which is deposited into the fetal skeleton (Pitkin 1985).

Maternal calcium metabolism during pregnancy undergoes dramatic changes to supply gestational needs. A substantial amount of the calcium required by the fetus is provided by increased maternal absorption of calcium from the intestine. Efficiency of intestinal absorption increases early in the second trimester, and by the fifth month, calcium absorption is double that in nonpregnant women (Institute of Medicine 1990). Although urinary calcium excretion also increases, calcium balance is strongly positive (Pitkin 1985). Elevated concentrations of $1,25\text{-}(OH)_2$ vitamin D in maternal blood are responsible for many of these adaptations (Cross et al. 1995). Prolactin secreted by the placenta may also stimulate calcium absorption in late pregnancy.

It is unclear whether increased calcium absorption from the diet during pregnancy prevents a loss of calcium from maternal bone. Several studies have shown that calcium absorbed from dietary sources, even in well-nourished women, is insufficient to cover the estimated needs of the fetus in the third trimester. This suggests that calcium is withdrawn from maternal bone. However, most studies have found no evidence of overall bone loss during a healthy pregnancy (Cross et al. 1995), (Sowers et al. 1995). Because calcium retention increases early in pregnancy before fetal needs become significant, calcium added to maternal bone earlier in gestation may be transferred to the fetus later in gestation.

Because about half of blood calcium is bound to albumin, and albumin concentrations decline in pregnancy, total plasma calcium falls about 5% by the 34th week of gestation. However, levels of free ionized calcium in the maternal blood remain constant, so adequate calcium is available for placental transport into the fetal blood (Mimouri and Tsang 1994).

The RDA for calcium during pregnancy is 1200 mg. This represents an increment of 400 mg/day after age 24—a 50% increase over suggested levels of intake for nonpregnant women. The basis of this recommendation is the estimated daily retention of about 300 mg of calcium that occurs during the third trimester (NAS 1989).

Magnesium

The normal fetus at term contains nearly 1 g of magnesium, most of which is acquired during the second and third trimesters at an average rate of about 6 mg/day (Mimouri and Tsang 1994). Concentrations of magnesium in maternal blood increase slightly in early pregnancy and return to nonpregnant levels by term (Reitz et al. 1975).

The RDA for pregnant women is 320 mg of magnesium. For women after age 18, this represents a 14% increase over nonpregnancy needs. This recommendation is based on both fetal needs and requirements for maternal tissue growth, and it assumes approximately 50% of dietary magnesium is absorbed by the mother (NAS 1989).

TRACE ELEMENTS

Iodine

The placenta readily transfers iodine from the mother to the fetus. An additional 25 mg of iodine per day is recommended to cover the extra needs of the developing fetus (NAS 1989).

Iron

Iron requirements increase dramatically during pregnancy. As shown in Figure 2.20, about 350 mg of iron are retained by the fetus and placenta, and 450 mg are required for the large increase in maternal red blood cell mass (Hallberg 1988). Normal losses of iron are about 240 mg over the course of a normal pregnancy. Thus, the total iron requirements during gestation average approximately 1 g—nearly half of an adult woman's total body iron pool of 2.2 g. The daily need for iron during pregnancy is over 4 mg, considerably greater than the typical daily iron absorption of 1.3 mg in nonpregnant women. These needs are concentrated in the second and third trimesters (Hallberg 1988).

The mother responds to these demands by sharply increasing intestinal absorption of iron (see Figure 2.21). Iron absorption from both supplements and food rises as pregnancy progresses. At term, the percent of iron absorbed from a 30–40 mg dose of ferrous iron is nearly 2.5 times that at 12 weeks gestation (Institute of Medicine 1990). Compared to nonpregnant women, pregnant women absorb three to five times the amount of iron from a mixed meal (Rosso 1990).

In most women increased absorption of dietary iron cannot cover the demands of pregnancy, and body stores of iron are mobilized. Serum iron levels fall progressively during gestation; values at term are 30%

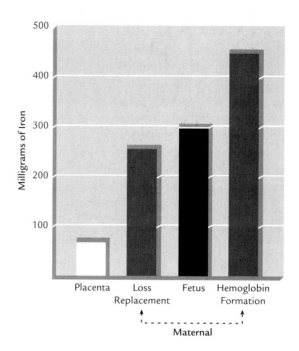

FIGURE 2.20

Distribution of iron needs during a term pregnancy

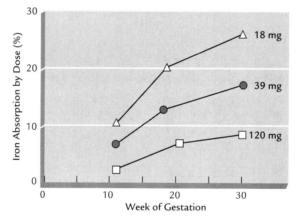

Source: Adapted from Institute of Medicine, 1990

FIGURE 2.21

Increase in iron absorption over the course of pregnancy

lower than prepregnancy levels (Institute of Medicine 1990). At the same time, transferrin levels at term are twice prepregnancy values (hepatic production of transferrin is stimulated by increased circulating estrogens). Consequently, transferrin saturation falls markedly. Concentrations of ferritin in the maternal blood are lower in pregnancy, particularly early in the second trimester as the red cell mass expands. Because of this dramatic mobilization of maternal iron stores, marrow iron stores are low or absent in most women in the third trimester.

During the two weeks preceding birth, the average fetus retains about 3 mg/day of iron. This is considerably more than the 0.6 mg/day needed

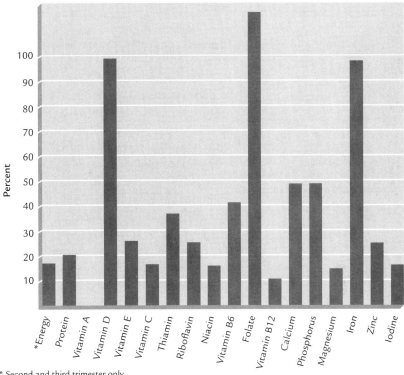

FIGURE 2.22

RDAs for pregnancy: percentage above nonpregnant RDAs for 25–50 years of age

* Second and third trimester only

to maintain normal iron status just after birth. It appears that the fetus draws extra iron from the mother before delivery, iron that can then be used by the infant to maintain normal status during lactation, to compensate for the low iron content of human milk (Rosso 1990).

For most nutrients, deficiency in the mother will produce a deficiency in the fetus. Maternal–fetal iron metabolism is unusual in that the iron status of the fetus is maintained near normal even if depletion of iron stores and anemia occur in the mother. In mothers who are iron deficient, fetal stores of iron, as measured by newborn hemoglobin concentrations or serum ferritin, are largely unaffected by iron supplementation. However, risk of low birthweight, prematurity, and perinatal mortality are increased in infants of iron-deficient mothers (Viteri 1994) (discussed in the next chapter).

To meet high gestational demands for iron, the RDA during pregnancy is set at 30 mg, a 100% increase over nonpregnancy levels. As Figure 2.22 shows, it is one of the highest increases (along with vitamin D and folate). This amount of iron is difficult to obtain even from normal balanced diets, and the iron intake of most pregnant women falls far short. Because of this, many expert groups recommend daily iron supplements for pregnant women (Institute of Medicine 1990), (American College of Obstetricians and Gynecologists 1992). Iron supplementation and iron-deficiency anemia during pregnancy are discussed in the next chapter.

Zinc

Maternal concentrations of plasma zinc decrease during gestation, with the average decline about 45% from prepregnancy levels (Hambidge et al. 1983). Several normal physiological changes during pregnancy may account for at least part of the decline, including lower plasma albumin concentrations (zinc is bound to albumin in plasma) and hemodilution. However, no significant differences in body retention of zinc are found when pregnant women in the third trimester are compared to control groups of nonpregnant women (Rosso 1990).

About 100 mg of zinc are retained in maternal and fetal tissues during gestation. Daily requirements during pregnancy average 0.2–0.3 mg during early pregnancy and 0.6–0.75 mg during the later stages (Swanson and King 1987). To cover these extra needs, and considering the percent absorption of zinc during pregnancy (about 25% in the third trimester), the RDA for zinc during pregnancy is set at 15 mg—25% higher than the RDA for nonpregnant women (NAS 1989).

Copper

Concentrations of copper in maternal plasma increase progressively during pregnancy. At term, values are twice those in nonpregnant women. This rise is due to a doubling of plasma levels of ceruloplasmin, the protein produced by the liver that binds over 90% of blood copper (Rosso 1990). Total copper retention during gestation is about 30 mg, of which 17 mg are retained in the fetus (Campbell 1988). Most of the copper is accumulated in the last trimester, when the average daily retention of copper is 0.28 mg. Because in 1989 (when the latest RDAs were published), information about copper needs during pregnancy was incomplete, there is no current RDA for copper during pregnancy.

Selenium

Selenium concentrations in maternal plasma decline during pregnancy (probably because of hemodilution), while levels in red blood cells do not change (Institute of Medicine 1990). The RDA for selenium during pregnancy is set at 65 mg, or 10 mg higher than that for nonpregnant women (NAS 1989). The basis of this recommendation is that about 5 kg of lean tissue is accumulated in maternal and fetal tissues during gestation, containing an average selenium content of 0.25 mg/kg (NAS 1989). The total selenium retained during pregnancy would therefore be

1.25 mg, or an average of about 5 mg/day. Allowing for variability and assuming an absorption rate of 80%, an additional 10 mg/day is recommended (NAS 1989).

Fluoride

The primary dentition begins to calcify in the fetus during the third trimester. Although one study found that fluoride supplements given to pregnant women reduced the incidence of dental caries in the primary dentition of their children (Glenn and Glenn 1987), the American Dental Association does not currently recommend fluoride supplementation during pregnancy. Intake of fluoride provided by fluoridated water supplies is encouraged (Institute of Medicine, 1990).

WATER AND ELECTROLYTES

Water

The need for water increases during pregnancy to support expansion of maternal plasma and extracellular volume, maintain the amniotic fluid, and provide for fetal needs. About 62% of gestational weight gain is water, corresponding to a daily increment in water during pregnancy of about 30 ml. It is recommended that pregnant women drink 8–10 cups of fluid per day, choose fluids that contribute to nutrient needs (such as fruit juices), and avoid beverages with caffeine, nonnutritive sweeteners, and alcohol (Institute of Medicine 1992). Adequate water intake also facilitates urinary excretion of maternal and fetal wastes.

Sodium and Potassium

The extracellular water gained during pregnancy contains about 150 mEq of sodium/liter, so the average daily sodium requirement increases during pregnancy by about 70 mg (NAS 1989). Because the RDA for sodium for adults is about 500 mg, and the typical U.S. diet contains greater than 2 g of sodium, the modest increase in sodium requirement during pregnancy is easily met by usual salt intake (NAS 1989). The small amounts of potassium accumulated in new tissue during gestation can also be easily supplied by typical adult diets.

EATING HABITS DURING PREGNANCY

NUTRITIONAL SURVEYS OF PREGNANT WOMEN

The National Academy of Sciences recently summarized the results of 17 large surveys of nutrient intake of pregnant women in the 1980s. In general, intakes of protein, thiamin, riboflavin, niacin, and vitamins A, C, and B_{12} met or exceeded the U.S. RDAs for pregnancy. Although mean daily energy intakes varied substantially, with reported values ranging from 1500 to 2800 kcal, most surveys found that energy intakes were well below current recommendations of 2500 kcal/day (Institute of Medicine 1990).

Consumption of many micronutrients was below current recommendations. In nearly all studies, intake of vitamins B_6, D, E, and folate, and the minerals iron, calcium, zinc and magnesium was well below the RDA (Institute of Medicine 1990). For example, the mean daily consumption of iron was between 10 and 17 mg— less than half the current RDA of 30 mg. Calcium consumption by pregnant African Americans, Hispanics, and Native Americans tended to be lower than that of Whites (Suiter et al. 1990), (Rush et al. 1988). Table 2.23 lists the nutrients that are most often lacking in diets of pregnant women.

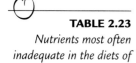

TABLE 2.23

Nutrients most often inadequate in the diets of pregnant women

Vitamin B_6
Vitamin D
Vitamin E
Folate
Iron
Calcium
Zinc
Magnesium

CHANGES IN APPETITE

Although changes in appetite are a characteristic of pregnancy, the popular image of the pregnant mother with a robust increase in appetite has not been supported by careful research. Most studies examining energy intake in well-nourished pregnant women have found only small increases during pregnancy. The increase in caloric intake averaged over the entire pregnancy has usually been in the range of 50–70 kcal/day—about the number of calories in an egg or ½ cup of whole milk (Institute of Medicine 1990).

Patterns of increased energy intake during pregnancy found in different studies vary considerably. A study in Scotland showed energy intake increased slowly through the second and third trimesters to about 150 kcal/day at term (Durnin 1987). In contrast, Dutch scientists found energy consumption did not change in the first two trimesters and rose only about 45 kcal/day in the final trimester (Von Raaij et al. 1987). A large recent Australian study found no increase in energy intake during pregnancy (Truswell et al. 1988). Changes in energy intake during pregnancy are difficult to detect because they are usually small, and available assessment measures are imprecise.

Food intake during pregnancy may be influenced by placental hormones. Maternal levels of progesterone and estrogen increase steadily through pregnancy, and high levels may stimulate appetite (Rosso 1990). Most women report hunger and food intake are greatest during the second trimester, and this is consistent with the usual pattern of gestational weight gain—daily gains are greatest in the second trimester.

FOOD PREFERENCES

More than three quarters of women report a shift in food preferences during pregnancy. A recent study found preferences for certain food flavors varies over the course of pregnancy (Bowen 1992). Food consumption patterns and taste perceptions were measured in 50 pregnant women. Although there were no changes in perception of sweet taste, women consumed substantially more sweet-tasting foods in the second trimester. In the third trimester, perception of salt taste was diminished, and women consumed more salty foods .

One common myth of pregnancy is that a woman's instincts will lead her to select appropriate, healthy foods during pregnancy. Although food preferences change during gestation, there is no evidence these changes instinctively allow a woman to correct or balance a poor diet. A more likely explanation for healthy changes in food preferences during pregnancy is the increased awareness of the importance of eating well.

Cravings and Aversions

Food craving is the compulsive urge for certain foods during pregnancy, for which there had been no great desire before pregnancy. *Aversions* are strong revulsions against food and drink that were not disliked before pregnancy.

Most women develop intense feelings about certain foods during pregnancy. In a study in Britain, over two thirds of pregnant women reported food cravings or aversions (Tierson et al. 1985). Studies in the United States and Britain have shown that certain food classes tend to arouse strong likes or dislikes in pregnant women. The most commonly craved products were sweets, fruits, and dairy products. The most frequent aversions were to high-protein foods, meats, alcohol, and caffeinated products (Worthington-Roberts et al. 1989), (Tierson et al. 1985).

The nutritional impact of cravings and aversions during pregnancy is usually minimal. In general, they tend to increase maternal intake of calcium and simple sugars and decrease intake of animal protein. They typically do not have detrimental effects on overall diet quality.

DIET FOR A HEALTHY PREGNANCY: REVIEW AND SUMMARY

Consumption of a high-quality diet before pregnancy is important to ensure optimal nutritional status and adequate prepregnancy weight. Once pregnant, women must carefully choose foods to cover increased gestational demands. The need for several micronutrients increases substantially in pregnancy: a 100% increase for folate, vitamin D, and iron, and more than 50% increase for many other nutrients. Yet daily energy needs are increased by only 300 kcal. Therefore, to meet increased nutrient needs without exceeding caloric needs, it is essential that pregnant women consume foods with high nutrient quality and density.

The Committee on Nutrition of the Institute of Medicine has published guidelines for healthy eating during pregnancy (Institute of Medicine 1992). Recommendations include:

- Eat a well-balanced and nutrient-dense diet. Include fruits, vegetables, grains, milk products, and protein foods (such as meat, eggs, legumes, nuts, and tofu) in meals and snacks every day. Emphasize foods that supply ample iron, calcium, vitamin D, zinc, fiber, and folate.

- Eat small to moderate-sized meals at regular intervals, and choose nutritious snacks.

- Eat enough food to gain weight at a desirable rate. Weight gain for most women should be 25–35 lb, or about 1 lb/week during the second and third trimesters. Weight gain for women who are underweight before pregnancy should be at the upper end of this range. Those who were overweight before pregnancy should gain 15–25 lbs.

- To obtain adequate calcium and vitamin D, take three or more servings of milk products daily (a serving is one cup), either with or between meals. Choose low-fat or skim milk products often. For women who are lactose intolerant, calcium-rich alternatives to dairy products are shown in Table 4.15.

■ To absorb more iron, include sources of heme iron (meat, poultry, fish) or vitamin C-rich foods (such as orange juice, broccoli, or strawberries) in meals.

■ Salt food moderately to taste. Healthy women do not need to restrict salt intake during pregnancy.

■ Abstain from or limit consumption of coffee or other caffeinated beverages to two servings or less each day.

■ The only sure way to avoid the possible harmful effects of alcohol on the fetus is to avoid drinking alcoholic beverages entirely.

Prenatal care, including nutrition assessment and counseling, should begin early in pregnancy, as a healthy lifestyle is important to both maternal and fetal health. Pregnant women should completely avoid smoking and illicit drugs and use medicinal drugs (even over-the-counter preparations) only with the approval of their physician. Healthy, pregnant women can benefit from a regular, low to moderate-intensity exercise program during pregnancy. (The impact of diet and lifestyle on pregnancy outcome is the focus of the next chapter).

The California Department of Health Services has developed a food guide to help women choose a nutritionally adequate diet (see Table 2.24). The guide focuses on the special nutritional needs of women during the childbearing years, both before and during pregnancy (California Dept. of Health Services 1990).

The guide specifies the minimum number of recommended servings of the various food groups needed to meet the RDAs, and women are encouraged to eat more food from the various food groups, depending on their appetite and energy needs. Using these guidelines, "best choice" sample menus for pregnant women have been devised (see Table 2.25).

It is recommended that pregnant women who are vegetarians supplement their diets with a vitamin and mineral supplement and consume ample amounts of complementary vegetable proteins, milk products, cereals, and vitamin C-rich fruits and vegetables (see Table 2.26). Finally, all pregnant women can ensure adequate intake of nutrients without consuming too many calories by following the guidelines in Table 2.27.

Food Group	One Serving Equals		Recommended Minimum Servings		
			Nonpregnant		Pregnant/
			11–24yrs	25+yrs	Lactating
	Protein	**Protein**			
Protein Foods Provide protein, iron, zinc and B-vitamins for growth of muscles, bone, blood, and nerves. Vegetable protein provides fiber to prevent constipation.	1 oz. cooked chicken or turkey 1 oz. cooked lean beef, lamb, or pork 1 oz. or ¼ cup fish or other seafood 1 egg 2 fish sticks or hot dogs 2 slices luncheon meat	½ cup cooked dry beans lentils, or split peas 3 oz. tofu 1 oz. or ¼ cup peanuts pumpkin, or sunflower seeds 1½ oz. or ⅓ cup other nuts 2 tbsp. peanut butter	5 A half serving of vegetable protein daily	5	7 One serving of vegetable protein daily
Milk Products Provide protein and calcium to build strong bones, teeth, healthy nerves and muscles and to promote normal blood clotting.	8 oz. milk 8 oz. yogurt 1 cup milk shake 1½ cups cream soup (made with milk) 1½ oz. or ½ cup grated cheese (like cheddar, monterey, mozzarella, or swiss)	1½ slices presliced American cheese 4 tbsp. parmesan cheese 2 cups cottage cheese 1 cup pudding 1 cup custard or flan 1½ cups ice milk, ice cream, or frozen yogurt	3	2	3
Breads, Cereals, Grains Provide carbohydrates and B-vitamins for energy and healthy nerves. Also provide iron for healthy blood. Whole grains provide fiber to prevent constipation.	1 slice bread 1 dinner roll ½ bun or bagel ½ English muffin 1 small tortilla ¾ cup dry cereal ½ cup granola ½ cooked cereal	½ cup rice ½ cup noodles or spaghetti ¼ cup wheat germ 1 4-inch pancake or 1 small muffin 8 medium crackers 4 graham cracker squares 3 cups popcorn	7	6	7

TABLE 2.24

Daily food guide for nonpregnant, pregnant, and lactating women

TABLE 2.24 *Continued*

Food Group	One Serving Equals		Recommended Minimum Servings		
			Nonpregnant 11–24yrs	25+yrs	Pregnant/ Lactating
	Protein	**Protein**			
Vitamin C-Rich Fruits and Vegetables Provide vitamin C to prevent infection and to promote healing and iron absorption. Also provide fiber to prevent constipation.	6 oz. orange, grapefruit, or fruit juice enriched with vitamin C 6 oz. tomato juice or vegetable juice cocktail 1 orange, kiwi, mango ½ grapefruit, cantaloupe ½ cup papaya 2 tangerines	½ cup strawberries ½ cup cooked or 1 cup raw cabbage ½ cup broccoli, brussels sprouts, or cauliflower ½ cup snow peas, sweet peppers, or tomato puree 2 tomatoes	1	1	1
Vitamin A-Rich Fruits and Vegetables Provide beta-carotene and vitamin A to prevent infection and to promote wound healing and night vision. Also provide fiber to prevent constipation.	6 oz. apricot nectar or vegetable juice cocktail 3 raw or ¼ cup dried apricots ¼ cantaloupe or mango 1 small or ½ cup sliced carrots 2 tomatoes	½ cup cooked or 1 cup raw spinach ½ cup cooked greens (beet, chard, collards, dandelion, kale, mustard) ½ cup pumpkin, sweet potato, winter squash, or yams	1	1	1
Other Fruits and Vegetables Provide carbohydrates for energy and fiber to prevent constipation.	6 oz. fruit juice (if not listed above) 1 medium or ½ cup sliced fruit (apple, banana, peach, pear) ½ cup berries (other than strawberries) ½ cup cherries or grapes ½ cup pineapple ½ cup watermelon	¼ cup dried fruit ½ cup sliced vegetable (asparagus, beets, green beans, celery, corn, eggplant, mushrooms, onion, peas, potato, summer squash, zucchini) ½ artichoke 1 cup lettuce	3	3	3
Unsaturated Fats Provide vitamin E to protect tissue	½ med. avocado 1 tsp. margarine 1 tsp. mayonnaise 1 tsp. vegetable oil	2 tsp. salad dressing (mayonnaise-based) 1 tbsp. salad dressing (oil-based)	3	3	3

Note: The Daily Food Guide for Women may not provide all the calories you require. The best way to increase your intake is to include more than the minimum servings recommended.

Adapted from California Department of Health Services, Maternal and Child Health Branch and WIC Supplemental Foods Branch, "Dietary Guidelines and a Daily Food Guide", in *Nutrition During Pregnancy and the Postpartum Period: A Manual for Health Care Professionals*

Breakfast:	1 cup oatmeal	1 sl. whole wheat toast
	⅛ cup raisins	1 tsp. margarine
	12 oz. nonfat milk	¾ cup shredded wheat
		12 oz. nonfat milk
		½ cup blueberries, raw
Snack:	6 oz. fresh orange juice	3 cups popcorn, plain
Lunch:	Chili made from:	
	3 oz. lean ground round	¾ cup lentil soup
	½ cup kidney beans,	1 sl. whole-grain rye bread
	cooked	2 oz. lean roast beef
	1/4 cup onions. cooked	½ cup beets, fresh cooked,
	2 sl. whole wheat bread	pickled
	1 cup lettuce	
	1 tbsp. Italian dressing	
Snack:	½ cup prunes cooked	½ cup prunes, cooked
Dinner:	3 oz. turkey, dark meat	4 oz. halibut, cooked
	1 cup brown rice, cooked	2 whole wheat rolls
	½ tsp. margarine	1 tsp. margarine
	½ cup peas, frozen	½ cup beet greens, fresh
	½ cup spinach, fresh cooked	cooked
	½ tsp. olive oil plus garlic	1 cup cabbage, raw, shredded
	powder (on spinach)	2 tsp. salad dressing
	1 whole wheat roll	(mayonnaise-type)
	1 tsp. margarine	
Snack:	12 oz. nonfat milk	12 oz. nonfat milk
		½ cup whole wheatflakes

TABLE 2.25

Two "best choice" sample menus for pregnant women

Adapted from K. Newman and D. Lee, "Developing a Daily Food Guide for Women", *J of Nut Educ 23* (1991): 76–82.

Food	Amino Acids Deficient	Complementary Protein Food Combinations
Grains	Isoleucine Lysine	Rice + legumes Corn + legumes Wheat + legumes Wheat + peanuts + milk Wheat + sesame + soybeans
Legumes	Tryptophan Methionine	Legumes + rice Beans + wheat Beans + corn Soybeans + rice + wheat Soybeans + corn + milk Soybeans + peanuts + seasame Soybeans + peanuts + wheat + rice Soybeans + sesame + wheat
Nuts and seeds	Isoleucine Lysine	Peanuts + sesame + soybeans Sesame + beans Sesame + soybeans + wheat Peanuts + sunflower seeds
Vegetables	Isoleucine Methionine	Lima beans Green beans Brussels sprouts + Sesame seeds or Brazil nuts or Cauliflower mushrooms Broccoli Greens + millet or rice

TABLE 2.26
Vegetarian food guide for protein combinations

Substitute similar food that provide more nutrients for the calories, for example:

Foods That Are Low in Vitamins and Minerals	Foods That Are Higher in Vitamins and Minerals
Bologna	Sliced chicken
Bacon	Ham
Cake	Ice milk, pudding
Fruit drinks	Fruit juices
	Fresh fruit

Choose lower-calorie versions of the same foods to make room for good nutrient sources, for example:

Commonly Eaten Food	Lower-Calorie Version
Ice cream	Ice milk or frozen yogurt
Whole milk	Low-fat or skim milk
Fried chicken	Grilled chicken or fried chicken without the skin
American cheese	Low-fat cheese
Cream soup	Vegetable soup
Choice or prime beef	Lean beef
Chips	Unbuttered popcorn, pretzels

Cut back on foods high in calories, such as those listed below, to make room for foods that are better sources of vitamins and minerals.

High-Calorie Foods That Are Low in Essential Nutrients

- Most deep-fried foods and fat-rich foods (such as onion rings, fried taco shells, chips, doughnuts, bacon and sausages)

- Soft drinks and fruit punch

- Salad dressings and mayonnaise

- Most desserts (such as cookies, cakes, and pies)

- Fats and oils

- Most sweets (such as candy, sugar, and honey)

Adapted from Institute of Medicine, Subcommittee for a Clinical Application Guide, *Nutrition During Pregnancy and Lactation: An Implementation Guide*, (Washington D.C.: National Academy Press, 1992).

TABLE 2.27
Getting adequate nutrients without too many calories

REFERENCES

Alexander, C.S. and Guyer, B., "Adolescent Pregnancy: Occurence and Consequences," *Pediatric Annals* 22 (1993):85–88.

American Academy of Pediatrics, *Pediatric Nutrition Handbook*, 3d ed. (Elk Grove, IL: American Academy of Pediatrics, 1994).

American College of Obstetricians and Gynecologists, American Academy of Pediatrics, *Guidelines for Perinatal Care*, (Washington D.C.: ACOG, 1992).

Apgar, J., "Zinc and Reproduction: An Update," *J Nutr Biochem* 3 (1992):266–278.

Baron, T.H., Ramirez, B., and Richter J.E., "Gastrointestinal Motility Disorders During Pregnancy," *Ann Int Med* 118 (1993):366–375.

Barr, S.I., Janelle, C., and Prior J.C., "Energy Intakes Are Higher During the Luteal Phase of Ovulatory Menstrual Cycles," *Am J Clin Nutr* 61 (1995):39–43.

Bates, G.W. and Whitworth, N.S., "Effect of Body Weight Reduction on Plasma Androgens in Obese Infertile Women," *Fertil Steril* 38 (1982):406.

Bendich, A., "Lifestyle and Environmental Factors That Can Adversely Affect Maternal Nutritional Status and Pregnancy Outcomes. *Ann NY Acad Sci* 678 (1993):255265.

Binkin, N.J., et al., "Birthweight and Childhood Growth," *Pediatrics* 82 (1988):828–34.

Block, G. and Abrams, B., "Vitamin and Mineral Status of Women of Childbearing Potential." *Ann NY Acad Sci* 678 (1993):245–54.

Bowen, D.J., "Taste and Food Preferencs Changes Across the Course of Pregnancy," *Appetite* 19 (1992):233–42.

Braverman, D.Z., Johnson, M.L., and Kern, F., "Effects of Pregnancy and Contraceptive Steroids on Gall Bladder Function," *N Engl J Med* 302 (1980):362–64.

Brown, J.E. and Toma, R.B., "Test Changes During Pregnancy," *Am J Clin Nutr* 43 (1986):414–418

Brown, P.F., Molloy, C.M., and Johnson, P.M., "Transferrin Receptor Affinity and Iron Transport in the Human Placenta," *Placenta* 3 (1982):21–28.

California Department of Health Services, Maternal and Child Health Branch and WIC Supplemental Foods Branch, "Dietary Guidelines and a Daily Food Guide," in *Nutrition during Pregnancy and the Postpartum Period: A Manual for Health Care Professionals,* (Sacramento: Department of Health Services, 1990), 59–92.

Campbell, D.M., "Trace Element Needs in Human Pregnancy," *Proc Nutr Soc* 47 (1988):45–53.

Centers for Disease Control, "Recommendations for the Use of Folic Acid to Reduce the Number of Cases of Spina Bifida and Other Neural Tube Defects," *MMWR* 41 (1992):RR–14.

Centers for Disease Control, "Use of Folic Acid for Prevention of Spina Bifida and Other Neural Tube Defects," *MMWR* 40 (1991):513–16.

Cetin, I. et al., "Umbilical Amino Acid Concentrations in Appropriate and SGA Infants: A Biochemical Difference Present in Utero," *Am J Obstet Gynecol* 158 (1988):120–126.

Coleman, R.A., "Placental Metabolism and Transport of Lipids," *Fed Proc* 45 (1986):2519–23.

Crawford, M.A., "The Role of Essential Fatty Acids in Neural Development: Implications for Perinatal Nutrition," *Am J Clin Nutr* 57 (1993):703S–10S.

Cross, N.A., et al., "Calcium Homeostasis and Bone Metabolism during Pregnancy, Lactation and Postweaning: A Longitudinal Study," *Am J Clin Nutr* 61 (1995):514–23.

Czeizel, A.E. and Dudas, I., "Prevention of First Occurrence of Neural Tube Defects by Periconceptional Vitamin Supplementation," *N Engl J Med* 327 (1992):1832–35.

Davison, J.S. and Dunlop, W., "Renal Hemodynamics and Tubular Function in Normal Human Pregnancy," *Kidney Int* 18 (1980):152–161.

De Luca, L.M. et al.,"Vitamin A in Epithelial Differentiation and Skin Carcinogenesis," *Nutr Rev* 52 (1994):45–53.

Durnin, J.V.G.A., "Energy Requirements in Pregnancy: An Integration of Longitudinal Data from the Five-Country Study," *Lancet* 2 (1987):1131–33.

Food and Agriculture Organization, "Requirements of Vitamin A, Iron, Folate and Vitamin B_{12}," *FAO Food and Nutrition Series* no.23 (Rome: FAO,.1988).

Friedman, S. et al., "A Saturable, High-affinity Binding Site for Transcobalamin-vitamin B_{12} Complexes in Human Placental Membrane Preparations," *J Clin Invest* 59 (1977):51–58.

Frisch, R.E. and McArthur, J.W., "Menstrual cycles: Fatness as a Determinant of Minimum Weight for Height Necessary for Their Maintenance or Onset," *Science* 185 (1974):949–951.

Garner, P.R., "The Effect of Body Weight on Menstrual Function," *Curr Probl Obstet Gynecol* 7 (1984):4–37.

Giugliani, E.R.J., Jorge, S.M. and Goncalves, A.L., "Serum Vitamin B_{12} Levels in Parturients, in the Intervillous Space of the Placenta, and in Full-term Newborns and Their Interrelationships with Folate Levels," *Am J Clin Nutr* 41 (1985):330–35.

Glenn, F.B. and Glenn, W.D., "Optimum Dosage for Prenatal Fluoride Supplementation," *J Dent Child* 54 (1987):445–50.

Gonzalez-Cossio, T. and Delgado, H., "Functional Consequences of Maternal Malnutrition," in Selected Vitamins, Minerals and the Functional Consequences of Maternal Malnutrition, ed. A.P. Simopoulos, *World Rev Nutr Diet* (1991).

Hallberg, L., "Iron Balance in Pregnancy," in *Vitamins and Minerals in Pregnancy and Lactation,* ed. H. Berger (New York: Raven Press, 1988).

Hambidge, K.M., et al., "Zinc Nutritional Status during Pregnancy: A Longitudinal Study," *Am J Clin Nutr* 37 (1983):429–442.

Hiriake, H., Kimura, M., and Itokawa, Y., "Distribution of K Vitamins in Human Placenta and Maternal and Umbilical Cord Plasma," *Am J Obstet Gynecol* 158 (1988):564–69.

Hytten, F.E. and Leitch, I., *The Physiology of Human Pregnancy,* 2d ed. (London: Blackwell Scientific,. 1971).

Institute of Medicine, *Nutrition During Pregnancy and Lactation: An Implementation Guide* (Washington D.C.: National Academy Press, 1992).

Institute of Medicine, *Nutrition During Pregnancy* (Washington, D.C.: National Academy Press, 1990).

Institute of Medicine, *Preventing Low Birthweight* (Washington, D.C.:National Academy Press, 1985).

Kirksey, A. and Wasynczuk, A.Z., "Morphological, Biochemical and Functional Consequences of Vitamin B$_6$ Deficits During Central Nervous System Development," *Ann NY Acad Sci* 678 (1993):62–80.

Landon, M.J. and Hytten, F.E., "The Excretion of Folate in Pregnancy," *J Obstet Gynecol Br Cwlth* 78 (1971):769–75.

Lechtig, A., et al., "Effect of Moderate Maternal Malnutrition on the Placenta," *Am J Obstet Gynecol* 123 (1975):191–201.

Lester, G.E., "Cholecalciferol and Placenta Calcium Transport," *Fed Proc* 45 (1986):2524–27.

Leturque, A., et al., "Glucose Metabolism in Pregnancy," *Biol Neonate* 51 (1987):64.

Liu, K. et al., "International Infant Mortality Rankings: A Look Behind the Numbers," *Health Care Fin Rev* 13 (1992):105–118.

Luke, B., Johnson, T.R.B., and Petrie, R., *Clinical Maternal-Fetal Nutrition* (Boston, MA: Little Brown and Co., 1993).

Martini, M.C., et al., "Effect of Menstrual Cycles on Energy and Nutrient Intake," *Am J Clin Nutr* 60 (1994):895–9.

Metzger, B.E., et al., "Accelerated Starvation and the Skipped Breakfast in Late Normal Pregnancy," *Lancet* 1 (1982):588.

Mimouni, F. and Tsang, R.C., "Perinatal Mineral Metabolism," in *Maternal-Fetal Endocrinology,* eds. D. Tulchinsky and A.B. Little (Philadelphia: W.B. Saunders Co., 1994).

Mimouni, F., et al., "Parathyroid Hormone and Calcitriol Changes in Normal and Diabetic Pregnancies," *Obstet Gynecol* 74 (1989):47.

Moore, K.L., *The Developing Human,* 4th ed., (Philadelphia, PA: W.B. Saunders, 1988).

MRC Vitamin Study Research Group, "Prevention of Neural Tube Defects," *Lancet* 338 (1991):131–7.

Murphy, J.F., O'Riordan, J., and Newcombe, R.G., "Relation of Hemoglobin Levels in the First and Second Trimesters of Pregnancy to Outcome of Pregnancy," *Lancet* 1 (1986):992–95.

National Academy of Sciences Food and Nutrition Board of the National Research Council, *Recommended Dietary Allowances,* 10th ed., (Washington D.C.:National Academy Press, 1989).

Newman, V. and Lee, D., "Developing a Daily Food Guide for Women," *J Nut Educ* 23 (1991):76–82.

Nilsen, S.T., et al., "Males with Low Birthweight Examined at 18 Years of Age," *Acta Ped Scan* 73 (1984):168–75.

Norton, M., Buchanan, T.A., and Kitzmiller, J.L., "Endocrine Pancreas and Maternal Metabolism," in *Maternal-Fetal Endocrinology*, eds. D. Tulchinsky and A.B. Little (Philadelphia: W.B. Saunders Co, 1994).

O'Connor, D.L., "Folate Status during Pregnancy and Lactation," in *Nutrient Regulation During Pregnancy, Lactation and Infant Growth*, eds. L. Allen, J. King, and B. Lonnerdal (New York: Plenum Press, 1994).

Oakley, N.W., Beard, R.W., and Turner, R.C. "Effect of Sustained Hyperglycemia on the Fetus in Normal and Diabetic Pregnancies," *Br Med J* 1 (1972):466–69.

Parham, E.S., Astrom, M.F., and King, S.H., "The Association of Pregnancy Weight Gain With the Mother's Postpartum Weight," *J Am Diet Assoc* 90 (1990):550–54.

Perez-Escamilla, R. and Pollitt, E., "Causes and Consequences of IUGR in Latin America," *Bull PAHO* 26 (1992):128–148.

Pitkin, R.M., "Calcium Metabolism in Pregnancy and the Perinatal Period: A Review," *Am J Obstet Gynecol* 151 (1985):99–109.

Prasad, A.S., "Clinical and Biochemical Spectrum of Zinc Deficiency in Human," in *Clinical, Biochemical and Nutritional Aspects of Trace Elements*, ed. A. Prasad (New York: Alan R. Liss, 1982)

Prentice, A.M., et al., "Energy Balance in Pregnancy and Lactation," in *Nutrient Regulation During Pregnancy, Lactation and Infant Growth*, eds. L. Allen, J. King, and B. Lonnerdal (New York: Plenum Press, 1994).

Rassin, D.K., "Amino Acid and Protein Metabolism in the Premature and Term Infant," in *Neonatal Nutrition and Metabolism*, ed. W.W. Hay - (St. Louis: Mosby-Year Book, 1991).

Reitz, R.E., et al., "Calcium, Magnesium, Phosphorus and Parathyroid Hormone Interrelationships in Pregnancy and Newborn Infants," *Obstet Gynecol* 50 (1977):701–705.

Robson, S.C., et al, "Serial Study of Factors Influencing Changes in Cardiac Output During Pregnancy," *Am J Physiol* 256 (1989):H1060–65.

Rosso, P. et al., "Physiological Adjustments and Pregnancy Outcome in Low-Income Chilean Women," *Fed Proc* 17 (1983):138.

Rosso, P., *Nutrition and Metabolism in Pregnancy* (New York: Oxford University Press, 1990).

Rush, D., et al., "The National WIC Evaluation: Longitudinal Study of Pregnant Women," *Am J Clin Nutr* 48 (1988):439–483.

Salmenpera, L., "Vitamin C during Prolonged Lactation: Optimal in Some Infants While Marginal in Others," *Am J Clin Nutr* 40 (1984):1050–56.

Schuster, K., Bailey, L.B., and Mahan, C.S., "Effect of Maternal Pyridoxine-HCl Supplementation on the Vitamin B$_6$ Status of the Infant and Mother and on Pregnancy Outcome," *J Nutr* 977 (1984):114–88.

Seely, E.W. and Moore, T.J., "The Renin-Angiotensin-Aldosterone System and Vasopressin," in *Maternal-Fetal Endocrinology,* eds. D. Tulchinsky and A.B. Little (Philadelphia: W.B. Saunders Co., 1994).

Shoupe, D., "Effect of Body Weight on Reproductive Function," in *Infertility, Contraception, and Reproductive Endocrinology,* eds. D.R. Mishell, V. Darajan, R. Lobo (Boston: Blackwell Scientific Publishers, 1991).

Silverman, B.L., Landsberg, L., and Metzger, B.E., "Fetal Hyperinsulinism in Offspring of Diabetic Mothers: Association with the Subsequent Development of Childhood Obesity," *Ann NY Acad Sci* 699 (1993):36–45.

Sowers, M., et al., "A Prospective Study of Bone Density and Pregnancy After an Extended Period of Lactation with Bone Loss," *Obstet Gynecol* 82 (1995):285–9.

Stein, Z. et al., *Famine and Human Development* (New York: Oxford University Press, 1975).

Suiter, C.W., et al., "Characteristics of Diet Among a Culturally Diverse Group of Low-Income Pregnant Women," *J Am Diet Assoc* 90 (1990):543–49.

Swanson, C.A. and King, J.C., "Zinc and Pregnancy Outcome," *Am J Clin Nutr* 46 (1987):763–771.

Tanner, J.M., *Fetus into Man: Physical Growth from Conception to Maturity* (Cambridge: Harvard University Press, 1990).

Teratology Society, "Recommendations for Vitamin A Use During Pregnancy," *Teratology* 35 (1987):267–75.

Tierson, F.D., Olsen C.L., and Hook E.B., "Influence of Cravings and Aversions on Diet in Pregnancy," *Ecol Food Nutr* 17 (1985):117–29.

Trotz, M., Gansau C.H., and Nau H., "Effect of Folic Acid Deficient Diet and Folinic Acid Treatment on the Embryo Toxicity of Valproic Acid in the Mouse," *Teratology* 32 (1985):35.

Truswell, A.S., Ash S., and Allen J.R., "Energy Intake during Pregnancy," *Lancet* 1 (1988):49.

US Human Nutrition Information Service, Nutrition Monitoring Division, *Nationwide Food Consumption Survey, Continuing Survey of Food Intakes by Individuals: Women 19–50 Years and Their Children 1–5 Years, 4 Days, 1985,* report no. 85–4 (Washington, D.C.: U.S. Govt. Printing Office, 1987).

van Raaij, J.M.A., et al., "Energy Requirements of Pregnancy in the Netherlands," *Lancet* 2 (1987):953–55.

Viteri, F.E., "The Consequences of Iron Deficiency and Anemia in Pregnancy," in *Nutrient Regulation During Pregnancy, Lactation and Infant Growth,* eds. L. Allen, J. King, and B. Lonnerdal B (New York: Plenum Press,1994).

Wallingford, J.C., and Underwood, B.A., "Vitamin A Deficiency in Pregnancy, Lactation and the Nursing Child," in *Vitamin A Deficiency and Its Control,* ed. J.C. Bauernfeind (New York: Academic Press, 1986).

Warren, M.P., et al., "Scoliosis and Fratures in Young Ballet Dancers. Relation to Delayed Menarche and Secondary Amenorrhea," *N Engl J Med* 315 (1986): 905.

Wegman, M.E., "Annual Summary of Vital Statistics," *Pediatrics* 94 (1994):792–803.

Whitsett, J.A. et al, "Synthesis of 1,25 Dihydroxyvitamin D_3 by Human Placenta in Vitro," *J Clin Endocrinol Metabol* 53 (1981):484–88.

Widdowsen, E.M., "Changes in Body Composition During Growth," in *Scientific Foundations of Pediatrics,* eds. J. A. Davis, and J. Dobbing (Baltimore: University Park Press, 1981).

Wise, P.H., "Poverty, Technology and Recent Trends in the U.S. Infant Mortality Rate," *Ped and Perinatal Epidemiol* 4 (1990): 390–401.

Wolf, H., "Hormonal Alterations of Efficiency of Conversion of Tryptophan to Urinary Metabolites of Niacin in Man," *Am J Clin Nutr* 42 (1971):792–99.

World Health Organization. 1985. "Energy and Protein Requirements." Report of a Joint FAO/WHO/UNO Expert Consultation. Technical Reports Series 724. World Health Organization, Geneva.

Worthington-Roberts, B. et al., "Dietary Cravings and Aversions in the Postpartum Period," *J Am Diet Assoc* 89 (1989): 647–651.

3

pregnancy and associated nutritional problems

▼

MALNUTRITION AND SUCCESSIVE PREGNANCIES
IN THE DEVELOPING WORLD

In impoverished areas of the developing world, many women enter their reproductive years without having achieved their full growth potential. Because of malnutrition and adverse conditions during their own development, these young women—generally from lower socioeconomic groups—tend to be of short stature and light weight. You have learned that maternal size (both height and prepregnancy weight) are important determinants of pregnancy outcome: mothers whose growth has been impaired because of chronic undernutrition tend to produce smaller, less healthy babies.

A classic study in rural Guatemala illustrates how the legacy of chronic undernutrition and poverty in one generation is passed to the next

(Lechtig et al. 1975). Guatemalan mothers of short stature tended to have lower birth-weight infants than mothers of normal stature. When women received food supplements during pregnancy, mothers of both short stature and normal stature delivered larger babies.

Although the food supplement increased birthweights more in infants born to shorter mothers, it did not eliminate the differences in infant size between the groups. The smaller mothers continued to deliver smaller, less healthy infants. Thus, despite adequate nutrition during pregnancy, two generations or more may be required to undo the effects of adverse environments on birthweight.

This chapter discusses how lower birthweight infants have higher morbidity and mortality in the newborn period. Many—no matter how well they are nourished after birth—cannot recover the growth potential lost in utero. Clearly, the nutritional health of childbearing women has profound and long-reaching repercussions on the health and well-being of future generations.

UNDERNUTRITION DURING PREGNANCY

UNDERNUTRITION AND INTRAUTERINE GROWTH RETARDATION

Impaired development of the fetus may be the result of a genetic defect in the fertilized ovum, a disorder of the placenta that limits the supply of nutrients and oxygen to the fetus, or disease or malnutrition in the mother. The multiple causes of intrauterine growth retardation (IUGR) are shown in Figure 3.1. There are two distinct patterns of IUGR:

- Type I (symmetric) IUGR is characterized by uniform growth impairment: height, weight, and head size of the affected infant are equally reduced. These babies have correct proportions but have low birthweights. Symmetric IUGR is thought to result from a disturbance of fetal growth that begins early in gestation and persists (Singer and O'Shea 1995).

- Type II (asymmetric) IUGR involves a greater reduction in weight relative to length and head size. These babies appear asymmetric—they have near normal length and head circumference, but are very thin and have little skeletal muscle or body fat. Asymmetric IUGR develops when placental function is impaired or the supply of nutrients or oxygen is curtailed later in gestation, usually in the third trimester (Singer and O'Shea 1995).

Although many factors can contribute to IUGR, one of the most important is nutrition. The rapidly growing fetus is sensitive to an inadequate supply of nutrients, and fetal development can be adversely affected even if the mother shows no visible signs of deficiency.

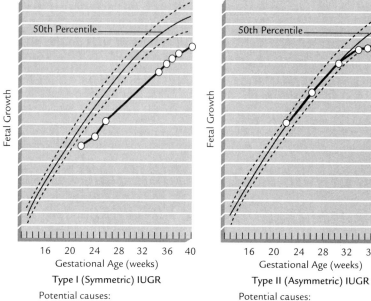

FIGURE 3.1

Growth patterns of infants with symmetric and asymmetric IUGR

DEFICIENCIES OF ENERGY AND PROTEIN DURING PREGNANCY

Because calories and protein supply the energy and building blocks for synthesis of new maternal, placental, and fetal tissue, deficiency during gestation has serious adverse effects on pregnancy outcome. Most studies examining the impact of dietary deficiency in pregnancy have focused on inadequate energy intake.

Insufficient calories to support adequate gestational weight gain will impair growth of the fetus and increase risk of low birthweight (Gonzalez-Cossio and Delgado 1991). The timing, severity, and duration of deficiency determines the pattern and severity of fetal growth retardation:

■ Maternal undernutrition before pregnancy and continuous fetal malnutrition throughout pregnancy causes symmetric IUGR. Cellular division and growth in all fetal tissues, including the brain, is reduced. Symmetric IUGR is often associated with poverty and chronic malnutrition and is common in the developing world (Perez-Escamilla and Pollitt 1992).

■ In contrast, fetal malnutrition confined to the third trimester produces asymmetric IUGR. By the end of the second trimester the fetus has already achieved about 75% of term head size and length, but only about a third of term weight (almost all the fat of a newborn is accumulated in the last 10 weeks of gestation).

Most low-birthweight infants in the industrialized countries, including the United States, are affected by asymmetric IUGR. This type is caused by a variety of factors, including diabetes, hypertension, and smoking (Perez-Escamilla and Pollitt 1992).

Fetal needs for energy and nutrients vary at different points in gestation. Early in pregnancy, total energy needs of the tiny embryo are small and usually easily met—even if the maternal diet is marginal. However, inadequate micronutrient supply during this period may disturb morphogenesis and produce birth defects (Keen and Zidenberg-Cherr 1994). Maternal energy deficit begins to significantly affect fetal growth only after about weeks 20–25, as the larger fetus demands a high quantity of daily energy.

Studies of the effects of famine on pregnant women during World War II illustrate how the timing and severity of maternal undernutrition affect fetal growth (Stein et al. 1975). The dramatic effects of the Dutch famine of 1944–45 on fertility were discussed in the previous chapter, but the famine also had severe adverse effects on pregnancy outcome.

Occurring in a previously well-nourished population, the famine in Holland lasted for a well-defined period of six months. Pregnant women were affected at different periods of gestation. In women exposed during the entire second half of pregnancy, mean birthweight fell over 300 g, or about 9% (see Figure 3.2). Exposure during only the first half of pregnancy had different effects: the number of congenital malformations and **spontaneous abortions** increased, but if the nutritional deficit was corrected in the second half of pregnancy (after the famine), mean birthweight was normal (Stein et al. 1975).

spontaneous abortion:
the premature expulsion from the uterus of the embryo or a nonviable fetus that occurs naturally, without external influence; a miscarriage

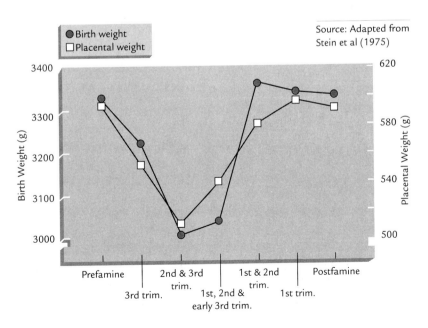

FIGURE 3.2
Birthweight and placental weight: effects of severe undernourishment at different trimesters during pregnancy

	Incidence (Percentage of Births)	
Condition	Prewar	Conceived during the famine
Abortion and miscarriage	1.67	8.3
Prematurity	5.27	8.4
Stillbirth	3.5	4.0
Neonatal death	1.55	5.1
Malformed	1.36	2.4

Adapted from Z. Stein, M. Susser, and G. Saenger, *Famine and Human Development: The Dutch Hunger Winter of 1944–45* (New York: Oxford University Press, 1975).

TABLE 3.1
Pregnancy outcome during the Dutch famine of 1944–45

The impact of the famine during the siege of Leningrad in 1942 has also been studied. In contrast to the Holland famine, most women who were pregnant in Leningrad had consumed marginal diets for a significant period before the famine. The famine lasted about 16 months and produced a period of severe starvation. Compared to the famine in Holland, it had much more devastating effects on pregnancy: babies born during this time weighed 500 g less than expected, and over 40% were born preterm. Nearly 50% of the term infants were LBW infants, and a third of the LBW infants died in the neonatal period (Antonov 1947).

There are several ways maternal undernutrition can impair fetal growth. A poor maternal diet can reduce the amount of nutrients available for transport to the fetus. Dietary restriction of energy and protein can reduce placental size and limit the capacity of the placenta to transfer nutrients to the fetus. Maternal malnutrition can also curtail the normal expansion of red cell mass and blood volume and limit blood flow to the placenta, thereby reducing nutrient availability. All of these mechanisms may play a role in fetal growth retardation in mothers who are undernourished (Rosso 1990).

newborn hypocalcemia: abnormally low blood concentrations of calcium during the newborn period

rickets: vitamin D deficiency in children characterized by poor mineralization of bone and defective skeletal growth

VITAMIN AND MINERAL DEFICIENCIES DURING PREGNANCY

Vitamin D Deficiency

Deficiency of vitamin D reduces availability of calcium during pregnancy and can adversely affect both mother and fetus. Marginal deficiency of vitamin D reduces bone density in the fetus. More severe deficiency can impair development of tooth enamel, cause **newborn hypocalcemia** or, in extreme cases, produce **rickets** in utero (Mimouri and Tsang 1994). If maternal intake of calcium is low, the calcium needs of the fetus will be partially met by increased resorption of maternal bone. Thus, poor vitamin D status can contribute to maternal **osteomalacia.**

In Britain, a high prevalence of vitamin D deficiency was found in pregnant Indian and Pakistani women. Overall, about a third of pregnant women and their infants had undetectable levels of vitamin D in the first week postpartum. Diets inadequate in vitamin D were the major cause of the widespread deficiencies (Rosso 1990). Women from these ethnic groups who were vegetarians were particularly at risk: over two thirds were markedly deficient in vitamin D.

Vitamin A Deficiency

In both animal and human studies, poor maternal vitamin A status is associated with intrauterine growth retardation and premature birth (Institute of Medicine 1990).

Vitamin E Deficiency

Several studies have suggested that congenital malformations and spontaneous abortions are more common in women with low vitamin E status during pregnancy. However, these are preliminary findings that need to be confirmed by further research.

Vitamin B$_6$ Deficiency

Low levels of pyridoxal phosphate (PLP) in maternal plasma have been linked to a variety of complications of pregnancy, including **preeclampsia**, low birthweight, and problems in newborn infants (Institute of Medicine 1990). However, findings have been inconsistent, and until further research is done, it is difficult to draw firm conclusions.

Folate Deficiency

The links between folate deficiency, neural tube defects, and anemia in pregnancy were discussed in Chapter 2, but studies have found associations between poor maternal folate status and other adverse birth outcomes. In one study, two groups of South African women were given a 0.5 mg folate supplement during pregnancy. One group was consuming a diet low in folate and the other a diet higher in folate. The supplement was associated with a 50% reduction in the incidence of small-for-gestational-age newborns in the group with the diet low in folate, but it had no effect

osteomalacia: impaired mineralization and softening of the bones caused by a deficiency of vitamin D; adult rickets

preeclampsia: a serious disorder of pregnancy characterized by hypertension, fluid retention, and protein in the urine, generally occurring after the 20th week of gestation

on the group whose folate intake was higher. The results suggested that inadequate folate status may impair fetal growth (Baumslag et al. 1970).

Thiamin Deficiency

Although the mother may have few or no symptoms, dietary deficiency of thiamin during pregnancy can reduce stores of fetal thiamin and contribute to the development of **beriberi** in the newborn. If not treated promptly, infantile beriberi can cause heart failure and death in the first few days after birth.

Calcium Deficiency

Similar to vitamin D deficiency, dietary deficiency of calcium during gestation may lead to decreased bone density in the newborn and, if severe, impair development of the fetal skeleton (Mimouri and Tsang 1994).

Iodine Deficiency

Because of increased requirements during pregnancy, intakes of iodine that are adequate in nonpregnant women can lead to **iodine-deficiency goiter** in pregnant women, particularly pregnant adolescents. Deficiency of iodine in the mother can have severe effects on fetal development, as adequate iodine is required during fetal life to support the developing thyroid gland (Fischer and Polk 1994).

A study from Zaire illustrates the consequences of iodine deficiency in utero. In a region where iodine deficiency was common, over 10% of babies had evidence of severe **hypothyroidism** at birth (DeLong 1990). Iodine deficiency in utero and early childhood can have devastating, long-term effects on mental and motor development and is a leading cause of mental retardation in the developing world (Dunn 1993) (discussed in Chapter 7).

Zinc Deficiency

Zinc is essential to normal cell replication and differentiation. In animal studies, maternal zinc restriction causes birth defects, premature birth, and fetal growth retardation. In humans, severe zinc deficiency in pregnant women with **acrodermatitis enteropathica** sharply increases the risk of pregnancy complications and birth defects (Jameson 1993).

beriberi: a thiamin deficiency disorder characterized by muscle weakness, heart abnormalities, edema and nerve degeneration

iodine-deficiency goiter: enlargement of the thyroid gland, characterized by a swelling in the lower front portion of the neck; caused by a lack of iodine in the diet

hypothyroidism: diminished production of thyroid hormone, leading to a low metabolic rate, a tendency to gain weight, and somnolence

acrodermatitis enteropathica: a rare disease resulting from an inherited inability to absorb dietary zinc; characterized by severe skin rash, hair loss, diarrhea and poor growth

The consequences of milder zinc deficiency in pregnancy are not completely clear. Although several studies have found associations between low maternal levels of zinc and congenital malformations, IUGR, and preterm birth (Adeniyi 1987), others have not shown adverse effects. Overall, study findings have been contradictory, and no firm conclusions can be drawn.

IRON DEFICIENCY AND ANEMIA

Among all the stages of the human life cycle, the pregnant woman is probably most vulnerable to iron deficiency. Both iron needs and the prevalence of iron deficiency increase markedly during pregnancy. The World Health Organization has estimated that over two-thirds of pregnant women worldwide are anemic, most because of iron deficiency (ACC/SCN 1993). Table 3.2 gives the definition for anemia in the three stages of pregnancy.

In the United States, national surveys indicate that certain factors increase the risk of iron deficiency in both pregnant and nonpregnant women. Prevalence of iron deficiency tends to be higher among adolescents, women living in poverty, and those with less than a high school education. Hispanic women are at greater risk of iron deficiency than non-Hispanic Whites, while rates are similar among White and Black women (Institute of Medicine 1990). Risk factors for iron deficiency are shown in Table 3.3.

Iron-deficiency anemia during pregnancy may be harmful to the fetus. In a recent study in Wales, outcomes of over 50,000 pregnancies were compared to maternal hemoglobin concentrations in the first half of pregnancy. The risk of low birthweight, preterm birth, and perinatal mortality were substantially higher when maternal hemoglobin levels were in the anemic range (Murphy et al. 1986). Although it is often difficult to distinguish the specific effects of iron deficiency from other, potentially confounding factors (education, poverty, diet), these findings have been supported by additional studies (Viteri 1994).

Considering the potential risks and the prevalence of iron deficiency in pregnancy, hemoglobin should be measured routinely during prenatal

TABLE 3.2

Definition of anemia during pregnancy

Trimester	Hemoglobin (g/dl)
First	<11.0
Second	<10.5
Third	<11.0

Source: Institute of Medicine (1990)

- Adolescence

- Poverty

- Limited education

- Hispanic

- History of heavy menstrual periods

- Diet low in meat and ascorbic acid

- Increasing parity

- Chronic use of aspirin

Adapted from Institute of Medicine, Subcommittee on Nutritional Status and Weight Gain during Pregnancy, *Nutrition during Pregnancy* (Washington D.C.: National Academy Press, 1990).

TABLE 3.3

Risk factors for iron deficiency in women

visits. Anemia accompanied by a low serum ferritin can be presumed to be iron deficiency. Treatment of iron-deficiency anemia includes daily iron supplements containing 60-120 mg of iron, along with advice on how to increase dietary sources of iron.

The current RDA for iron during pregnancy is 30 mg. Because the typical mixed diet in the United States provides only about 6 mg of iron/1000 kcal (NAS 1989), consuming adequate dietary iron during pregnancy is difficult. Expert opinion is divided on the issue of whether iron supplements should be given to all pregnant women to prevent iron deficiency. Several groups, including the U.S. Public Health Service, think the available evidence is insufficient to recommend routine supplementation (U.S. Preventive Services Task Force 1993). Other groups, including the American College of Obstetricians and Gynecologists (American College of Obstetricians and Gynecologists 1992), the Food and Nutrition Board of the National Academy of Sciences (Institute of Medicine 1990), and an expert panel convened by the U.S. Food and Drug Administration (Federation of the American Society for Experimental Biology 1991), believe supplementation for all pregnant women is warranted. See Figure 3.3.

To prevent iron deficiency, most experts recommend routine daily supplementation of 30 mg of ferrous iron for all pregnant women (Institute of Medicine 1990). Supplements should be taken in conjunction with a well-balanced diet that contains enhancers of iron absorption (meat, ascorbic acid). Although high-dose iron supplements may cause nausea, constipation, and reduced appetite, these side effects are rare with doses of 30 mg/day.

Source: Institute of Medicine (1990)

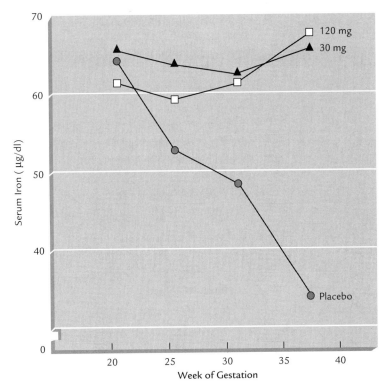

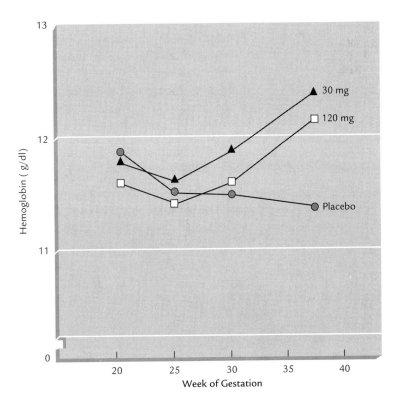

FIGURE 3.3

Iron status of pregnant women receiving iron supplementation vs. those receiving placebos

LOW-BIRTH-WEIGHT INFANTS AND THE DEVELOPING WORLD

The World Health Organization has estimated that nearly one in five infants born in the developing countries are LBW infants. Compare the mean birthweight of infants in various populations shown in Table 3.4. As stated earlier, the majority of LBW infants in the developing world suffer from symmetric IUGR caused by chronic maternal undernutrition throughout pregnancy (Perez-Escamilla and Pollitt 1992). Rates of IUGR are as high as 35% in some impoverished regions (Uvin 1994), and IUGR is a major cause of infant mortality worldwide (ACC/SCN 1992).

There are several aspects of undernutrition in developing populations that influence fetal growth and birthweight:

- Many women in developing countries are **nutritionally stunted** because of chronic undernutrition and other adverse conditions in earlier life. Smaller mothers tend to have less healthy babies despite nutritional adequacy during pregnancy (Lechtig et al. 1975).

- When added to the burden of chronic maternal undernutrition, deficits in energy and protein during gestation sharply reduce average birthweight (often by 500–700 grams) and greatly increase the incidence of LBW infants. In a study of poorly nourished women in Central America, those who had low prepregnancy weights (less than 45 kg) and who gained less than 3.5 kg during pregnancy had a 60% chance of delivering an LBW infant (Gonzalez-Cossio and Delgado 1991).

- Insufficient energy intake during pregnancy is the result of multiple factors, including low income, food scarcity, limited education, and increased energy expenditure in daily maternal work. Moreover, the prepregnancy weight and nutritional status of the mother is influenced by additional factors, including parity, interval since last pregnancy, and chronic infections and diseases.

Nutrition interventions in undernourished, pregnant women in developing countries have produced encouraging results. A recent study reported that a nutritional supplementation program increased mean

nutritionally stunted: impaired growth due to chronic malnutrition resulting in small stature

TABLE 3.4

Mean birthweight of full-term infants in various populations around the world

Population	Country	Year	Birthweight (g)
Lumi	New Guinea	1962–67	2400
Indian	India	1947–63	2740
Yomba	Nigeria	1949–57	2980
Caucasian	Australia	1955	3440
Caucasian	Sweden	1956–57	3502

Adapted from P. Rosso, *Nutrition and Metabolism in Pregnancy* (Oxford: Oxford University Press, 1990).

birthweight by 225 g in undernourished Gambian women (Prentice et al. 1987). In Chile, researchers supplemented the diets of low-income, underweight mothers with a milk-based food supplement containing vitamins and minerals. Their infants had mean birthweights significantly higher than a group of women not receiving the supplement (Mardonese-Santander et al. 1988). As shown in Table 3.5, the percentage of low-birth-weight babies in the group that took supplements was less than half that in the group that did not.

TABLE 3.5

Effects on birthweight of food supplement program for pregnant women in Chile

| Group | Food Supplement | |
	Consumers	Nonconsumers
Birthweight (g)	3178 ± 484	2991 ± 340
SGA babies (%)	32.5	84.2

Adapted from F. Mardones-Santander et al. "Effect of Milk-Based Food Supplement on Maternal Nutritional Status and Fetal Growth in Underweight Chilean Women," *Am J Clin Nutr* 47 (1988): 413–19.

DIGESTIVE DISORDERS IN PREGNANCY

Gastrointestinal complaints are common in normal, healthy pregnancies. Increased levels of circulating hormones, particularly progesterone, relax smooth muscle throughout the gastrointestinal tract. Muscle contractions that normally move food and liquids through the stomach and intestines diminish, and transit time is prolonged. These changes in motility and the symptoms they produce can have a significant impact on nutrition during pregnancy. Diet often plays a central role in managing these conditions (see Table 3.6).

HEARTBURN

Heartburn is caused by **reflux** of food, fluid, or gastric acid from the stomach into the lower esophagus, causing a burning feeling and pain. Thirty to fifty percent of pregnant women are bothered by heartburn, especially in the later months of pregnancy (Baron et al. 1993). High progesterone and estrogen levels cause the lower esophageal sphincter to relax, and the enlarging uterus increases pressure on the stomach.

The pain and discomfort of heartburn may cause some pregnant women to restrict their meals and reduce food intake at a time when nutritional requirements are high. Heartburn can be minimized by eating small, frequent meals. Spicy foods should be avoided, and meals should not be eaten just before physical activity or exercise. Because reflux is

reflux: a backward or return flow of substances in the digestive tract; for example, reflux of stomach contents back into the esophagus

Heartburn

- Eat small, low-fat meals, and eat slowly.

- Take low-fat snacks such as toast or fruit as needed for extra energy and nutrients.

- Drink fluids mainly between meals.

- Go easy on spices.

- Avoid lying down for 1 to 2 hours after eating or drinking, especially before going to bed.

- Wear loose-fitting clothing.

Constipation

- Drink 2 to 3 quarts of fluids daily. This includes water, milk, juice, and soup.

- Eat high-fiber cereals and generous amounts of other whole grains, legumes, fruits, and vegetables.

- Take part in physical activities such as walking and swimming.

- Avoid taking laxatives unless recommended by your health-care provider.

Nausea and Vomiting

- Keep crackers, toast, or dry cereal within reach of your bed. Eat some before getting up.

- Eat frequent small meals.

- Try to take adequate fluids even if you can't handle solids—for example, try clear juices and flat sugar-sweetened soft drinks.

- Avoid drinking coffee and tea. Avoid drinking citrus fruit juices and water upon arising. Drink liquids mainly between meals.

- Try to avoid cooking odors that make you feel ill.

- Avoid or limit your intake of high-fat and spicy foods.

TABLE 3.6
Dietary strategies for managing digestive disorders during pregnancy

usually worse when lying down, elevating the head of the bed and not eating or drinking within three hours of bedtime can be helpful.

NAUSEA AND VOMITING

Nausea is the most common gastrointestinal complaint of pregnancy, occurring in 50–90% of pregnancies; associated vomiting occurs in 25–55% (Baron et al. 1993). Nausea and vomiting often begin in the first weeks of pregnancy, peak in the third month, and usually disappear during the second trimester. It is generally worse early in the day (morning sickness), when even strong odors of foods can precipitate nausea. Both physiological and psychological factors contribute to the problem. Elevated levels of placental hormones play a role, as well as tensions and anxiety relating to pregnancy.

Simple changes in diet can help confront the problem. Eating small, frequent meals high in easily digestible carbohydrates and low in fat can help. Reducing meal volume by consuming liquids between meals may be of benefit. Many women report that eating dry toast and crackers helps settle their stomachs when nauseated. Mothers should be reassured that mild nausea and vomiting will not harm their fetus. Very rarely (about 1 in 300 pregnancies), symptoms become severe (a condition termed *hyperemesis gravidarum*), and require intensive medical and nutritional care (Baron et al. 1993).

CONSTIPATION

Another common disorder of pregnant women is constipation. Several factors contribute to the problem, including prolonged stool transit time, compression of the colon by the enlarging uterus, and decreased physical activity. Proper diet during pregnancy, including ample fluid and fiber intake, can minimize symptoms.

In a report from Israel, women who ate a diet high in fruits and vegetables reported very few problems with constipation (Levy et al. 1971). British researchers found that increasing fiber intake from 20 g/day to 27 g/day reduced the frequency of constipation in pregnant women (Anderson 1984). Hemorrhoids, another common complaint during pregnancy, can also be reduced by including ample fiber and fluids in the diet.

PICA

The compulsion to eat substances that have little or no nutritional value is called *pica*. This eating disorder has been reported in all ages and races, in both sexes, and in many cultures worldwide (Lackey 1982). Pregnant women with pica most often eat clay, dirt, or laundry starch, and other substances such as ice, gravel, charcoal, hair, mothballs, and baking soda.

The cause of pica is not known. Social, psychological, and physiological factors may be involved. Pica has been related in several studies to poor iron status (Rosso 1990), but the practice is more likely due to local habits and traditional culture than to a need for specific nutrients. Among low-income women in the southern United States, incidence of pica during pregnancy has been reported to be 10–40%, mostly involving laundry starch (Lackey 1982).

Depending on the severity of the condition and the substance ingested, pica can be extremely harmful. Lead poisoning from pica for wall plaster, severe anemia from ingested mothballs, and fatal infections from the eating of clay have been reported (Rosso 1990). More commonly, pica can displace nutritional foods from the diet or interfere with absorption of nutrients. It is uncertain how best to manage this baffling disorder, but if pica is interfering with maternal nutrition, every effort should be made to discourage it.

POTENTIALLY HARMFUL DIETARY SUBSTANCES IN PREGNANCY

ALCOHOL

Consumption of alcohol during pregnancy has potentially devastating effects on the fetus. Alcohol is teratogenic; heavy alcohol use during pregnancy causes a distinct constellation of abnormalities in newborns, termed the *fetal alcohol syndrome* (FAS) (Beattie 1992).

FAS has the following three primary characteristics (a detailed list is given in Table 3.7):

- *Prenatal and/or postnatal growth retardation.* The most common sign of FAS is retarded growth in height, weight, and head circumference both in utero and during infancy and childhood. Growth and development during childhood may be impaired even if adequate nutrition and home support are provided.

- *Nervous system abnormalities.* Children with FAS have persistent developmental delays and intellectual impairment.

- *Abnormal facial features.* Infants born with FAS have small eyes, short **palpebral fissures,** and a poorly developed nose, upper jaw, and lip (see Figure 3.4).

Heavy alcohol use during pregnancy also causes a variety of other defects as shown in Table 3.8. Children who do not have full FAS but show some of the symptoms are described as having fetal alcohol effects (FAE).

palpebral fissures: the folds or grooves in the skin around the eyes formed by the eyelids

Symptom	Description
Growth deficiency	Prenatal and/or postnatal growth retardation; failure to thrive (weight, length, and/or head circumferences < 10th percentile)
Central nervous system involvement	Signs of neurological, developmental, and/or behavioral abnormality (developmental delay with intellectual impairment, irritability in infancy or hyperactivity/easy distractibility in childhood)
Characteristic facial dysmorphism (at least 2 of 3)	1. Microcephaly (head circumference < 3rd percentile) 2. Microphthalmia (small eyes) 3. Poorly developed upper lip and flattened profile of the upper jaw

Adapted from J.O. Beattie, "Alcohol Exposure and the Fetus," *Eur J Clin Nutr* 46 (1992): S7–17.

TABLE 3.7
Clinical criteria for the diagnosis of FAS

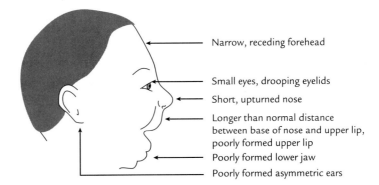

FIGURE 3.4

Features of a child with FAS

In the United States, over two thirds of newborns will have been exposed to alcohol at some point in pregnancy (Institute of Medicine 1990). About 40% of women consume alcohol regularly during pregnancy, and about 3% consume an average of more than three drinks a day (Prager et al. 1984). However, alcohol use among pregnant women is decreasing in the U.S. (Centers for Disease Control 1992). Pregnant women who drink tend to be older, single, or divorced women, who are cigarette smokers and have had several previous pregnancies (Beattie 1992).

The worldwide incidence of FAS is about 1 per 500 live births, while that of FAE is 1 per 200–300 births (Luke et al. 1993). In the U.S., the incidence of FAS is approximately 1 in 10,000 births (Wegman 1994). Among alcoholic women, incidence of FAS is estimated to be 2–3% . FAS is the leading cause of mental retardation in the developed countries of the world, and the yearly cost of providing treatment for children with FAS in the United States has been estimated to be nearly $320 million (Luke et al. 1993).

The exact mechanism by which alcohol damages the fetus has not been established. Both alcohol and its major metabolite, acetaldehyde, readily cross the placenta into the fetus and have been clearly shown to be teratogenic. The fetal brain is exposed to particularly high concentrations of alcohol, and, as the fetus does not have fully developed enzyme systems to break it down, most of the alcohol reaching the fetus must later diffuse back through the placenta to the mother for metabolism (Beattie 1992).

Heavy alcohol use may have detrimental effects on maternal nutrition. In heavy drinkers, intakes of protein, vitamins, and minerals are often inadequate, as ethanol becomes a major source of energy in the diet and displaces more nutritious food.

TABLE 3.8

Alcohol-related birth defects

- Cardiac abnormalities
- Defects of the joints: elbow, hip, digits
- Genital abnormalities
- Skin hemangiomas (benign tumors composed of blood vessels)

Narrow, receding forehead

Small eyes, drooping eyelids

Short, upturned nose

Longer than normal distance between base of nose and upper lip, poorly formed upper lip

Poorly formed lower jaw

Poorly formed asymmetric ears

Alcohol can cause intestinal malabsorption and produce zinc, thiamin, and folate deficiencies in heavy drinkers (discussed in Chapter 9). Zinc deficiency, common in chronic alcoholism, has been associated with congenital abnormalities in humans (Beattie 1992). Placental transfer of certain nutrients, particularly amino acids, may be impaired by alcohol. Therefore, along with the direct toxicity of alcohol, the malnutrition associated with heavy alcohol use may harm the fetus.

Although full FAS is thought to occur only in offspring of heavy, chronic users (more than six drinks a day), smaller and less frequent alcohol use during pregnancy may also have detrimental effects on the fetus (Institute of Medicine 1990). Two to four drinks a day (1–2 oz of pure ethanol) increases the risk of intrauterine growth retardation (Wright et al. 1983). Impairments in intelligence, attention span, and reaction time at ages 4 to 7 are correlated with estimates of alcohol intake in early pregnancy (Streissguth et al. 1989).

The effects of even lower levels of prenatal alcohol exposure are uncertain. Several studies have found no or inconsistent associations between low levels of consumption and fetal effects (Institute of Medicine 1990). However, one study found a significantly increased risk of delivering a growth-retarded infant in women who consumed only one to two drinks per day (Mills et al. 1984).

Because no absolutely safe level of alcohol consumption has yet been established, most expert groups recommend complete abstinence during pregnancy (American College of Obstetricians and Gynecologists 1992), (The Surgeon General's Report on Nutrition and Health 1988). A multivitamin–mineral supplement should be taken by those women who choose to continue to drink alcohol during pregnancy. Routine prenatal care should include information on the dangers of alcohol, and every effort should be made to encourage women to limit or eliminate alcohol intake during pregnancy.

CAFFEINE

Over three fourths of pregnant women in the United States consume significant amounts of caffeine—average daily intake is about 100–150 mg. Tea, chocolate, colas, and various medications contribute to caffeine intake but, in most women, coffee accounts for the vast majority of caffeine ingested. Table 3.9 gives the caffeine content of beverages and chocolate.

In 1980, the FDA cautioned pregnant women to limit their intake of caffeine (Food and Drug Administration 1980). Research in animals had linked caffeine to birth defects and spontaneous abortions. Pregnant rats given high doses of caffeine were more likely to deliver offspring with deformed limbs and other malformations. Although many studies have looked at the effects of caffeine during pregnancy, the safety of caffeine in human pregnancy remains unresolved (Shiono and Klebanooff 1993).

Beverages and Foods	Average (mg)	Range (mg)
Coffee (5-oz cup)		
Brewed, drip method	130	110–150
Brewed, percolator	94	64–124
Instant	74	40–108
Decaffeinated, brewed or instant	3	1–5
Tea (5-oz cup)		
Brewed	40	20–90
Instant	30	25–50
Iced (12-oz glass)	70	67–76
Soft drinks (12-oz can)		
Colas		
Regular		30–46
Diet		2–58
Caffeine-free		0–trace
Cocoa beverage (5-oz cup)	4	2–20
Chocolate milk beverage (8-oz)	5	2–7
Milk chocolate candy (1 oz)	6	1–15
Dark chocolate, semisweet (1 oz)	20	5–35
Baker's chocolate (1 oz)	26	26
Chocolate flavored syrup	4	4

TABLE 3.9
Caffeine content of beverages

Caffeine is rapidly absorbed and distributed throughout the body and readily crosses the placenta. Pregnancy alters the metabolism of caffeine, as the activity of liver enzymes that break down caffeine is slowed by high levels of circulating estrogens. This causes slower elimination of caffeine, and its **half-life,** which is 2 to 6 hours in nonpregnant women, increases to 7 to 11 hours during pregnancy (Nehlig and Debry 1994). Thus caffeine levels in the maternal blood are elevated for longer periods. In the fetus, the enzymes that break down caffeine are not fully mature until about six months after birth. In newborns, the metabolism of caffeine is very slow, with the half-life of caffeine lasting 40 to 130 hours (Nehlig and Debry 1994).

Even at low levels of intake, caffeine constricts the blood vessels in the placenta. After two cups of coffee, the blood flow through the placenta is significantly reduced. A reduction in placental blood flow can potentially reduce the supply of oxygen and nutrients to the fetus.

Studies in humans looking for associations between caffeine intake and pregnancy outcomes have produced inconsistent findings (Institute of Medicine1990), (Nehlig and Debry 1994), (Shiono and Klebanoff 1993):

■ *Preterm birth.* In most surveys, caffeine intake has not been associated with increased incidence of preterm birth.

half-life: the time required by an organism or tissue to eliminate one half of a drug or radioactive substance that has been introduced into it

- *Spontaneous abortion (SAB).* Caffeine appears to slightly increase the risk of SAB, although not all studies agree. In a recent Canadian study, risk of SAB was significantly increased by caffeine consumption of 165 mg/day during the first trimester (Infante-Rivard et al. 1993).

- *Birth defects.* Although caffeine at high doses does appear to be teratogenic in animals, there is no convincing evidence that it causes birth defects in humans.

- *Fetal growth.* Daily intakes of more than 300 mg caffeine a day are associated with increased risk for IUGR. Women who smoke and consume caffeine are at particular risk. Compared to women who smoke but do not consume caffeine, the risk for IUGR is significantly increased in women who smoke during pregnancy and consume more than 300 mg caffeine/day. In a recent prospective study in the United States, caffeine consumption of less than 300 mg/day did not increase risk for IUGR (Mills et al. 1993).

- *Neonatal health.* Heavy coffee consumption just before term can increase caffeine levels in newborns and produce irritability and vomiting in the first week after birth.

Overall, the weight of evidence suggests ingestion of more than 300 mg caffeine/day (more than three cups of coffee a day) during pregnancy may be harmful to the fetus. Lower levels of consumption have not been clearly shown to be safe. Because caffeine is widely consumed by pregnant women, even small adverse consequences would have major implications for public health. Most experts agree with the 1980 FDA advisory: pregnant women should limit their consumption of caffeine to a minimum (Eskanazi 1993).

FOOD ADDITIVES

Before being added to the food supply, new food additives are rigorously tested in animal studies, and proof of safety during pregnancy must be provided to the FDA. Many additives have been used in foods for years, and although they have not been thoroughly tested, they are labeled "generally recognized as safe" by the FDA. Despite these safeguards, our knowledge of the potential adverse effects of many food additives remains incomplete. Several substances that were found to be harmful in pregnant animals have been banned, including the food coloring red dye no. 2, and cyclamate, a nonnutritive sweetener (Bopp et al. 1986). The safety of other nonnutritive sweeteners during pregnancy has been extensively studied, but data on the effects of chronic consumption during human pregnancy remain limited. While these compounds continue to be studied, limiting consumption to moderate levels during pregnancy would seem prudent.

Aspartame

The body breaks down aspartame into aspartate, phenylalanine, and very small amounts of methanol. Aspartame and its metabolites have been extensively studied in pregnancy. No evidence of fetal risk has been found, even with large doses (six times the 99th percentile of projected daily intake, or 200 mg/kg) over long periods (American Dietetic Association 1993).

Aspartate does not cross the placenta in appreciable amounts. Although little is known about the effects of methanol on the fetus, the amount of methanol ingested is small (Butchko and Kotsonis 1991), and it does not significantly raise maternal blood levels of formic acid (a toxic compound formed from methanol in severe methanol poisoning). Phenylalanine is actively transported by the placenta, but even high intakes do not raise fetal levels near the toxic range. With the exception of women with **phenylketonuria,** aspartame appears to be safe for use in pregnancy (American Dietetic Association 1993).

Saccharin

Although saccharin can cross the placenta, there is no evidence that it is harmful to the fetus. However, high doses of saccharin are weakly carcinogenic in rats, particularly when exposure begins in utero and continues into adult life (Hoover 1980). Excessive use during pregnancy should be avoided (American Dietetic Association 1993).

FOOD CONTAMINANTS

Evidence from animal studies suggests that several heavy metals found in the environment, including mercury, lead, cadmium, and nickel, are potentially harmful to the fetus (Persaud 1991). Most laboratory studies looking at the effects of these substances in animal pregnancies have used very high doses, and the potential effects of the trace levels found in our food and water on fetal health are mostly unknown.

Lead

phenylketonuria: an inherited disease caused by a defect in the liver's ability to metabolize phenylalanine to tyrosine; if untreated it results in mental retardation and other abnormalities

In the past decade, the adverse effects of low body levels of lead, particularly in infants and children, have become a major public health concern. Recently, several studies have found associations between lead exposure in utero and poor pregnancy outcomes (Andrews et al. 1994). Lead is transported by the placenta into the developing fetus, and high levels of maternal lead increase the risk of prematurity and low birthweight.

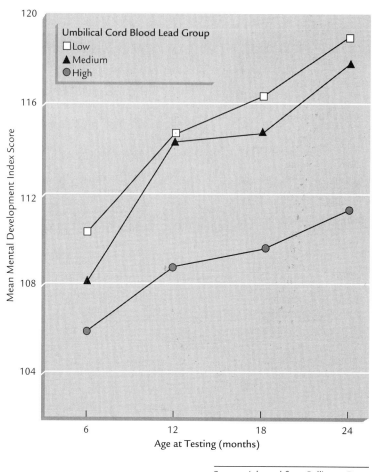

FIGURE 3.5
*In utero lead exposure:
its effects on mental
development*

Source: Adapted from Bellinger, D.
et al. *N Engl J Med* 316 1987) : 1037.

Prenatal exposure to lead also carries the risk of long-term intellectual impairment in childhood. As shown in Figure 3.5, children exposed to low levels of lead in utero have reduced mental development during infancy and childhood (Baghurst et al. 1995). Low-level contamination of food and water supplies is a major source of lead during pregnancy. (A complete discussion of the health risks of lead can be found in Chapter 7).

Mercury

Methylmercury is a fungicide that has been implicated in several outbreaks of mercury poisoning around the world, with particularly devastating effects in pregnant women. For example, in Iraq in 1971–72, 31 pregnant women ate barley and wheat that had been contaminated with methyl-mercury. Many of the women died, and infants born to the survivors suffered from severe neurological and ocular abnormalities. Similar tragic incidents have occurred in Russia and Sweden (Koos and Longo 1976).

Pesticides and Chemicals

The effects of exposure to small amounts of pesticides in food eaten during pregnancy are mostly unknown. Polychlorinated biphenyls (PCBs) are industrial chemicals that are teratogenic. In Japan in 1968, consumption of cooking oil that was contaminated with high levels of PCBs caused intrauterine growth retardation and other congenital abnormalities (Kuwabara et al. 1978).

HIGH-DOSE VITAMIN SUPPLEMENTS

In the United States, because of perceived health benefits, many people regularly consume vitamin and mineral supplements. The indiscriminate use of supplements should be avoided during pregnancy. High doses of vitamins, particularly vitamins A and D, and possibly vitamins C and B_6, can harm the fetus.

Effects of High Doses of Vitamin A

Large doses of vitamin A during pregnancy are potent teratogens. Characteristic fetal abnormalities produced by high doses of vitamin A include malformations of the skull and face, heart, and central nervous system (Teratology Society 1987). Although the minimum harmful dose during human pregnancy is not known, most studies have shown intake of more than 6000 RE (20,000 IU) is required for teratogenicity. This level of intake is eight to ten times the current RDA for vitamin A in pregnancy. Pregnant women should be careful to maintain total vitamin A intake at levels near the RDA of 800 RE (Teratology Society 1987).

The drug isotretinoin (Accutane®), a synthetic analog of vitamin A used to treat severe acne, is a potent teratogen. Before the drug's teratogenicity was known, many women who took it gave birth to babies with the isotretinoin teratogen syndrome. This characteristic group of birth defects, shown in Table 3.10, is similar to those associated with vitamin A toxicity (Lammer et al. 1985). Although the drug is now contraindicated during pregnancy, its toxicity demonstrates the potential teratogenicity of vitamin A-like compounds.

Effects of High Doses of Vitamin D

Very high doses of vitamin D consumed during pregnancy are potentially harmful to the fetus. In both animals and humans, high doses have produced fetal hypercalcemia, abnormal bone and dental development, and malformation of the aorta (Luke and Johnson 1993).

- Craniofacial malformations:
  malformed ears
  small jaws
  cleft palate

- Cardiac abnormalities

- Neurological abnormalities

- Abnormal brain development

- Hydrocephalus (abnormal accumulation of fluid inside the skull)

TABLE 3.10
Vitamin A-associated birth defects

Effects of High Doses of Vitamins C and B$_6$

High doses of vitamin C during pregnancy may be harmful. Two infants born to mothers who had consumed 400 mg of ascorbic acid/day during gestation developed scurvy during the first postnatal month. It was hypothesized that the large maternal doses caused fetal metabolism and excretion of vitamin C to increase, so that the abrupt fall in vitamin C supply after birth produced signs of deficiency in the newborns (Cochrane 1965).

Similarly, high doses of vitamin B$_6$ (50 mg/day) during pregnancy have also been suspected of causing deficiency in newborns (Institute of Medicine 1990). Although there was no direct evidence that the infants' symptoms in these reports were caused by the high maternal intakes, high doses of these vitamins should be avoided during pregnancy.

LIFESTYLE AND ENVIRONMENTAL INFLUENCES ON MATERNAL NUTRITIONAL STATUS AND PREGNANCY OUTCOME

SMOKING

Despite the widely publicized harmful effects of smoking during pregnancy, about one in five pregnant women continue to smoke. The highest rates occur among White women, followed by Black and Hispanic women. Smoking during pregnancy is more common among teenagers and women in their early 20s than among older women (Centers for Disease Control 1992). Table 3.11 gives more detailed data on the rates of smoking.

Cigarette smoking has major adverse effects on maternal and fetal health. In the industrialized countries, it is one of the most important risk factors for IUGR, and smokers are twice as likely to deliver a low-birth-weight infant (Floyd et al. 1993). See Table 3.12.

Mothers who smoke are at greater risk for preterm delivery, perinatal mortality, and spontaneous abortion. Also, children of mothers who

	Smoked during the 12 Months before Birth (%)		Quit after Learning of Pregnancy (%)		Smoked after Learning of Pregnancy (%)	
	1985	1990	1985	1990	1985	1990
All women	31.8	23.7	21.2	22.6	25.1	18.3
18–24 years	40.1	29.3	21.1	23.4	31.6	22.4
25-29 years	34.4	25.7	23.6	26.3	26.3	18.9
30–34 years	24.3	21.1	18.9	19.4	19.7	17.0
35–44 years	23.4	17.7	16.7	17.4	19.5	14.6
Education						
Less than 12 years	46.0	35.5	14.8	13.5	39.2	30.7
12 years	35.8	28.1	20.2	24.2	28.6	21.3
13–15 years	24.0	19.7	29.4	28.3	16.9	14.1
16 years or more	13.4	7.8	37.5	32.2	8.4	5.3
Race/Ethnicity						
White	33.2	25.3	21.7	22.3	26.0	19.7
Black	27.5	19.0	17.9	25.8	22.6	14.1
Hispanic	16.8	12.1	38.9	34.1	10.3	8.0

Adapted from Centers for Disease Control, "CDC Surveillance Summaries," *MMWR* 41 (27 Nov 1992): SS-7.

TABLE 3.11
Rates of smoking before and during pregnancy by age, education, and ethnicity

	Percentage of Women Giving Birth to Low Birthweight Infant	
	Nonsmokers	Smokers
White	4.3%	8.6%
African American	9.0%	15.1%
Hispanic	5.5%	9.4%
Native American	3.3%	6.1%

TABLE 3.12
Smoking, ethnicity, and the incidence of low birthweight

From Centers for Disease Control, "CDC Surveillance Summaries," *MMWR* 41 (27 Nov 1992): SS-7.

Nutrient	Smokers	Nonsmokers
Umbilical Cord Blood vitamin C (mg/dl)	0.61	1.68
Placental vitamin C (mg/100g)	10.1	20.9
Maternal plasma vitamin E (mg/dl)	0.4	0.8
Umbilical Cord Blood vitamin E (mg/dl)	0.2	0.3
Maternal plasma beta-carotene (µg/dl)	19	44
Umbilical Cord Blood beta-carotene (µg/dl)	7	20

Adapted from A. Bendich, "Lifestyle and Environmental Factors That Can Adversely Affect Maternal Nutritional Status and Pregnancy Outcomes," *Ann NY Acad of Sci* 678 (1993): 255-65.

TABLE 3.13

Effects of maternal smoking on newborn status of vitamins C, E, and beta-carotene

smoked during pregnancy may have long-term impairments in physical growth and intellectual performance (Institute of Medicine 1990). These adverse effects are dose dependent: the greater number of cigarettes smoked during pregnancy, the greater the likelihood of harm (Luke and Johnson 1993).

Smoking impairs maternal and fetal nutrition in several ways. Smokers tend to have more irregular eating habits than nonsmokers and smokers consume more alcohol and coffee than nonsmokers (Morabia and Wynder 1990). Intakes of protein, zinc, riboflavin, thiamin, and iron are lower in pregnant women who smoke, compared to nonsmoking pregnant women, and are associated with greater risk of having LBW infants (Haste et al. 1991).

Smoking alters the metabolism of certain micronutrients and increases requirements for nutrients (discussed in Chapter 8). Smoking can have negative effects on the status of zinc, vitamin C, vitamin B_6, folate, and vitamin B_{12} in both pregnant and nonpregnant women. Although dietary intakes of vitamins C, E, and beta-carotene are similar in smoking and nonsmoking pregnant women, concentrations in the fetus, maternal plasma, and placenta are markedly lower in smoking women (Bendich 1993). See Table 3.13.

Smoking may reduce blood flow through the placenta and restrict oxygen and nutrient flow to the fetus. Compared to nonsmokers, smokers tend to have lower prepregnancy weights and lower gestational weight gains, even though surveys show energy intakes of smoking women during pregnancy tend to be greater than those of nonsmokers. The increase in metabolic rate that accompanies smoking may reduce calories available for weight gain during pregnancy (Institute of Medicine 1990).

Cessation of prenatal smoking could save the health-care system millions of dollars and substantially improve maternal and child health. Stopping smoking during pregnancy is one of the six key objectives for reducing maternal and infant health risks outlined by the federal government in the report Healthy People 2000 (see Chapter 1) (USDHHS 1990).

JOB STRESS

A recent study examined the effect of increased job stress—defined as increased demands of work and decreased control at work—on pregnancy outcome in over 200,000 Danish women. Greater job stress significantly increased the risk of spontaneous abortions and delivering an infant of below normal weight, and it doubled the risk of infant mortality (Brandt and Nielson 1992). Along with other lifestyle factors, these findings could reflect the dietary patterns of pregnant women who work at demanding jobs outside the home. Because of inadequate time for shopping and food preparation, many pregnant women who work outside the home have irregular eating habits and may not consume nutritionally adequate diets.

ILLICIT DRUGS

Illicit drug use is associated with poor dietary quality, and impaired nutritional status in drug-abusing pregnant women may contribute to poor pregnancy outcome. Serum folate, vitamin C, and ferritin concentrations are lower in pregnant women who use illicit drugs than in women who do not, and infants exposed in utero have lower birthweights than non-exposed infants (Knight et al. 1990).

Marijuana

It is currently estimated that about one in seven newborns in the United States are exposed in utero to one or more illicit drugs (Chasnoff et al. 1990), and marijuana is the most commonly used illicit drug in women of childbearing age. Estimates of marijuana use during pregnancy have ranged from 10–27% (Institute of Medicine 1990), and in certain high-risk groups, rates may be higher: a recent report indicated about a third of women from poor urban areas used marijuana in the first trimester of pregnancy (Bendich 1993).

The main active ingredient in marijuana, tetrahydrocannabinol, crosses the placenta. Exposure of the fetus may be prolonged because of slow clearance of the drug by the fetus and placenta. Adverse effects of marijuana in pregnancy include an increased risk for premature delivery and lower birthweight (Zuckerman et al. 1989).

Cocaine

Surveys of inner city populations have found that 15–18% of pregnant women used cocaine at least once during their pregnancy (Zuckerman et al. 1989). Cocaine users are more likely to deliver an infant of low birthweight, and they have a five times greater risk of premature delivery, than pregnant women who do not use cocaine. Moreover, mental and psychomotor development are significantly impaired in infants of mothers who used cocaine and other drugs during pregnancy (Griffith et al. 1994).

Cocaine causes constriction of placental blood vessels and can impair delivery of nutrients and oxygen to the fetus. Cocaine is an appetite suppressant, and regular use during pregnancy is associated with lower gestational weight gain (Zuckerman et al. 1989).

MEDICINAL DRUGS

Many medicines used regularly without ill effects by nonpregnant women are dangerous in pregnancy. The fetus is extremely vulnerable to untoward drug effects: even commonly used drugs such as aspirin and antibiotics can harm the fetus (Jones 1994). Pregnant women should always seek medical advice before using any drugs.

PHYSICAL EXERCISE

Regular aerobic exercise for weight control and general health has become important to many women of childbearing age. About 15–20% of women considering pregnancy participate in strenuous aerobic exercise programs (Clapp 1994). Most physically active women who become pregnant wish to continue exercising during pregnancy, but the question of how active a women can be without compromising the health and development of her fetus remains largely unanswered. Exercise during pregnancy raises the following concerns:

- *Elevated body temperature.* The developing embryo is sensitive to elevated temperatures, and increased maternal core temperature during the first trimester has been associated with fetal growth retardation and congenital abnormalities (Lotgering et al. 1987). The threshold for adverse effects is thought to be about 39.2°C, and sustained exercise can raise maternal temperature above this level (Clapp 1994).

- *Decreased blood glucose.* During exercise, blood glucose is rapidly taken up by muscle for use as fuel, and circulating levels of glucose fall. The fall in blood glucose during exercise tends to be greater in pregnant women than in nonpregnant women (Clapp et al. 1987). Glucose is the major source of energy for the fetus, and the fall in blood glucose during exercise may limit the amount available to the fetus.

- *Decreased blood flow to the fetus.* During exercise, as blood flow increases to muscles and skin, blood is diverted from the vessels supplying the uterus and placenta. This normal physiological response to exercise could potentially reduce oxygen and nutrient availability to the fetus.

- *Increased biomechanical stress.* Weight gain, changes in posture, and relaxation of connective tissue in ligaments and tendons (caused by high progesterone levels) may increase the likelihood of musculoskeletal injury during weightbearing exercises (such as running and aerobics). Also, bouncing movements could increase stress on the large, mobile, pregnant uterus and potentially tear the membranes surrounding the fetus or cause premature labor.

In response to these concerns, research on exercise during pregnancy has increased over the past several decades. Overall, studies have found that low- to moderate-intensity exercise regimens do not have adverse effects on pregnancy in healthy women (Paisley and Mellion 1988), (American College of Obstetricians and Gynecologists 1994). They may actually improve maternal well-being and decrease problems during labor and delivery. Recent studies have shown:

- Moderate exercise during pregnancy does not appear to increase the risk of spontaneous abortion, congenital abnormalities, or preterm labor (Clapp 1994).

- Most studies that have looked at the impact of moderate exercise on birthweight have found no effect. However, studies of women who continue rigorous aerobic exercise through the third trimester have found significant decreases in birthweight (Clapp and Capeless 1990). In a recent study, the median birthweight of infants born to mothers who exercised vigorously in later pregnancy was 400 g less than women in a control group who did not exercise (3200 g versus 3600 g) (Clapp 1994). However, the median birthweight in the exercising group was still within the normal range for healthy newborns. See Figure 3.6.

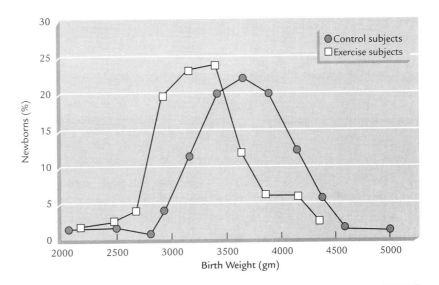

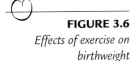

FIGURE 3.6

Effects of exercise on birthweight

Source: Adapted from Clapp (1994).

The following exercise recommendations are for women who do not have any additional risk factors for adverse maternal or perinatal outcome.

- Women can continue to exercise and derive health benefits during pregnancy even from mild to moderate exercise routines. Regular exercise (at least three times per week) is better than intermittent activity.

- Women should avoid exercise in the supine position after the first trimester. Such a position is associated with decreased cardiac output in most women; such regimens are best avoided during pregnancy. Pregnant women should also avoid prolonged periods of motionless standing.

- Women should be aware of decreased oxygen levels available for aerobic exercise during pregnancy. They should be encouraged to modify the intensity of their exercise according to symptoms. Pregnant women should stop exercising when fatigued and not exercise to exhaustion. Weight-bearing exercises may, under some circumstances, be continued during pregnancy at intensities similar to those prior to pregnancy. Nonweight-bearing exercises such as cycling or swimming will minimize the risk of injury and facilitate the continuation of exercise during pregnancy

- Morphologic changes should serve as a relative contraindication to types of exercise in which loss of balance could be detrimental to maternal or fetal well-being, especially in the third trimester. In addition, any type of exercise with the potential for even mild abdominal trauma should be avoided.

- Women who exercise during pregnancy should be particularly careful to follow an adequate diet, since pregnancy requires an additional 300 kcal per day to maintain metabolic homeostasis.

- Pregnant women who exercise in the first trimester should augment heat dissipation by ensuring adequate hydration, appropriate clothing and optimal environmental surroundings during exercise.

- Many of the physiologic and morphologic changes of pregnancy continue four to six weeks postpartum. Thus, prepregnancy exercise routines should be resumed gradually based on the woman's physical capability.

From American College of Obstetricians and Gynecologists, "Exercise during Pregnancy and the Postnatal Period," *Technical Bulletin* no.189 (February 1994).

TABLE 3.14
Guidelines for exercise in pregnancy and postpartum

- Moderate, regular exercise does not appear to significantly reduce oxygen or nutrient supply to the fetus, as measured by changes in fetal heart rate or blood flow (Clapp 1994). However, concerns were raised by two studies in sedentary pregnant women performing very strenuous cycling. Both studies found that vigorous exercise slowed fetal heart rates significantly—an indication that oxygen availability to the fetus was impaired during exercise (Artal and Posner 1991).

Guidelines for exercise during pregnancy are shown in Table 3.14. Optimal nutrition for women who exercise during pregnancy is important. Caloric intake should be adequate to cover both the extra needs of pregnancy and energy expended in exercise, so that desirable gestational weight gain is achieved. Pregnant women should consume ample carbo-hydrates during exercise to support blood glucose. Liberal intake of cool fluids can prevent dehydration and reduce the rise in maternal body temperature during exercise.

NUTRITION AND HIGH-RISK PREGNANCY

TEENAGE PREGNANCY

In the U.S., approximately 1 million women under age 20 become pregnant each year (10% of all women 15–19 years old), and 500,000 give birth (more than 12% of all births) (Centers for Disease Control 1993). Twenty-five percent of pregnant teenagers become pregnant again within one year of their first delivery. Although the majority of teen pregnancies occur in women 17 and older, recent statistics show an increasing rate of pregnancy among adolescents less than 15 years old (Alexander and Guyer 1994).

Pregnancy during the teenage years can have detrimental consequences for the health and nutrition of both mother and infant. Compared to older women, teenage mothers are more likely to develop iron-deficiency anemia and high blood pressure during pregnancy, and they suffer from higher rates of postpartum obesity. Younger teenage mothers (under 15 years old) are at particularly high risk for these complications (Stevens-Simon and McAnalney 1992).

As shown in Table 3.15, pregnant teens under 15 years of age have the highest rate of low-birth-weight infants of any age group, and they are twice as likely to deliver a low-birth-weight infant than older mothers. Teenage mothers are also at greater risk of prolonged and difficult labor. Infants of adolescent mothers are more likely to be preterm than infants of adult mothers, and they are three times more likely to die in the first month of life (Stevens-Simon and McAnalney 1992).

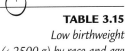

TABLE 3.15

Low birthweight (< 2500 g) by race and age in the United States

Age (Yr)	Race		
	All (%)	White (%)	Black (%)
All ages	6.9	5.6	11.5
<15	13.6	11.0	15.2
15–19	9.9	7.5	12.6

Adapted from "Advance Report of Final Natality Statistics," *Monthly Vital Statistics Report* 39, no. 4, Supplement, (15 Aug 1990).

In the United States, the economic costs of teenage pregnancy are staggering: from 1985 through 1990, government costs related to teenage childbearing totaled $120 billion (Centers for Disease Control 1993). The psychological costs of teenage pregnancy are also high, because most teenage parents do not have the opportunity to obtain the education, skills, and maturity necessary to support a healthy family. Children of teenage mothers are more likely to have health problems and learning difficulties at school (Hoekelman 1994).

There are several reasons why teenage pregnancies have less favorable outcomes. Biological factors—such as reproductive immaturity—may be partially responsible, particularly among very young adolescents. Social and environmental conditions also play a major role, as teenage mothers are often poor, unmarried, come from single-parent families, and have limited education (Stevens-Simon and McAnarney 1992). They are more likely to have low prepregnancy weights and less likely to receive adequate prenatal care. Many also continue to smoke and drink alcohol during pregnancy (Kokotailo 1992). Statistics on prenatal care and tobacco and drug use are shown in Tables 3.16 and 3.17.

One of the most important environmental determinants of maternal and fetal health during teenage pregnancy is nutritional status. The nutritional demands of the developing fetus and placenta, added to the requirements for the mother who is still growing herself, are extremely

TABLE 3.16

Use of prenatal care by age and ethnicity in the U.S. (1992)

	% Starting Care in First Trimester			% No or Late Prenatal Care		
	All Births	<15	15 to 19	All Births	<15	15 to 19
All races	77.7	42.9	59.5	5.2	17.2	9.7
White	80.8	47.5	62.4	4.2	15.8	8.6
Black	63.9	39.2	53.2	9.9	18.2	12.1

Source: Adapted from Center for Disease Control (1993).

TABLE 3.17

Tobacco, alcohol, and drug use during teenage pregnancy

	Use (% Reporting at Initial Prenatal Visit)
Tobacco	22.2
Alcohol	6.6
Marijuana	7.5
Other drugs	6.1

Adapted from P. K. Kokotailo, H. Adger and A. K. Duggan, "Cigarette, Alcohol, and Other Drug Use by School-Age, Pregnant Adolescents: Prevalence, Detection and Associated Risk Factors," *Pediatrics* 90 (1992):328.

high. In nonpregnant women between 11 and 17 years of age, height typically increases 15%, and weight 69%. Skeletal maturation continues, and calcium requirements remain high.

Poor dietary habits are common in adolescence, making the pregnant teenager vulnerable to nutritional deficiencies at a time of increased need. Because of irregular meals, food preferences, and dieting and weight concerns, many adolescents do not obtain adequate micronutrients in their diets (American Dietetic Association 1994) (discussed in Chapter 8).

Surveys in pregnant teens have found widespread dietary deficiencies: zinc intakes average only 6 mg (40% of the RDA), iron about 15 mg (half the RDA), and calcium about 870 mg (72% of the RDA) (Guitierrez and King 1992). Inadequate micronutrient intake may have detrimental effects on long-term health of the mother. For example, inadequate dietary calcium can impair bone mineralization in pregnant teenagers. Because optimal bone density achieved during early life may protect against osteoporosis in later years, inadequate calcium caused by the high demands of teenage pregnancy may increase risk.

Because of smaller prepregnancy size and immaturity, adolescents less than 15 years of age (or less than two years postmenarche) tend to have lower gestational weight gains than older women. Moreover, even when nutritional status and gestational weight gain are comparable to older women, teenagers are at higher risk of delivering a smaller infant (Institute of Medicine 1990). Because of this, women in this age group should strive to gain weight in the upper end of the recommended ranges shown in Table 3.18.

Inadequate weight gain early in teenage pregnancy increases the risk of poor pregnancy outcome. In a recent study of nearly 1800 pregnant teenagers in the United States, weight gain of less than 4.3 kg by 24 weeks gestation was associated with increased risk of low birthweight, even if later gains brought total weight gain at term up to recommended levels (Hediger et al. 1989).

Prepregnant Weight Categories	Recommended Total Gain	
	kg	lb
Underweight (BMI<19.8)	12.5–18	28–40
Normal weight (BMI=19.8–26)	11.5–16	25–35
Overweight (BMI>26–29)	7.0–11.5	15–25
Very overweight (BMI>29)	7.0–9.1	15–20

Very young adolescents (≤14 years of age or <2 years postmenarche should strive for gains at the upper end of the range.

Adapted from Institute of Medicine, Subcommittee on Nutritional Status and Weight Gain during Pregnancy, *Nutrition during Pregnancy* (Washington D.C.: National Academy Press 1990).

TABLE 3.18
*Gestational weight
gain recommendations
for adolescents*

Prenatal nutritional assessment and counseling are important factors in helping pregnant teens to eat a healthy, well-balanced diet, and following accepted guidelines for pregnancy. The diet should emphasize ample intake of energy, calcium, iron, zinc, and vitamins C, B_6, and folic acid (American Dietetic Association 1994). Weight gain should be monitored to ensure that energy and protein intakes are sufficient to support a gain of about 0.4 kg (1 lb) per week in the last two trimesters (Institute of Medicine 1990). If nutritional intake from food sources is inadequate, a daily vitamin–mineral supplement should be recommended. Pregnant adolescents may particularly benefit from support through programs such as WIC (discussed later in this chapter).

MULTIPLE PREGNANCY

In the United States, about 2% of pregnancies are **multiple pregnancies**. In 1990, there were nearly 100,000 multiple births, and nearly all (97%) were twin births. The incidence of twin gestation is steadily increasing—from 1980 to 1990, the number of twin births increased by 35% (Luke and Johnson 1993). The increase is due in part to the increasing numbers of women who are delaying childbirth: twin births are more common among mothers over 30 years old.

Compared with single pregnancies, twins are over eight times more likely to be LBW infants and five times as likely to be born preterm (Luke and Johnson 1993). Nutritional demands of twin pregnancy are high, and inadequate maternal nutrition is a contributing factor in many LBW twins.

Maternal adaptation to a twin pregnancy is in many ways a heightened version of the response to a single pregnancy. Compared to single pregnancies, twin pregnancies have higher circulating levels of placental hormones, 25% greater blood volume expansion, and greater placental and renal blood flow (Campbell 1988). Women with twin pregnancies gain weight earlier and have larger total weight gains than women with single pregnancies. Several studies have found weight gains of at least 24 lb by 24 weeks and 40–45 lb at term to be associated with better outcomes, including a decreased incidence of low birthweight and decreased perinatal mortality (Institute of Medicine 1990).

Women pregnant with twins should carefully select foods using the guidelines for single pregnancy in Table 2.24 (in Chapter 2), but increase the number of servings from each food group. Optimal prenatal care and nutritional counseling are essential during twin pregnancy, and goals for gestational weight gain should be 40 to 45 lb (Luke et al. 1992). The National Academy of Sciences recommends a vitamin and mineral supplement for women carrying twins (Institute of Medicine 1990). There are no RDAs for women pregnant with twins. Considering the established need for significantly higher weight gains, the guidelines for single pregnancy have been recently modified for women with twin pregnancy (Luke and Johnson 1993). (See Table 3.19).

multiple pregnancy: pregnancy resulting in the birth of more than one infant

Nutrient	RDA for Single Pregnancy	Modified RDA for Twin/Triplet Pregnancy
Folic acid	400 µg	800 µg
Vitamin D	10 µg	15 µg
Iron	30 mg	50 mg
Calcium	1200 mg	1800 mg
Phosphorus	1200 mg	1800 mg
Pyridoxine	2.2 mg	4.0 mg
Thiamin	1.5 mg	3.0 mg
Zinc	15 mg	30 mg
Riboflavin	1.6 mg	3.0 mg
Protein	60 mg	120 mg
Iodine	175 µg	300 µg
Vitamin C	70 mg	150 mg
Energy	2500 kcal	3000 kcal
Magnesium	320 mg	450 mg
Niacin	17 mg	25 mg
Vitamin B_{12}	2.2 µg	3.0 µg
Vitamin A	800 µg	1000 µg

TABLE 3.19

Daily nutrient allowances for twin pregnancies vs. single pregnancy

Adapted from B. Luke, T. R. B. Johnson and R. Petrie, *Clinical Maternal-Fetal Nutrition* (Boston: Little, Brown and Co. 1993).

GESTATIONAL DIABETES

Pregnancy increases levels of several hormones—estrogen, progesterone, human placental lactogen, prolactin, and cortisol—that reduce the ability of insulin to control blood sugar. Most pregnant women secrete more insulin to balance these effects, and glucose metabolism remains normal. However, in about 3 to 5% of pregnant women, insulin is insufficient to control blood sugar, and diabetes develops (Hadden 1985). Although symptoms may go unnoticed by the mother, diabetes can harm the developing fetus. Risk factors for gestational diabetes are shown in Table 3.20.

1. Previous history of gestational diabetes

2. Glycosuria (sugar in the urine) or symptoms of diabetes

3. Fasting plasma glucose ≥105 mg/dl or 2-hr postmeal plasma glucose level of ≥120 mg/dl

4. Obesity (>150 lb or body mass index (kg/m^2) of >27 before pregnancy)

5. Previous infant >9 lb (4100 g)

TABLE 3.20

Major risk factors for gestational diabetes

Adapted from P. Rosso, *Nutrition and Metabolism in Pregnancy* (Oxford: Oxford University Press, 1990).

If gestational diabetes is not recognized and blood sugar is not carefully controlled, pregnancy can be adversely affected in several ways. Risks of perinatal morbidity and mortality are increased, and birth defects are more common (Norton et al. 1994). Infants born to diabetic mothers tend to have very high birthweights and are at increased risk for complications during delivery. Because of this, all pregnant women should be screened for diabetes at 24–28 weeks gestation (American Diabetes Association 1994).

Nutrition and moderate exercise are the cornerstones of diabetic management during pregnancy, and women with gestational diabetes should be provided an individualized dietary program consistent with current expert recommendations (American Diabetes Association 1994). Current guidelines suggest dietary protein should contribute 10–20% of calories; less than 10% of calories should come from saturated fat; up to 10% from polyunsaturated fat, and the remaining 60–70% of calories should come from carbohydrates and monounsaturated fats.

Limiting carbohydrate intake at breakfast and dividing food intake during the day into frequent, smaller meals often improves blood glucose control. In combination with a moderate exercise program, dietary adjustments allow many women to avoid having to take insulin during pregnancy. After delivery, glucose metabolism returns to normal in nearly all of women with gestational diabetes (American Diabetes Association 1994).

OBESITY

Problems associated with obesity and pregnancy are discussed in detail in Chapter 2. Women who enter pregnancy significantly overweight are at increased risk for complications shown in Table 3.21. Obese women also have higher rates of complications during labor and delivery (Institute of Medicine 1990).

Obese women should not try to lose weight during pregnancy. Low-calorie diets can impair growth of the fetus and are usually low in micronutrients important during pregnancy—including iron, calcium, and folate. In pregnancy, fasting or very low calorie diets markedly increase levels of ketone bodies in maternal blood, and high levels of ketones may be harmful to the developing fetus (Norton et al. 1994).

Overweight women should follow current guidelines for gestational weight gain provided by the National Academy of Sciences (see Chapter 2) (Institute of Medicine 1990). Prenatal care and regular checkups during pregnancy are particularly important for obese mothers.

	Obese (%)	Normal Weight (%)
Gestational hypertension	38.0	19.0
Edema	56.4	47.2
Albumin in the urine	4.3	1.3
Gestational diabetes	1.8	1.2
Anemia	6.7	8.6
Cholelithiasis (gallstones)	3.7	3.1
Postpartum infection	6.1	0.6

Adapted from Z. Tilton et al. "Complication and Outcome of Pregnancy in Obese Women," *Nutrition* 5 (1989):95–99.

TABLE 3.21

Complications during pregnancy and in the postpartum period in obese vs. normal weight women

PREGNANCY-INDUCED HYPERTENSION (PIH)

In the United States, 10 to 15% of women develop high blood pressure during pregnancy (Saftlas et al. 1990). For many women, hypertension is mild and transient, but a significant number develop *preeclampsia*— a potentially serious disorder of pregnancy characterized by an elevation in blood pressure with **proteinuria.** Loss of plasma proteins leads to edema, and if preeclampsia becomes severe, it can lead to convulsions and coma in the mother *(eclampsia)* (Saftlas et al. 1990). Table 3.22 gives diagnostic criteria for the various forms of hypertension in pregnancy.

Pregnancy-induced hypertension (PIH) is the second leading cause of maternal death in the United States (Saftlas et al. 1990), and is also harmful to the fetus. Warning signs of the condition are listed in Table 3.23. The fetus in a hypertensive pregnancy is at greater risk for low birthweight and preterm birth, and the perinatal mortality rate in hypertensive pregnancy is more than twice that of normal pregnancy (Roberts 1994).

proteinuria: the abnormal excretion of protein in the urine

Diagnosis	Criteria
Hypertension	Systolic BP ≥ 140 mm Hg Diastolic BP ≥ 90 mm Hg
Preeclampsia	Hypertension plus proteinuria
Eclampsia	Gestational hypertension and convulsions or coma

From National high blood pressure education program working group on high blood pressure in pregnancy, "Consensus Report," *Am J Obstet Gynecol* 163 (1990):1689–712.

TABLE 3.22

Diagnostic criteria for hypertensive disorders of pregnancy

- Severe and constant headaches
- Swelling, especially of the face
- Dizziness
- Blurred vision
- Sudden weight gain (1 lb/day)

TABLE 3.23

Warning signs for PIH

- Women with history of hypertension when not pregnant
- Primigravidas (6–8 x risk)
- Hypertension in a previous pregnancy
- Women with diabetes
- Women with kidney disease
- Multiple pregnancy (5x risk)
- African American

Adapted from J. Roberts, "Current Perspectives on Preeclampsia," *Nurse-Midwifery* 39 (1994):70-90.

TABLE 3.24

Women at risk for PIH

PIH is more common in women who were hypertensive before pregnancy or in a previous pregnancy. Risk is increased in Black women, teenagers (50% higher risk), older mothers (over 35 years), those with multiple pregnancy, and in **primigravidas.** See Table 3.24.

Both high and low gestational weight gains have been linked to PIH. Large weight gains, particularly in the second and third trimesters, have been associated with increased risk of PIH. However, because edema and increased body water are common features of preeclampsia, it has been difficult to determine whether weight gain causes PIH or merely accompanies it. Other studies have found that poor weight gain in early pregnancy increases risk of PIH, and that most women with PIH have total gestational weight gains below average (Institute of Medicine 1990). Thus, the relationship between PIH and weight gain remains unclear. Women with significant risk factors for PIH should strive to achieve gestational weight gains within the recommended ranges (Institute of Medicine 1990).

Studies have shown that PIH is not related to high salt intakes during pregnancy, as low salt diets have actually been found to aggravate high

primigravida: a woman who is pregnant for the first time

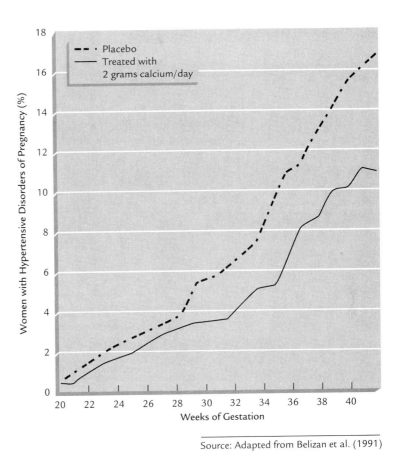

FIGURE 3.7

Calcium supplementation during pregnancy and reduced risk of PIH

Source: Adapted from Belizan et al. (1991)

blood pressure during pregnancy. Women should not attempt to restrict salt intake during pregnancy (Newman and Fuller 1990).

Several epidemiological studies have found an inverse relationship between calcium intake and PIH. Clinical trials have shown that calcium supplementation during pregnancy can reduce the risk of preeclampsia (Anonymous 1992). Recently, a large, placebo-controlled study found that supplementation with 2 g of calcium/day from the 20th week of gestation to term reduced the rate of PIH (Belizan et al. 1991). As shown in Figure 3.7, the incidence of PIH in women who took supplements was one third that of women who took the placebo. The mechanism responsible for this beneficial effect is unknown.

The doses used in these clinical trials (1.5–2 g/day) are higher than the current RDA for calcium in pregnancy (1200 mg) (NAS 1989). Women, particularly those at risk for PIH, should strive to obtain at least the RDA for calcium during pregnancy. The National Academy of Sciences recommends calcium supplementation for pregnant women who have low intakes of dietary calcium (Institute of Medicine 1990).

VITAMIN AND MINERAL SUPPLEMENTS IN PREGNANCY

As discussed earlier in this chapter, the Food and Nutrition Board of the National Academy of Sciences recommends routine iron supplementation (30 mg/day) for all pregnant women. However, it does not recommend routine vitamin and mineral supplementation for well-nourished pregnant women (Institute of Medicine 1990). For women at risk of nutritional deficiency, including women with poor eating habits, women with multiple pregnancy, and those who use alcohol or cigarettes, a daily supplement is recommended. It should contain the nutrients shown in Table 3.25.

In special circumstances, the Institute of Medicine recommends supplementation with other nutrients during pregnancy (Institute of Medicine 1990):

- *Vitamin D:* 10 mg daily for complete vegetarians (vegans) and women with low intake of vitamin D-fortified milk. Women at northern latitudes in winter and those with minimal exposure to sunlight may also require supplemental vitamin D.

- *Vitamin B_{12}:* 2.0 mg daily for vegans.

- *Calcium:* 600 mg daily for women under age 25 whose dietary intake of calcium is less than 600 mg/day.

- *Zinc:* 15 mg daily for women taking therapeutic levels of iron. Large amounts of supplemental iron are given to women who develop iron-deficiency anemia during pregnancy. Prenatal iron supplements, especially doses greater than 30 mg of elemental iron/day, can lower zinc concentrations in maternal plasma, probably by interfering with zinc

TABLE 3.25

Recommended vitamin and mineral supplements for women at risk of nutrient deficiency during pregnancy

Nutrient	Amount
Iron	30 mg
Zinc	15 mg
Copper	2 mg
Calcium	250 mg
Vitamin B_6	2 mg
Vitamin C	50 mg
Vitamin D	5 μg
Folic acid	300 μg

Adapted from Institute of Medicine, Subcommittee on Nutritional Status and Weight Gain during Pregnancy, *Nutrition during Pregnancy* (Washington D.C.: National Academy Press, 1990).

absorption. The NAS recommends zinc supplementation when more than 30 mg/day of supplemental iron is administered during pregnancy (Institute of Medicine 1990).

■ *Copper:* 2 mg daily for women taking therapeutic levels of iron (mor than 30 mg/day). Both iron and zinc supplementation can interfere with copper absorption and metabolism (Institute of Medicine 1990).

THE ROLE OF NUTRITION IN PRENATAL CARE

Prenatal care should begin as soon as a woman has missed one menstrual cycle—the earlier prenatal care begins, the greater the benefit for the mother and fetus. Prenatal care usually includes a medical examination, laboratory tests, and counseling. Nutrition is a critical component of prenatal care, and all women can potentially benefit from prenatal nutritional counseling, regardless of income, education, or lifestyle.

Prenatal health-care providers should consider the nutritional status of the newly pregnant woman and risk factors for inadequate nutrition in the upcoming pregnancy (see Table 3.26). Initial considerations are prepregnancy weight, gestational weight gain, and iron status, as well as recommendations for an optimal diet, education about risk factors (such as alcohol use and smoking), and determination of the need for vitamin or mineral supplementation.

Obtaining prenatal care early in pregnancy is associated with better pregnancy outcome and sharply decreased risk for delivering a low-birth-weight infant (Murray and Bernfield 1988).

A study in Montreal demonstrated the effectiveness of adequate prenatal nutritional care. A group of poor pregnant women were given nutrition education and provided with supplementary foods (extra milk, eggs, and oranges), as well as a vitamin and mineral supplement. As shown in Table 3.27, in the women who received prenatal nutritional care, the incidence of stillbirths, neonatal mortality, and perinatal mortality was substantially lower than in the province of Quebec and in Canada as a whole (Primrose and Higgins 1971).

In the United States, more than three quarters of pregnant women receive prenatal care during the first trimester, many through the efforts of government programs such as the Supplemental Food Program for Women, Infants, and Children (WIC).

- Age (<15 years or >35 years)

- Recent pregnancies with close birth intervals

- Poor outcome in a previous pregnancy

- Low income

- Limited education

- Inadequate access to food

- Restrictive diets (including fad diets, certain cultural diets, or vegan diets)

- Avoidance of milk products because of lactose intolerance

- Abuse of alcohol, drugs, or caffeine

- Smoking

- Prepregnancy underweight (<85% of standard weight)

- Prepregnancy overweight (>120% of standard weight)

- Failure to gain weight according to recommended patterns

- Diet restriction in an attempt to lose weight

- Multiple pregnancy (twins or triplets)

- Severe nausea and vomiting

- Chronic systemic disease (diabetes, hypertension, liver or kidney disease, anemia)

- Extreme food cravings or aversions, pica

- Lifestyle likely to interfere with ability to obtain adequate nutrition, for example, that of a busy professional person or that of a poor woman living alone

- Unhappiness with being pregnant

TABLE 3.26
*Factors that impose
nutritional risks on the
outcome of pregnancy*

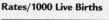

	Rates/1000 Live Births		
	Women Provided Prenatal Nutritional Care	Quebec Province	Canada
Stillbirths	8.7	11.8	11.4
Neonatal mortality in first week	5.7	16.2	14.4
Perinatal mortality	14.4	28.0	25.8

Adapted from T. Primrose and A. Higgins, "A study in human antepartum nutrition," *J Reprod Med* (1971):257–64

TABLE 3.27
*Influence of nutritional
intervention on birth
outcome*

WIC: SUPPLEMENTAL FOOD PROGRAM FOR WOMEN, INFANTS, AND CHILDREN

The U.S. Congress created the WIC program in 1972 to address the need of "substantial numbers of pregnant, postpartum and breastfeeding women, infants, and children from families of inadequate income—those at special risk with regard to their physical and mental health by reason of inadequate nutrition or health care or both" (Abrams 1993). The major goals of WIC include improving the outcome of pregnancy, preventing low birthweight, and ensuring adequate nutrition during important stages of growth in high-risk individuals. Components for pregnant women include:

1. Assessment of nutritional and medical risk status

2. Regular nutrition education and specialized education tailored to high-risk status.

3. Supplementary foods

4. Referral to and encouraged use of prenatal services

WIC is currently the biggest public health nutrition program in the United States. Federal funding for the program has increased from $20 million in 1974 to $2.6 billion in 1992, and WIC now provides assistance to nearly five million women and children, including over 700,000 pregnant women (Burdich and Murray 1992). See Figure 3.8. About two thirds of WIC-eligible women in the United States are being served by the program (Abrams 1993). Table 3.28 lists the criteria for eligibility.

WIC is administered federally by the United States Department of Agriculture and locally through health centers, clinics, and hospitals. Eligibility for the program is based on income (most states use a cutoff of 185% of the U.S. poverty guideline) and nutritional risk. Pregnant, lactating, and postpartum women, and children under 5 years old are eligible, and those who qualify are provided with nutrition education and counseling, health and social services, and supplementary foods.

A typical package of food for a pregnant woman includes eggs, milk, cheese, fruit juice, dry legumes, iron-fortified cereal, and peanut butter. These foods are selected to supply protein, calcium, iron, and vitamins A and C—all nutrients often absent from the diets of low-income women.

Does WIC work? In 1992, the U.S. government comprehensively reviewed the program and evaluated the impact and cost-benefit of prenatal WIC participation (General Accounting Office 1992). Findings suggested that WIC reduced the incidence of low birthweight by 25% and very low birthweight by 44% in women who participated in the program. The earlier in pregnancy and the longer a woman receives WIC support, the more effective the outcome.

WIC was also found to be cost effective. Because the costs of treating low-birth-weight infants are so high, the investment in WIC actually saves money by preventing poor pregnancy outcome (General Accounting Office 1992). It is estimated that every dollar invested in WIC saves over three dollars in potential costs to governments and private payers. More importantly, WIC provides an opportunity for many infants to begin life with greater potential for future health.

- Detrimental or abnormal nutritional conditions detectable by biochemical or anthropometric measurement, such as anemia, extremes of prepregnancy weight, or low prenatal weight gain

- Nutritionally related medical conditions, such as diabetes, vitamin or mineral deficiencies, or lead poisoning

- Dietary deficiencies that impair or endanger health

- Conditions that predispose toward inadequate nutritional intake or status, such as substance abuse, short interconceptional periods, chronic infections, or extremes of maternal age

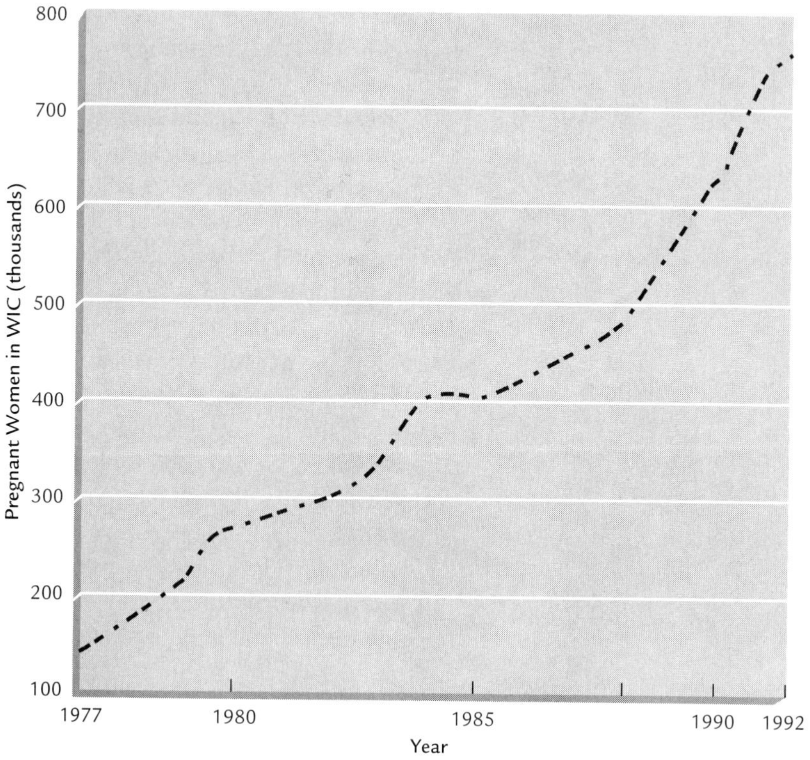

TABLE 3.28
Nutritional criteria for WIC

FIGURE 3.8
Estimated participation of pregnant women in WIC 1977–1992

REFERENCES

Abrams, B., "Preventing Low Birthweight: Does WIC Work?" *Ann NY Acad Sci* 678 (1993):306–17.

ACC/SCN, "Volume I: Global and Regional Results," second report on the world nutrition situation (Geneva: ACC/SCN, 1992).

ACC/SCN, "Volume II: Country Trends, Methods and Statistics," second report on the world nutrition situation (Geneva: ACC/SCN, 1993).

Adeniyi, F. A. A., "The Implications of Hypozincemia in Pregnancy," *Acta Obstet Gynecol Scand* 66 (1987):579–82.

Alexander, C. S. and Guyer, B., "Adolescent Pregnancy: Occurrence and Consequences," *Pediatric Ann* 22 (1994):85–88.

American College of Obstetricians and Gynecologists, "Exercise during Pregnancy and the Postnatal Period," *Technical Bulletin* no.189 (February 1994).

American College of Obstetricians and Gynecologists, American Academy of Pediatrics, *Guidelines for Perinatal Care* (Washington D.C.: ACOG, 1992).

American Diabetes Association, "Nutrition Recommendations and Principles for People with Diabetes Mellitus," *Diabetes Care* 17 (1994):519–22.

American Dietetic Association, "Nutrition Care for Pregnant Adolescents," *J Amer Diet Assoc* 94 (1994):449–50.

American Dietetic Association, "Use of Nutritive and Nonnutritive Sweeteners," *J Am Diet Assoc* 93 (1993):816–21.

Anderson, A. S., "Constipation during Pregnancy: Incidence and Methods Used in Treatment in a Group of Cambridgeshire Women," *Health Visit* 12 (1984):363–4.

Andrews, K. W., Savitz D. A. and Hertz-Piciotto I., "Prenatal Lead Exposure in Relation to Gestational Age and Birthweight: A Review," *Am J Indust Med* 26 (1994):13–32.

Anonymous, "Calcium Supplementation Prevents Hypertensive Disorders of Pregnancy," *Nutr Rev* 50 (1992):233–36.

Antonov, A., "Children Born during the Seige of Leningrad," *J Pediatr* 30 (1947):250–59.

Artal, R. M. and Posner, M. D., "Fetal Response to Maternal Exercise," in *Exercise in Pregnancy*, eds. R. A. Mittlemark, R. A. Wiswell and B. L. Drinkwater, 2d ed. (Baltimore: Williams and Wilkins, 1991).

Baghurst, P. A., et al., "Exposure to Environmental Lead and Visual-Motor Integration at Age Seven Years: The Port Pirie Cohort Study," *Epidemiol* 6 (1995):104–09.

Baron, T. H., Ramirez, B., and Richter, J. E., "Gastrointestinal Motility Disorders during Pregnancy," *Ann Int Med* 118 (1993):366–75.

Baumslag, N., Edelstein, T., and Metz, J., "Reduction of Incidence of Prematurity by Folic Acid Supplementation during Pregnancy," *Br Med J* 1 (1970):16–17.

Beattie, J. O., "Alcohol Exposure and the Fetus," *Eur J Clin Nutr* 46 (1992):S7–S17.

Belizan, J. M., et al., "Calcium Supplementation to Prevent Hypertensive Disorders of Pregnancy," *N Engl J Med* 325 (1991):1399–405.

Bendich, A., "Lifestyle and Environmental Factors that Can Adversely Affect Maternal Nutritional Status and Pregnancy Outcomes," *Ann NY Acad Sci* 678 (1993):255–65.

Bopp, B. A., et al., "Toxicological Aspects of Cyclamates and Cyclohexamine," *CRC Crit Rev Toxicol* 16 (1986):216–306.

Brandt, L. P. A., and Nielson C. V., "Job Stress and Adverse Outcome of Pregnancy: A Causal Link or Recall Bias?" *Am J Epidemiol* 135 (1992):302–11

Burdich, M., and Murray, J., "Study of WIC Participants and Program Characteristics," (Washington D.C.: USDA, Food and Nutrition Service, Office of Analysis and Evaluation, 1992).

Butchko, H. H., and Kotsonis, F. N., "Acceptable Intake vs. Actual Intake: The Aspartame Example," *J Am Coll Nutr* 10 (1991):258–66.

Campbell, D. M., "Physiological Changes and Adaptation," in *Twinning and Twins*, eds. I. MacGillivray, D. M. Campbell, and B. Thompson (New York: Wiley, 1988).

Centers for Disease Control, "CDC Surveillance Summaries," *MMWR* 41 (1992):25–43.

Centers for Disease Control, "Teenage Pregnancy and Birth Rates," *MMWR* 42 (1993):733–37.

Chasnoff, I. J., Landress, H. J., and Barrett, M. E., "The Prevalence of Illicit-Drug or Alcohol Use during Pregnancy and Discrepancies in Mandatory Reporting in Pinellas County, Florida," *N Engl J Med* 322 (1990):1202–06.

Clapp, J. F., "A Clinical Approach to Exercise in Pregnancy," *Clin Sports Med* 13 (1994):443–58.

Clapp, J. F., Wesley, M., and Sleamaker, R. H., "Thermoregulatory and Metabolic Responses to Jogging Prior to and during Pregnancy," *Med Sci Sports Exerc* 19 (1987):124–30.

Clapp, J. F., and Capeless, E. L., "Neonatal Morphometrics Following Endurance Exercise during Pregnancy," *Am J Obstet Gynecol* 163 (1990):1805–11.

Cochrane, W. A., "Overnutrition During Fetal Life—A Problem?" *Can Med Assoc J* 93 (1965):893–99.

DeLong, G. R., "The Effect of Iodine Deficiency on Neuromuscular Development," *1990 IDD Newslet* 6 (1990):1–9.

Dunn, J. T., "Iodine Supplementation and the Prevention of Cretinism," *Ann NY Acad Sci* 678 (1993):159–68.

Eskanazi, B., "Caffeine during Pregnancy: Grounds for Concern?" *JAMA* 270 (1993):2973–74.

Federation of the American Society for Experimental Biology, *Guidelines for Assessment and Management of Iron Deficiency in Women of Childbearing Age* (Bethesda, Md: FASEB, 1991).

Fischer, D. A., and Polk, D. H. "The Ontogenesis of Thyroid Function and Actions," in *Maternal-Fetal Endocrinology,* eds. D. Tulchinsky, A. B. Little (Philadelphia: W. B. Saunders Co., 1994).

Floyd, R. L., et al., "A Review of Smoking in Pregnancy: Effects on Pregnancy Outcomes and Cessation Efforts," *Annu Rev Pub Health* 14 (1993):379–411.

Food and Drug Administration, "Caffeine and Pregnancy," *FDA Drug Bull* 10 (1980):19–20.

General Accounting Office, *Early Intervention: Federal Investments Like WIC Can Produce Savings* (Washington D.C.: U.S. General Accounting Office, 1992).

Gonzalez-Cossio, T., and Delgado, H., "Functional Consequences of Maternal Malnutrition," in *Selected Vitamins, Minerals and the Functional Consequences of Maternal Malnutrition,* ed. A. P. Simopoulos, *World Rev Nutr Diet* (Basel: Karger, 1991).

Griffith, D. R., Azuma, S. D., and Chasnoff, I. J., "Three Year Outcome of Children Exposed Prenatally to Drugs," *J Am Acad Child Adolescent Psych* 33 (1994):20–7.

Guitierrez, Y., and King, J. C., :Nutrition during Teenage Pregnancy," *Pediatric Ann* 22 (1992):99–108.

Hadden, D. R., "Geographic, Ethnic and Racial Variations in the Incidence of Gestational Diabetes Mellitus," *Diabetes* 34 (1985):8–12.

Haste, F. M., et al., "The Effect of Nutritional Intake on Outcome of Pregnancy in Smokers and Nonsmokers," *Br J Nutr* 65 (1991):347–54.

Hediger, M. L., et al., "Patterns of Weight Gain in Adolescent Pregnancy: Effects on Birth Weight and Preterm Delivery," *Obstet Gynecol* 74 (1989):6–12.

Hoekelman, R. A., "Teenage Pregnancy—One of Our Nation's Most Challenging Dilemmas," *Pediatric Ann* 22 (1994):81–2.

Hoover, R., "Saccharin—A Bitter Aftertaste?" *N Engl J Med* 302 (1980):573–74.

Infante-Rivard, C., et al., "Fetal Loss Associated with Caffeine Intake before and during Pregnancy," *JAMA* 270 (1993):2940–43.

Institute of Medicine, *Nutrition during Pregnancy and Lactation: An Implementation Guide* (Washington D.C.: National Academy Press, 1992).

Institute of Medicine, *Nutrition during Pregnancy* (Washington D.C.: National Academy Press, 1990).

Institute of Medicine, *Nutrition Services in Perinatal Care* (Washington D.C.: National Academy Press, 1992).

Jameson, S., "Zinc Status in Pregnancy: The Effect of Zinc Therapy on Perinatal Mortality, Prematurity and Placental Ablation," *Ann NY ACad Sci* 678 (1993):178–93.

Jones, K. L., "Effects of Therapeutic, Diagnostic and Environmental Agents," in *Maternal-Fetal Medicine*, 3d ed. Eds. R. K. Creasy, R. Resnik (Philadelphia: W. B. Saunders, 1994).

Keen, C. L., and Zidenberg-Cherr, S., "Should Vitamin-Mineral Supplements Be Recommended for All Women of Childbearing Potential?" *Am J Clin Nutr* 59 (1994):532S–9S.

Kjos, S. L., et al., "Gestational Diabetes Mellitus: The Prevalence of Glucose Intolerance and Diabetes Mellitus in the First Two Months Postpartum," *Am J Obstet Gynecol* 163 (1990):93–98.

Knight, E. M., et al., "Illicit Drug Use in Pregnancy: Effect of Maternal Nutritional Status and Birthweight," *FASEB J* 6 (1990):A1960.

Kokotailo, P. K., et al., "Cigarette, Alcohol, and Other Drug Use by School-Age Pregnant Adolescents: Prevalence, Detection and Associated Risk Factors," *Pediatrics* 90 (1992):328–34.

Koos, B. J., and Longo L. B., "Mercury Toxicity in the Pregnant Woman, Fetus, and Newborn Infant: A Review," *Am J Obstet Gynecol* 126 (1976):390–8.

Kuwabara, K., et al., "Relationship between Breast-Feeding and PCB in Blood of Children whose Mothers were Occupationally Exposed to PCBs," *Int Arch Occup Environ Health* 41 (1978):189.

Lackey, C. J., "Pica-Pregnancy's Etiological Mystery," in *Alternative Dietary Practices and Nutritional Abuses in Pregnancy*, Committee on Nutrition of the Mother and School-Age Child (Washington D.C.: National Academy of Sciences, 1982).

Lammer, E. J., et al., "Retinoic Acid Embryopathy," *N Engl J Med* 313 (1985):837–41.

Lechtig, A., et al., "Influence of Maternal Nutrition on Birthweight," *Am J Clin Nutr* 28 (1975):1223–33.

Levy, N., Lemberg, E., and Sherf, M., "Bowel Habits in Pregnancy," *Digestion* 4 (1971):216–22.

Lotgering, F. K., Gilbert, R. D., and Longo, L. D., "Maternal and Fetal Responses to Exercise during Pregnancy," *Physiol Rev* 65 (1985):1–36

Luke, B., et al., "The Association Between Maternal Weight Gain and the Birthweight of Twins," *J Maternal-Fetal Med* 1 (1992):267–76.

Luke, B., Johnson, T. R. B., and Petrie, R., *Clinical Maternal-Fetal Nutrition*, (Boston: Little Brown and Co., 1993).

Mardones-Santander, F., et al., "Effect of a Milk-Based Food Supplement on Maternal Nutritional Status and Fetal Growth in Underweight Chilean Women," *Am J Clin Nutr* 47 (1988):413–19.

Metzger, B. E., et al., "Summary and Recommendations of the Third International Workshop-conference on Gestational Diabetes Mellitus," *Diabetes* 40 (1991):197.

Mills, J. L., et al., "Maternal Alcohol Consumption and Birthweight. How Much Drinking during Pregnancy Is Safe?" *JAMA* 252 (1984):1875–79.

Mills, J. L., et al., "Moderate Caffeine Use and the Risk of Spontaneous Abortion and IUGR," *JAMA* 269 (1993):593–97.

Mimouni, F., and Tsang, R. C., "Perinatal Mineral Metabolism," in *Maternal-Fetal Endocrinology*, eds. D. Tulchinsky, A. B. Little (Philadelphia: W. B. Saunders Co., 1994).

Morabia, A., and Wynder, E. L., "Dietary Habits of Smokers, People Who Never Smoked, and Nonsmokers," *Am J Clin Nutr* 52 (1990):933–37.

Murphy, J. F., et al., "Relation of Hemoglobin Levels in the First and Second Trimesters to Outcome of Pregnancy," *Lancet* 1 (1986):992–95.

Murray, J. L., and Bernfield, M., "The Differential Effect of Prenatal Care on the Incidence of Low Birth Weight Among Blacks and Whites in a Prepaid Health Plan," *N Engl J Med* 319 (1988):1385–91.

National Academy of Sciences, Food and Nutrition Board of the National Research Council, *Recommended Dietary Allowances*, 10th ed (Washington D.C.: National Academy Press, 1989).

Nehlig, A., and Debry, G., "Consequences on the Newborn of Chronic Maternal Consumption of Coffee during Gestation and Lactation: A Review," *J Am Coll Nutr* 13 (1994):6–21.

Newman, V., and Fullerton, J. T., "Role of Nutrition in the Prevention of Preeclampsia—Review of the Literature," *J Nurse Midwifery* 35 (1990):282–91.

Norton, M., Buchanan, T. A., and Kitzmiller, J. L., "Endocrine Pancreas and Maternal Metabolism," in *Maternal-Fetal Endocrinology*, eds. D. Tulchinsky and A. B. Little (Philadelphia: W. B. Saunders Co., 1994).

Paisley, J., and Mellion, M. B., "Exercise during Pregnancy," *Am Fam Physician* 38 (1988):143–50.

Perez-Escamilla, R., and Pollitt, E., "Causes and Consequences of IUGR in Latin America," *Bull PAHO* 26 (1992):128–48.

Persaud, T. V. N., *Environmental Causes of Birth Defects*, (Springfield, Il., Charles C. Thomas Publ., 1991).

Prager, K., et al., "Smoking and Drinking Behavior before and during Pregnancy of Married Mothers of Live-Born Infants and Stillborn Infants," *Public Health Rep* 99 (1984):117–27.

Prentice, A. M., et al., "Increased Birthweight after Prenatal Dietary Supplementation of Rural African Women," *Am J Clin Nutr* 46 (1987):912–25.

Primrose, T., and Higgins, A., "A Study in Human Antepartum Nutrition," *J Reprod Med* 7 (1971):257–64.

Roberts, J., "Current Perspectives on Preeclampsia," *J Nurse Midwifery* 39 (1994):70

Rosso, P., *Nutrition and Metabolism in Pregnancy* (New York: Oxford University Press, 1990).

Saftlas, A. F., et al., "Epidemiology of Preeclampsia and Eclampsia in the U.S.," *Am J Obstet Gynecol* 163 (1990):460–5.

Shiono, P. H., and Klebanof, F. M. A., "Caffeine and Birth Outcomes," *Am J Epidemiol* 37 (1993):951–54.

Singer, D. B., and O'Shea, P. A., "Fetal and Neonatal Pathology," in *Human Reproduction: Growth and Development*, ed. D. R. Coustan (Boston: Little Brown, 1995).

Stein, Z., et al., *Famine and Human Development: The Dutch Hunger Winter of 1944–45*, (New York: Oxford University Press, 1975).

Stevens-Simon, C., and McAnarney, E. R., "Adolescent Pregnancy," in *Textbook of Adolescent Medicine*, eds. E. R. McAnarney, et al., (Philadelphia: W. B. Saunders, 1992).

Streissguth, A. P., et al., "Attention, Distraction and Reaction Time at Age Seven Years and Prenatal Alcohol Exposure," *Neurobehav Toxicol Teratol* 8 (1986):717–25.

Tanner, J. M., *Fetus into Man: Physical Growth from Conception to Maturity* (Cambridge: Harvard University Press, 1990).

Teratology Society, "Recommendations for Vitamin A Use During Pregnancy," *Teratology* 35 (1987):267–75.

The Surgeon General's Report on Nutrition and Health: Summary and Recommendations. DHHS (PHS) publication no. 88-50211 (Washington D.C.: US Govt Printing Office, 1988).

U.S. Dept. of Health and Human Services, *Healthy People 2000: National Health Promotion and Disease Prevention Objectives* (Washington D.C.: USDHHS, Govt. Printing Office, 1990).

U.S. Preventive Services Task Force, "Routine Iron Supplementation during Pregnancy," *JAMA* 270 (1993):2846–48.

Uvin, P., "The State of World Hunger," *Nutr Rev* 52 (1994):151–62.

Viteri, F. E., "The Consequences of Iron Deficiency and Anemia in Pregnancy," in *Nutrient Regulation during Pregnancy, Lactation and Infant Growth*, eds. L. Allen, J. King, and B. Lonnerdal (New York: Plenum Press, 1994).

Wegman, M. E., "Annual Summary of Vital Statistics," *Pediatrics* 94 (1994):792–803.

Wright, J. T., et al., "Alcohol Consumption, Pregnancy and Low Birthweight," *Lancet* 1 (1983):663–65.

Zuckerman, B., et al., "Effects of Maternal Marijuana and Cocaine Use on Fetal Growth," *N Engl J Med* 320 (1989):762–768.

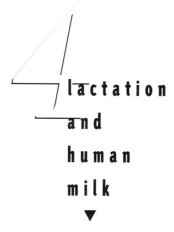

lactation and human milk
▼

THE MAMMARY GLAND

EARLY DEVELOPMENT OF THE BREAST

The growth and development of the female breast, or *mammogenesis,* occurs in a series of steps that culminate in lactation. Mammary gland development begins **in utero**, and the infant is born with a small amount of rudimentary mammary tissue. Maternal and placental hormones near term entering the fetal bloodstream stimulate these immature mammary glands, and newborns often secrete a thin, milklike fluid (called witches milk) for several days after birth. The breast then remains quiescent through childhood until puberty.

in utero: within the uterus; not yet born

In females, high levels of estrogen and progesterone during puberty stimulate rapid development of the glands and growth of the breast. During adolescence and young adulthood, secretion of progesterone during each menstrual cycle stimulates further growth of the mammary glands and ducts.

Stimulated by high levels of circulating ovarian hormones, final development and maturation of the breast occurs during pregnancy and, by term, the glands are fully formed and capable of producing breast milk to feed the newborn.

Pregnant adolescents who have had periods of amenorrhea (missed menstrual periods) or late **menarche**, followed by early pregnancy, may have underdeveloped glands and ductwork. They may have insufficient mammary tissue to support breast-feeding. However, the majority of women who enter adulthood and prepare for childbearing have ample mammary tissue.

ovarian steroid hormones: the female sex hormones produced by the ovary, which contain carbon atoms in a complex structure of four interlocking rings; the two major hormones are estrogen and progesterone

menarche: the beginning of menstrual function

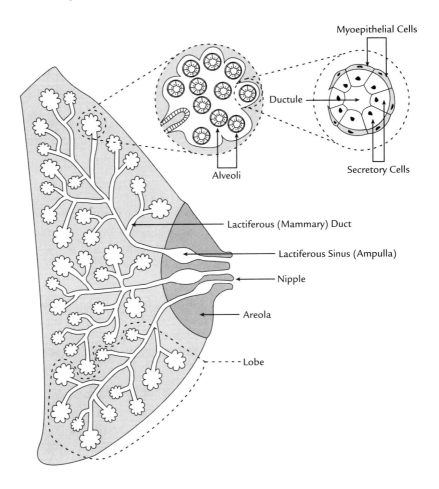

FIGURE 4.1
Anatomy of the breast and the alveolus

THE ANATOMY OF THE BREAST

The mature mammary gland is a **compound tubuloalveolar gland** embedded in a cushion of adipose tissue. Small clusters of secretory cells are surrounded by a contractile network of myoepithelial cells forming alveoli, the basic secretory units of the breast. The alveoli secrete milk into a branched system of small ducts that merge to form larger collecting ducts. See Figure 4.1.

The ductal systems terminate in the lactiferous sinuses, a series of reservoirs beneath the nipple and areola, and these sinuses empty through 15 to 20 small pores in the nipple. Large numbers of fat cells, nerves, and blood vessels surround the mammary glands and each breast is supported and subdivided by connective tissue into 15 to 20 lobes. Although variable in size (mainly due to differences in adipose tissue content), the average breast of an adult woman before pregnancy weighs about 150–200 g.

compound tubulo-alveolar gland: a gland made up of a number of smaller units containing tubules whose excretory ducts combine to form ducts of progressively higher order

prolactin: a hormone secreted by the anterior pituitary gland that promotes the growth of breast tissue and stimulates milk production postpartum

human placental lactogen (hPL: a hormone produced by the placenta toward the end of pregnancy that stimulates lactogenesis

parturition: labor leading to childbirth

anterior pituitary gland: the anterior lobe of the pituitary gland, an endocrine gland located at the base of the brain

BREAST DEVELOPMENT DURING PREGNANCY

The female breast undergoes dramatic changes in structure, size, and function during pregnancy, typically more than doubling in size and increasing weight to between 400 and 500 g. Large numbers of new adipose cells, supporting connective tissue cells, and blood vessels are formed. Development of the ductal system is completed, and large numbers of new secretory cells and alveoli—the sites of future milk production—are formed.

These changes are the result of the unique hormonal environment of pregnancy. Progesterone promotes an increase in the number and size of the alveoli, while the ductal system proliferates under the influence of estrogen. **Prolactin** and **human placental lactogen (hPL)** also play central roles in increasing breast mass: the ducts and alveoli mature as prolactin and hPL levels sharply rise in later pregnancy (Tay et al. 1992). See Figure 4.2.

During the second half of pregnancy, secretory activity in the alveolar cells steadily increases. The alveolar cells produce a thin, yellowish liquid called colostrum and, in the weeks before **parturition**, the alveoli become distended by accumulated colostrum.

The mammary glands are fully mature and capable of lactation by the beginning of the third trimester, but secretion of mature milk during pregnancy does not occur because lactation is inhibited by high levels of estrogen and progesterone produced by the placenta (Tay et al. 1992). These hormones inhibit lactation by slowing the release of prolactin by the **anterior pituitary gland**. Prolactin is the major hormone controlling lactation and is essential for milk synthesis.

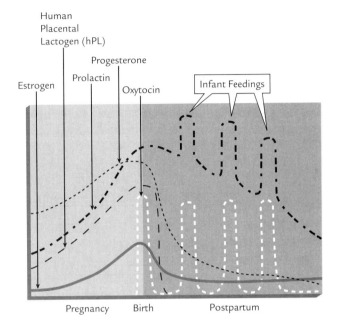

FIGURE 4.2

Hormonal changes of pregnancy, delivery, and lactation

BIRTH AND THE EARLY HORMONAL PROCESSES OF LACTATION

The maternal hormonal environment changes abruptly after delivery of the newborn. With loss of the placenta, progesterone and estrogen levels fall sharply, and the anterior pituitary gland, no longer inhibited by these hormones, releases large amounts of prolactin (Tay et al. 1992). Prolactin stimulates the final differentiation of alveolar cells in the breast to mature, milk-secreting cells.

The other major hormone involved in the onset and maintenance of lactation is **oxytocin**. Like prolactin, it is produced by the pituitary but is produced in the posterior portion of the gland. There is also an abrupt increase in oxytocin release by the posterior pituitary at birth (Tulchinsky 1994). Oxytocin has two major functions **postpartum**.

First, oxytocin circulates to the breast and stimulates milk secretion from the alveoli. Second, oxytocin travels to the uterus and stimulates the uterus to contract. Uterine contractions help control postpartum bleeding (detachment of the placenta during delivery causes the lining of the uterus to bleed), and they aid in **uterine involution**.

oxytocin: a hormone secreted by the pituitary gland that stimulates release of milk from the mammary glands

postpartum: after childbirth

uterine involution: the process of reduction of the size of the uterus to its nonpregnant size and state following childbirth

PHYSIOLOGY OF LACTATION

The breast is much more than a passive reservoir of milk. It is a dynamic organ uniquely capable of **lactogenesis**. Highly specialized cells in the mammary glands extract water, amino acids, lactose, fats, vitamins, minerals and other substances from the maternal blood. These cells package these substrates, synthesize many new nutrients and compounds, and secrete a unique and complex fluid specifically tailored to the needs of the infant. Unlike many other mammals, such as the cow with her udder, the human mammary gland has only limited storage capacity. Thus mammary epithelial cells must continuously synthesize new milk.

SYNTHESIS OF MILK

As the mammary glands mature during pregnancy, the appearance of the epithelial cells lining the alveoli changes dramatically. They develop the organization and morphology of active secretory cells and become crowded with **ribosome**s, **endoplasmic reticulum**, **lipid droplets**, and **secretory granules**.

The constituents of milk are derived from three main sources:

- Most of the components of milk are synthesized in the secretory cells of the mammary gland, from precursors derived from the maternal bloodstream.

- Some substances are produced by other cells in the breast, passed to the secretory cells, and then secreted into the milk.

- Other components are pulled intact from the maternal plasma and transferred directly into the milk.

Because the components of milk are ultimately derived from substrates in the maternal plasma, factors that affect the composition of the plasma also influence the composition of the milk. Therefore, the nutritional state of the breast-feeding mother can have a substantial influence on the nutritional character of her milk.

Figure 4.3 shows a cross-section of a mammary secretory cell, and Figure 4.4 illustrates the five principal secretory pathways within the mammary epithelial cells (Neville 1990):

1. *Exocytosis of Golgi-derived vesicles.* Many of the water-soluble components of breast milk are synthesized on ribosomes and packaged in the **Golgi apparatus** of mammary cells. They are then transported to the alveolar surface of the cell and secreted by **exocytosis**. The milk sugar lactose (synthesized from glucose and galactose in the Golgi) is secreted via this pathway. The principal milk proteins are synthesized

---

lactogenesis: milk production

ribosome: intracellular particles involved in protein synthesis; they occur either bound to cell membranes or free in the cytoplasm

endoplasmic reticulum: an intracellular network of tubules or flattened sacs important in protein and fatty acid synthesis

lipid droplets: intracellular drops of fats or fat-soluble substances

secretory granules: intracellular particles, often containing proteins that are formed in the endoplasmic reticulum and later secreted from the cell

Golgi Apparatus: a network of flattened vesicles responsible for secretion of the protein synthesized in the endoplasmic reticulum and ribosomes

exocytosis: the discharge from the cell of particles too large to diffuse through the cell membrane

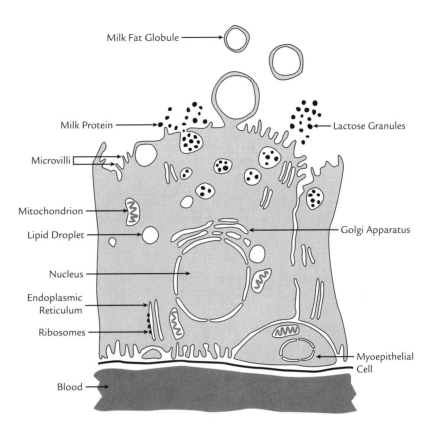

FIGURE 4.3
Mammary secretory cell

from amino acids derived from maternal blood and secreted via exocytosis. The water-soluble vitamins and many minerals, including calcium and phosphorus (as phosphate), are also secreted into breast milk via exocytosis.

2. *Apocrine secretion of milk-fat globules.* The fats present in milk are derived from both *de novo* fatty acid synthesis from glucose within the mammary cell, and fatty acids taken up from maternal plasma. Triglycerides, fatty acids, and fat-soluble vitamins accumulate in lipid droplets in the cytoplasm of the mammary epithelial cells. The droplets migrate to the cell surface where they are pinched off in small envelopes of cell membrane, forming milk-fat globules.

3. *Transport through channels in the cell membrane.* Many ions, including sodium, potassium, and chloride, pass through channels in the mammary cell membrane into the milk. The movement of these ions into the milk is carefully regulated by the secretory cells. Breast milk is over 90% water and is **isoosmotic** with maternal plasma. Water diffuses across the alveolar cell membrane into the milk following the osmotic pull of several components secreted into the milk.

4. ***Transcytosis*** *of intact proteins.* Immunoglobulin A (IgA), the principal immunoglobulin of milk, is produced by white blood cells in the breast tissue. IgA is released into the bloodstream and then taken up

isoosmotic: a state of having the same total osmotic pressure as another fluid

transcytosis: a mechanism of transcellular transport of large particles

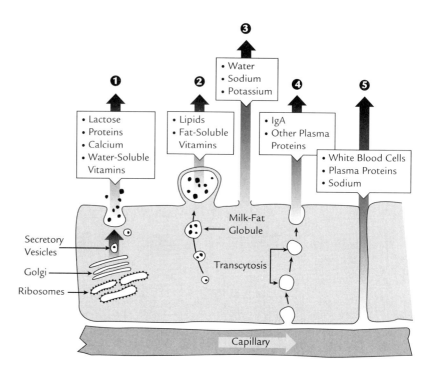

FIGURE 4.4
Five major pathways of milk synthesis

by receptor-mediated pinocytosis into the mammary secretory cells. It is repackaged into secretory IgA (discussed later in this chapter), transferred across the secretory cell, and secreted into the milk.

5. *The Pericellular Pathway.* In certain situations, such as in the early stages of lactation or if inflammation is present, substances can permeate the mammary epithelium by passing between loosely joined cells and find their way into the breast milk. This pericellular pathway allows intact plasma proteins, such as albumin, electrolytes, and maternal blood cells to pass into the breast milk. As the mammary epithelium matures, the junctions between the cells tighten, and there is less movement of substances via pericellular pathways.

Milk production is a two-step process. First, milk is synthesized and secreted by the mammary epithelium and accumulates in the alveoli and the small ducts that empty the alveoli. The second step is the ejection of the synthesized milk from the alveoli toward the nipple.

MILK EJECTION

neuroendocrine reflex:
a pathway whereby cells release a hormone into the blood in response to a nerve stimulus

Milk ejection (also called milk letdown) is the process that moves the milk from the alveoli through the ducts to the nipple, where it becomes available for the infant. See Figure 4.5. Suckling by the infant initiates a **neuroendocrine reflex.** During nipple stimulation, sensory nerves in the breast send signals to the brain. In response to this signal, the posterior pituitary releases oxytocin into the bloodstream. When the oxytocin

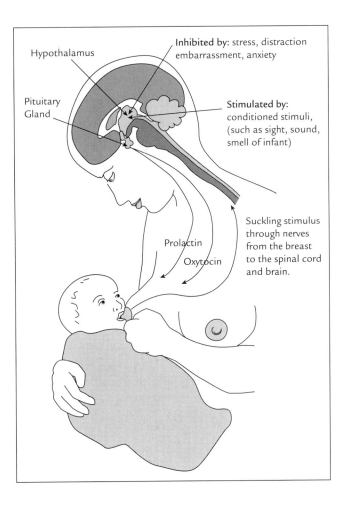

FIGURE 4.5
*Milk letdown reflex
and the suckling—
prolactin reflex*

reaches the breast, it interacts with receptors on the myoepithelial cells surrounding the alveoli, causing them to contract. As the myoepithelial cells contract, they squeeze the alveoli, pushing milk out into the ducts and toward the nipple.

The milk-ejection reflex can be felt by the mother as a tingling or tightening sensation in the breasts soon after the infant begins to suckle. It is an essential step in the production of milk, and is sensitive to physiological and psychological influences. Maternal embarrassment, stress, anxiety, or pain can interfere with the reflex and inhibit milk letdown (Newton 1992). The reflex is also easily conditioned, occurring even when the mother thinks about the infant or hears the infant cry.

HORMONAL CONTROL OF MILK SECRETION AND EJECTION

The postpartum fall in progesterone in the presence of high levels of prolactin triggers the onset of mature milk secretion (Tay et al. 1992). Full lactation does not begin immediately after birth but, on average, is delayed about 48 to 72 hours after parturition. Why? Because progesterone levels decline gradually, taking about four to five days to return to normal. Most women sense the onset of milk secretion as a feeling of fullness or engorgement of their breasts in the first few days postpartum, and the milk is said to have "come in".

Prolactin, produced by the anterior pituitary, stimulates milk synthesis and secretion. Blood levels of prolactin rise to 200–300 µg/mL (20 to 25 times nonpregnancy levels) and they remain elevated for two to three weeks postpartum—even in women who are not breastfeeding.

Prolactin stimulates protein synthesis and lactose formation by the mammary epithelium and also increases the fat content in the milk. Prolactin stimulates **lipoprotein lipase** in the mammary secretory cells, thereby increasing uptake of fatty acids from the maternal plasma. At the same time, it slows lipoprotein lipase activity in adipose cells elsewhere in the body, so triglycerides are available for mammary gland uptake (Williamson and Lund 1994).

Similar to the milk ejection reflex mediated by oxytocin, a neuroendocrine pathway initiated by infant suckling stimulates the pituitary to release prolactin (refer to Figure 4.5). Prolactin then circulates to the breast and stimulates the secretory cells in the alveoli to produce milk. This reflex is rapid and efficient (prolactin levels in blood are doubled after 10 to 15 minutes of suckling). During lactation, there is a rise and fall of prolactin levels in maternal blood proportional to the intensity and frequency of nipple stimulation. Even in women who have not been pregnant, regular nipple stimulation can cause prolactin-mediated synthesis of breast milk (this will be discussed in Chapter 5).

Lactogenesis, under the control of prolactin, occurs regardless of whether the mother is breast-feeding: milk production in the first few days postpartum is similar in both breast-feeding and nonbreast-feeding women. However, once lactogenesis is established, regular milk removal from the breast becomes crucial for continued milk production. If suckling is not begun by the fourth postpartum day, milk production slows and the milk composition reverts quickly to an immature, prepartum type. If a woman decides not to breast-feed her infant, the mammary gland involutes. Connective tissues and fat gradually replace many of the alveoli, and the remaining mammary epithelium regresses to the quiescent state of prepregnancy (Gould 1983).

Similar changes in the breast occur during weaning, as the infant is shifted to solid foods and suckles less frequently. Prolactin levels and milk production fall, and the breast gradually involutes.

lipoprotein lipase:
an enzyme on the surface of cells; it splits triglycerides passing by in the blood into fatty acids and glycerol; these can then be absorbed by the cells for metabolism or storage

BALANCING MILK PRODUCTION WITH INFANT DEMAND

Soon after birth, demand for milk by the infant becomes the primary determinant of milk production. The removal of milk stimulates continued milk production; conversely, stasis of milk in the breast inhibits synthesis. Both pituitary hormones and local factors within the breast participate in the supply-demand response that balances milk secretion with infant intake during lactation. Prolactin secretion in response to the infant's suckling continues to play an important role; however, the level of prolactin in most women gradually falls during lactation—even as milk secretion increases (Battin et al. 1985).

Several local factors within the breast are thought to play roles in balancing milk production with demand. In animal studies, scientists have found a milk protein that can inhibit secretion from the mammary epithelium in a dose-dependent manner (Wilde et al. 1988). If milk is not regularly removed from the alveoli, the protein builds up and begins to inhibit secretion by the epithelial cells. Another local regulator of secretion is distension of the alveoli by accumulated milk—pressure and distension of the secretory epithelium inhibits milk production.

MATURATIONAL CHANGES IN HUMAN MILK

Breast milk is a remarkably complex substance, with over 200 recognized components. See Table 4.1. Although milk is fundamentally a solution of sugar, protein, and salts containing a suspension of fat, it is much more:

- It contains enzymes to help the newborn digest and absorb nutrients.

- It contains immune factors to protect the infant from infection.

- It contains hormones and growth factors that influence infant growth.

Many individual milk constituents serve dual roles, both as nutrients and as functional components.

Although the basic components of breast milk are the same in all women, the concentration of the individual components varies considerably from mother to mother. Even in milk from the same woman, the level of many of the constituents in milk changes at different stages of lactation, from day to day, diurnally, and from the beginning of a single feeding to the end.

During the first two to three weeks postpartum, as the secretion of the alveolar cells evolves into fully mature milk, marked changes in the composition of breast milk occur. In general, the mammary secretion changes from a solution high in protein and electrolytes and low in fat and lactose, to one lower in protein and electrolytes but high in lactose and fat. See Table 4.2.

Nitrogen Compounds	Carbohydrates
Proteins	Lactose
Caseins	Oligosaccharides
α-Lactalbumin	Bifidus factors
Lactoferrin	Glycopeptides
Secretory IgA and other	
immunoglobulins	Lipids
ß-Lactoglobulin	Triglycerides
Lysozyme	Fatty acids
Enzymes	Phospholipids
Hormones	Sterols and hydrocarbons
Growth factors	Fat-soluble vitamins
Nonprotein Nitrogen Compounds	A and carotene
Urea	D
Creatine	E
Creatinine	K
Uric acid	
Glucosamine	Minerals
Nucleic acids	Calcium
Nucleotides	Phosphorus
Polyamines	Magnesium
	Potassium
Water-Soluble Vitamins	Sodium
Thiamin	Chlorine
Riboflavin	Sulfur
Niacin	
Pantothenic acid	Trace Elements
Biotin	Iodine
Folate	Iron
Vitamin B_6	Copper
Vitamin B_{12}	Zinc
Vitamin C	Manganese
	Selenium
	Chromium
Cells	Cobalt
Leukocytes	
Epithelial cells	

TABLE 4.1

Classes of constituents in human milk

From Institute of Medicine, Subcommittee on Nutrition during Lactation, *Nutrition during Lactation* (Washington D.C.: National Academy Press, 1991).

Constituent (per 100 ml)		Colostrum 1–5 days	Mature Milk >30 days
Energy	kcal	58	70
Total solids	g	12.8	12.0
Lactose	g	5.3	7.0
Total nitrogen	mg	360	171
Protein nitrogen	mg	313	129
NPN	mg	47	42
Total protein	g	2.3	0.9
Casein	mg	140	187
α-Lactalbumin	mg	218	161
Lactoferrin	mg	330	167
IgA	mg	364	142
Amino acids (total)			
Alanine	mg	—	52
Arginine	mg	126	49
Aspartate	mg	—	110
Cysteine	mg	—	25
Glutamate	mg	—	196
Glycine	mg	—	27
Histidine	mg	57	31
Isoleucine	mg	121	67
Leucine	mg	221	110
Lysine	mg	163	79
Methionine	mg	33	19
Phenylalanine	mg	105	44
Proline	mg	—	89
Serine	mg	—	54
Threonine	mg	148	58
Tryptophan	mg	52	25
Tyrosine	mg	—	38
Valine	mg	169	90
Taurine	mg	—	8
Urea	mg	10	30
Creatine	mg	—	3.3
Total fat	g	2.9	4.2
Fatty acids (% total fat)			
12:0 tauric		1.8	5.8
14:0 myristic		3.8	8.6
16:0 palmitic		26.2	21.0
18:0 stearic		8.8	8.0
18:1 oleic		36.6	35.5
18:2, n-6 linoleic		6.8	7.2
18:3, n-3 linolenic		—	1.0
C_{20} and C_{22} polyunsaturated		10.2	2.9

TABLE 4.2

Composition of breast milk: Colostrum and mature milk

TABLE 4.2 *Continued*

Constituent (per 100 ml)		Colostrum 1–5 days	Mature Milk >30 days
Cholesterol	mg	27	16
Vitamins			
Fat soluble			
Vitamin A (retinol equivalents)	µg	189	60
ß-Carotene	µg	112	23
Vitamin D	µg	—	0.05
Vitamin E (total tocopherols)	µg	1280	315
Vitamin K	µg	0.23	0.21
Water soluble			
Thiamin	µg	1.5	14
Riboflavin	µg	25	35
Niacin	µg	75	150
Folic acid	µg	—	8.5
Vitamin B_6	µg	12	18
Biotin	µg	0.1	0.6
Pantothenic acid	µg	183	240
Vitamin B_{12}	ng	200	45
Ascorbic acid	mg	4.4	4.0
Minerals			
Calcium	mg	23	28
Magnesium	mg	3.4	3.0
Sodium	mg	48	18
Potassium	mg	74	58
Chlorine	mg	91	42
Phosphorus	mg	14	15
Sulphur	mg	22	14
Trace elements			
Chromium	ng	—	50
Cobalt	µg	—	1
Copper	µg	46	25
Fluorine	µg	—	16
Iodine	µg	12	11
Iron	µg	45	40
Manganese	µg	—	0.6
Nickel	µg	—	2
Selenium	µg	—	2.0
Zinc	µg	540	120

Adapted from R. A. Lawrence, *Breastfeeding: A Guide for the Medical Profession,* 4th ed. (St. Louis: Mosby, 1994) and American Academy of Pediatrics, Committee on Nutrition, *Pediatric Nutrition Handbook,* 3rd ed. (Elk Grove, IL: American Academy of Pediatrics, 1993).

COLOSTRUM

colostrum: the fluid secreted by the mammary gland during the first few days postpartum

Produced during the first three to seven days postpartum, **colostrum** is thicker and more viscous, slightly yellow and not as milky as mature milk. The yellow tint is due to a high concentration of **carotenoids**. The carotene content of colostrum is about ten times higher than mature milk (Patton et al. 1990). It contains less fat and sugar than mature milk, is lower in calories (58 versus 65 kcal/100ml) but has more electrolytes, ash, and protein. The principal protein in colostrum is **immunoglobulin A (IgA)**, which helps protect the newborn from infections in the gastrointestinal tract (discussed later in this chapter). **Multiparous** women, particularly those who have previously breast-fed infants, usually produce greater volumes of colostrum than **primiparous** women.

carotenoids: generic term for the carotenes and their oxygenated derivatives

immunoglobulin A: a protein that functions as an antibody and is the primary immunoglobulin in breast milk; it binds to and inactivates foreign antigens (such as viruses and bacteria)

TRANSITIONAL MILK

Colostrum evolves into *transitional* milk at about one week postpartum. The composition of transitional milk is midway between colostrum and mature milk, having more protein and less lactose and fat than mature milk (Hibberd et al. 1982). By around 21 days postpartum, most women are producing fully mature milk. However, there can be wide variation in the rate at which maturational changes in milk composition occur.

multiparous: having given birth at least two times or more to viable offspring

A recent study found about 40% of women produce transitional milk at day 15, while a few women secrete mature milk as early as day 5 postpartum (Humenick 1987). Multiparous women who have breast-fed infants previously often produce mature milk sooner than first-time mothers.

primiparous: having given birth to one viable offspring

COMPOSITION OF MATURE HUMAN MILK

ENERGY CONTENT OF HUMAN MILK

Reported values for energy content of mature milk vary between 58 and 72 kcal/100 ml, primarily because of variation in fat content. The generally accepted value is around 65 kcal/100 ml, with approximately 50% of the energy from fat, about 40% from lactose, and the remainder from protein (Institute of Medicine 1991).

LIPID CONTENT OF HUMAN MILK

The lipids in milk are carried in the milk-fat globule, which contains a rich core of triglycerides surrounded by a membrane. The globule structure, contained in mammary cell membranes pinched off during apocrine secretion, enables the fat to remain in suspension in the milk

(Jensen 1989). The average fat content in milk from well-nourished mothers is 4.2 g/100 ml.

Over 95% of the fat is in the form of triglycerides. Lipases present in the newborn's small intestine, as well as in the breast milk itself (discussed later in this chapter), break down the triglycerides to provide energy for the growing infant. Small amounts of phospholipids, glycolipids, and free fatty acids are also present in milk, and about 0.5% of milk fat is cholesterol—a precursor for synthesis of **myelin,** cell membranes, and steroid hormones in the newborn.

Variations in Fat Content

Fat is one of the most variable components of milk, varying from 2 g to over 5 g/100 ml, and is influenced by several factors. The fat content of milk increases steadily during the first week postpartum and then plateaus. There is diurnal variation in the fat content: it tends to be lowest at about 6:00 A.M. and steadily rises to its highest level at 2:00 P.M. (Brown et al. 1982). At any one feeding, the fat content of milk gradually increases from the beginning of the feed to the end. After an average feeding, about 20% of the milk remains in the gland, and this "hindmilk" is particularly rich in fat (Jensen 1989). See Figure 4.6.

myelin: the fatty substance forming a sheath around certain nerve fibers

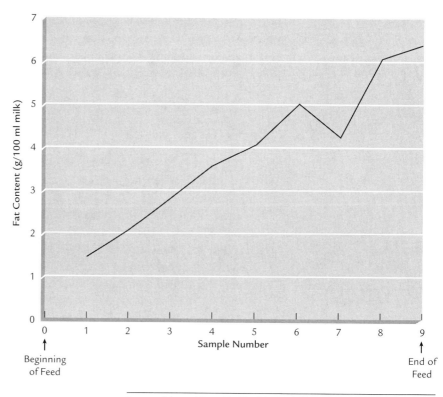

FIGURE 4.6

Variations in fat content of milk during a single feed

Source: Adapted from Neville, M. C., Neifert, M. R. *Lactation: Physiology, Nutrition and Breastfeeding* (New York: Plenum Press, 1983)

The parity of the mother may also influence milk-fat content, as multiparous women generally have less fat in their milk in later pregnancies. Maternal nutrition can markedly affect the milk-fat concentration and the type of fats present in the breast milk (discussed later in this chapter).

CARBOHYDRATE CONTENT OF HUMAN MILK

The major carbohydrate in breast milk is lactose, a milk-specific disaccharide synthesized in the alveolar cells of the breast. Figure 4.7 shows the synthesis of lactose. It is the second most abundant substance in milk (after water). About 90% of the total carbohydrate content in milk is lactose; the remaining 10% is a variety of mono- and oligosaccharides, including glucose, galactose, fructose, and glucosamines (Lawrence 1994).

Newborn intestinal epithelial cells have ample amounts of lactase (Riby 1994), which readily hydrolyses the lactose to glucose and galactose. Lactose enhances the absorption of calcium from breast milk (Lawrence 1994). In contrast to the variability of milk-fat content, the content of lactose in milk is fairly constant after the first postpartum week at about 7 g/100 ml, and levels are constant throughout the day.

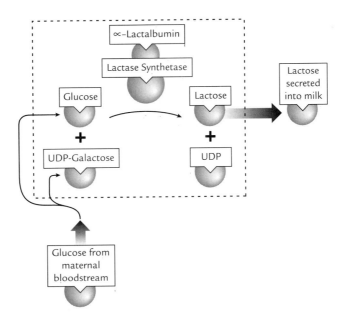

FIGURE 4.7
Synthesis of lactose

The protein content of mature breast milk from well-fed mothers is 0.7–0.9 g/100 ml (WHO 1985). However, not all of the protein is nutritionally available for infant needs. Certain proteins, such as IgA, have immunological functions and are poorly digested. Because of the high content of IgA in early milk and the immaturity of the newborn's digestive tract, up to 10% of milk protein in the first few postpartum weeks is not digested (Williams 1991). Digestion and absorption of milk protein increases as the infant grows, and by four months postpartum only about 3% of dietary protein goes unabsorbed.

The milk-specific proteins are broadly divided into *whey* and *casein* fractions, based on their different physical structures and distinct patterns of digestion. During digestion in the infant gastrointestinal tract, the caseins (or "curd" proteins) are **precipitated** into sizable clumps in the newborn's stomach. The casein-derived curds are complexes of calcium caseinate and calcium phosphate, which are tougher and less readily digestible than the softer, more flocculent clumps formed when whey proteins are acidified in the newborn's stomach.

Casein Proteins

Casein proteins make up 40% of the total protein in milk (Lonnerdal 1985). Caseins are linear phosphoproteins synthesized by the mammary epithelium, which have a charged end that avidly binds calcium, phosphate, and magnesium. Caseins, in combination with phosphates and calcium, form stable **micelles** which are secreted into the milk. The micelles are efficient packages for delivering protein and minerals to the infant, allowing milk to carry greater amounts of calcium, magnesium, and phosphate than could be carried in a simple aqueous solution.

Whey Proteins

The three major whey proteins are alpha-lactalbumin, lactoferrin, and secretory immunoglobulin A. Alpha-lactalbumin makes up 30% of the total protein in milk, lactoferrin 10–20%, and secretory immunoglobulin A 10%. Most whey proteins can be broken down to provide amino acids for the infant, but they also have important functional roles.

The mammary epithelium synthesizes large amounts of alpha-lactalbumin during lactation, and it has multiple functions. It is a component of the enzyme lactase synthetase—the enzyme that synthesizes lactose from glucose and galactose in the alveolar cells of the mammary gland. See Figure 4.7. It is then secreted into the milk and has calcium and zinc binding sites that help carry these minerals in the milk. Once in

precipitate: to cause a substance in solution to settle down into solid particles

micelle: a stable spherical structure formed by lipids, with the charged, polar heads of the lipid molecules oriented outward and the hydrophobic tails on the inside

the intestine of the infant, it is readily digested and provides a rich source of amino acids for the newborn.

Lactoferrin, found only in human milk, is an iron-binding protein that transports iron in milk and has several other important functions. It has antibacterial properties, in that it inhibits the growth of certain iron-requiring bacteria in the intestines, thus protecting the infant from infection. Lactoferrin is also a stimulus for growth and development of **lymphocytes** in the newborn's intestine (Hashizume et al. 1983). Lactoferrin can be digested by the newborn and, like alpha-lactalbumin, it is an important source of amino acids for growth.

Secretory immunoglobulin A (SIgA) is synthesized by mammary secretory cells from immunoglobulin A and other proteins (Goldman and Goldblum 1989). SIgA is found only in human milk, and colostrum is particularly rich in SIgA, with concentrations five times higher than in mature milk. Although IgA concentrations decrease in mature milk, SIgA remains a major and important milk protein.

In older children and adults, immune cells produce IgA in response to foreign antigens, and IgA is a central component of mucosal immunity. However, the newborn is deficient in IgA and vulnerable to infection from pathogens in the gastrointestinal tract. SIgA entering the infant's gastrointestinal tract resists hydrolysis by enzymes and acid and blankets the epithelium of the intestine, blocking adhesion and invasion by bacteria and viruses (Goldman and Goldblum 1989). This protective effect provides a transient immunological defense while the infant's own immune system is maturing.

The Protein Quality of Human Milk

The protein content of breast milk is considered ideal for the human infant. Milk proteins are complete and balanced, supplying optimal amounts of the essential and nonessential amino acids. For example, breast milk is particularly rich in **taurine**, which is the second most abundant free amino acid in human milk and is found at levels 30 times that in cow's milk. In the infant, taurine is not incorporated into proteins but is important in bile acid metabolism and may function as a neurotransmitter (Lawrence 1994).

Because the enzyme systems that normally synthesize taurine are still developing in early infancy, taurine is poorly synthesized by the newborn (Sturman 1988). While the enzyme systems mature, the ample supply in breast milk covers the newborn's needs.

Similarly, the enzyme systems that metabolize phenylalanine, tyrosine, and methionine mature late in fetal life and, particularly in preterm infants, overconsumption of these amino acids can increase blood levels to potentially detrimental levels. Human milk contains only moderate amounts of phenylalanine, tyrosine, and methionine, compared to the much higher content in cow's milk (AAP 1993).

lymphocyte: a type of white blood cell formed in lymphatic tissues throughout the body

taurine: an amino acid important in the conjugation of bile; also found in small quantities throughout the body, including in the lung, muscle, and nervous tissues

NONPROTEIN NITROGEN CONTENT OF HUMAN MILK

Small amounts of free amino acids, peptides, and urea are present in breast milk, but overall, they do not significantly contribute to nutritional needs. Certain of these compounds have other important functions:

- **N-acetyl glucosamine,** an amino sugar, may promote the growth of *Lactobacillus bifidus* in the colon (Institute of Medicine 1991). This bacteria is not **pathogenic**, and colonization of the colon by lactobacilli is considered beneficial, as it discourages the replication of pathogenic bacteria and helps protect the infant.

- **Carnitine** is essential in the oxidation of fatty acids for energy, and newborns particularly need carnitine because fat provides a large percentage of the energy in milk. Human milk provides amounts sufficient to meet the needs of most infants (Sandor et al. 1982) (discussed in Chapter 6).

VITAMIN CONTENT OF HUMAN MILK

Fat-Soluble Vitamins in Human Milk

Vitamin A About 90% of the vitamin A content of human milk is in the form of retinyl esters carried in the milk-fat globules; smaller amounts of retinol and beta-carotene are also present. The retinyl esters are hydrolyzed by lipases in the intestinal tract of the infant and the retinol is efficiently absorbed. The amount of vitamin A in breast milk declines over the course of lactation, from 200 µg/100ml in the first weeks postpartum to 40–60 µg/100 ml after several months (Chappell et al. 1986).

Beta-carotene is stored in the developing mammary gland during gestation and secreted into colostrum at high levels (34– 750 µg/100 ml); amounts in mature milk are reported to range from 10–30 µg/100 ml (Patton et al. 1990).

Vitamin D Breast milk contains small amounts of vitamin D; reported values range from 0.05 to 0.15 µg/100ml. About 75% of the vitamin D activity in the milk is accounted for by 25-OH vitamin D. Because the activity of the hepatic hydroxylase enzymes that convert vitamin D to 25-OH vitamin D may be low in the newborn, provision of most of the vitamin in milk in the 25-OH form may be advantageous during early infancy (Lawrence 1994).

Vitamin K The vitamin K content of mature milk is low (only about 0.2 µg/100 ml). Although vitamin K synthesized by intestinal bacteria contributes to the needs during later infancy, in the newborn the

N-acetyl glucosamine: an amino derivative of glucose occurring in many glycoproteins and mucopolysaccharides

pathogenic: able to produce disease

carnitine: an amino acid important in fatty acid metabolism in cells

intestine is sterile. Therefore, during the first weeks postpartum as the newborn establishes a stable population of microflora, vitamin K must be obtained from dietary sources. Because the recommended intake of 12 μg/day during infancy (NAS 1989) cannot be met by levels in breast milk, it is recommended that all infants receive supplementation with vitamin K during the newborn period (AAP 1993). Breast-feeding without supplementation has been associated with hemorrhagic disease of the newborn (discussed in Chapter 6).

Vitamin E Vitamin E in milk is a mixture of several **tocopherols**, with α-tocopherol accounting for 83% of the total. Levels of tocopherols are two to three times higher in colostrum (0.8–1.0 mg/100 ml) than in mature milk (0.2–0.3 mg/100 ml) (Jannsen et al. 1981). The high levels of tocopherols and beta-carotene in colostrum provide antioxidant protection during the newborn period.

Water-Soluble Vitamins in Human Milk

Vitamin C Vitamin C is avidly transferred into milk by the mammary secretory cells, and the level of vitamin C in milk is about ten times that in maternal plasma (Bates et al. 1983). In well-nourished women, mature milk contains 4–6 mg/100 ml.

Thiamin The thiamin content of milk varies considerably over the course of lactation and between individual women. Colostrum contains low levels (1–2 μg/100 ml), and there is a sharp increase (seven to tenfold) in secretion of thiamin into mature milk—with maximum levels at two to three months postpartum.

Riboflavin and Niacin The amount of riboflavin in milk is high in colostrum and declines gradually over the course of lactation; well-fed women produce milk containing about 35 μg/100 ml. The niacin content of human milk increases from 75 μg/100 ml in colostrum to 100–200 μg/100 ml in mature milk.

Vitamin B$_6$ Levels of vitamin B$_6$ in milk are low in colostrum and increase markedly during the first two weeks postpartum. By three weeks, levels in milk plateau at around 18 μg/100 ml and remain constant with little day-to-day or diurnal variation for the remainder of lactation.

Folate Folate in milk is bound to specific folate-binding proteins in the whey fraction, and this association may enhance infant absorption of the bound folate from milk. Unlike most other water-soluble vitamins, even if maternal dietary intake is low, folate continues to be secreted into the milk—maintaining milk folate content at the expense of maternal folate status (Institute of Medicine 1991). A large portion of the folate present in milk is in the form of **folylpolyglutamates**. Folate levels in milk generally increase over the course of lactation and range from 5–14 μg/100 ml.

tocopherols: a generic term for vitamin E and chemically related compounds

folylpolyglutamates: compounds formed when folic acid combines with glutamate

Vitamin B$_{12}$ Levels of vitamin B$_{12}$ in mature milk vary widely between women, ranging from 0.03–0.32 μg/100 ml. Women who are complete vegetarians produce milk with much lower levels (Gambon et al. 1986), levels as low as 0.005 μg/100 ml (discussed later in this chapter).

Biotin and Pantothenic Acid   The mammary secretory cells actively secrete biotin into breast milk at levels several hundred times greater than the level in plasma (Mock 1994). The level of biotin in breast milk is usually about 0.6 μg/100 ml, and levels gradually increase over the course of lactation. The pantothenic acid content of mature milk is 0.22–0.26 mg/100 ml— about 20% higher than levels in colostrum (Johnston et al. 1981).

MINERAL CONTENT OF HUMAN MILK

Mature breast milk contains all of the essential minerals needed for infant growth and development. The mineral content of human milk is well-balanced and the bio-availability to the infant is high, making milk a high-quality source of these nutrients. The most abundant minerals are the electrolytes (potassium, sodium, and chloride) and the major minerals (calcium, phosphorus, sulphur, and magnesium). Smaller amounts of zinc, iron, chromium, and copper, and the other trace minerals are also present.

The mineral content of breast milk varies significantly with the stage of lactation. Colostrum is richer in minerals than mature milk, and levels of many of the minerals gradually decline over the course of lactation. In addition, during a single feeding the mineral content of breast milk may vary significantly. Table 4.3 shows the changes in foremilk versus hindmilk, and Figure 4.8 shows the decline in zinc and calcium content of milk over the course of lactation.

Time of Sampling	Calcium (mg/100 ml)	Sodium (mg/100 ml)	Iron (μg/100 ml)	Zinc (μg/100 ml)
Foremilk				
Midmorning	28.5	13.4	23	102
Evening	30.7	11.8	34	96
Hindmilk				
Midmorning	26.9	14.4	25	89
Evening	29.0	15.7	51	91

The mineral content of breast milk can vary by time of day and over the course of a single feeding (foremilk=milk produced at the beginning of a feeding; hindmilk=produced toward the end of the feeding).

Adapted from M. F. Picciano, "Mineral Content of Human Milk During a Single Nursing," Nutr Rep Int 18 (1978):5

TABLE 4.3

Mineral content of foremilk vs. hindmilk at different times of the day

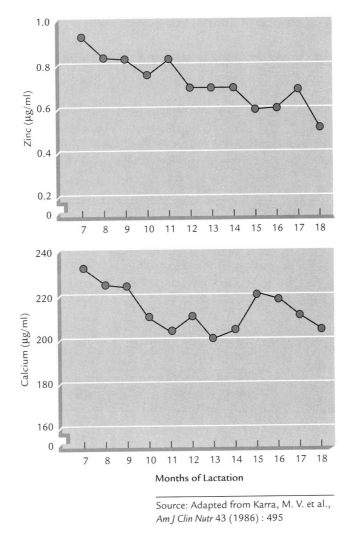

FIGURE 4.8

Zinc and calcium in breast milk, from 7 to 18 months lactation

Months of Lactation

Source: Adapted from Karra, M. V. et al., *Am J Clin Nutr* 43 (1986) : 495

Major Minerals in Human Milk

Calcium, phosphorus, and magnesium are carried in milk largely bound to casein proteins. About two thirds of the calcium in breast milk is in casein micelles, and the remainder forms a soluble complex with citrate. Although amounts of calcium are small (20–35 mg/100 ml), the calcium is highly bioavailable (infants absorb about two thirds of the calcium in breast milk, compared with 25 to 40% in bovine milk) (Fomon 1993).

Electrolytes in Human Milk

The sodium, potassium, and chloride concentrations of colostrum are high, but as the mammary secretion matures, levels fall and concentrations in mature milk are one to two thirds lower than in colostrum.

Trace Minerals in Human Milk

Iron Breast milk is low in iron, providing only about 0.5 mg/day or less to the breast-feeding infant. Concentrations are highest immediately after birth, and levels in mature milk decline with duration of lactation (from 2 weeks to 9–12 months postpartum, levels fall 20–50%). The level of iron in breast milk varies widely among women and from day to day in the same woman. Values reported for mature milk range from 20–90 µg/100 ml, with levels slightly higher in the evening than in the morning—and higher in women who are multiparous (Casey and Hambidge 1983).

Lactoferrin, synthesized by the mammary epithelium, avidly binds iron and transports it in the milk. The iron is highly bioavailable: the infant absorbs 50–70% of the iron in breastmilk compared with less than 10% from cow's milk or formula (Dallman 1986). Several factors present in the milk may increase the availability of iron, including ascorbic acid, lactoferrin, and inosine and its metabolites.

Zinc The most abundant trace element in human milk, zinc is actively transported into the milk by the mammary epithelium. Concentrations vary substantially over the course of lactation, with levels in colostrum four to five times higher than those in mature milk. Levels fall rapidly during the first postpartum month and then decline more slowly for the remainder of lactation. Concentrations at 1, 3, and 12 months postpartum are 0.4, 0.1, and 0.05 mg/100 ml, respectively (Casey et al. 1984).

There is little day-to-day or diurnal variation in the zinc content of milk (Lawrence 1994). Bioavailability is high, and several factors may facilitate zinc absorption from human milk—including picolinic acid, citrate, and zinc-binding proteins. (Johnson and Evans 1978). In one study (see Figure 4.9), 25 mg of oral zinc was given to infants with either breastmilk or cow's milk. The absorption of the zinc given with breastmilk was significantly higher (Casey et al. 1981).

Copper Concentrations of copper are 25 to 30% higher in colostrum than in mature milk, with levels gradually declining over the first four months of lactation and then becoming stable up to the 12th month (Casey et al. 1989). Reported values range from 10-60 µg/100 ml.

Manganese The level of manganese in mature milk declines gradually from about 0.6 µg/100 ml in the first month of lactation to 0.3 µg/100 ml by the third month (Institute of Medicine 1991).

Selenium Selenium in breast milk is primarily bound to casein and other proteins. Concentrations are highest in colostrum (3 µg/100 ml) and decline over the course of lactation. Values reported for mature milk, ranging from 1–3 µg/100 ml, vary with geographic location and selenium content of the diet (Kumpulanien 1989). Levels of selenium in milk are directly related to maternal plasma levels, with values in milk generally five to seven times those in maternal plasma (Mannan and Picciano

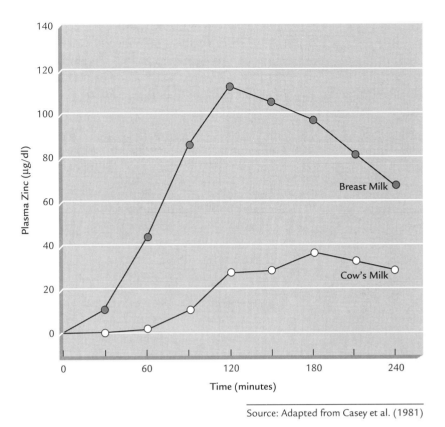

FIGURE 4.9

Absorption of 25 mg of oral zinc given with either breastmilk or cow's milk

Source: Adapted from Casey et al. (1981)

1987). The selenium-containing enzyme glutathione peroxidase is present in milk and may help protect the milk from oxidative damage. The activity of the enzyme in milk is positively correlated with the selenium content of the milk (Mannan and Picciano 1987).

Iodine During lactation, iodine metabolism is unique among the trace elements in that the mammary gland avidly accumulates iodine. The concentration of iodine in milk is 20 to 30 times higher than that in the maternal plasma. Iodine levels in milk are strongly correlated with maternal intake of iodine from the diet (discussed later in this chapter).

DIGESTIVE ENZYMES IN HUMAN MILK

There are a variety of enzymes secreted into human milk that have digestive functions. These enzymes help the immature gastrointestinal tract of the newborn digest and absorb nutrients in the milk.

Lipases

lipase: a fat-splitting or lipolytic enzyme

The mammary epithelium synthesizes several **lipases** that are carried in the milk and are inactive until they contact bile salts in the upper small intestine of the newborn. Once activated, the so-called bile-salt stimulated lipases augment the activity of infant **pancreatic lipase** (Hamosh 1989).

Lipases act on the milk-fat globules, breaking down triglycerides and retinyl esters. They increase the absorption of fatty acids, glycerol, and retinol.

Amylases and Proteases

In early infancy, activity of pancreatic amylase is low. The mammary cells secrete an amylase into the milk that is resistant to proteolysis in the newborn's stomach (Hamosh 1989). It passes into the small intestine of the infant and aids in the digestion of nonlactose carbohydrates in milk. Breast milk also contains proteases, which aid in protein digestion in the infant intestine.

ANTI-INFECTIVE FACTORS IN HUMAN MILK

A complex variety of anti-infective substances in milk, including immune cells, immunoglobulins, enzymes, and other protective factors, are produced throughout lactation (they are particularly abundant in colostrum). They enter the infant's gastrointestinal tract during feedings and are relatively resistant to digestion. They interact to protect the newborn from disease, particularly from infections of the gastrointestinal tract and infant diarrheal disease.

Immune Cells

Several types of white blood cells are found in human milk, particularly in the early stages of lactation. About 90% are phagocytic cells that engulf and destroy pathogenic bacteria and viruses, and the remainder are lymphocytes that produce antibodies and other protective factors—including lysozyme, lactoferrin, and complement.

Immunoglobulins

Antibodies to various antigens (including bacteria, viruses, and toxins) are produced by white blood cells in the mammary glands. They are then transferred into the milk and provide **passive immunity,** temporarily protecting the infant while its own immune system matures. See Figure 4.10. Although small amounts of immunoglobulins G and M are present in milk, the major immunoglobulin is secretory immunoglobulin A (SIgA) (Goldman and Goldblum 1989).

pancreatic lipase: an enzyme secreted by the pancreas that catalyzes the hydrolysis of triglycerides and phospholipids

passive immunity: immune protection provided by immune proteins obtained from an outside source and not produced by the organism itself

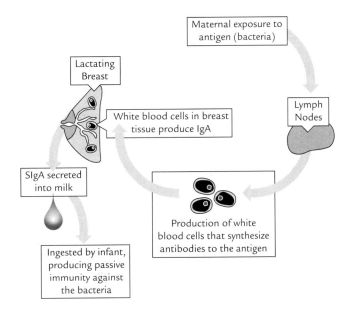

FIGURE 4.10

Maternal exposure to bacteria, production of IgA, transfer of passive immunity to the infant

Lactoferrin

In addition to transporting iron in milk, lactoferrin has protective effects against bacteria and viruses in the digestive tract of the newborn (Sanchez et al. 1992). Lactoferrin inhibits growth of certain bacteria by complexing with iron. Because iron is needed by bacteria for growth, and iron is bound to lactoferrin and unavailable, multiplication of the bacteria is inhibited. Lactoferrin, in conjunction with antibodies and immune cells, also has antiviral activity in the gastrointestinal tract.

Lysozyme

This enzyme is present in large amounts in human milk throughout lactation. Lysozyme breaks down the cell walls of pathogenic microbes and is a potent antibacterial factor (Lawrence 1994).

Nonlactose Carbohydrates

As mentioned earlier in this chapter, glucosamines in milk promote the growth of *Lactobacillus bifidus* in the lower intestinal tract of the infant. Lactobacilli normally are the dominant bacteria in the colon during infancy. They break down lactose to acetic acid and other metabolites—metabolites that inhibit the growth of pathogenic bacteria. A variety of moderate chain-length oligosaccharides are also present in small amounts in human milk, and many of them have anti-infective properties, preventing the attachment of pathogenic bacteria in the intestine (Goldman et al. 1986) and neutralizing toxins secreted by certain bacteria.

GROWTH FACTORS AND HORMONES IN HUMAN MILK

Human milk contains several substances, called growth factors, that promote cell growth and differentiation (Iacopetta et al. 1992). Although their biological significance is not yet completely understood, they may stimulate growth of intestinal epithelium in the newborn and strengthen the integrity of the mucosal barrier, protecting the infant from pathogens and antigens. They may also promote growth, differentiation, and repair in the mammary gland during lactation. Cortisol, insulin, and thyroid hormones are also present in milk in small amounts; their functional significance is uncertain.

CONTAMINANTS IN HUMAN MILK

PESTICIDES, CHEMICALS, AND HEAVY METALS

Worldwide, contamination of our air, water, and food supply with chemicals, heavy metals, and radioactivity continues to be a major public health concern. Although many of these substances find their way into breast milk, usually in trace amounts, there is little information available on their potential short or long-term effects on infant health.

The major chemical contaminants are pesticides, such as DDT, and organohalogen compounds, including dioxin, polybrominated biphenyls (PBBs), and polychlorinated biphenyls (PCBs) (Jensen 1991). Because humans are at the top of the food chain, levels of pesticides in human milk tend to be higher than those in cow's milk. DDT is one of the most widespread contaminants in breast milk. In general, levels of DDT in breast milk are higher in developing countries, where DDT still is, or recently has been, used extensively in agriculture or disease prevention. Figure 4.11 shows levels of DDT found in human milk from several countries in the Western Hemisphere.

Trace amounts of other chemicals have been identified in milk. In one study, breast milk samples from mothers in four cities in the United States were tested for chemicals (Pellizari et al, 1982). **Cyclic hydrocarbons** (such as **benzene** and **toluene**), halocarbons (including perchloroethylene and freon), and other chemicals were detected in nearly two thirds of the samples.

The significance of trace levels of chemicals in breast milk to maternal and infant health is unknown. However, findings from a recent study are encouraging. Researchers studied 850 children whose mothers had been exposed to high levels of chemicals during lactation and whose milk contained PCBs and dichlorodiphenylethane (DDE), a metabolite of

cyclic hydrocarbons: substances containing a ring of carbon atoms

benzene and toluene: liquid hydrocarbons, often used as solvents

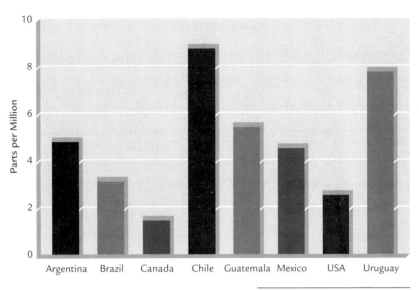

FIGURE 4.11

*DDT in human milkfat:
the Americas*

Source: Adapted from Jensen (1991)

DDT. There were no adverse effects on growth or general health of the children, who were followed from birth to 5 years of age (Gladen 1988).

Maternal exposure to heavy metals such as lead, mercury, cadmium, and arsenic can result in contamination of breast milk. Usually only trace amounts of these metals find their way into human milk, and even in regions where the environment contains large amounts of these metals, levels in human milk tend to be much lower than levels in water or cow's milk (Dabeka 1986). Infant poisoning from lead transferred through breast milk has been reported, although maternal exposure must be high for this to occur (maternal serum lead levels greater than 40 mg/dl) (Institute of Medicine 1991).

Fortunately, most nursing mothers are at minimal risk for contamination of their milk. However, risk is higher in certain areas, particularly in regions near contaminated water, agricultural areas with regular aerial spraying of crops, and industrial regions where use of chemicals is heavy. Lactating women can reduce their exposure to environmental contaminants by following the guidelines in Table 4.4. Concerned mothers should contact their local Department of Health or the Human Effects Monitoring Branch of the Environmental Protection Agency for more information.

TABLE 4.4

*Guidelines to reduce
exposure to chemicals
and pesticides
during lactation*

- Wash thoroughly or peel vegetables and fruits before eating.
- Discard the outside leaves of foods such as lettuce and cabbage.
- Avoid fish from waters known to be high in contaminants.
- Eat low-fat foods and trim the fat from meats before cooking.

RADIOACTIVITY

In 1986 an explosion at the Chernobyl nuclear power plant in Russia released large amounts of radioactivity into the environment, and fallout occurred throughout Western Europe. In many areas, levels of radioactivity in cow's milk and leafy vegetables increased markedly, and pregnant and lactating women were advised to avoid these foods.

Although increased levels of radioactivity were found in breastmilk samples from women throughout Europe, the amounts were very small. For example, in Austria, levels in human milk were less than one tenth those in cow's milk (Lindemann and Christensen 1987). In reports from Italy, Austria, Norway, and Sweden, scientists concluded that concentrations of radioactivity were so low they did not pose a threat to nursing infants (Lindemann and Christensen 1987).

VIRUSES AND BACTERIA

Ordinarily, human milk provides substantial immunological protection to the newborn. However, if the nursing mother has certain serious infections, viruses and bacteria can be transmitted to the infant in the breast milk. For example, newborn infection with tuberculosis, hepatitis B, herpes simplex virus, and the human immunodeficiency virus (HIV) can occur via human milk (Goldfarb 1993).

MILK VOLUME

MEASURING MILK OUTPUT

Once lactation is established, extraction of milk from the breast by the suckling infant is the primary stimulus for milk production, and the mammary gland carefully balances production with infant demand. Therefore, for most women who breast-feed, the volume of milk produced during lactation is determined by infant need. Unless extra milk is removed from the breast (and not consumed by the newborn) or the infant regurgitates milk, milk intake by the infant will equal milk production by the mother.

The most commonly used method of measuring milk intake and milk production is test weighing of the infant: the infant is carefully weighed before and after each feeding, and the amount of weight gained is used to determine milk intake.

Source: Adapted from
Neville et al. (1988)

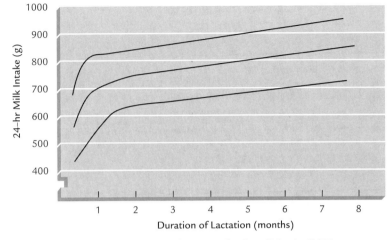

FIGURE 4.12

*Milk intake during
first eight months
of lactation*

The lines show the mean value for milk intake (±SD).

Milk Output during Lactation

The milk intake of healthy, exclusively breast-feeding infants from several studies around the world is shown in Figure 4.12. Once lactation is established, milk intake in the first four to five months averages 750–800 g/day, but there is wide variability, with ranges reported from 400 and 1200 g/day (Neville et al. 1988).

After the first 6 months, as solid foods are introduced to the breast-feeding infant, milk consumption typically falls. A recent study measured the milk intake of U.S. infants who were breast-fed for 12 months, and introduced to solid foods between 4 and 7 months. Milk intake averaged 770 g/day at 6 months, 635 g/day at 9 months, and 445 g/day at 12 months (Dewey et al. 1992).

Breast-feeding mothers have the potential to produce far more milk than is needed to meet usual infant demands. For example, mothers who breast-feed twins or triplets can produce 2000–3000 g milk/day. In women with single infants, using a **breast pump** to extract additional milk after feedings can increase milk production by 20 to 40% (Dewey and Lonnerdal 1986).

breast pump:
a mechanical aid used
to express milk from
the breast

MATERNAL METABOLISM DURING LACTATION

SUBSTRATES FOR MILK SYNTHESIS

During lactation, profound alterations in maternal metabolism enable the nursing mother to produce large quantities of milk for her infant while supporting her own nutritional needs. However, the metabolism of nutrients during lactation is altered to give priority to milk production: energy and nutrients from the diet and in maternal stores are channeled into the mammary gland. If maternal nutritional status is marginal, available substrates will be used for milk synthesis, and maternal stores will be depleted.

During lactation, the secretory cells of the mammary epithelium require an abundant and continuous supply of triglycerides, amino acids, glucose, and other substrates for milk synthesis.

Lipid and Protein Metabolism

Lactation alters lipid and lipoprotein metabolism. Cholesterol and high-density lipoprotein levels tend to be higher in lactating women, while triglyceride levels are lower. See Table 4.5. High prolactin levels during lactation modify fat metabolism, increasing mobilization of fats from adipose tissue and directing dietary lipids toward the mammary gland (Willamson and Lund 1994).

Protein metabolism is also altered during lactation. Compared to non-lactating women, well-nourished women who are breast-feeding tend to have lower rates of protein turnover and lower rates of muscle protein breakdown, as measured by urinary 3-methylhistidine excretion (Motil et al. 1990).

Type of Lipoprotein	Mean Value, mg/dl	
	Lactating Women	Nonlactating Women
Total		
Triglycerides*	92	112
Cholesterol*	207	188
Phospholipids	227	217
Low-density lipoprotein		
Triglycerides	26	24
Cholesterol	129	121
Phospholipids	70	70
High-density lipoprotein		
Triglycerides*	12	10
Cholesterol*	65	51
Phospholipids*	141	123

*Significant difference (p < .05) between lactating and nonlactating subjects.

Adapted from R. H. Knopp et al., "Effect of Postpartum Lactation on Lipoprotein Lipids and Apoproteins" *J Clin Endocrinol Metab* 60 (1985): 542–47

TABLE 4.5

Lipids and lipoprotein levels during lactation, compared with nonlactating women six weeks postpartum

Micronutrient Metabolism

As maternal requirements for most micronutrients increase during lactation, enhanced absorption from dietary sources occurs. For example, calcium absorption from dietary sources increases during lactation (King et al. 1994). Lactational amenorrhea reduces maternal losses of iron, preserving maternal iron stores for use in milk synthesis. Other alterations in micronutrient metabolism occur; for example, between weeks 1 and 20 of lactation, zinc concentrations in blood steadily rise, while copper levels fall (Van der Elst et al. 1986).

WEIGHT LOSS AND ENERGY EXPENDITURE

Patterns of Weight Loss during Lactation

Most women are concerned about returning to their prepregnancy weight during the postpartum period. Breast-feeding can help women lose weight because milk production requires significant energy (about 500–650 kcal/day) and, if the mother is in negative energy balance during lactation, breast-feeding can facilitate loss of body fat gained during gestation. However, breast-feeding is no guarantee that weight will be lost effortlessly postpartum.

In general, a woman who gains about 13 kg during pregnancy loses about 5 kg during delivery. About another 2 kg will be lost during the first week, mostly through diuresis of body water. During lactation, most women continue to lose weight, as body fat accumulated during gestation is mobilized to supply the needs of the breast-feeding infant.

During the first four to six months of lactation, most women are in negative energy balance. In well-nourished women, the reported range of mean daily energy deficit is between 110 and 343 kcal/day (Institute of Medicine 1991). These deficits would be expected to produce a weight loss of between 6 and 16 pounds over six months, and most women do gradually lose weight during lactation.

Average weight loss during the first six months postpartum in breast-feeding women is 0.6–0.8 kg per month, or a total of between 9 and 10 pounds. For many women who breast-feed past six months, weight loss continues during months 6 through 12, but typically at a slower rate (0.2–0.4 kg per month) than in the first six months (Heinig et al. 1990).

Among individual women, weight changes postpartum vary widely. Studies have shown that the intensity and duration of breast-feeding does not predict maternal weight loss (Ohlin and Rossner 1990), and many women gain weight during lactation.

Weight Loss: Breast-feeding versus Formula Feeding

Do breast-feeding women lose weight more readily than women who feed their babies formula? A recent study examined patterns of weight loss and changes in body composition in 56 new mothers. The women were divided into three groups: exclusively breast-feeding, formula feeding, and combination breast and formula feeding. All three groups lost weight (about 8 kg) in the six months after delivery. There were no significant differences in weight loss or loss of body fat between the groups (Brewer et al. 1989).

Weight loss in the postpartum period is influenced by several factors. Mothers who have gained more weight during pregnancy tend to lose more postpartum, and in general, multiparous women lose less of their gestational weight gains than first-time mothers (Institute of Medicine 1991).

Changes in Body Fat

Women who lose weight during lactation metabolize fat stores to provide energy for milk synthesis. One study, using skinfold measurements to estimate changes in body fat percentage during lactation, found that suprailiac and subscapular measurements typically decrease during the first four to six months postpartum—indicating body fat percentage is decreasing as fat is mobilized to meet the energy needs of lactation (Butte et al. 1984). In a second study, weight and triceps skinfold thickness were measured for 2 years postpartum in two groups of women: those who breastfed for at least 12 months; and those who breastfed for less than 3 months. Weight loss and loss in skinfold thickness were significantly greater in those who breastfed for at least 12 months (Dewey et al. 1993). See Figure 4.13. In lactating women, fat is more readily mobilized from femoral adipose tissue (in the hips and thighs), compared to nonlactating women (Stoneham et al 1988).

Energy Expenditure during Lactation

In most women, weight loss of 0.6–0.8 kg per month (beginning in the second postpartum month) does not affect milk production. Women who are significantly overweight can lose up to 2 kg per month without adversely affecting lactation, but more rapid weight loss can reduce a mother's ability to produce adequate milk (Institute of Medicine 1991). All lactating mothers should maintain adequate daily energy intake (a minimum of 1500 calories should be consumed daily) and avoid strict diets and weight loss medications.

Many new mothers respond to the increased demands of lactation by decreasing energy expenditure. In the United States, compared to the

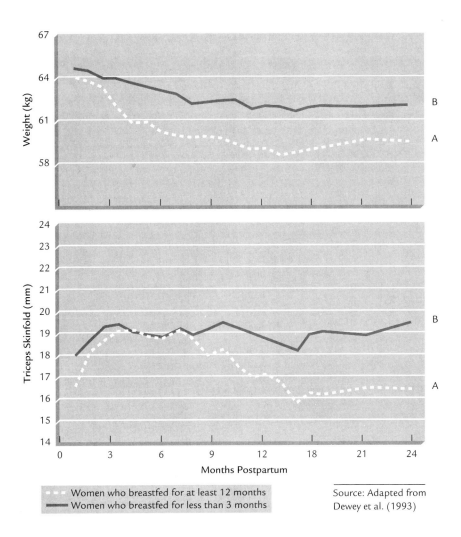

FIGURE 4.13

Weight loss and skinfold measurements: Breast-feeding for at least 12 vs. less than 3 months

Women who breastfed for at least 12 months
Women who breastfed for less than 3 months

Source: Adapted from Dewey et al. (1993)

energy expenditure of light to moderately active nonlactating women (2200 kcal/day), the total energy expenditure of lactating women (not including milk production) averages about 1800–1900 kcal/day (Lovelady et al. 1990).

Several energy-sparing adaptations have been reported in lactating women. Lactating women have lower basal metabolic rates than nonlactating women (Lovelady et al. 1990), (Blackburn and Calloway 1985). The **thermic effect of food** is reduced during lactation, resulting in a small reduction in daily energy expenditure (Illingworth et al. 1986). See Figure 4.14.

Although many women reduce their activity during lactation and thereby reduce energy expenditure, the energy expenditure varies widely among nursing women. For some women, activity may actually increase, as the care of the infant adds to the demands of caring for older children and energy spent at work. Many women pursue exercise programs postpartum to increase energy expenditure.

thermic effect of food: energy required to digest food and to absorb, metabolize, and store nutrients

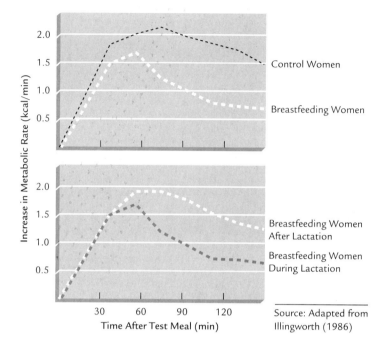

FIGURE 4.14
Reduced thermic effect of food during lactation

Source: Adapted from Illingworth (1986)

DIETARY SURVEYS OF LACTATING MOTHERS

A limited number of studies have examined the dietary intake and nutritional status of lactating women. The women surveyed have been mostly well nourished, college-educated, and Caucasian, and most surveys have looked at only the first few months of lactation.

Although energy intakes are often below recommended amounts, in general, these surveys have found intake of protein, calcium, iron, vitamins A and C, thiamin, riboflavin, and niacin to be at least 80% of the RDAs for lactation, and in many cases, intakes exceed the RDAs (Institute of Medicine 1991).

The nutrient content of diets eaten by lactating women differs from those typically eaten by adult women in the United States. For example, the densities of calcium and vitamin A are 40 to 50% higher in the diets of nursing mothers compared with non-nursing women (Institute of Medicine 1991).

There could be several reasons for this difference. Mothers who choose to breast-feed could be women eating diets rich in these nutrients; or women may eat more foods rich in calcium and vitamin A because they are breast-feeding. Either way, in these relatively affluent women, the dietary intake of most nutrients appeared to be adequate.

In contrast, little information is available on the intakes of lactating women from other socioeconomic levels or minority groups. Data that are available indicate dietary intakes of many nursing mothers may be suboptimal. For example, in a nationally representative study, data from the 1977–78 Nationwide Food Consumption Survey were analyzed to determine the dietary intakes of lactating women in the United States. Nutrients for which dietary intake of nursing women was most likely to be low were vitamins A and C, calcium, magnesium, and iron (Krebs-Smith and Clark 1989), (Sneed et al. 1981).

In a survey of American Indians, intake of calcium, magnesium, zinc, and vitamins D, E, and folate in nursing mothers was low (Butte and Calloway 1981).

Although knowledge of the dietary intakes and nutritional profile of lactating women is incomplete, dietary surveys of women in their childbearing years have indicated certain groups may be at high risk for nutrient inadequacy. Diets of adolescent women typically contain less iron and vitamin A than is recommended during lactation. Also, diets of Black women, particularly those living below the poverty line, are often low in calcium, magnesium, iron, and vitamin A. Vegetarian women tend to have low intakes of vitamin B_{12}, iron, and zinc. Breast-feeding women from these groups may be at high risk for nutrient inadequacy.

MATERNAL NUTRITIONAL NEEDS DURING LACTATION

NUTRITIONAL DEMANDS OF LACTATION

Although the nutritional demands of pregnancy have traditionally attracted more attention than nutrition during lactation and breast-feeding, the stress of breast-feeding on the mother is substantial. A healthy infant doubles its weight in the first four to six months postpartum, and a mother who chooses to exclusively breast-feed her infant must provide all the energy, protein, and micronutrients to support this rapid growth.

With a few exceptions, maternal needs during breast-feeding are greater than those during pregnancy and, in fact, are typically higher than in any other period of a woman's life. Several factors determine maternal nutritional requirements during lactation: the volume and composition of milk produced, maternal nutritional status when beginning lactation, and underlying maternal needs.

Synthesis and secretion of milk by the nursing mother require ample energy, protein, and micronutrients. These nutrients are normally supplied by increased food intake but are also drawn from maternal stores—particularly when dietary intakes are inadequate.

RECOMMENDED DIETARY ALLOWANCE FOR LACTATING MOTHERS

Table 4.6 gives the RDAs for lactating women nursing a single infant (NAS 1989). Current RDAs are presented as absolute amounts (not increments to be added to the allowances for nonlactating women, as in previous editions) and are divided into recommendations for 0–6 months and 6–12 months postpartum. The differences between the RDAs for the two periods reflect the differences in milk production during early and late lactation—which are estimated to be 750 and 600 ml/day, respectively. The RDAs for lactating women are aimed at maintaining adequate amounts of nutrients in breast milk and preventing depletion in the mother (NAS 1989). Note the difference in RDAs for vitamins A, D, and C, calcium, and zinc in lactating versus nonlactating women, shown in Figure 4.15.

Milk production among lactating women varies widely, and the RDAs reflect average values for milk output (an individual's needs may be significantly higher or lower than the RDAs depending on the intensity of breast-feeding). For example, for a woman who introduces substantial

	First 6 Months	Second 6 Months
Energy (kcal)*	+500	+500
Protein (g)	65	62
Vitamin A (retinol equivalents)	1300	1200
Vitamin D (µg)	10	10
Vitamin E (α- tocopherol equivalents)	12	11
Ascorbic acid (mg)	95	90
Folacin (µg)	280	260
Niacin (mg)	20	20
Riboflavin (mg)	1.8	1.7
Thiamin (mg)	1.6	1.6
Vitamin B_6 (mg)	2.1	2.1
Vitamin B_{12} (µg)	2.6	2.6
Calcium (mg)	1200	1200
Phosphorus (mg)	1200	1200
Iodine (µg)	200	200
Iron (mg)	15	15
Magnesium (mg)	355	340
Zinc (mg)	19	16

*Added to usual recommended daily energy intake for adolescent and adult females (2200 kcal).

Adapted from National Academy of Sciences, Food and Nutrition Board of the National Research Council, "Recommended Dietary Allowances," 10th ed. (Washington D.C.: U.S. Government Printing Office, 1989).

TABLE 4.6
RDAs for lactation

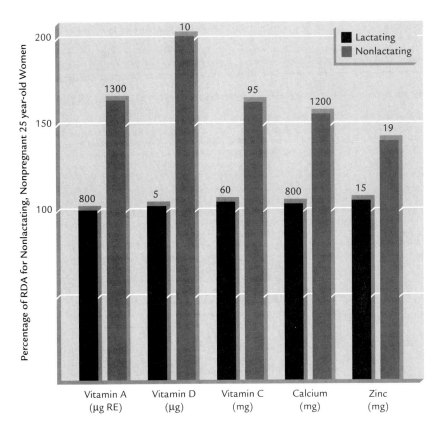

amounts of solid foods to her infant early in lactation and continues limited breast-feeding, the RDAs may overestimate maternal nutrient needs. The following discussion of nutrient needs during lactation is based on milk outputs of 750 ml/day over the first 6 months postpartum and 600 ml/day from 6 to 12 months.

RDAs for Energy

The average energy content of human milk is 65–70 kcal/100 ml (WHO 1985), and it is estimated that the efficiency at which maternal energy (from dietary sources or body stores) is converted into milk energy is about 80% (WHO 1985). The nursing mother therefore requires about 85 kcal for every 100 ml of milk she produces. Thus, the average nursing mother needs an extra 640 kcal/day during the first 6 months and 510 kcal/day during the second six months of lactation.

A portion of this extra energy for milk production can be supplied by fat stored during pregnancy: women who gain about 12 kg during gestation store 2–3 kg of adipose tissue (discussed in Chapter 3). Mobilization of these stores could provide 100–150 kcal/day over six months of breast-feeding.

The RDAs for energy are calculated assuming normal postpartum weight loss and utilization of body fat will provide 150 kcal/day over the first six months of lactation. Thus, the RDAs call for an extra 500 kcal/day throughout lactation (NAS 1989). Women who have not gained sufficient weight during pregnancy and have not stored sufficient body fat, and women who are thin (weight below standards for height and age), should consume an additional 650 kcal during the first six months of lactation (NAS 1989).

RDA for Protein

The average protein content of human milk is 1.3 g/100 ml during the first few weeks postpartum and falls to 1.1 g/100 ml in mature milk. The conversion of maternal protein to milk protein is estimated to be 70% efficient. Using these estimates, the additional protein requirement during lactation is calculated to be 15 g/day during the first six months postpartum, and 12 g/day thereafter (NAS 1989).

RDAs for Fat-Soluble Vitamins

Vitamin A The vitamin A content of breast milk from well-nourished mothers ranges from 40–70 μg/100 ml (Wallingford and Underwood 1986), and daily losses of vitamin A in milk during the first six months postpartum are between 300–525 μg/day. The maternal liver normally contains about 200 mg of stored vitamin A. If all of the extra vitamin A in milk came from maternal stores, six months of breast-feeding would deplete liver stores by 25 to 50%. Thus, to conserve maternal stores, the RDA for lactating women for the first six months postpartum is set at 500 RE. As milk output declines, the requirement is reduced to 400 RE from six to twelve months postpartum (NAS 1989).

Vitamin D Breast milk contains small amounts of vitamin D (average values are about .05–.075 μg/100 ml), and ample vitamin D during lactation is important to maintain calcium balance. Calcium requirements of lactation are high, and to minimize potential mineral loss from the maternal skeleton, optimal calcium absorption from dietary sources is important. The RDA for vitamin D for all lactating women is set at 10 μg/day for the duration of lactation—an increment of 5 μg/day for women beyond age 24 (NAS 1989).

Vitamins E and K About 3 mg of α-tocopherol is secreted into the milk each day during the first six months of lactation, and 2–2.5 mg is secreted from six to twelve months. Considering the incomplete absorption of vitamin E from the diet, the increment of intake for vitamin E during lactation is set at 4 mg for the first six months and 3 mg thereafter (NAS 1989). Because the extra requirement for vitamin K during lactation is low, and the vitamin K intake from the typical U.S. diet normally exceeds the RDA for adult women, no additional vitamin K is recommended during lactation (NAS 1989).

RDAs for Water-Soluble Vitamins

Vitamin C The daily increment for vitamin C is set at 35 mg and 30 mg during the first and second six months of lactation, respectively. These recommendations are based on average maternal losses of 22 mg/day in milk in the first six months postpartum and 18 mg/day from six to twelve months, and they assume the efficiency of intestinal absorption for vitamin C is 85% (NAS 1989).

Thiamin, Riboflavin, and Niacin Because of losses into milk and increased energy consumption during lactation, maternal requirements for thiamin, riboflavin, and niacin increase 30 to 50% (NAS 1989). Losses of thiamin into milk average 0.2 mg/day, and losses of niacin average 1.0–1.3 mg/day. Riboflavin secretion into milk is approximately 0.26 mg/day during the first six months of lactation and 0.21 mg/day thereafter. Considering extra energy requirements and the absorption efficiency for these vitamins, the RDAs for lactation are set at 1.6 mg for thiamin and 20 mg for niacin. The riboflavin allowance is for 1.8 mg and 1.7 mg during the first and second six months postpartum, respectively (NAS 1989).

Vitamin B_6 Vitamin B_6 plays a central role in the metabolism of amino acids. The extra protein intake during lactation should be accompanied by additional vitamin B_6—an increment of 0.016 mg per extra gram of protein is considered optimal. Also, losses of the vitamin in milk average 0.15 mg/day, once lactation is established. To cover these needs, the RDA is set at 2.1 mg/day during lactation, an increment of 0.5 mg/day over values for nonlactating women (NAS 1989).

Folate Folate reserves in many women are depleted during pregnancy, and poor folate status postpartum is common. Daily losses of folate in milk are estimated to be approximately 50 µg/day and 40 µg/day during the first and second six months of lactation. Absorption of folate from dietary sources is estimated to be about 50%. Thus, the RDA calls for 280 µg/day (an increment of 100 µg/day) during the first six months of lactation and 260 µg/day (an increment of 80 µg/day) thereafter (NAS 1989).

However, because of links between folate intake and neural tube defects (discussed in Chapter 3), more recent guidelines from the U.S. Public Health Service recommend 400 µg/day of folate for all women capable of becoming pregnant (see discussion in Chapter 2). Although lactational amenorrhea provides some contraceptive protection during lactation, many women become pregnant while breast-feeding. Thus, lactating women should consume 400 µg of folate/day. (Center for Disease Control 1992).

Vitamin B_{12} Based on an average level in milk of 0.06 µg/100 ml, daily losses of vitamin B_{12} in milk are estimated to be about 0.4–0.5 µg. The absorption of vitamin B_{12} from normal diets is approximately 70%, thus the current RDA calls for an increment of 0.6 µg/day throughout lactation (NAS 1989).

Biotin and Pantothenic Acid   The RDAs for biotin and pantothenic acid for lactating women are identical to those for nonlactating women (NAS 1989).

RDAs for Major Minerals

Calcium   Daily losses of calcium into milk are approximately 250–320 mg/day. In non-lactating adult women, at dietary intakes of around 800 mg/day, calcium absorption averages 15 to 20%, and studies suggest calcium absorption from dietary sources increases during lactation (King et al. 1994). To cover increased needs during lactation, the RDA for calcium and phosphorus is set at 1200 mg/day (NAS 1989).

Particularly if dairy products are not a significant part of the diet, many lactating women do not obtain the RDA for calcium. Because calcium levels in milk are maintained even if maternal intake is low, calcium from the skeleton may be mobilized during periods of low intake.

Inadequate intake during lactation may have adverse affects on bone density, particularly in younger women who need calcium for bone mineralization. Although rates of bone mineralization are highest in the adolescent years (discussed in Chapter 8), the calcium content of the skeleton continues to increase into the late 20s.

Magnesium   The concentration of magnesium in human milk from well-nourished mothers is 3–4 mg/100 ml, so during the first six months of lactation about 25 mg/day would be lost in milk, and about 20 mg/day would be lost thereafter. Assuming absorption of magnesium from the diet averages 50%, the current RDA calls for an increment of 75 mg/day and 60 mg/day during the first and second six months of lactation, respectively (NAS 1989).

RDAs for Trace Minerals

Iron   Daily losses of iron in breast milk are estimated to be about 0.3 mg. Many women experience a period of lactational amenorrhea during breastfeeding and avoid the usual monthly losses of iron in the menses (about 0.5 mg/day when averaged over one month) (Habicht et al. 1985). Therefore, the needs of lactating, nonmenstruating women are not substantially different from those of nonlactating, menstruating women. For this reason, there is no increase in the current RDA for iron intake during lactation (NAS 1989). Lactating women who resume their menses will require significantly higher amounts than the RDA (an additional 0.5 mg/day).

Zinc   Daily zinc losses into milk are 1.2 mg during the first six months of lactation, and 0.6 thereafter. The RDA is calculated assuming zinc

absorption from the diet is 20%; thus, recommendations are for increments of 7 and 4 mg/day during the first and second six months of
breast-feeding (NAS 1989).

Selenium. Although the level of selenium in human milk varies considerably in different populations around the world, and loss of selenium
in milk in women in the U.S. is typically about 13 µg/day. Assuming
absorption of dietary selenium is approximately 80%, the RDA throughout lactation is set at 20 µg/day (NAS 1989).

Manganese, Iodine and Other Trace Elements   Lactation is not
thought to increase manganese requirements appreciably. Manganese
losses into milk are very low, so the current RDA does not call for additional manganese during breast-feeding. The current RDA for iodine
during lactation calls for an extra 50 µg/day. This recommendation is
based on the estimated requirement of the infant for iodine during the
first year. The RDAs for copper, fluoride, chromium and molybdenum
during lactation are the same as those for adults (NAS 1989).

RDA for Water

Human milk is 87% water. Based on an average milk production of 750
ml and 600 ml/day during the first and second six months of lactation,
nursing mothers require about 650 ml of extra fluid each day during the
first six months postpartum and about 530 ml thereafter (NAS 1989).

RDAs for Electrolytes

Requirements for sodium and potassium increase during lactation.
Sodium losses into milk average about 135 mg/day, while potassium
losses are about 375 mg/day—requirements easily met by current adult
intakes of sodium and potassium. Therefore, no additional intake of
these electrolytes is recommended during lactation (NAS 1989).

EFFECTS OF MATERNAL NUTRITION ON MILK VOLUME AND COMPOSITION

MILK VOLUME: BODY FAT AND ENERGY BALANCE

During pregnancy, well-nourished women store significant energy as
adipose tissue—stores they later draw upon to provide energy for milk
production. However, in most women, the amount of body fat during
lactation does not significantly affect milk volume (overweight and
normal-weight women do not produce more milk than thinner

women). Studies in industrialized countries have found that maternal weight for height and amount of body fat, when measured in either the prenatal or postpartum period, have little effect on milk volume. Only when maternal malnutrition becomes severe is milk production impaired. Several studies in developing countries have found that very low maternal weight for height is associated with lower infant milk intakes (Brown and Dewey 1992).

Also, neither higher energy intake nor positive energy balance during lactation greatly influences milk volume. When well-nourished women are compared with women who are marginally malnourished, despite substantial differences in energy and nutrient intake, milk outputs during lactation are similar. In a study of marginally malnourished mothers in Bangladesh, there was no association between maternal weight or body fat and milk volume. However, women with greater body fat produced milk with higher fat and total energy content than thinner women (Brown et al. 1986). Several reports have suggested that higher circulating prolactin levels allow marginally malnourished women to maintain milk synthesis (Institute of Medicine 1991).

Although milk output is generally maintained even if maternal food intake is restricted, studies suggest that severe food restriction can decrease milk production. In a study in lactating baboons, when ad libitum diets were cut 20%, there was no effect on milk volume. But when diets were reduced 40%, milk output fell 20% (Roberts et al. 1985).

One study examined the effects of energy restriction for one week among previously well-nourished mothers. In women who substantially reduced their energy intake but maintained intakes of at least 1500 kcal/day, there was no effect on milk volume. However, intakes below 1500 kcal/day reduced prolactin levels and milk production by 15% (Strode et al. 1986).

MILK VOLUME: PROTEIN INTAKE

Studies in developing countries have suggested that protein supplementation in malnourished mothers increases milk output (Institute of Medicine 1991). In well-nourished women, increasing protein intake for short periods—from four to ten days—does not affect milk volume (Forsum and Lonnerdal 1980), (Motil et al. 1986).

MILK VOLUME: FLUID INTAKE

Women who are breast-feeding should obtain ample fluids; however, consuming fluids at levels higher than daily needs, or in excess of thirst, does not increase milk volume. Women who are breast-feeding can tolerate a significant amount of water restriction without affecting milk volume. In a group of women who fasted and drank no fluids for 12 to 14 hours and continued to breastfeed their infants, although signs of mild dehydration were noted, there was no effect on milk output (Prentice et al. 1984).

MILK VOLUME: AGE AND PARITY

Mothers of all ages can successfully breast-feed, as maternal age has little effect on milk production. Most well-nourished adolescent mothers can produce ample milk, and among women between the ages of 20 and 40, there is no association between age and infant milk intake.

Although milk production in the first week postpartum may be higher in multiparous women, once lactation is established, most studies have shown no associations between parity and milk production among well-fed women. In highly multiparous Gambian women, milk production diminished only in women who had breastfed seven to ten or more children (Prentice 1986).

MATERNAL SUPPLEMENTATION IN THE DEVELOPING WORLD

Studies in developing countries examining the effect of food supplementation on milk production in malnourished mothers have found that programs aiming to "feed the nursing mother and thereby the infant" have had mixed results. In a study in Gambia, 130 nursing women were given a supplement containing 725 kcal/day and 57 g of protein. There was no effect on milk production at any stage in lactation (Prentice et al. 1983). In contrast, a study in Burma found that nursing women supplemented with 900 kcal and 39 g protein per day for two weeks significantly increased milk production from 660 to 790 g/day (Naing and Oo 1987).

Although providing malnourished women who are breast-feeding with food supplements has had inconsistent effects on milk volume, supplementation can clearly provide significant maternal health benefits and improve milk composition.

MILK COMPOSITION

In general, the nutritional composition of human milk is stable. The mammary epithelium maintains a steady secretion of many nutrients into the milk despite fluctuations in maternal intake. Most of the energy, proteins, carbohydrates, electrolytes, and minerals in milk are present in adequate amounts even when maternal dietary supply is limited (Institute of Medicine 1991).

During periods of inadequate intake, the concentrations of these constituents are maintained in the milk at the expense of maternal reserves, thus protecting the nutritional status of the breast-feeding infant. The demand for energy and nutrients is very high during early infancy; this mechanism ensures the steady delivery of nutrients during this critical period of development.

However, levels of other nutrients do respond to changes in maternal diet or nutritional status. Concentrations of the vitamins, several minerals, and the types of fat in milk can be markedly altered by maternal nutrition (see Table 4.7).

Nutrient or Nutrient Class	Intake on Milk Composition
Macronutrients	
Proteins	+/0
Lipids	+*
Lactose	0
Minerals	
Calcium	0
Phosphorus	0
Magnesium	0
Sodium	0
Potassium	0
Chlorine	0
Iron	0
Copper	0
Zinc	+
Manganese	+
Selenium	+
Iodine	+
Fluoride	+
Vitamins	
Vitamin C	+
Thiamin	+
Riboflavin	+
Niacin	+
Pantothenic acid	+
Vitamin B_6	+
Biotin	+
Folate	+
Vitamin B_{12}	+
Vitamin A	+
Vitamin D	+
Vitamin E	+
Vitamin K	+

+   Denotes a positive effect of intake on nutrient content of milk. The magnitude of the effect varies widely among nutrients.

0   Denotes no known effect of intake on nutrient content of milk.

*   Effect appears to be on type of fatty acids present but not on total content of triglycerides or cholesterol in the milk.

Adapted from Institute of Medicine, Subcommittee on Nutrition during Lactation, *Nutrition during Lactation* (Washington D.C.: National Academy Press, 1991).

TABLE 4.7
Influence of maternal intake in nutrient content of milk

PROTEIN AND LACTOSE

Although there have been a few reports of low concentrations of protein in milk from undernourished women in developing countries, there is no firm evidence that maternal body size and composition or maternal diet influence the total concentration of milk protein (Institute of Medicine 1991). The concentration of lactose in human milk is not influenced by dietary factors.

LIPIDS

In well-nourished women, changes in maternal fat intake generally have little effect on the total quantity of fat in the milk. Mothers who consume low-fat diets do not secrete less fat in their breast milk, but both the nature of the fat eaten by the nursing mother and her energy balance influence the composition of the milk fat.

Many of the fatty acids present in the milk are derived directly from the maternal diet. For example, although the total fat content is similar, the types of fatty acids in the milk from vegetarian and nonvegetarian mothers reflect their different diets (Jensen 1989): vegetarians produce milk with greater amounts of the fatty acids present in plant foods (such as linoleic acid); nonvegetarians produce milk with more animal fatty acids (such as palmitic and stearic acid). (See Table 4.8). Women who consume more partially hydrogenated fats and oils secrete greater amounts of trans fatty acids in their milk (Chappell et al. 1985).

The energy balance of the lactating mother also influences the source of the fatty acids in her milk. If the mother is in energy balance, about 30% of the fatty acids found in milk come directly from the diet (Hachey et al. 1989). If she is in negative balance, more fatty acids will originate from adipose stores.

	Vegans	Nonvegetarians
Lauric ($C_{12.0}$)	39	33
Myristic ($C_{14.0}$)	68	80
Palmitic ($C_{16.0}$)	166	276
Stearic ($C_{18.0}$)	52	108
Oleic ($C_{18.1}$)	313	353
Linoleic ($C_{18.2}$)	317	69
Linolenic ($C_{18.3}$)	15	8

Mean values expressed as milligrams per gram total methyl esters.

Adapted from T. A. B. Sanders et al. *Am J Clin Nutr* 31 (1978):805.

TABLE 4.8
Concentrations of fatty acids in milk in vegetarians and nonvegetarians

Concentrations of cholesterol in breast milk are not influenced by maternal diet. Women who consume low-cholesterol diets do not secrete less cholesterol in their milk, as the cholesterol level remains constant at 10–15 mg/100 ml.

In undernourished women, milk-fat concentrations may be related to the percentage of maternal body fat. In studies in Gambia and Bangladesh, women with very low body fat produced milk with less milk fat, and concentrations of fat in the milk decreased as lactation progressed (Institute of Medicine 1991).

MILK COMPOSITION: FAT-SOLUBLE VITAMINS

Levels of the fat-soluble vitamins in milk are sensitive to maternal nutritional status, and levels generally fall during maternal deficiency.

Vitamin A

Lactating mothers with poor vitamin A stores secrete less vitamin A in their milk. In a study of Ethiopian women, poorly nourished women had lower levels of retinyl esters in their milk compared with well-nourished women (Gebre-Medhin et al. 1976). In nursing women with poor vitamin A status, milk levels of vitamin A increase if dietary intake from food or supplements is increased. In well-nourished populations, results of maternal supplementation with vitamin A have been equivocal: several studies have found supplementation increased the amount of vitamin A in the milk; other studies have found no effects (Institute of Medicine 1991).

Vitamin D

In women deficient in vitamin D (from little sunlight exposure and poor dietary supply), vitamin D levels in breast milk fall markedly, sometimes to undetectable levels (Hollis et al. 1983). In such cases, exposure to ultraviolet light and supplementation promptly increase the amount of vitamin D in breast milk (Greer et al., 1984). See Figure 4.16.

Vitamins E and K

Levels of vitamin E and vitamin K in breast milk are responsive to maternal dietary intake. In one study, nursing women with low vitamin K intakes were given 20 mg of phylloquinone; concentrations of vitamin K in breast milk increased significantly (Von Kries et al. 1988). High maternal intake of vitamin E can substantially increase amounts secreted in breast milk (Anderson and Pittard 1985).

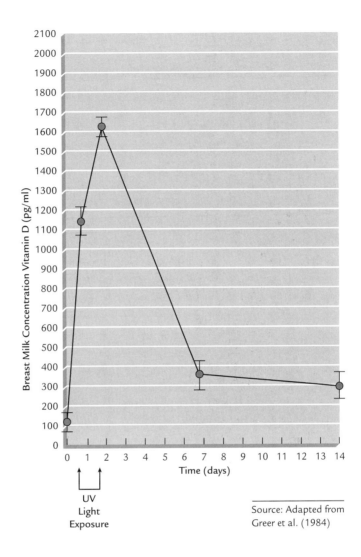

FIGURE 4.16

UV exposure in vitamin D-deficient lactating mothers increases vitamin D in breast milk.

UV
Light
Exposure

Source: Adapted from
Greer et al. (1984)

MILK COMPOSITION: WATER-SOLUBLE VITAMINS

In general, levels of water-soluble vitamins in breast milk are sensitive to changes in maternal status. Maternal deficiency will reduce the amounts in the milk, while supplementation will increase levels—but generally only up to a certain point. After this point, even with increased intake, the amount secreted in milk plateaus.

Vitamin C

As maternal intake of vitamin C increases, greater amounts of the vitamin appear in the milk. However, this effect plateaus at intakes of 100 mg/day, as the vitamin C content of the milk levels off at 5–6 mg/100 ml.

Thiamin

Mothers with beriberi produce milk with very low levels of thiamin, and their infants develop beriberi by two to four weeks postpartum. Supplementation with thiamin can readily increase levels in breast milk up to a plateau value of around 20 mg/100 ml (Institute of Medicine 1991).

Vitamin B$_6$

Dietary intake has a marked effect on levels of vitamin B$_6$ in human milk. A study compared vitamin B$_6$ levels in milk from two groups of nursing mothers: one group received a 2.5 mg supplement each day, and one group received no supplementation. Levels of vitamin B$_6$ in breast milk from the first group were more than twice those in the second group (Styslinger and Kirksey 1985).

Vitamin B$_{12}$

Researchers have found extremely low levels of vitamin B$_{12}$ in the milk of several groups of women with low intakes of vitamin B$_{12}$—malnourished women, complete vegetarians, and women with pernicious anemia (Gambon et al. 1986).

Folate

Folate is unique among the vitamins in that adequate levels in milk are generally maintained in the face of maternal deficiency. In nursing women with marginal dietary intakes of folate, levels in milk gradually increase over the course of lactation even as blood levels of the vitamin fall (O'Connor 1994). During lactation, available folate is preferentially directed toward the mammary gland, but this mechanism may not be adequate to maintain milk folate levels if maternal deficiency is extreme. Women with megaloblastic anemia due to severe folate deficiency have low levels of folate in their breast milk (less than 0.5 μg/100 ml); supplementation returns levels in the milk to near normal values (Matz et al. 1968) (Cooperman et al. 1982). See Figure 4.17.

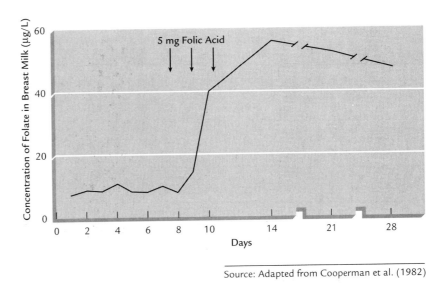

FIGURE 4.17

Breast milk folate: Response to maternal supplementation with folate in folate-deficient mothers

Source: Adapted from Cooperman et al. (1982)

Other Water-Soluble Vitamins

Milk content of the remainder of the water-soluble vitamins, such as biotin, pantothenic acid, riboflavin, and niacin, are correlated with maternal intake, and supplementation markedly increases amounts secreted in breast milk. Low concentrations of riboflavin and niacin in milk rise promptly during maternal supplementation. Similarly, nursing women who received a daily supplement of 1 mg of pantothenic acid had higher mean values of the vitamin in their breast milk compared with women who did not receive a supplement (0.48 mg/100 ml versus 0.26 mg/100 ml) (Institute of Medicine 1991).

MILK COMPOSITION: MAJOR MINERALS AND ELECTROLYTES

Concentrations of calcium, magnesium, and phosphorus in human milk show little variability among women and do not appear to be influenced by maternal nutritional status or diet. Similarly, the levels of the major electrolytes in milk (sodium, potassium, and chloride) are not altered by changes in maternal nutrition.

MILK COMPOSITION: TRACE MINERALS

Maternal nutritional status or diet do not influence the concentrations of iron, manganese, and copper in breast milk. Most studies have shown that the mother's zinc intake does not influence concentrations in milk—zinc levels are maintained even when the dietary zinc is low. However, two recent studies have found that zinc supplementation during lactation (13 mg/day for six months and 50 mg/day for one month) produced small but significant increases in milk concentrations (Institute of Medicine 1991).

Maternal fluoride intake has minimal effects on the fluoride level in milk. One study found only about 0.2% of a large dose of fluoride (11 mg) was transferred into the milk (Ekstrand et al. 1984). There are no significant differences in milk fluoride levels among women consuming a wide range (from 0.2 to 1.0 ppm) of fluoride from drinking water. Only when fluoride concentrations in the water exceed 1.4–1.7 ppm do fluoride levels in human milk increase, and even then only slightly.

In contrast to the other minerals, maternal intake of iodine and selenium markedly affects the levels found in breast milk. The mammary gland avidly accumulates iodine, and milk levels increase as dietary intake by the nursing mother increases. In northwestern Zaire, where iodine deficiency is widespread, maternal intakes are typically very low and milk levels average only 2 µg/100 ml (Delange 1985). In contrast, in the United States, iodine concentrations in milk average nearly 18 µg/100 ml (Institute of Medicine 1991).

Similar to iodine, selenium values in breast milk reflect the selenium content of the food consumed by the nursing mother. There are substantial differences in the content of selenium in the soil worldwide and therefore in the local food supply (Kumpulanien 1989).

MATERNAL HEALTH DURING LACTATION

NUTRITIONAL HEALTH OF THE MOTHER

We have seen that levels of protein, carbohydrate, folate, and most of the minerals in human milk are maintained even if maternal dietary intake is low. Yet many lactating women fall short of their requirements for several of these nutrients. If dietary intake of a nutrient is inadequate, and the nutrient continues to be secreted into the milk, maternal stores must provide the difference to ensure the mother's good health. Prolonged dietary inadequacy will deplete maternal stores. The demands of breast-feeding can therefore have a major impact on the nutritional health of the mother.

A recent report estimated the total output in milk of selected nutrients during periods of partial and exclusive breast-feeding and calculated the potential impact of these demands on body stores (Institute of Medicine 1991). The results are shown in Table 4.9. To support six months of exclusive breast-feeding, a mother who did not consume the suggested increment of protein in the RDAs for lactation (an extra 15–19 g protein/day) would need to mobilize nearly 20% of her lean tissue.

The average adult woman has about 1000 g of calcium in her body, almost all of it in her bones. If extra calcium is not consumed to cover losses into the milk, during six months of breast-feeding, nearly 5% of her total calcium would need to be used for milk production. Folate

Source of Demand	Nutrient Output				
	Protein, g (% of total body content)	Calcium, g (% of total body content)	Iron, mg (% of stores)	Vitamin A, mg (% of stores)	Folate, mg (% of stores)
Estimated body content	11,250	1,035	300	209	6–7
Exclusive breast-feeding for 6 mo	1,512 (19)	40 (4)	43 (14)	96 (46)	12 (~100)
Exclusive breast-feeding for 6 mo, plus partial breast-feeding for 6 mo	2,268 (29)	60 (6)	65 (22)	144 (69)	18 (>100)

For a 60 kg woman, based on average milk composition values and a volume of 800 ml/day for exclusively breast-feeding and 400 ml/day for partial breast-feeding.

Adapted from Institute of Medicine, Subcommittee on Nutrition during Lactation, *Nutrition during Lactation* (Washington D.C.: National Academy Press, 1991).

TABLE 4.9

Estimated long-term demands of lactation for selected nutrients

stores in the body are small, and unless additional folate is consumed to cover the extra needs of milk production, body reserves of folate would be depleted in just a few months of breast-feeding. Clearly, the potential consequences of nutrient inadequacy during lactation are substantial.

INVOLUTION OF THE UTERUS

Oxytocin released by the posterior pituitary in response to infant suckling causes contractions of the uterus, and these contractions help the uterus shrink and return to normal prepartum size—as well as decrease the risks of uterine bleeding during the postpartum period.

LACTATIONAL AMENORRHEA AND CONTRACEPTION

Breast-feeding prolongs the time between delivery of a baby and the return of the menses. As part of the normal female menstrual cycle, pulsatile release of gonadotropin-releasing hormone (GnRH) stimulates the pituitary gland to produce luteinizing hormone (LH). LH then circulates to the ovary and promotes maturation of the ovarian follicle and ovulation.

Regular stimulation of the breast by the nursing infant disrupts these neuroendocrine pathways. Suckling inhibits the release of GnRH by the hypothalamus, and this reduces LH secretion by the pituitary (Tulchinsky 1994). LH levels in women who are breast-feeding are much lower than in postpartum, nonlactating women. Reduced LH secretion removes the stimulus on the ovary, and produces postpartum anovulation and amenorrhea, as shown in Figure 4.18. The duration of postpartum anovulation varies widely among women: some women who choose not

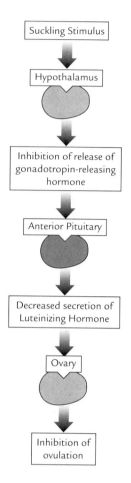

FIGURE 4.18
*Physiological mechanisms
of lactational infertility*

to breast-feed wait months before ovulating, while others resume menses in the first several months postpartum—even while breast-feeding (Diaz et al. 1991). Suckling is more effective at inhibiting GnRH release in some women than in others, but in general, the greater the intensity and frequency of breast-feeding, the more ovulation is inhibited.

For most women, breast-feeding provides contraceptive protection during the postpartum period, but it is not an entirely dependable method of contraception. Although lactating women who are not ovulating are not capable of becoming pregnant, some women do ovulate during breast-feeding. Indeed, many women have begun a second pregnancy while still nursing. However, during the first six months postpartum, if the mother is fully breast-feeding and not menstruating, breast-feeding provides more than 98% protection from pregnancy (Short et al. 1991). See Figure 4.19.

The nutritional status of the mother during lactation can also influence the duration of postpartum amenorrhea, as there is a clear association between undernutrition in the mother and prolonged postpartum anovulation. Several mechanisms for this effect have been proposed. It may be that a critical proportion of body fat is necessary for the return of

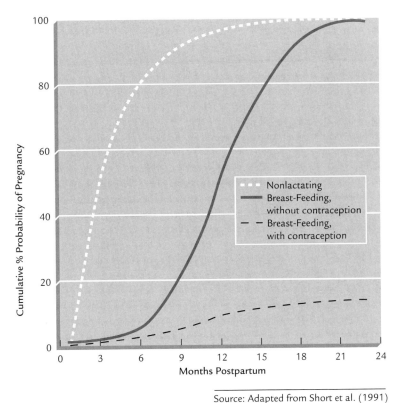

FIGURE 4.19

Probability of pregnancy while breast-feeding

Source: Adapted from Short et al. (1991)

ovulation after delivery. Also, inadequate food intake may alter maternal hormonal status and inhibit ovulation. Finally, undernutrition may influence the composition of breast milk by reducing fat content and increasing the amount of suckling required to transfer sufficient nutrients to the infant—thereby prolonging anovulation.

Intervention programs in developing countries that improve maternal nutritional status have produced substantial reductions in the duration of postpartum amenorrhea. In a supplementation program in Gambia, the length of postpartum amenorrhea was shortened from 66 to 42 weeks (Delange 1985). Nutritional supplementation in undernourished mothers may have important public health implications because, worldwide, breast-feeding has important contraceptive benefits. Postpartum anovulation contribute to child spacing by providing a period during which the mother can recover from her previous pregnancy before beginning a new one.

In developing countries, breast-feeding continues to provide more woman-months of contraception than all other modern methods of birth control combined (Wray 1991). In many areas the birth interval provided by lactation is the major determinant of fertility—the total number of pregnancies a woman will have. This may have important implications in regions of the world where overpopulation is severe and food is scarce. (Thapa and Short 1988).

LONG-TERM EFFECTS OF BREAST-FEEDING ON MATERNAL HEALTH

Obesity

There have been few long-term studies of lactation and body weight, and it is not clear if women who breastfeed are more or less likely to become obese when compared to women who feed their babies formula. Milk synthesis and secretion require significant energy, and if energy intake is not increased during lactation, negative energy balance and weight loss should occur. However, although most women lose weight during the postpartum period, some nursing mothers do not lose weight, and some gain weight during breast-feeding. Studies in animals have generally found that pregnancy without breast-feeding results in increased adipose tissue and accumulation of excess fat, compared with pregnancy followed by breast-feeding (Jen et al. 1988), but studies in humans have provided equivocal results (Institute of Medicine 1991).

A recent study found that a group of women who breast-fed had actually gained more weight than a group that did not, when measured at nine months postpartum (Rookus et al. 1987). Clearly, more research is needed to clarify the consequences of lactation on long-term energy balance.

Osteoporosis

The average 55 kg woman stores 900–1000 g of calcium in her body, nearly all of it in her bones. During the first six months of lactation, losses of calcium into the milk total approximately 50 g, or about 5% of calcium stores. Added to the 30 g of calcium required by the fetus during gestation, the equivalent of 8% of maternal body stores must be transferred in 15 months. Unless dietary intake can increase to offset these potential losses, successive pregnancies could reduce mineral content of the bones and increase the risk of osteoporosis later in life.

Many women do not consume the RDA for calcium (1200 mg) during pregnancy and lactation. In these women, do calcium losses to the fetus and newborn increase the likelihood of osteoporosis? Studies have shown that bone mineral is lost during pregnancy and lactation (King et al. 1994), (Sowers et al. 1993). Blood osteocalcin levels of lactating women are higher than those in nonlactating women, indicating turnover of bone is greater during lactation (King et al. 1994). In a study of 98 healthy lactating women, bone mineral density was measured across twelve months postpartum. Women who lactated for greater than five months had losses of bone mineral of approximately 5% at six months postpartum (Sowers et al. 1993). See Figure 4.20.

However, these losses are usually replenished during the post-lactation period. A longitudinal study showed that recovery of bone density begins in the first four months after lactation (Hreshchyshyn et al. 1988). Studies

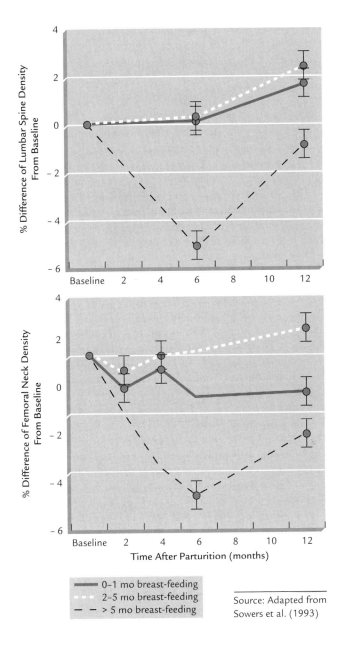

FIGURE 4.20

Bone mineral density during lactation

Legend:
— 0–1 mo breast-feeding
····· 2–5 mo breast-feeding
– – – > 5 mo breast-feeding

Source: Adapted from Sowers et al. (1993)

that have examined women before pregnancy and at intervals after lactation is complete have found that overall bone density has not changed or is positive in the post-lactation period (Sowers et al. 1993).

One study looked at the relationships between parity, lactation, and bone density. Researchers measured the bone mineral density of nearly 600 women who had been previously pregnant—some of the women had breast-fed their infants, and some had not. Density of the spine and the femoral head were measured, and no differences were found in bone density in the femur between the two groups. In the spine, bone density was higher in the women who had breast-fed, particularly those who had nursed more than one infant. The researchers estimated lumbar spine density was increased by about 1.5% per breast-fed child compared with women who had chosen not to breast-feed (Hreshchyshyn et al. 1988). Other studies have found no association between the number of pregnancies and duration of lactation and the occurrence of osteoporotic fractures in later life (Alserman et al. 1986).

Although a recent expert panel cautioned that "the data are not conclusive" (Institute of Medicine 1991), most studies indicate that breast-feeding is not detrimental to long-term bone health. Further research is needed to confirm and clarify these findings.

Breast Cancer

Breast-feeding may be protective against breast cancer. Most—but not all—epidemiological studies have found that a positive history of lactation is associated with a reduced risk of breast cancer in later life (Newcomb et al. 1994). In general, protection increases with increasing duration of lactation. That is, the more time a woman spends breast-feeding, the less her risk of developing breast cancer in later life. In one study, women who lactated for 36 months during their lifetime had a 30% reduction in risk compared to women who never lactated (Newcomb et al. 1994). See Figure 4.21.

Most scientists believe the protective effect is due to the effect of lactation on levels of sex hormones. Exposure of the breast to estrogens and other steroid hormones during normal ovulatory cycles is thought to increase the chance of developing breast cancer. Breast-feeding modifies the maternal neuroendocrine system and produces a period of postpartum anovulation. Because ovulation is inhibited, the amount of circulating estrogen is markedly reduced, as the developing ovarian follicle normally produces estrogen. Thus, the protective effect of lactation may

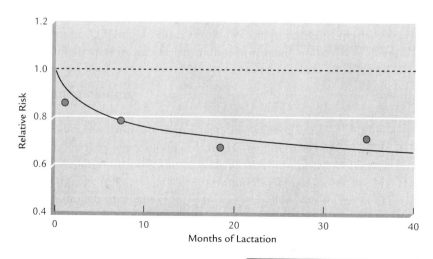

FIGURE 4.21

Risks of breast cancer according to cumulative number of months of lactation

Source: Adapted from Newcomb (1994)

be due to a reduction in estrogen exposure during the postpartum months (Kelsey and John 1994).

DRUG USE DURING LACTATION

EFFECTS OF MEDICINAL DRUGS ON THE INFANT

Most drugs taken by the nursing mother, whether taken by mouth, injection, or inhalation, pass into the breast milk in small quantities (usually less than 1 percent of the mother's dose). Although only a few drugs are **contraindicated** during lactation, women who are breast-feeding should always consult their physicians before taking any over-the-counter or prescription drugs.

Several factors influence the potential toxicity of drugs taken during lactation:

- *Maturity of the infant.* Preterm infants and newborns are particularly vulnerable to adverse drug effects, for several reasons. The enzyme systems needed to metabolize drugs are still developing during early infancy, particularly in the first month postpartum. Also, excretory systems in the liver and kidney are not fully mature. Therefore drugs may reach high levels and have a long half-life in the infant.

- *Drug characteristics.* Drugs that are smaller, more lipid soluble, and less ionized pass more readily into the milk. Insulin, for example, because it is a relatively large protein that is not lipid soluble, does not cross into the milk. Compounds given in high doses or with prolonged duration of effects accumulate in greater amounts in the milk.

- *Intensity of breast-feeding.* The greater the milk intake, the greater the potential dose to the infant. Infants who are exclusively breast-fed and nurse eight to ten times a day will be more vulnerable to a maternal drug than an infant who is taking supplementary foods and nursing only once or twice a day.

- *Timing of the drug dose relative to breast-feeding.* Most milk is produced by the mammary gland during feedings (only small amounts of milk are stored in the breast). If the infant nurses while maternal blood levels of a drug are high, more of the drug will be present in the milk. For many medications, particularly those taken orally, levels in the blood will be highest 30 to 60 minutes after taking the drug. Levels in milk will tend to be lowest if the mother takes the medicine right after nursing, and sufficient time is available to metabolize the drug before the next feeding.

Many commonly used drugs can affect the breast-feeding infant. In general, effects are similar in both mother and infant, but may be more severe

contraindicated: when the use of a drug becomes inadvisable, usually because of risks

in the infant. For example, sedatives (such as Valium) can cause profound lethargy and drowsiness in infants. Anticoagulants (including aspirin in large doses) can produce abnormal bleeding in infants (Lawrence 1994). Although many antibiotics are safe during lactation, the nursing infant may develop an allergy to the antibiotic that the mother is taking.

DRUGS THAT INFLUENCE MILK PRODUCTION

A variety of commonly used drugs inhibit milk production (see Table 4.10). For example, popular cold preparations contain antihistamines and **sympathomimetics**, and nursing women who take these preparations often notice a marked falloff in milk production. Sympathomimetics may inhibit prolactin release, and antihistamines "dry" all body secretions, including breast milk.

Inhibit Milk Production	Enhance Milk Production
Alcohol (heavy use)	Many antidepressants (such as Elavil)
Antihistamines (including OTC cold and flu preparations)	Many antipsychotics (such as Thorazine)
	Insulin
Estrogens (including birth control pills)	
Barbiturates (sedatives)	

Adapted from M. C. Neville and M. Neifert, *Physiology, Nutrition and Breastfeeding* (New York: Plenum Press, 1983),367–403.

TABLE 4.10
Drugs that affect milk volume

ORAL CONTRACEPTIVES

Many women begin taking oral contraceptives in the postpartum period to prevent conception. In the United States, about one in six women use oral contraceptives at some point during lactation (Ford and Labbok 1987).

There are several different preparations available, and the composition and dosage of the pill determines its effect on milk production. The use of pills combining estrogen and progesterone (even low-dose preparations) is associated with significant reductions in milk volume (Koetsawang 1987). Because of this, combined estrogen-progesterone preparations are not recommended, particularly in the first few months postpartum.

In contrast, several studies have found that progestin-only preparations (progestin is a form of progesterone) have no effect on milk volume or composition. Consequently, the World Health Organization now rec-

sympathomimetic: an agent that produces effects similar to those produced by stimulation of the sympathetic nervous system

ommends progestin-only pills for women who use oral contraceptives during breast-feeding (WHO 1988).

HERBS

Many herbs contain drugs and druglike substances. Although they are in a natural rather than a purified form, they can have significant effects on the infant if used by the nursing mother. For example, senna and cascara are herbal remedies that can cause diarrhea in the infant, and they should not be used by the breast-feeding mother. Mothers should be as cautious with herbal remedies as they are with drugs during breast-feeding.

DECISIONS ABOUT DRUGS DURING LACTATION

The decision to use a drug while breastfeeding must be carefully considered, and medication should be used only if absolutely necessary. A large amount of research has been done to clarify the safety of many commonly used drugs, but the potential effects of many drugs on infant health are still not well understood. The final decision should be made by the mother in consultation with her physician. See Table 4.11 for some general guidelines.

- Use a medication only if it is absolutely necessary.
- Delay starting the medication (if there is a choice) until the infant is more mature and better able to detoxify and metabolize drugs that might be transported through the milk.
- Take the lowest dose possible for the shortest time possible.
- Choose a drug that transfers the least amount into breastmilk.
- Avoid a drug with a long half-life, if possible.
- Avoid sustained release preparations.
- Schedule taking the medication so that the lowest amount gets into the milk—usually immediately after a feeding or before the infant has a long sleep period.
- Observe for any untoward reaction, such as fussiness, rash, colic, or change in feeding or sleeping habits. If any of these occur, the physician should be notified and given information about all of the drugs used by the mother.
- Teach manual expression or provide a breast pump and instructions for its use if the mother must take a contraindicated drug for a short time.

Adapted from J. Riordan and K. G. Auerbach, *Breastfeeding and Human Lactation* (Boston: Jones and Bartlett, 1993).

TABLE 4.11
Guidelines for breastfeeding and use of medications

ILLICIT DRUGS

Marijuana

Marijuana appears to be the most commonly used illegal drug among pregnant and lactating women. A recent study found that 5 to 15% of pregnant women use the drug (Braude et al. 1987). Tetrahydrocannabinol (THC), the active ingredient in marijuana, passes into breast milk and is poorly absorbed from the infant's intestinal tract. It can cause drowsiness and decrease the frequency of infant feeding if exposure is prolonged.

Cocaine

Cocaine is also secreted in human milk and can be absorbed by the infant. Although absorption of cocaine from the milk is low, several reports indicate cocaine exposure through breast-feeding can be toxic for the infant (Chasnoff et al. 1987). Infants exposed to cocaine through breast-feeding exhibit signs of cocaine intoxication, including abnormal heartbeat patterns and repeated vomiting (Chasnoff et al. 1987). Metabolism of cocaine by the infant is slow and exposure is prolonged. Because of the potential toxicity, use of cocaine by breast-feeding mothers can be the basis for legal action of child abuse.

Heroin

Heroin passes into the breast milk and is absorbed by infants. They can become physically addicted to the drug if breast-feeding mothers are regular heroin users.

Amphetamines

Amphetamines are readily secreted into breast milk and levels in milk can be several times higher than levels in maternal plasma. Maternal use during breastfeeding can cause irritability and disturb sleep in infants. Table 4.12 shows the effects of several illicit drugs on infants who are being breastfed.

Drug	Effect
Amphetamine	Irritability, poor sleep pattern
Cocaine	Cocaine intoxication
Nicotine (smoking)	Shock, vomiting, diarrhea, rapid heart rate, restlessness; decreased milk production
Phencyclidine	Potent hallucinogen

Adapted from American Academy of Pediatrics, Committee on Nutrition, *Pediatric Nutrition Handbook*, 3d ed. (Elk Grove, IL: American Academy of Pediatrics, 1993)

TABLE 4.12
Illicit drugs and their effects on infants during lactation

A HEALTHY LIFESTYLE AND DIET DURING LACTATION

EXERCISE DURING LACTATION

Many new mothers are eager to begin or continue exercise programs after delivery, and physical activity during lactation can provide significant health benefits (Lovelady et al. 1990). Exercise in the postpartum period can help women lose weight gained during pregnancy. Increased energy expenditure allows the breast-feeding mother to eat more food, increasing the likelihood of adequate protein and micronutrient intake. Also, for many women, exercise provides a psychological lift.

Breast-feeding mothers who follow a moderate exercise program have greater daily energy intakes, lower levels of body fat, and produce equal amounts of breast milk, compared with nonexercising mothers (Dewey et al. 1994). The potential impact of strenuous exercise during lactation is unknown. Intense exercise can increase lactate concentrations in the mother's blood and breast milk. Because lactate may impart a sour taste to the milk, infant acceptance of breast milk may be diminished if feeding occurs after strenuous exercise. Infant acceptance of breast milk collected before intense exercise is significantly greater than acceptance of milk collected 10 to 30 minutes after exercise (Wallace et al. 1992). Lactate levels are much higher in post-exercise milk and remain elevated for up to 90 minutes after exercise.

The American College of Obstetrics and Gynecology has issued guidelines for exercise in both the pre- and postpartum periods (presented in Table 3.14) (American College of Obstetrics and Gynecologists 1994). By following these guidelines, breast-feeding mothers who wish to exercise can safely pursue a fitness program. To protect the breasts, a sturdy support brassiere should be worn and jerking or bouncing motions should be minimized. Mothers should also be sure to drink generous amounts of fluid before and during exercise, as lactation increases fluid needs 500–650 ml/day.

SMOKING DURING LACTATION

Effects on Milk Volume

Smoking during breast-feeding reduces milk volume in several ways. Infants of mothers who smoke have average birth weights 200 g lower than those of nonsmokers and a smaller infant usually has reduced demands for milk (Hopkinson et al. 1992). In addition, heavy smoking (more than 15 cigarettes/day) lowers blood levels of prolactin 30 to 50% during the early postpartum period (Anderson et al. 1982). Smoking also stimulates adrenaline release, which is known to inhibit release of oxytocin by the pituitary. See Figure 4.22.

Reduced volume may contribute to the increased incidence of early weaning in breast-feeding mothers who smoke. Within the same socio-economic group, there is a lower prevalence of breast-feeding at two to three months postpartum among smoking mothers compared to mothers who don't smoke (Matheson and Rivrud 1989).

Source: Adapted from Hopkinson et al. (1992)

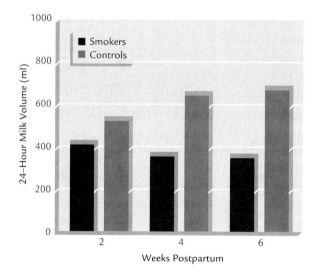

FIGURE 4.22
Smoking and lactation

Adverse Effects on the Infant

Cigarette smoke is rich in nicotine, and nicotine and its metabolite cotinine pass readily from the maternal blood into the milk (Dahlstrom et al. 1990). After a cigarette, levels of nicotine can be higher in the mother's milk than in her blood, and although both levels fall over time, the half-life of nicotine is much longer (more than 90 minutes) in milk than in blood (Lawrence 1994). Heavy smoking during breast-feeding can cause nicotine poisoning in infants. See Table 4.12. Carcinogens in cigarette smoke can pass into the milk during lactation.

Because of the potential adverse health effects for both the mother and infant, current recommendations are for women to stop smoking completely during lactation (Lawrence 1994). For women who continue to smoke, smoking should be done away from the infant, and the time between smoking and nursing should be as long as possible to reduce exposure of the infant to nicotine.

COFFEE AND CAFFEINE CONSUMPTION DURING LACTATION

About one percent of a maternal dose of caffeine (whether from coffee, tea, soft drinks, chocolate, or medicines) is transported into the milk, and levels in milk are typically 50 to 80% of levels in maternal plasma (Lawrence 1994). Metabolism of caffeine by infants is much slower than in adults, and frequent doses can cause caffeine to accumulate in the newborn—producing symptoms of irritability and wakefulness.

Breast-feeding mothers should avoid overconsumption of caffeine during lactation, but consumption of one to two cups of coffee or one to two soft drinks per day will not produce unacceptable levels of caffeine in the milk (Committee on Drugs 1989).

A study in Costa Rican women found that mothers who drink three or more cups of coffee per day during lactation had reduced iron concentrations in their breast milk compared with breast-feeding mothers who did not drink coffee. The reduction in iron in the milk had detrimental effects on iron status in early infancy (Munoz et al. 1988).

ALCOHOL CONSUMPTION DURING LACTATION

Traditionally, small amounts of alcohol during lactation have been suggested as beneficial—allowing the mother to relax and facilitating the milk ejection reflex. However, alcohol actually inhibits milk ejection (Institute of Medicine 1991). The effect of ethanol is dose dependent and involves inhibition of oxytocin release from the pituitary. In one study, there was no inhibition of oxytocin at doses of ethanol less than 0.5 g/kg body weight. At doses greater than 1.0 g/kg, the ejection reflex was completely inhibited in nearly half the women (Cobo 1973). In a woman of average size, 0.5 g/kg corresponds to about 24 oz of beer or 2–2.5 oz of liquor.

Alcohol exposure during breast-feeding may have serious adverse effects on infant development. Although acetaldehyde (the major toxic metabolite of ethanol) cannot pass through the mammary epithelium into the milk, ethanol itself readily passes into the milk at concentrations approaching those in maternal blood. It can produce lethargy and drowsiness in the breast-feeding infant.

A recent report found that heavy alcohol use by nursing mothers increased the risk of retarded psychomotor development in their infants at one year of age (Little et al. 1989). There were no signs of impairment in infants of breast-feeding mothers who drank occasionally—defined as one to two drinks per week.

Clearly, heavy alcohol use during breast-feeding can harm both the infant and mother. Although the level of alcohol consumption that may harm the nursing infant is not known, occasional light use of alcohol during lactation appears to be reasonably safe. A recent report from the National Academy of Sciences advises that occasional consumption of small amounts of alcohol is not contraindicated during lactation (Institute of Medicine 1991).

VITAMIN AND MINERAL SUPPLEMENTATION DURING LACTATION

Routine vitamin and mineral supplementation is not recommended for breast-feeding mothers (Institute of Medicine 1991). Most mothers can obtain adequate amounts of vitamins and minerals from a balanced and varied diet. However, certain groups of women eating restricted diets may benefit from supplementation.

Women who are lactose intolerant or do not consume dairy products should strive to obtain alternate sources of calcium through low-lactose dairy products. If dietary sources of calcium are inadequate, consumption of a daily supplement containing 600 mg of elemental calcium is advisable. Breast-feeding women whose food intake is sharply limited because of low income, poor access to nutritious foods, or excessive dieting, may benefit from a vitamin and mineral supplement at the level of the RDAs. Women who have limited exposure to sunlight and who avoid vitamin D-enriched milk products (for example, complete vegetarians and those who are lactose intolerant) should consume a daily supplement containing 10 µg of vitamin D. Complete vegetarians should consume a daily supplement containing 2.6 µg vitamin B_{12} (Institute of Medicine 1991).

Breast-feeding women should avoid nutritional supplements containing amounts of iodine and vitamin D in excess of the RDAs. Large doses of these micronutrients can result in toxic levels in breast milk and potentially harm the infant. For women who choose to take a vitamin and mineral supplement during lactation, an appropriate formulation is shown in Table 4.13. Infants who are exclusively breast-fed may require supplementation with iron, vitamin K, and fluoride during the first year (AAP 1993). In special situations, vitamin D and vitamin B_{12} supplementation can be beneficial during infancy. Vitamin and mineral supplementation during infancy is discussed in detail in Chapter 6.

Iron	30–60 mg
Zinc	15 mg
Copper	2 mg
Calcium	250 mg
Vitamin D	10 μg (400 IU)
Vitamin C	50 mg
Vitamin B_6	2 mg
Folate	300 μg
Vitamin B_{12}	2 μg

If vitamin A is included, beta-carotene is preferred over retinol to reduce the risk of toxicity or other adverse reactions. Since calcium and magnesium may interfere with iron absorption, upper limits of 250 and 25 mg/dose, respectively, are recommended as a part of vitamin–mineral supplements.

From Institute of Medicine, Subcommittee for a Clinical Application Guide, *Nutrition during Pregnancy and Lactation: An Implementation Guide* (Washington D.C.: National Academy Press, 1992).

TABLE 4.13
Appropriate vitamin and mineral supplements for lactating mothers

VEGETARIAN DIETS DURING LACTATION

Breast-feeding mothers who consume high-quality lacto-ovo-vegetarian diets (eggs and milk in addition to plant foods) obtain adequate nutrition to support maternal and infant health during lactation. In a study in the United States, breast-feeding mothers who were partial vegetarians (they ate no meat or poultry but did consume milk products) were surveyed at 2 to 18 months postpartum (Finley et al. 1985). Mean daily intakes of calcium, iron, thiamin, riboflavin, niacin, and vitamins A and C met or exceeded the RDAs for these nutrients. Mean daily protein intake (78 g) was lower than those reported for nonvegetarian mothers, but still exceeded the RDA.

To meet the increased demands of breast-feeding, ovo-lacto-vegetarians should consume a balanced diet containing a variety of grains, legumes, eggs, nuts, dairy products, fruits, and vegetables. Vegetarian mothers should emphasize foods rich in high-quality protein, calcium, iron, zinc, and vitamins D, B_6, and B_{12}. Table 4.14 provides a basic diet guide for lactating vegetarians.

Women eating vegan diets (no meat, fish, poultry, eggs, or milk) should pay special attention to vitamin B_{12} intake. Several studies have found the content of vitamin B_{12} in breast milk of vegans is low, and severe vitamin B_{12} deficiency has been reported in infants of vegan mothers. Lactating vegans can obtain adequate vitamin B_{12} from special foods, such as vitamin B_{12}-fortified soy milk and fortified yeast. If

- Grains, legumes, nuts, and seeds: Six or more servings including several slices of whole grain bread, beans, and nuts or seeds

- Vegetables: Three servings or more, including one or more servings of dark leafy greens

- Fruit: One to four pieces, including citrus fruits for a source of vitamin C

- Milk and eggs: Dairy products and eggs to meet protein requirements.

Adapted from J. Riordan and K. G. Auerbach, *Breastfeeding and Human Lactation* (Boston: Jones and Bartlett, 1993), 135–152.

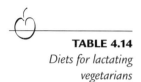

TABLE 4.14
Diets for lactating vegetarians

vitamin B_{12} intake from these sources is inadequate, breast-feeding mothers should take a daily supplement of vitamin B_{12} at the level of the RDA (Institute of Medicine 1991).

Levels of vitamin D are low in the diets of many vegans. Vegan mothers should obtain frequent exposure to sunlight to maintain vitamin D status. It is recommended that vegans who consume low amounts of vitamin D and do not obtain adequate sunlight take a daily supplement of 10 µg of vitamin D (Institute of Medicine 1991).

LACTATION AND MATERNAL HEALTH: REVIEW AND SUMMARY

The importance of a healthy diet while breast-feeding cannot be over-emphasized. If a mother is exclusively breast-feeding, the rapid growth of early infancy depends entirely on a continuous supply of energy, high-quality protein, and micronutrients in the breast milk. Moreover, a carefully chosen, nutrient-dense diet protects maternal health by allowing the breast-feeding mother to lose weight gained during pregnancy, replenish nutrient stores depleted by the demands of pregnancy, and maintain nutrient stores to support milk production.

The levels of a few of the nutrients in breast milk (most vitamins and several minerals) are responsive to changes in maternal status. If the mother's diet is inadequate, levels in the milk will fall and may imperil the growth and health of the infant. However, the levels of most nutrients in breast milk (energy, protein, most minerals, and folate) are maintained even if dietary intake is inadequate, but at the expense of maternal reserves. This may have detrimental short- and long-term effects on maternal health. For example, inadequate folate can cause maternal anemia during breast-feeding and poor calcium and vitamin D status during lactation may compromise bone health in later life. Clearly, both infant and maternal health depend on the quality of the mother's diet during lactation.

Nutrients most often lacking in the diets of breast-feeding women are the minerals calcium, zinc, magnesium, and iron, and the vitamins folate, thiamin, B_6 and A. Breast-feeding mothers should emphasize foods rich in these nutrients, particularly calcium. Calcium-rich foods, and strategies to increase the calcium intake of women with lactose intolerance, are shown in Table 4.15.

Here are good dietary sources of the other nutrients many breast-feeding mothers need most:

- *Zinc:* meat, seafood, poultry, eggs, seeds and legumes, and whole grains

- *Magnesium:* nuts, seeds, legumes, whole grains, and green vegetables

- *Iron:* fortified cereals, meat, and spinach

- *Folate:* leafy vegetables, green beans, legumes, fruit, liver, fortified cereals, and whole grains

- *Vitamin B_6:* poultry, meat, fish, potatoes, bananas, prunes, watermelon, spinach, nuts, and fortified cereal

- *Thiamin:* meats, fish, whole grains, legumes, corn, seeds and nuts, and fortified cereals

- *Vitamin A (or its precursor beta-carotene):* vitamin A-enriched dairy products, liver, carrots, and sweet potatoes

Foods equal to about 1 cup of milk in calcium content:
  3 oz sardines (if the bones are eaten)

Foods equal to about ½ cup of milk in calcium content:
  3 oz canned salmon (if the bones are eaten)
  4 oz tofu (if it has been processed with calcium sulfate)
  4 oz collards
  4 corn tortillas (if processed with calcium salts)

Foods equal to about ⅓ cup of milk in calcium content:
  1 cup cooked dried beans
  4 oz bok choy, turnip greens, or kale
  1 medium square of cornbread
  7 to 9 oysters
  3 oz shrimp

Foods that can be made high in calcium:
  Soups made from bones cooked with vinegar or tomato

TABLE 4.15

Sources of calcium for women with lactose intolerance and those avoiding milk products

Adapted from Institute of Medicine, Subcommittee for a Clinical Application Guide, *Nutrition during Pregnancy and Lactation: An Implementation Guide* (Washington D.C.: National Academy Press, 1992).

Fluid needs are also increased during lactation, and mothers should drink about two quarts of fluids (milk, juice, soup, or water) each day. A good rule of thumb is to drink a glass of fluid with each meal and with each feeding.

A goal for many new mothers is the loss of extra weight gained during pregnancy. We have seen that gradual, steady weight loss during lactation does not impair breast milk production and can reduce the likelihood of later obesity—particularly in women with repeated pregnancies. In most women, weight loss of 1 to 2 lb per month (beginning in the second postpartum month) will not affect milk production. Women who are significantly overweight can lose up to 4½ lb per month without adversely affecting lactation.

Therefore, a healthy diet for lactation should allow for weight loss of 1 to 2 lb per month while providing adequate amounts of protein and micronutrients. The breast-feeding mother should be careful not to overeat. The RDAs during lactation call for a modest increase of only 500 kcal/day (NAS 1989) during lactation, or about the amount in an additional two cups of whole milk, an apple, and one cup of rice. For the average mother, this increase does not completely cover the increased energy costs of lactation, and fat gained during pregnancy will be mobilized to supply the difference.

A modest program of exercise during the postpartum period will not affect milk production and can help in postpartum weight loss. Rapid weight loss can impair milk production and is not recommended. Lactating mothers should maintain adequate daily energy intake (a minimum of 1500 calories should be consumed daily). Strict diets and weight loss medications should be avoided.

Younger breast-feeding women, particularly those in their teens and early 20s, should emphasize foods rich in calcium and vitamin D and strive to maintain intakes near the RDAs for these nutrients. The skeleton is still developing at this age, and rates of bone mineralization are high. Poor calcium status during adolescence and young adulthood may jeopardize long-term bone health and increase the risk of osteoporosis.

The Institute of Medicine's Committee for Nutrition during Pregnancy and Lactation has published guidelines to help breast-feeding women choose foods to meet their nutritional needs (Institute of Medicine 1991). By following these recommendations (Tables 4.16 and 4.17), a new mother can eat a balanced diet that meets the RDAs for lactation—and provide optimal nutrition for both herself and her infant.

Type of Restrictive Eating Pattern	Corrective Measures
Excessive restriction of food intake, i.e., ingestion of < 1,800 kcal of energy per day, which ordinarily leads to unsatisfactory intake of nutrients compared with the amounts needed by lactating women	Encourage increased intake of nutrient-rich foods to achieve energy intake of at least 1800 kcal/day; if the mother insists on curbing food intake sharply, promote substitution of foods rich in vitamins, minerals, and protein for those lower in nutritive value; in individual cases, it may be advisable to recommend a balanced multivitamin–mineral supplement; discourage use of liquid weight loss diets and appetite suppressants
Complete vegetarianism, i.e., avoidance of all animal foods, including meat, fish, dairy products, and eggs	Advise intake of a regular source of vitamin B_{12} such as special vitamin B_{12}-containing plant food products or a 2.6 µg vitamin B_{12} supplement daily
Avoidance of milk, cheese, or other calcium-rich dairy products	Encourage increased intake of other culturally appropriate dietary calcium sources; provide information on the appropriate use of low-lactose dairy products if milk is being avoided because of lactose intolerance; if correction by diet cannot be achieved, it may be advisable to recommend 600 mg of elemental calcium per day taken with meals
Avoidance of vitamin D-fortified foods, such as fortified milk or cereal, combined with limited exposure to ultraviolet light	Recommend 10 µg of supplemental vitamin D per day

TABLE 4.16

Food choices to boost nutrient intake in women with restrictive eating patterns during lactation

Adapted from the National Academy of Sciences, Institute of Medicine, Food and Nutrition Board, Committee on Nutritional Status during Pregnancy and Lactation, Subcommittee on Nutrition during Lactation, *Nutrition during Lactation*, (Washington D.C.: National Academy Press, 1991).

- Avoid diets and medications that promise rapid weight loss.

- Eat a wide variety of breads and cereal grains, fruits, vegetables, milk products, and meats or meat alternates each day.

- Take three or more servings of milk products daily.

- Choose a diet low in fat, saturated fat, and cholesterol. This reduces the risk of chronic disease and may help you manage your weight. Keeping fat intake under control also helps make room for foods that are rich in vitamins and minerals. Aim to have 2 servings daily of low-fat meat, fish, or poultry, or of legumes. Also aim to have 2 to 3 servings of low-fat, calcium-rich milk products such as low-fat milk, cheese, or yogurt. One cup of milk or yogurt is an example of one serving.

- Choose a diet with plenty of vegetables, fruits, juices, and grain products. Choose whole-gain products at least part of the time. Aim to have 2 or more servings of fruit or juice, 3 or more servings of vegetables, and 6 to 11 servings of grains each day. One slice of bread and ½ cup of rice, fruit, or vegetables are examples of one serving. Use sweets, sugars, and soft drinks only in moderation. ·Make a greater effort to eat vitamin-A-rich vegetables or fruit often. Examples of vitamin-A-rich foods include carrots, spinach or other cooked greens, sweet potatoes, and cantaloupe. Take enough fluids (especially milk, juice, water, and soup) to keep from getting thirsty.

- You need enough food (at least 1800 kcal/day) to help maintain milk production and to provide the nutrients that you and your baby need.

- For the first 6 weeks, the best guide to how much you should be eating is your own appetite.

- Try to keep your intake of coffee, cola, or other sources of caffeine to 2 servings or less per day.

- For those who choose to take alcoholic beverages: It is best to avoid drinking alcoholic beverages, but certainly have no more than 2 to 2.5 oz of liquor, 8 oz of table wine, or 2 cans of beer on any one day (less for small women).

- If environmental contaminants (e.g., heavy metals such as mercury and organic chemicals such as pesticides) are a potential problem in the area, be on the alert for official advisories concerning foods or areas to avoid.

Adapted from Institute of Medicine, Subcommittee for a Clinical Application Guide, *Nutrition during Pregnancy and Lactation: An Implementation Guide* (Washington D.C.: National Academy Press, 1992).

TABLE 4.17

Dietary recommendations for lactating women

REFERENCES

Alderman B. W., et al., "Reproductive History and Postmenopausal Risk of Hip and Forearm Fracture," *Am J Epidemiol* 124 (1986):262–67.

American Academy of Pediatrics, *Pediatric Nutrition Handbook* (Elk Grove Il, American Academy of Pediatrics, 1993).

American College of Obstetricians and Gynecologists, "Exercise during Pregnancy and the Postnatal Period," *Technical Bulletin* no. 189 (February 1994).

Anderson, A. N., et al., "Suppressed Prolactin but Normal Neurophysin Levels in Cigarette Smoking Breastfeeding Women," *Clin Endocrinol* 17 (1982):363.

Anderson, D. M., and Pittard W. B., "Vitamin E and C Concentrations in Human Milk with Maternal Megadosing: A Case Report," *J Am Diet Assoc* 85 (1985):715–17.

Bates, C. J., et al., "The Effect of Vitamin C Supplementation on Lactating Women in Kenaba, a West African Rural Community," *Int J Vitam Nutr Res* 53 (1983):68–76.

Battin, D. et al., "Effect of Suckling on Serum Prolactin, Luteinizing Hormone, Follicle-Stimulating Hormone and Estradiol during Prolonged Lactation, *Obstet Gynecol* 65 (1985):785–88.

Blackburn, M. W., and Calloway, D. H., "Heart Rate and Energy Expenditure of Pregnant and Lactating Women," *Am J Clin Nutr* 42 (1985):1161–69.

Braude, M. C., et al., "Perinatal Affects of Drugs of Abuse," *1987 Fed Proc* 46 (1987):2446–53.

Brewer, M. M., Bates, M. R., and Vannoy, L. P., "Postpartum Changes in Maternal Weight and Body Fat Depots in Lactating vs. Nonlactating Women," *Am J Clin Nutr* 49 (1989):259–65.

Brown, K. H., and Dewey, K. G., "Relationships Between Maternal Nutritional Status and Milk Energy Output of Women in Developing Countries," in *Mechanisms Regulating Lactation and Infant Nutrient Utilization*, eds. M. F. Picciano and B. Lonnerdal (New York: Wiley-Liss, 1992).

Brown, K. H., et al., "Clinical and Field Studies of Human Lactation: Methodological Considerations," *Am J Clin Nutr* 35 (1982):742.

Brown, K. H., et al., "Lactational Capacity of Marginally Nourished Mothers: Relationships Between Maternal Nutritional Status and Quantity and Proximate Composition of Milk," *Pediatrics* 78 (1986):909–19.

Butte, N. F., and Calloway, D. H., "Evaluation of Lactational Performance of Navajo Women," *Am J Clin Nutr* 34 (1981):2210–15.

Butte, N. F., et al., "Effect of Maternal Diet and Body Composition on Lactation Performance," *Am J Clin Nutr* 39 (1984):296–306.

Casey, C. E. and Hambidge, K. H., "Nutritional Aspects of Human Lactation: Minerals," in *Lactation: Physiology, Nutrition and Breastfeeding*, eds. M. C. Neville and M. R. Neifert (New York: Plenum Press, 1983).

Casey, C. E., Neville, M. C. and Hambidge, K. M., "Studies in Human Lactation: Secretion of Zinc, Copper and Manganese in Human Milk," *Am J Clin Nutr* 49 (1989):773–85.

Centers for Disease Control, "Recommendations for the Use of Folic Acid to Reduce the Number of Cases of Spina Bifida and Other Neural Tube Defects," *MMWR* 41 (1992):RR–14.

Chappell, J. E., Clandinin, M. T., and Kearney-Volpe, C., "Trans Fatty Acids in Human Milk Lipids: Influence of Maternal Diet and Weight Loss," *Am J Clin Nutr* 42 (1985):49–56

Chappell, J. E., Francis, T., and Clandinin, M. T., "Simultaneous High Performance Chromatography Analysis of Retinol Esters and Tocopherol Isomers in Human Milk," *Nutr Res* 6 (1986):849–52.

Chasnoff, I. J., Lewis D. E., and Squires, L., "Cocaine Intoxication in a Breastfed Infant," *Pediatrics* 80 (1987):836–38.

Cobo, E., "Effect of Different Doses of Ethanol on the Milk-Ejecting Reflex in Lactating Women," *Am J Obstet Gynecol* 115 (1973):817–21.

Committee on Drugs, "Transfer of Drugs and Other Chemicals into Human Milk," *Pediatrics* 84 (1989):924–36.

Cooperman, J. M., et al., "The Folate in Human Milk," *Am J Clin Nutr* 36 (1982): 576-80.

Dabeka, R. W., et al., "Survey of Lead, Cadmium and Fluoride in Human Milk and Correlation Levels with Environment and Food Factors," *Food Chem Toxicol* 24 (1986):913–21.

Dahlstrom, A., et al., "Nicotine and Cotinine Concentrations in the Nursing Mother and her Infant," *Acta Pediatr Scand* 79 (1990):142–7.

Dallman, P. R., "Iron Deficiency in the Weanling: A Nutritional Problem on the Way to Solution," *Acta Pediatr Scand Suppl* 323 (1986): 59–67.

Delange, F., "Physiopathology of Iodine Nutrition," in *Trace Elements in Nutrition of Children*, ed. R. K. Chandra, *Nestle Nutrition Workshop Series*, Vol. 8 (New York: Raven Press, 1985).

Dewey, K. G. and Lonnerdal, B., "Infant Self-regulation of Breastmilk Intake," *Acta Pediatr Scand* 75 (1986):893–98.

Dewey, K. G., et al., "A Randomized Study of the Effects of Aerobic Exercise by Lactating Women on Breastmilk Volume and Composition," *N Engl J Med* 330 (1994):449–53.

Dewey, K. G., et al., "Maternal vs Infant Factors Related to Breastmilk Intake and Residual Milk Volume: The DARLING Study," *Pediatrics* 87 (1992):829–37.

Diaz, S., et al., "Early Difference in the Endocrine Profile of Long and Short Lactational Amenorrhea," *J Clin Endocrinol Metab* 72 (1991):196.

Ekstrand, J., et al., "Distribution of Fluoride to Human Breast Milk: Following High Doses of Fluoride," *Caries Res* 18 (1984):93–5.

Finley, D. A., et al., "Food Choices of Vegetarians and Nonvegetarians during Pregnancy and Lactation," *J Am Diet Assoc* 85 (1985):678–685.

Ford, K., and Labbok, M., "Contraceptive Usage During Lactation in the US: An Update," *Am J Pub Health* 77 (1987):79–81

Fomon, S. J., *Nutrition of Normal Infants* (St. Louis: Mosby, 1993).

Forsum, E., and Lonnerdal, B., "Effect of Protein Intake on Protein and Nitrogen Composition of Breast Milk," *Am J Clin Nutr* 33 (1980):1809–13.

Gambon, R. C., Lentze, M. J., and Rossi, E., "Megaloblastic Anemia in one of Monzygous Twins Breast Fed by Their Vegetarian Mother," *Eur J Pediatr* 145 (1986):570–571.

Gebre-Medhin, M., al., "Breastmilk Composition in Swedish and Ethiopian Mothers," *Am J Clin Nutr* 29 (1976):441–51.

Gladen B. C., et al., "Development After Exposure to PCB and DDE Transplacentally and Through Human Milk," *J Pediatr* 113 (1988):991–95.

Goldfarb, J., "Breastfeeding: AIDS and Other Infectious Diseases," *Clin Perinatol* 20 (1993):225–44.

Goldman, A. S., and Goldblum, R. M., "Immunoglobulins in Human Milk," in *Protein and Nonprotein Nitrogen in Human Milk.* eds. S. A. Atkinson and B. Lonnerdal (Boca Raton: CRC Press,1989).

Goldman, A. S., et al., "Anti-Inflammatory Properties of Human Milk," *Acta Pediatr Scand* 75 (1986):689–95.

Gould, S. F., "Anatomy of the Breast," in *Lactation: Physiology, Nutrition and Breastfeeding*, eds. M. C. Neville and M. R. Neifert (New York: Plenum Press, 1983).

Greer, F. R., et al., "Effects of Maternal Ultraviolet B Irradiation on Vitamin D Content of Human Milk," *J Pediatr* 105 (1984): 431.

Habicht, J. P., et al., "The Contraceptive Role of Breastfeeding," *Popul Stud* 39 (1985):213–32.

Hachey, D. L., et al., "Human Lactation II: Endogenous Fatty Acid Synthesis in the Mammary Gland," *Pediatr Res* 25 (1989):63–68.

Hamosh, M., "Enzymes in Human Milk: Their Role in Nutrient Digestion, Gastrointestinal Function and Nutrient Delivery to the Newborn Infant," in *Textbook of Gastroenterology and Nutrition in Infancy*, ed. E. Lebenthal, 2d ed. (New York: Raven Press, 1989).

Hashizume, S., Kuroda, K., and Murakami, H., "Identification of Lactoferrin as an Essential Growth Factor for Human Lymphocyte Lines in Serum-Free Media," *Biochem Biophys Acta* 763 (1983):377.

Heinig, M. J., Nommsen, L. A., and Dewey, K. G., "Lactation and Postpartum Weight Loss," *FASEB J* 4 (1990):362

Hibberd, C. M., et al., "Variation in the Composition of Breastmilk During the First Five Weeks of Lactation: Implications for the Feeding of Preterm Infants," *Arch Dis Child* 57 (1982):658.

Hollis, B. W., Lambert, P. W., and Horst, R. L., "Factors Affecting the Antirachitic Sterol Content of Native Milk," in *Perinatal Calcium and Phosphorus Metabolism* eds. M. F. Holick, T. K. Gray, and C. S. Anast (Amsterdam: Elsevier, 1983).

Hopkinson, J. M., et al., "Milk Production by Mothers of Premature Infants: Influence of Cigarette Smoking," *Pediatrics* 90 (1992):934.

Hreshchyshyn, M. M., et al., "Associations of Parity, Breastfeeding and Birth Control Pills with Lumbar Spine and Femoral Neck Bone Densities," *Am J Obstet Gynecol* 159 (1988):318–22.

Humenick, S. S., "The Clinical Significance of Breastmilk Maturation Rates," *Birth* 14 (1987):174–79.

Iacopetta, B. J., et al., "Epidermal Growth Factor in Human and Bovine Milk," *Acta Pediatr Scand* 81 (1992):287.

Illingworth, P. J., et al., "Diminution in Energy Expenditure During Lactation," *Br Med J* 292 (1986):437–41.

Institute of Medicine, *Nutrition during Pregnancy and Lactation: An Implementation Guide* (Washington D.C.: National Academy Press, 1992).

Institute of Medicine, *Nutrition during Lactation* (Washington D.C.: National Academy Press, 1991).

Institute of Medicine, *Nutrition during Pregnancy* (Washington D.C.: National Academy Press, 1990).

Jannsen, L., Akesson, B., and Holmberg, L., "Vitamin E and Fatty Acid Composition of Human Milk," *Am J Clin Nutr* 34 (1981):8.

Jen, K. L. C., Juuhl, N., and Lin, P. K. H., "Repeated Pregnancy without Lactation: Effects on Carcass Composition and Adipose Tissue Cellularity in Rats," *J Nutr* 118 (1988):93–8.

Jensen, A. A., "Levels and Trends of Environmental Chemicals in Human Milk," in *Chemical Contaminants in Human Milk*, eds. A. A. Jensen and S. A. Slorach (Boca Raton: CRC Press, 1991).

Jensen, R. G., *The Lipids of Human Milk* (Boca Raton, CRC Press, 1989).

Johnson, P. E., and Evans, G. W., "Relative Zinc Availability from Human Milk, Infant Formulas and Cow's Milk," *Am J Clin Nutr* 31 (1978):416.

Johnston, L., Vaughn, L., and Fox, H. M., "Pantothenic Acid Content of Human Milk," *Am J Clin Nutr* 34 (1981):2205.

Kelsey, J. L., and John, E. M., "Lactation and the Risk of Breast Cancer," *N Engl J Med* 330 (1994):136–7.

King, J. C., et al., "Calcium Metabolism during Pregnancy and Lactation," in *Nutrient Regulation during Pregnancy, Lactation and Human Growth*, eds. L. Allen, J. C. King, and B. Lonnerdal (New York: Plenum Press, 1994).

Koetsawang, S., "The Effects of Contraceptive Methods on the Quality and Quantity of Breastmilk," *Int J Gynecol Obstet* 25 (Suppl) (1987):115–27.

Krebs-Smith, S. M., and Clark, L. D., "Validation of a Nutrient Adequacy Score for Use with Women and Children," *J Am Diet Assoc* 89 (1989):775–83.

Kumpulanien, J., "Selenium: Requirement and Supplementation," *Acta Pediatr Scand Suppl* 351 (1989):114–7.

Lawrence, R. A., "Breastfeeding: A Guide for the Medical Profession," (St. Louis: Mosby Year-Book, 1994).

Lindemann, R., and Christensen G. C., "Radioactivity in Breastmilk After the Chernobyl Accident," *Acta Pediatr Scand* 76 (1987):981–2.

Little, R. E., et al., "Maternal Alcohol Use During Breastfeeding and Infant Mental and Motor Development at One Year," *N Engl J Med* 321 (1989):425–30.

Lonnerdal, B., "Casein Content of Human Milk," *Am J Clin Nut* 41 (1985):113.

Lovelady, C. A., Lonnerdal, B., and Dewey, K. G., "Lactation Performance of Exercising Women," *Am J Clin Nutr* 52 (1990):103–09.

Mannan, S., and Picciano, M. F., "Influence of Maternal Selenium Status on Human Milk Selenium Concentration and Glutathione Peroxidase Activity," *Am J Clin Nutr* 46 (1987):95–100.

Matheson, I., and Rivrud, G. N., "The Effect of Smoking on Lactation and Infantile Colic," *JAMA* 26 (1989):42–43

Metz, J., Zalusky, R., and Herbert, V., "Folic Acid Binding by Serum and Milk," *Am J Clin Nutr* 21 (1968):289–97.

Mock, D. M., "Biotin in Human Milk: When, Where and in What Form?" in *Nutrient Regulation during Pregnancy, Lactation and Human Growth*, eds. L. Allen, J. C. King and B. Lonnerdal (New York: Plenum Press, 1994).

Motil, K. C., et al., "Dietary Protein and Nitrogen Balance in Lactating and Nonlactating Women," *Am J Clin Nutr* 51 (1990):378–84.

Motil, K. J., Montandon, C. M., and Garza C., "Effect of Dietary Protein Intake on Milk Production in Lactating Women," *Am J Clin Nutr* 43 (1986):677.

Munoz L. M., et al., "Coffee Consumption as a Factor in Iron Deficiency Anemia Among Pregnant Women and Their Infants in Costa Rica," *Am J Clin Nutr* 48 (1988):645–51.

Naing, K. M., and Oo, T. T., "Effect of Dietary Supplementation on Lactation Performance of Undernourished Burmese Mothers," *Food Nutr Bull* 9 (1987):59–61

National Academy of Sciences, Food and Nutrition Board of the National Research Council, *Recommended Dietary Allowances*, 10th ed. (Washington D.C.: National Academy Press, 1989).

Neville, M. C., "The Physiological Basis of Milk Secretion. Part 1: Basic Physiology," *Ann NY Acad Sci* 586 (1990):1.

Neville, M. C., et al., "Studies in Human Lactation: Milk Volumes in Lactating Women During the Onset of Lactation and Full Lactation," *Am J Clin Nutr* 48 (1988): 1375-86.

Newcomb, P. A., et al., "Lactation and a Reduced Risk of Breast Cancer," *N Engl J Med* 330 (1994):81–87.

Newton, N., "The Relation of the Milk-ejection Reflex to the Ability to Breastfeed," *Ann NY Acad Sci* 652 (1992):484.

O'Connor, D. L., "Folate Status during Pregnancy and Lactation," in *Nutrient Regulation during Pregnancy, Lactation and Human Growth*, eds. L. Allen, J. C. King, and B. Lonnerdal (New York: Plenum Press, 1994).

Ohlin, A., and Rossner, S., "Maternal Body Weight Development after Pregnancy," *Int J Obes* 14 (1990):159–73.

Patton, S., et al., "Carotenoids in Human Colostrum," *Lipids* 25 (1990):159–65.

Pellizari, E. D., et al., "Purgeable Organic Compounds in Mother's Milk," *Bull Environ Contam Toxicol* 28 (1982):322.

Prentice, A., "The Effect of Maternal Parity on Lactational Performance in a Rural African Community," *in Human Lactation 2: Maternal and Environmental Factors*, eds. M. Hamosh and A. S. Goldman (New York: Plenum Press, 1986).

Prentice, A., et al., "Breastmilk Fatty Acids of Rural African Women: Effect of Diet and Maternal Parity," J *Pediatr Gastroenterol Nutr* 8 (1989):486–90.

Prentice, A. M., et al., "Dietary Supplementation of Lactating Gambian Women," *Hum Nutr:Clin Nutr* 37C (1983):53–64.

Prentice, A. M., et al., Dietary Supplementation of Gambian Women, Effect on Maternal Health, Nutritional Status, and Biochemistry," *Hum Nutr: Clin Nutr* 37C (1983):65–74.

Prentice, A. M., Lamb, W. H., and Prentice, A., "The Effect of Water Abstention on Milk Synthesis in Lactating Women," *Clin Sci* 66 (1984):291–98.

Riby, J. "Perinatal Development of Digestive Enzymes," in *Infant Nutrition*, eds. A. F. Walker and B. A. Rolls (London: Chapman and Hall, 1994).

Roberts, S. B., Cole, T. J., and Coward, W. A., "Lactational Performance in Relation to Energy Intake in the Baboon," *Am J Clin Nutr* 41 (1985):1270–76.

Rookus, M. A., et al., "The Effect of Pregnancy on the Body Mass Index 9 Months Postpartum in 49 Women," *Int J Obesity* 11 (1987):609–18.

Sanchez, L., Calvo, M., and Brock, J. H., "Biological Role of Lactoferrin," *Arch Dis Child* 67 (1992):657.

Sandor, A. et al., "On the Carnitine Content of Human Breast Milk," *Pediatr Res* 16 (1982):89.

Short, R. V., et al., "Contraceptive Effects of Extended Lactational Amenorrhea: Beyond the Bellagio Conference," *Lancet* 337 (1991):715.

Sneed, S. M., Zane, C., and Thomas, M. R., "The Effects of Ascorbic Acid, Vitamin B_6, Vitamin B_{12} and Folic Acid Supplementation on the Breast Milk and Maternal Nutritional Status of Low Socioeconomic Lactating Women," *Am J Clin Nutr* 34 (1981):1338-1346.

Sowers, M. F., et al., "Changes in Bone Density with Lactation." *JAMA* 269 (1993):3130.

Stoneham, S., et al., "Adenosine and the Regional Difference in Adipose Tissue Metabolism in Women," *Acta Endocrinol* 327 (1988):118–31.

Strode, M. A., Dewey, K. G., and Lonnerdal, B., "Effects of Short Term Caloric Restriction on Lactational Performance of Well-Nourished Women," *Acta Pediatr Scand* 75 (1986):222–29.

Sturman, J. A., "Taurine in Development," *J Nutr* 118 (1988):1169.

Styslinger, L., and Kirksey, A., "Effects of Different Levels of Vitamin B_6 Supplementation on Vitamin B_6 Concentrations in Human Milk and Vitamin B_6 Intakes of Breastfed Infants,"*Am J Clin Nutr* 41 (1985):21–31.

Tay, C. C. K., Glasier, A., and McNeilly, A. S., "Endocrine Control of Lactation," in *Mechanisms Regulating Lactation and Infant Nutrient Utilization*, eds. M. F. Picciano and B. Lonnerdal (New York: Wiley-Liss, 1992).

Thapa, S., Short, R. V., and Potts, M,. "Breastfeeding, Birthspacing and Their Effects on Child Survival," *Nature* 335 (1988):679–82.

Tulchinsky, D., "Postpartum Lactation and the Resumption of Reproductive Functions," in *Maternal-Fetal Endocrinology*, eds. D. Tulchinsky and A. B. Little, 2d ed. (Philadelphia: W. B. Saunders, 1994).

van der Elst, C. W., et al., "Serum Zinc and Copper in Thin Mothers, Their Breast Milk and Infants," *J Trop Pediatr* 32 (1986):111–14

von Kries, R., Shearer, M. J., and Gobel, U., "Vitamin K Deficiency in Infancy," *Eur J Pediatr* 147 (1988):106–12

Wallace, J. P., Inbar, G., and Ernsthausen, K., "Infant Acceptance of Post-exercise Breast Milk," *Pediatrics* 89 (1992):1245.

Wallingford, J. C., and Underwood, B. A., "Vitamin A Deficiency in Pregnancy, Lactation and the Nursing Child," in *Vitamin A Deficiency and Its Control*, ed. J. Bauerenfeind (New York: Academic Press, 1986).

WHO, Task Force on Oral Contraceptives, Effects of Hormonal Contraceptives on Breastmilk Composition and Infant Growth," *Std Fam Plan* 19 (1988):361–69.

WHO, "Energy and Protein Requirements," *Technical Report Series* 724 (Geneva: World Health Organization, 1985).

WHO, "The Quantity and Quality of Breastmilk," *Report on the WHO Collaborative Study on Breastfeeding* (Geneva: WHO, 1985).

Wilde, C. J., et al., "Feedback Inhibition of Milk Secretion: The Effect of a Fraction of Goat Milk on Milk Yield and Composition," *Q J Exp Physiol* 73 (1988):391–97.

Willams, A. F., "Lactation and Infant Feeding," in *Textbook of Pediatric Nutrition*, eds. D. S. McLaren, et al., 3d ed. (Edinburgh: Churchill Livingstone, 1991).

Williamson, D. H., and Lund, P., "Cellular Mechanisms for the Regulation of Adipose Tissue Lipid Metabolism in Pregnancy and Lactation, in *Nutrient Regulation during Pregnancy, Lactation and Human Growth*, eds. L. Allen, J. C. King, and B. Lonnerdal (New York: Plenum Press, 1994).

Wray, J. D., "Breastfeeding: An International and Historical Review," in *Infant and Child Nutrition Worldwide: Issues and Perspectives*, ed. F. Falkner (Boca Raton: CRC Press, 1991).

5 breast-feeding

▼

CULTURE AND BREAST-FEEDING

Culture has a major influence on breast-feeding practices and patterns: a mother's attitude toward breast-feeding is largely determined by the attitudes of her society (Baumslag 1987). Beliefs and cultural norms concerning breast-feeding vary substantially, but the vast majority of infants worldwide are breast-fed. In most cultures, breast-feeding a newborn infant is perceived as the traditional, natural order of life. Indeed, a rich history of ceremonies and rituals may surround the onset of lactation and be directed toward ensuring an ample supply of milk (Guthrie et al. 1983). A new mother may even be allowed a one-month period of seclusion to become acquainted with her infant and establish her milk supply (Jimenez and Newton 1979).

In the United States, the cultural context of breast-feeding is very different. Although most women view breast-feeding as the optimal way to feed their infants, societal norms and customs discourage it. Many women work outside the home and are unable to take extended maternal leave from their jobs. High-quality infant formula is widely available and affordable. Also, many women have, as role models, mothers and grandmothers from generations that mostly formula-fed their infants. Moreover, breast-feeding is seen as an intimate and private activity not to be practiced in public. All of these influences are part of the cultural context of infant feeding in the United States, and they help define our society's attitudes and beliefs about the practice.

The powerful influence of this cultural context can be seen in the patterns of breast-feeding in women who immigrate to the United States from Southeast Asia. Although nearly all mothers in Vietnam and Indochina breast-feed their infants, very few Vietnamese and Indochinese women living in the United States breast-feed. In a recent survey of low-income Southeast Asian families in the United States, over 90% of the infants born in their native country had been breast-fed, compared with only 10% of those born in the United States (Serdula et al. 1991).

HISTORICAL TRENDS IN BREAST-FEEDING

Female homo sapiens, along with other members of the class Mammalia, synthesize and secrete breast milk that, for a time, can provide the sole nourishment for their newborns: for thousands of years, human mothers have nursed their infants at the breast.

Breast-feeding practices of modern hunter–gatherer societies, such as the !Kung people of the Kalahari Desert, are thought to be similar to the practices of our early prehistoric ancestors. Nursing consists of frequent, short feeds (about two minutes per feed and four feeds per hour) that are equally distributed throughout the day; nursing continues until the child is 2 to 4 years old (Konner and Worthman 1980). Such patterns of breast-feeding have existed for millennia, and breast milk continues to be the only infant food for most of the world's peoples.

Although formula feeding has only become common in the twentieth century, the search for alternatives to maternal breast-feeding goes back to ancient times. Writings from ancient Egypt and Greece contain records of infants consuming prepared foods or animal milk. Before the twentieth century, the most common alternative to maternal breast-feeding was use of a **wet nurse**. Wet nursing existed in ancient Babylonia and India and is described in the Old Testament of the Bible and in the Koran (Coates 1993).

Throughout Europe from 1500 into the 1700s, wet nurses were widely employed by women of the middle and upper classes. They often

wet nurse: a woman who breastfeeds the child of another at her own breast

sent a newborn away to a wet nurse to be returned to its parents only after it had been **weaned**. But by the late 1700s, the use of wet nurses was declining in England and North America because of public concern over the quality of care. Issues such as the moral character of the wet nurse (many earned money as prostitutes) and their health (tuberculosis and syphilis were common) caused many mothers to return to breast-feeding.

In the late nineteenth and early twentieth centuries, the impact of the Industrial Revolution changed the lifestyles of many families. The extended family began to disintegrate as people migrated in large numbers to the big cities. Many women entered the workforce and were no longer able or willing to breast-feed for long periods.

At the same time, the development and promotion of manufactured baby milks made formula feeding easy, cheap, and safe. Technological innovations, including the availability of glass bottles with rubber nipples (which could be easily cleaned) and refrigeration, were introduced. Sanitary water supplies were widely available, and the pasteurization process for milk was developed. Knowledge about infant and child nutrition expanded. For example, supplements of cod liver oil were found to prevent rickets in children.

Advances in disease prevention and medical knowledge led to an increasing faith in modern science and medicine. The food science industry began promoting infant formula to both mothers and physicians, using as major themes the difficulty of breast-feeding and the perfection of the manufactured formula. By the early 1900s, formulas based on cow's milk were widely available. Many physicians believed formula was comparable to human milk, and the use of formula in place of breast-feeding was often encouraged. At the same time, prevailing Victorian ideas of modesty discouraged breast-feeding in public.

Formula feeding was appealing to many women who wanted freedom from traditional sexual, social, and political roles. These women were influential—many were educated and from the upper socioeconomic classes—and they served as role models for other women in society (Coates 1993).

All of these factors contributed to the decline in breast-feeding in the United States from 1900 to 1970. At the turn of the century, about nine out of ten infants were breast-fed, and until World War II there were only modest declines: in 1936–40 about 75% of infants were still being breast-fed. However, between 1940 and 1970 the number of mothers choosing to breast-feed fell dramatically (Institute of Medicine 1991). Duration of breast-feeding fell as well, from an average of over 4 months in the 1930s to 2.2 months in the late 1950s. Breast-feeding in the United States reached its nadir in 1972, with only about two out of every ten infants receiving breast milk (see Figure 5.1).

In the 1970s, the trend reversed itself. The number of women breast-feeding their infants increased sharply in the United States from

wean: to discontinue breastfeeding an infant, with substitution of other foods

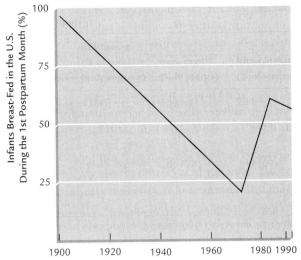

Adapted from Fomon
(1993)

FIGURE 5.1

*Percentage of U.S infants
breast-fed during the first
month (1900–1992)*

1972–82, with over sixty percent of women breast-feeding during early
infancy. What prompted this rapid turnaround? Research began to
identify the complex and potent immunological properties of human
milk, and scientists became increasingly aware of the biological specificity of breast milk and its unique benefits to infant health (Coates
1993). Also, many women's groups began to promote breast-feeding as
part of a move back to a simpler, more natural lifestyle.

Educated women with high incomes were the first to return to breast-
feeding (Grossman et al. 1990), just as they had been the first to embrace
formula feeding in the first half of the century. However, the increase was
not limited to women of higher socioeconomic levels: among women
whose education did not go beyond high school, rates of breast-feeding
increased more than five times (from 10 to 51%) from 1971 to 1981, and

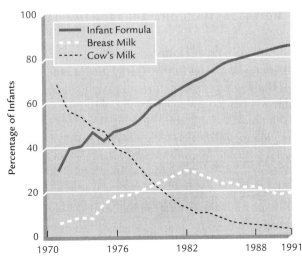

Adapted from Fomon
(1993)

FIGURE 5.2

*Percentage of U.S. infants
receiving breast milk,
cow's milk, or formula at
five to seven months
(1971–1991)*

in mothers from lower income families (less than $15,000/year), rates more than tripled (Institute of Medicine 1991). Despite these trends, in the United States and other industrialized countries, the incidence of breast-feeding among poor and disadvantaged women continues to be low.

The apex of breast-feeding was in 1982, when 61% of mothers were initiating breast-feeding, and 40% were nursing for three months or longer. There has been a persistent, slow decline since then, with the latest figures showing about 52% of women in the United States initiating breast-feeding, but only 20% continuing at five to seven months (Fomon 1993). See Figure 5.2.

THE DEMOGRAPHICS AND DETERMINANTS OF BREAST-FEEDING

ETHNICITY AND BREAST-FEEDING IN THE UNITED STATES

As Table 5.1 shows, there are marked differences in breast-feeding among mothers of different ethnic backgrounds. Far fewer Black mothers initiate breast-feeding (23%) and continue it at six months (7%), compared with White mothers (58 and 23%, respectively). Why Black mothers breast-feed less frequently than Whites is not clear— the differences remain when income, parity, and marital status are taken into account.

Until the 1950s, the reverse was true: more Black women than White women were breast-feeding, but the fall in breast-feeding during the 1960s was greater in Blacks than in Whites. By 1970 the rate of breast-feeding in Black women was half that in White women (14% versus 29%, respectively), and although rates have increased in both groups, the trend toward greater breast-feeding has been stronger in Whites (Fomon 1993). See Table 5.2.

Hispanic mothers breast-feed at intermediate rates (48% initiate nursing, with 15% continuing at six months), and mothers born in their native country are more likely to breast-feed than Hispanic women born in the United States. Studies have found that Hispanic mothers who intend to work are much less likely to initiate breast-feeding than those who do not (Institute of Medicine 1991).

Rates of breast-feeding in Southeast Asian mothers in the United States are declining and are even lower than the reported rates for other minority groups. Southeast Asian women tend to consider formula feeding the norm in the United States. They associate it with the general good health of infants in this country (Coates 1993).

Category	Total %		White %		Black %		Hispanic %	
	New-borns	5–6 mo Infants	New-borns	5–6 mo Infants	New-borns	5–6 mo Infants	New-borns	5–6 mo Infants
All mothers	52.2	19.6	58.5	22.7	23.0	7.0	48.4	15.0
Parity								
Primiparous	52.6	16.6	58.3	18.9	23.1	5.9	49.9	13.2
Multiparous	51.7	22.7	58.7	26.8	23.0	7.9	47.2	16.5
Marital status								
Married	59.8	24.0	61.9	25.3	35.8	12.3	55.3	8.8
Unmarried	30.8	7.7	40.3	9.8	17.2	4.6	37.5	8.6
Maternal age								
<20 yr	30.2	6.2	36.8	7.2	13.5	3.6	35.3	6.9
20–24 yr	45.2	12.7	50.8	14.5	19.4	4.7	46.9	12.6
25–29 yr	58.8	22.9	63.1	25.0	29.9	9.4	56.2	19.5
30–34 yr	65.5	31.4	70.1	34.8	35.4	13.6	57.6	23.4
≥35 yr	66.5	36.2	71.9	40.5	35.6	14.3	53.9	24.4
Maternal education								
No college	42.1	13.4	48.3	15.6	17.6	5.5	42.6	12.2
College	70.7	31.1	74.7	34.1	41.1	12.2	66.5	23.4
Family income								
<$7,000	28.8	7.9	36.7	9.4	14.5	4.3	35.3	10.3
$7,000–$14,999	44.0	13.5	49.0	15.2	23.5	7.3	47.2	13.0
$15,000–$24,999	54.7	20.4	57.7	22.3	31.7	8.7	52.6	16.5
≥$25,000	66.3	27.6	67.8	28.7	42.8	14.5	65.4	23.0
Maternal employment								
Full-time	50.8	10.2	54.8	10.8	30.6	6.9	50.4	9.5
Part-time	59.4	23.0	63.8	25.5	26.0	6.6	59.4	17.7
Not employed	51.0	23.1	58.7	27.5	19.3	7.2	46.0	16.7
U.S. census region								
New England	52.2	20.3	53.2	21.4	35.6	5.0	47.6	14.9
Middle Atlantic	47.4	18.4	52.4	21.8	30.6	9.7	41.4	10.8
East North Central	47.6	18.1	53.2	20.7	21.0	7.2	46.2	12.6
West North Central	55.9	19.9	58.2	20.7	27.7	7.9	50.8	22.8
South Atlantic	43.8	14.8	53.8	18.7	19.6	5.7	48.0	13.8
East South Central	37.9	12.4	45.1	15.0	14.2	3.7	23.5	5.0
West South Central	46.0	14.7	56.2	18.4	14.5	3.8	39.2	11.4
Mountain	70.2	30.4	74.9	33.0	31.5	11.0	53.9	18.2
Pacific	70.3	28.7	76.7	33.4	43.9	15.0	58.5	19.7

From Institute of Medicine, Subcommittee on Nutrition during Lactation, *Nutrition during Lactation* (Washington D.C.: National Academy Press, 1991).

TABLE 5.1

Demographic profile of breast-feeding in the United States:
% of infants breast fed at birth and 5–6 months

	1955	1960	1965	1970	1988	1992
White	49	43	39	29	58	58
Black	59	42	24	14	25	28
Hispanic	58	55	39	35	51	52

Adapted from Institute of Medicine, Subcommittee on Nutrition during Lactation, *Nutrition during Lactation* (Washington D.C.: National Academy Press, 1991) and NHANES III (1988–1991).

TABLE 5.2
Percentage of newborn infants breast-fed in the United States

SOCIOCULTURAL DETERMINANTS OF BREAST-FEEDING IN THE UNITED STATES

Other than ethnic background, the major determinants of breast-feeding in the United States include:

- *Age.* Younger mothers breast-feed much less frequently than older mothers, and rates are particularly low among teenage mothers. Also, older mothers tend to breast-feed for longer periods than younger mothers.

- *Parity.* Multiparous and primiparous women initiate breast-feeding at similar rates, while the former tend to breast-feed for longer periods.

- *Geographic differences.* Women from western states are much more likely to breast-feed than women from southern or eastern states. The lowest rates are found in Alabama, Mississippi, and Tennessee.

- *Marital status.* Unmarried women are much less likely to breast-feed than married women. Differences are particularly marked in Black women, where the rate of breast-feeding in unmarried women is less than half that of married women.

- *Education.* At least one year of college-level education is a strong positive predictor of breast-feeding.

- *Income.* Poverty is associated with decreased rates of breast-feeding. Women from households with the lowest family income are only half as likely to nurse their infants as those with family incomes more than $25,000/year.

- *Employment.* Maternal employment outside the home influences breast-feeding: mothers working full-time are less likely to initiate breast-feeding and much less likely to be breast-feeding at six months, compared with women working part-time or not working. The difficulties encountered by breast-feeding mothers who work outside the home are later discussed in this chapter.

Similar variations in breast-feeding rates among women of different socioeconomic backgrounds also occur in other industrialized countries. For example, in England, where about 60% of all women breast-feed, only 43% from lower socioeconomic levels initiate breast-feeding, compared with 87% of mothers from the highest socioeconomic levels (Editorial 1988).

BREAST-FEEDING IN THE DEVELOPING WORLD

Breast-feeding was nearly universal in the developing regions of the world until the middle of the twentieth century. However, from 1940 to 1970, as formula feeding displaced breast-feeding in the industrialized countries, greater contact was being established between Western health-care workers and the developing countries. Along with primary health care, vaccination programs, and improvements in sanitation, Western attitudes toward infant feeding were introduced to the developing world.

By direct recommendations and through training of indigenous health-care workers, infant formula was promoted as an alternative to breast-feeding, and many personnel in hospitals and clinics encouraged use of the manufactured formula (Winikoff and Laukaran 1989). As part of Western relief projects, large amounts of surplus dairy products, including cow's milk-based infant formula, were shipped to undernourished areas in the developing countries (Wade 1974).

Increasing modernization and urbanization disrupted the extended family, and more mothers began working away from home. In many areas, cultural attitudes were changed by contact with the outside developed world. Western lifestyles, including formula feeding, were perceived as role models.

Transnational food companies were eager to provide infant formula to a potentially huge market, and free or subsidized formula was widely distributed to hospitals and clinics. By the mid-1970s, a World Health Organization survey in four developing countries (India, Nigeria, Ethiopia, and the Philippines) found over 40 international corporations manufacturing and distributing infant milk products (WHO 1981).

Consequently, from 1950 to 1970, breast-feeding rates sharply declined in many developing regions, particularly among the urban poor (Helsing 1983). See Figure 5.3. In countries such as Malaysia, Chile, and the Philippines—where modernization was more advanced—breast-feeding rates fell more than the undeveloped countries (such as Zaire and Nigeria) (WHO, Part I 1989), (WHO Part II 1989). The decline in breast-feeding had severe adverse effects on infant morbidity and mortality.

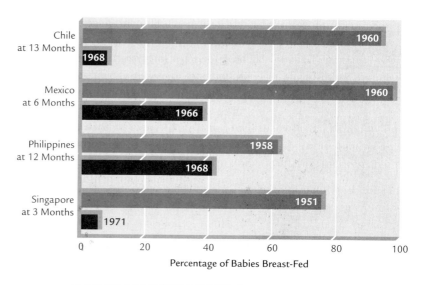

FIGURE 5.3

Decline in breastfeeding in the developing world

Adapted from Helsing and King (1983)

In impoverished developing countries, formula feeding is associated with sharply higher rates of infection and lower rates of infant survival, compared with breast-feeding. Manufactured formula lacks the immunological and other health-promoting factors of breast milk (discussed in Chapter 4). Also, in the poor countries of the developing world, water supplies are often contaminated, and unsanitary water used to clean bottles and mix formula carries pathogenic bacteria, viruses, and parasites. In these environments, bottle feeding can be a dangerous source of infection. In particular, babies fed formula are more susceptible to infectious diarrheal illnesses—a leading cause of death among newborns in many regions.

A report in 1981 documented the high costs to infant health: of the 100 million infants born each year in the developing countries in the late 1970s, 5 million were thought to suffer from diarrhea associated with bottle feeding, and over 1 million infant deaths were attributable to contaminated infant formula (WHO 1981).

Formula feeding in the developing world has other adverse effects. Formula is expensive, and widespread illiteracy and unfamiliarity with preparation methods often result in the feeding of **dilute formula**. Infants fed dilute formula suffer from undernutrition and poor growth. Also, breast-feeding is the most important form of contraception in many highly populated developing countries. Anovulation during breast-feeding contributes to child spacing by providing a period during which the mother can recover from her previous pregnancy before beginning a new one (discussed in Chapter 4).

By the late 1970s, as the importance of breast-feeding to maternal and infant health was increasingly recognized, earlier policies promoting formula feeding were reconsidered, and a renewed emphasis was

dilute formula: formula that has been prepared with too much water so that it is reduced in concentration

placed on encouraging breast-feeding. As part of the WHO/UNICEF program identifying 1979 as the International Year of the Child, the American Academy of Pediatrics and the Canadian Pediatric Society jointly issued guidelines endorsing breast-feeding (AAP 1978). In 1981, in response to concerns about widespread advertising of formula, WHO/UNICEF approved the International Code for Marketing of Breast Milk Substitutes (Armstrong 1988). The Code shown in Table 5.3 provides guidelines for the marketing of formula and prohibits direct advertising to consumers.

Many developing countries have taken firm steps to discourage use of breast milk substitutes. In the mid-1970s, Papua New Guinea enacted laws forbidding the advertisement of feeding bottles and formulas and made both available only by prescription. The government widely publicized the benefits of breast-feeding to health-care providers and the general population. Consequently, the prevalence of breast-feeding among infants younger than 2 years increased from 65% in 1976 to 88% in 1979 (Biddulph 1981).

In 1989, WHO reported on rates of breast-feeding in 15 developing countries in Asia, Africa, and Latin America (WHO, Part I 1989), (WHO, Part II 1989). During the 1980s, although breast-feeding initiation rates had stabilized and slightly increased in many countries, the median duration of breast-feeding had decreased in two thirds of the countries surveyed. A survey in Nicaragua found that the percentage of infants breast-fed at six months fell from 58% to 33% between 1977 and 1988 (Sandiford et al. 1991). Worldwide, rates of breast-feeding at birth and at

- No advertising of these products to the public.
- No free samples to mothers.
- No promotion of products in health-care facilities.
- No company nurses to advise mothers.
- No gifts or personal samples to health workers.
- No words or pictures idealizing artificial feeding, including pictures of infants, on the products.
- Information to health workers should be scientific and factual.
- All information on artificial feeding, including the labels, should explain the benefits of breast-feeding, and the costs and hazards associated with artificial feeding.
- Unsuitable products, such as condensed milk, should not be promoted for babies.
- All products should be of a high quality and take into account the climatic and storage conditions of the country where they are used.

TABLE 5.3
WHO code for marketing breast milk substitutes

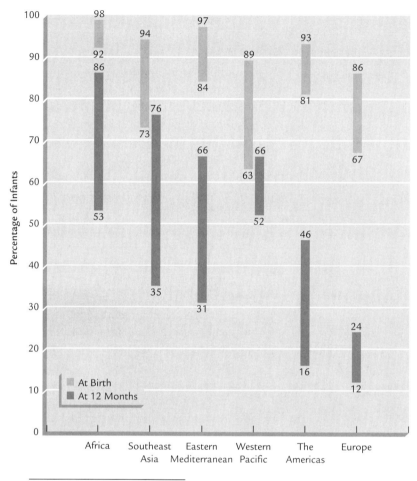

FIGURE 5.4

Percentage of infants breast-fed at birth and 12 months around the world
Source: Adapted from WHO (1989)

Source: Adapted from WHO (1989)

12 months vary considerably (see Figure 5.4). To reaffirm the importance of breast-feeding, WHO and UNICEF issued the Innocenti Declaration in 1990 (UNICEF 1990). This report put forth major goals designed to protect, promote, and support breast-feeding around the world. See Table 5.4.

MATERNAL EMPLOYMENT AND BREAST-FEEDING

Over half of all women in the United States with children under 3 years old are in the workforce, and most work full-time (U.S. Bureau of Labor Statistics 1989). Many of these women return to work during lactation, and some continue to successfully breast-feed while working. However, depending on the life situation and type of job, breast-feeding while working can be a challenge.

There are many barriers, and many working women choose either not to breast-feed at all, or to wean the infant before returning to work. The employed mother is less likely to breast-feed than the mother who stays at

Ten Steps to Successful Breast-feeding

Every facility providing maternity services and care for newborn infants should:

1. Have a written breast-feeding policy that is routinely communicated to all health-care staff.
2. Train all health-care staff in skills necessary to impliment this policy.
3. Inform all pregnant women about the benefits and management of breast-feeding.
4. Help mothers initiate breast-feeding within 30 minutes after birth.
5. Show mothers how to breast-feed, and how to maintain lactation even if they should be separated from their infants.
6. Give newborn infants no food or drink other than breast milk, unless *medically* indicated.
7. Practice rooming-in—allow mothers and infants to remain together—24 hours a day.
8. Encourage breast-feeding on demand.
9. Give no artifical teats or pacifiers to breast-feeding infants.
10. Foster the establishment of breast-feeding support groups and refer mothers to them on discharge from the hospital or clinic.

Adapted from UNICEF, Nutrition Cluster (H-8F), "Inocenti Declaration on the Protection, Promotion and Support of Breastfeeding," (Florence, Italy, 1 August 1990).

TABLE 5.4
Guidelines to support breastfeeding

home (Gielen et al. 1991). Ten percent of mothers with jobs breast-fed at six months, compared with nearly 25% of those not working (Institute of Medicine 1991).

Women who wish to breast-feed after returning to work have several options. When the mother is away, the infant can be fed stored milk—milk that is expressed from the breast and frozen for that purpose (discussed later in this chapter). Or the mother can continue to breast-feed when she is home but have her infant consume formula when she is away at work. Less commonly, she can have the baby brought to her at work for feedings, or she can return home throughout the day for feedings. All of these options have been used successfully by women choosing to breast-feed while working outside the home.

To encourage breast-feeding, several European countries have national policies allowing paid time off from work after birth of a baby. In Sweden, new mothers can take up to one year of paid leave after the birth of a child, and their job will be held for them. Danish women are entitled to six months of paid leave after the birth of a child (Auerbach 1993).

The Family and Medical Leave Act (FMLA), signed into law in the United States in 1993, makes it easier for a new mother to breast-feed her infant during early pregnancy. The FMLA says that workers at larger firms have a right to take up to 12 weeks of unpaid leave to care for a new baby or a close family member with a serious health condition (Zuffoletto 1994). The plan would also allow leave for any incapacity due to pregnancy.

When possible, the working mother should have a minimum of four to six weeks at home with her baby after birth. This period allows the mother and infant to develop a close attachment and enables the mother who chooses to breast-feed time to fully establish lactation.

A STEP-BY-STEP APPROACH TO BREAST-FEEDING

THE DECISION TO BREAST-FEED

Expectant mothers and their families are faced with a choice between breast-feeding and formula feeding. This important decision is usually made early: about 30% of families decide before pregnancy and over 90% by the end of the first trimester of pregnancy (Post and Singer 1983). A mother's choice of feeding method is determined by a number of individual, family, and sociocultural factors:

- The emotional motivations of the mother appear to be of primary importance. Motherhood and breast-feeding are potentially important opportunities for psychological growth. Many women are able to consolidate their self-concept as adults and develop increased confidence and feelings of competence and responsibility. However, many women considering breast-feeding have psychological concerns, including anxiety about being able to adequately nourish their baby, feelings of incompetence, and modesty and embarrassment about exposing their breasts.

- The attitudes of family members and friends have a modest influence on an expectant mother's decision to breast-feed. In most families, the opinion of the father of the baby has the greatest influence, but guidance from female relatives, particularly the maternal grandmother, are also important (see Table 5.5). When making a choice, women from

Ethnic Group	Influential Person Regarding Breast-Feeding
Caucasian	Male partner/baby's father Grandmother (maternal and paternal) Mother's best friend
Hispanic	Male partner/baby's father Grandmother (maternal and paternal)
African American	Peers Maternal Grandmother

Adapted from J. Riordan and K. G. Auerbach, *Breastfeeding and Human Lactation* (Boston: Jones and Bartlett, 1993).

TABLE 5.5
Whose advice mothers consider important in choosing to breast- or bottle feed by ethnic group

different ethnic groups consider advice from a variety of sources. Health-care providers—doctors and nurses—appear to have only minimal influence on the decision (Aberman and Kichoff 1985). Studies have found no correlations between a woman's history of being breast- or bottle-fed as an infant and her subsequent feeding preference as a mother (Switzky et al. 1979).

■ Social trends and role models have an important impact on many women's choice of feeding. As discussed earlier in this chapter, the increase in breast-feeding in the United States and other countries in the 1970s and early 1980s was influenced by the growth of advocacy groups, such as La Leche League International. Increased societal concerns about health, fitness, and mother–infant attachment also played a role.

When a mother makes her decision to bottle- or breast-feed, it is important that her choice be an informed one. Unfortunately, many women make this decision based on only limited information. Families should be fully aware of the advantages and disadvantages of breast- and bottle-feeding. Classes or books that explain breast-feeding, as well as informal discussions with women who have breast-fed babies and can share their experiences, are all valuable.

Breast-Feeding Education

Traditionally, inexperienced mothers have turned to their own mothers and other female relatives and friends for advice about whether to breast-feed, and for guidance and support during lactation. However, because of the marked decline in breast-feeding over the first 70 years of this century, expectant mothers today may be less familiar with breast-feeding, and there are fewer women who can share their breast-feeding experiences. Increasing geographic mobility has further isolated many expectant mothers from family support networks (Bocar and Shrago 1993).

In response, many hospitals and clinics have developed breast-feeding education programs. These programs usually have three main goals:

1. To help the expectant mother make an informed choice about infant feeding and, when appropriate, to promote breast-feeding.

2. To provide practical information and assistance in breast-feeding at the onset of lactation. Individualized guidance in the first few days of breast-feeding is important to initiate lactation, establish healthy patterns, and answer questions.

3. To provide continuing support after the establishment of lactation. Many women who are breast-feeding when they leave the hospital give up in the first month postpartum. Ongoing contact and support of the nursing mother is a strong positive influence on breast-feeding duration.

How effective is breast-feeding education? Both prenatal counseling and telephone and home follow-up visits after birth are effective in increasing duration of breast-feeding (Bocar and Shrago 1993). As an example, a program at inner-city hospitals in Arizona providing early postpartum education and support increased the mean duration of breast-feeding in the population by over 50% (Neifert 1983).

PREPARATION FOR BREAST-FEEDING

Nipple Conditioning

For some women prenatal preparation of the breasts may reduce nipple soreness during later breast-feeding. Although elaborate exercises are unnecessary, a moderate amount of nipple conditioning, such as periodic rolling of the nipples between the fingers, or rubbing the nipples to reduce sensitivity, may be helpful.

In two studies, pregnant women used nipple conditioning exercises on one breast, while the other was left alone to serve as a control. Nipple tenderness during subsequent lactation was significantly reduced in the conditioned breast (Atkinson 1979), (Storr 1988). However, too much friction can remove the oils that keep the areola and nipple moist. Often, simply going without a brassiere for periods in late pregnancy allows the nipples to rub against clothing and is adequate preparation.

Mothers should be reassured that breast size does not influence lactation. Almost all women have ample glandular tissue to support breast-feeding, and they should be reminded that variation in breast size reflects adipose tissue content, not milk-producing tissue.

Nipple Inversion

As part of the preparation for lactation, the breasts should be examined for problems with the nipple or areola. Rarely, a nipple will be inverted. An inverted nipple that everts and protrudes when it is squeezed between the thumb and finger should be left alone, as suckling by the infant will evert the nipple during feedings, and normal milk flow will occur. If the nipple does not protrude when squeezed, it is truly inverted, and it should be treated with massage or with a breast shield.

When using a breast shield, the central rubber tip is removed, and the shield is placed over the **areola**. A breast shield can be worn during the last weeks of pregnancy and early in lactation under a well-fitting, supportive brassiere. The steady, gentle pressure transmitted through the shield often everts the nipple (see Figure 5.5).

areola: the darkened ring of skin surrounding the nipple of the breast

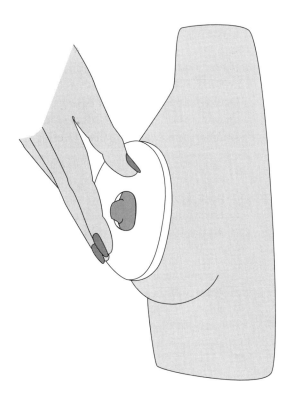

FIGURE 5.5
Using a breast shield to evert an inverted nipple

EARLY BREAST-FEEDING

The Postpartum Environment

Breast-feeding can be intimidating for the first-time mother. Ideally, during the first few feedings, an experienced health-care provider should be nearby to guide the process, answer questions, and provide support. When possible, the father should be included during this early period, as his knowledge and understanding may be important when the mother and baby are at home.

One of the major goals of postpartum confinement in the hospital or clinic is the promotion and initiation of breast-feeding in those mothers wishing to breast-feed. A supportive environment to encourage breast-feeding should include:

- Early first feeding at the breast

- Frequent feedings (at least every three hours)

- Feedings through the night

- Allowing the baby to stay in the mother's room, if desired

- Supportive nursing staff to answer questions and provide guidance

- Avoiding routine supplementary formula feeding

Initiating Breast-Feeding

If the birth has been uncomplicated and the condition of the mother and infant allow it, the mother should be encouraged to begin breast-feeding immediately after delivery. The mother should be as comfortable as possible, and the infant should be positioned so suckling and swallowing are not impaired (see Figure 5.6):

1. The most common position for breast-feeding is sitting upright in a comfortable chair with wide arms and good back, leg, and foot support. The mother cradles the baby in her arms and turns his head toward her at breast level, bringing the infant's mouth close to the nipple. The mother's other hand guides the nipple into the infant's mouth.

2. Particularly if the mother has had a cesarean delivery, or has been given anesthesia, another comfortable position is lying down on her side with pillows supporting her back and knees. The infant should be placed on his side on a pillow facing the breast while the mother's arm supports the infant's head.

The mother should support her breast by holding it in her hand, with the nipple pinched gently between the first and second fingers, and guide the nipple to the infant's mouth. The nipple should be slipped into the infant's mouth when it is open its widest, as suckling will be more effective if the nipple and a portion of the areola are grasped by the infant (see Figure 5.7). Enough of the areola should be grasped so that the infant can properly empty the collecting ducts at the base of the nipple. The correct suckling technique contributes significantly to successful breast-feeding, as shown in Table 5.6.

Although newborn infants often lick and nuzzle the nipple at first, they quickly begin to suckle vigorously. Suckling and swallowing are inborn, instinctive reflexes that appear during fetal life, usually at the beginning of the third trimester. The reflex becomes well established by 32–37 weeks in utero (Bu'Lock et al. 1990). Feeding reflexes are described in Table 5.7.

The structure of the newborn's lips, mouth, and pharynx are well-adapted for breast-feeding. The lips are slightly everted and can effect an airtight seal around the nipple and areola. The shape of the pharynx channels the incoming milk into the esophagus and away from the opening of the trachea, preventing choking on the milk. Infants breast-feed using a coordinated series of rhythmical mouth movements called the suckling cycle (Woolridge 1986). During a feeding, the infant suckles and swallows about once per second, as long as milk is present and the infant is hungry.

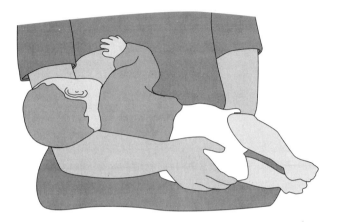

Cradle Position

Side-Lying Position

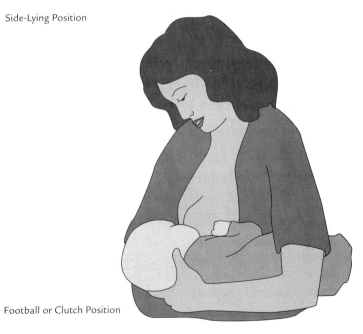

FIGURE 5.6
*Positions for
breast-feeding*

Football or Clutch Position

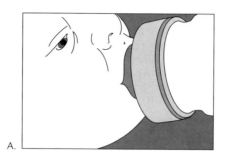

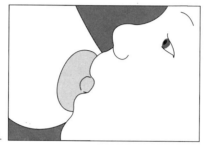

A.

B.

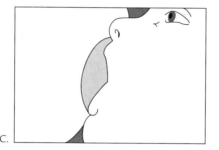

C.

(a) The position of the infant's lips and sucking action are different when bottle feeding. (b) If the infant sucks the nipple as if it was a bottle, poor milk flow and nipple soreness often result. (c) The infant is sucking in good position. It has most of the areola in it's mouth, and is very close to the breast.

Adapted from Helsing and King (1983)

FIGURE 5.7
Proper grasp of the areola and nipple by the suckling infant

Sucking Technique	Exclusively Breast-Feeding 5 days	Exclusively and Partly Breast-Feeding			
		1 mo	**2 mo**	**3 mo**	**4 mo**
No. (%) incorrect at discharge*	25 (100)	16 (64)	12 (48)	11 (44)	10 (40)
No. (%) correct at discharge	57 (100)	55 (96)	48 (84)	45 (79)	42 (74)

* Incorrect sucking consisted of the infant positioned as if bottle-feeding, sucking only at the nipple.

Adapted from L. Righard and M. O. Alade, "Sucking technique and its effect on successs of breastfeeding," *Birth* 19 (1992):185.

TABLE 5.6
Duration of breast-feeding with different sucking techniques

TABLE 5.7
Feeding reflexes in the newborn

Reflex	Stimulus/Description	Appears/Disappears
Rooting	Stroke infant's cheek/head will turn toward stimulus	Birth/disappears by 3–4 mo
Sucking	Stroke infant's lips; place finger in infant's mouth/elicits strong sucking movement	Birth/12 mo
Swallowing	Place liquid on posterior tongue/swallow follows	Persists throughout life
Gag	Stimulate posterior pharynx	Persists throughout life
Extrusion	Touch infant's tongue/tongue thrusts outward	Birth/2–4 mo

Adapted from J. Riordan and K. G. Auerbach, *Breastfeeding and Human Lactation* (Boston: Jones and Bartlett, 1993).

The First Feedings

During the first feedings, even if the infant's appetite is not large, the infant should be offered both breasts at each feeding to stimulate milk production. After a three- to four-minute suckling period on the first breast, the baby should be repositioned to the second breast. If the newborn suckles only one breast and then falls asleep, the next feeding should begin with the other breast (Woolridge et al. 1990).

Suckling should be long enough to initiate the "letdown" reflex (Mulford 1990). Prolonged feedings should be avoided at the beginning of lactation, as they can cause sore or cracked nipples (AAP 1993). A reasonable length is five minutes per breast per feeding the first day and ten minutes per breast per feeding on the second day (AAP 1993).

The newborn often suckles unevenly during the first few days. At times, the infant will not readily take the breast, while at others he will suckle vigorously. Both the mother and infant are learning a new skill, and some feeding sessions will be more successful than others. Usually, the mother and infant should remain together for the first 24 hours after delivery. Breast-feeding can continue when the infant is wakeful, as frequently as every two hours, or when the infant is ready to suckle. Feedings should occur at least every four to five hours to provide the full benefits of early and frequent feeding (AAP 1993).

Advantages of Early Feedings

Early and frequent breast-feeding has advantages for both the infant and the mother. For the infant:

- The infant's instinctive suckling reflex is most intense in the first 20–30 minutes after birth. If feeding is delayed, the newborn may have more difficultly learning to suckle properly (Anderson et al. 1982). When possible, the first feeding should occur within two hours of birth.

- The newborn is provided the immunological protection of colostrum soon after birth.

- Suckling stimulates **peristalsis** in the newborn intestine, facilitating the excretion of **meconium** (the substance that coats the intestine in utero), and preparing the intestine for digestion and absorption of milk.

- Peristalsis also increases elimination of bile products in the intestine produced by hemoglobin breakdown. Infant jaundice is less likely to occur if feeding and peristalsis begin early (discussed later in this chapter).

- Frequent and early milk intake slows the loss of weight that typically occurs in infants during the first week of life (discussed in Chapter 6).

For the mother there are also several advantages to early feeding:

- Suckling-induced release of oxytocin stimulates uterine contractions (discussed in Chapter 4), which help to control uterine bleeding after birth.

- Early and frequent feedings hasten the onset of lactation and the maturation of the milk, enabling the mother to provide the newborn with the full nutritional benefits of mature milk sooner.

- Breast engorgement (discussed later in this chapter) is less likely if feedings are frequent and begin early.

- Studies have repeatedly shown that the duration of breast-feeding is usually lengthened if feedings begin early (Winikoff et al. 1986). Early nursing facilitates attachment and bonding between the mother and infant, and successful and positive early feedings help the mother to relax and gain confidence in her ability to feed her infant.

THE FIRST TWO MONTHS OF BREAST-FEEDING

Feeding Patterns

By the third or fourth postpartum day, an infant typically breast-feeds about seven to eight times a day, and consume 300–400 ml of milk/24 hrs (Casey 1996). Feeding periods average about ten minutes each breast. The small size of the infant's stomach, and the large nutritional needs of the newborn dictate that the infant will need to be fed frequently, with some infants feeding up to 12–14 times a day (Butte et al. 1985). See Figure 5.8.

peristalsis: wavelike contractions by the muscular wall of the digestive tract that propel contents forward

meconium: dark green mucuslike material in the intestine of the newborn; it constitutes the first stools passed by the infant

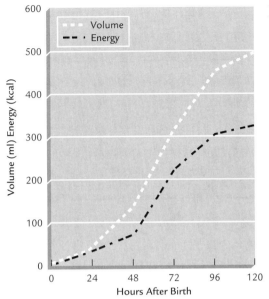

Adapted from C.E. Casey et al. (1986)

FIGURE 5.8
Daily intakes of milk during the first five days

Maternal–Infant Bonding

The infant derives a sense of belonging, security, and attachment from the warmth of the mother's body and many other maternal-infant interactions while feeding. The closeness and time spent together in the first month postpartum are important factors in material-infant bonding. (Richards 1982).

Importance of Support for the Mother

Today, mothers are often promptly discharged from the hospital after an uncomplicated delivery. This reduces the time hospital personnel have to provide instruction in breast-feeding and support during the initial stages of lactation. To prevent the potential negative effects of early discharge, an experienced nurse, physician, or lactation counselor should be available by phone during the first few weeks of lactation.

Support and guidance for the mother during this period are critical to continued breast-feeding, especially for young, first-time mothers. New mothers struggling with sore nipples or breast engorgement need to be reassured that frequent feedings are not a sign of insufficient milk and infant hunger, but the normal breast-feeding pattern.

Stooling Patterns of the Breast-Feeding Infant

A series of predictable changes occur in stooling patterns of the breast-feeding infant over the first several months postpartum. Stooling begins shortly after birth, and during the first several days, meconium is passed, and the stool appears black and tarry. With each successive feeding, the stool lightens in color and becomes more liquid. After the first week, the breast-feeding infant's stools are a mustard yellow liquid with a yeasty odor. The mother of the breast-feeding infant should be aware that many small, yellow, liquid stools are not diarrhea, but the normal pattern for breast-fed infants (Weaver et al. 1988).

BREAST-FEEDING FROM TWO TO SIX MONTHS

Feeding Patterns

Once lactation is established, the volume of milk produced during lactation is determined primarily by the infant's needs. The average milk intake of healthy infants during the first four to five months is about 750 ml/day (about 3 to 3.5 cups of breast milk) (Institute of Medicine 1991). However, milk intake varies widely between infants, and from day to day in the same infant, and ranges from 400 to 1200 ml/day.

After the first few months, although infants sleep longer at night, night-time feedings are usually still necessary. By two months, the milk letdown reflex is well conditioned and efficient, and feedings may take less time: an infant may be able to empty each breast in four to five minutes (Neifert 1983).

Appetite Spurts

During lactation, a nursing mother will often notice times when her infant's appetite suddenly increases. During these periods, termed *appetite spurts*, an infant will demand to be fed more frequently. These episodes may coincide with growth spurts and, although they can occur anytime, are often seen at about six weeks of age, and again at three and six months (Neifert 1983). Increasing the frequency of nursing for 24–48 hours will increase breast milk output sufficiently to cover infant needs during these periods.

Potential Problem Foods in the Mother's Diet

Foods with strong odors and spicy flavors, such as garlic, onion, and peppers, can influence the taste or odor of the mother's breast milk and may decrease the acceptability of the milk to the infant (Mennella and Beauchamp 1991). Also, certain foods eaten by the mother may produce allergic reactions in the breast-feeding infant, a subject discussed later in this chapter.

BREAST-FEEDING SIX MONTHS AND BEYOND

Adding Solid Foods

Most mothers begin to offer solid food to their infants at about six months. Infants are ready to accept solids when they begin to show interest in adult foods, show better hand–mouth coordination, and begin to lose their vigorous tongue-thrust during nursing (AAP 1993). Solid food should be given at the time of day when the mother's output of milk is lowest. For most women, this is late in the afternoon or evening. Chapter 6 contains a detailed discussion of introducing solid foods.

Teething

The primary teeth begin to erupt at around six to seven months. During teething, many infants will bite and chew at the nipple. This can be painful, and teething prompts many women to begin weaning. However, breast-feeding can comfortably continue during teething, as most infants quickly learn not to bite if reprimanded with a stern tone of voice. Biting can be minimized if the infant is quickly removed from the breast when active suckling slows and not allowed to play and nuzzle at the nipple. (Neifert 1983)

Psychological Benefits of Nursing

As the infant begins to eat more solid food and feedings become more sporadic, the nutritional importance of breast-feeding diminishes. However, the psychological support of nursing may continue to be important for many infants, particularly during periods of stress, illness, or fatigue.

WEANING

How Long Should a Mother Breast-Feed?

The Committee on Nutrition of the U.S. National Academy of Sciences recommends exclusive breast-feeding for normal, full-term infants from birth to age four to six months (Institute of Medicine 1991). Breast-feeding, along with supplemental feedings, is recommended for the remainder of the first year, and can be continued after the first year if desired. To provide the immunological benefits and growth factors in human milk during the newborn period, mothers should be strongly encouraged to breast-feed for at least the first two to three months (AAP 1993).

The Process of Weaning

Although the timing of *weaning* (the transition from the breast to other foods) varies widely among different populations, studies of cultures worldwide have found the mean age for weaning is around 2½ years (Neifert 1983). Breast-feeding is usually supplemented with other foods after four to six months, but breast milk continues to provide a valuable portion of infant needs. In New Guinea, where infants are normally nursed for three to four years, mothers who feed their infants mostly breast milk continue to have good milk output—over 500 ml/day—two years after delivery (Ford 1945).

In contrast, in the United States nearly two thirds of mothers who begin to breast-feed their infants stop before five to six months, and almost all women who breast-feed do so for less than a year. Because of cultural attitudes, many women in the United States who breast-feed past the first year have difficulty with acceptance from peers and family (Morse and Harnson 1987).

All children wean themselves eventually, but usually the mother is ready to wean the infant before the infant is ready to leave the breast. Many infants spontaneously give up breast-feeding at the beginning of the second year, particularly if they have been provided supplemental foods since age four to six months (AAP 1993). There is no ideal duration for breast-feeding for all families. Each mother–infant couple decides when to wean based on individual factors: maternal convenience, infant needs, and cultural customs all play a role.

Maternal Health during Weaning

Particularly during weaning, problems with plugged ducts and breast discomfort can be minimized by regular soaking of the breasts in warm water, manual expression to reduce breast fullness, and wearing a comfortable support brassiere. Women should examine their breasts carefully for signs of plugged ducts or mastitis (discussed later in this chapter). During weaning, to maintain maternal–infant attachment and closeness, the baby should be given extra cuddling and holding to replace the time formerly spent nursing. Table 5.8 lists some recommendations for weaning mothers.

- To help relieve breast soreness or fullness, shower and allow the warm water to run over the breasts, or soak the breasts by lying down in the tub.
- Use a breast pump or manual expression to relieve fullness.
- Wear a supportive, comfortable bra.
- Observe for signs of plugged ducts or breast infection.

TABLE 5.8
Recommendations for the weaning mother

Weaning and Infant Health in the Developing World

In regions of the developing world where food is scarce, sanitation poor, and fertility rates high, a child is often breast-fed until another child is born. In these areas, weaning is a period when the child is at sharply increased risk for illness and death (Wharton 1991). During the breast-feeding period, most infants grow well and stay relatively healthy. But as supplemental foods begin to replace breast milk, growth often falters and illness is more common. The infant is no longer provided the immunological benefits of milk, supplemental foods have less nutritional value, and water, food, and bottles are frequently contaminated because of poor sanitation. Diarrheal disease during weaning is common. In regions where food is scarce, kwashiorkor—a form of severe protein deficiency—appears as the infant is weaned off the breast. The term *kwashiorkor* is derived from the Ga language of Ghana, where it means "the disease of the baby displaced from the breast". Kwashiorkor is described in more detail in Chapter 7.

RELACTATION AND INDUCED LACTATION

Relactation is the process of restimulating lactation in a new mother after it has been interrupted. Relactation can occur days, weeks, or months after lactation has ended, and there are a number of circumstances in which relactation is desirable. Some women develop medical problems that require temporary cessation of breast-feeding. A woman who delivers a low-birth-weight infant (many of whom require hospitalization and specialized nutrition), and is separated from her infant for several weeks, will often notice a drop-off in milk output in the absence of suckling. When the baby returns home, the mother often wishes to relactate. Some women who chose to wean their infants early or had problems with nursing and gave up, change their minds, and wish to begin breast-feeding again.

In general, the shorter the time between stopping and attempting to relactate, the easier it is to do. Compared to a mother who has stopped lactating entirely, a woman who is still producing even small amounts of milk is much more likely to be able to reestablish her milk supply. The success of relactation also depends on the age and the infant's willingness to suckle. Ordinarily, the younger the baby, the more willing it will be to suckle.

Induced lactation is the stimulation of lactation in a woman who has not been recently pregnant. Increasingly, mothers of adopted infants are choosing to attempt induction, and many are at least partially successful in breast-feeding. (Auerbach and Avery 1981).

Induction is more likely to be successful in women who have given birth or nursed in the past. A great deal depends on the feeding style of the baby—how vigorously and frequently it is willing to suckle. In both relactation and induced lactation, the major stimulus to milk production is effective, regular suckling of the breasts, together with breast massage and nipple stimulation. As an adjunct, medications and hormones can be used to help promote lactation.

If the mother's milk output is low, the infant can be encouraged to suckle by use of one of several nursing supplementers. These devices, as shown in Figure 5.9, allow formula from a bag to flow through a small tube, the end of which is taped to the nipple. When the baby suckles, it receives both breast milk from the nipple and formula from the tube. As milk production increases, the mother can gradually wean the infant from the supplementer.

EXTRACTION AND STORAGE OF BREAST MILK

Manual Expression of Milk

Most mothers will be away from their infants for brief periods during lactation. These periods may include a feeding, and the infant can be provided with expressed breast milk that has been stored.

Manual expression of breast milk is a useful and simple technique to learn. A warm, damp cloth is applied to the breast, and the breast is mas-

Adapted from Lawrence (1994)

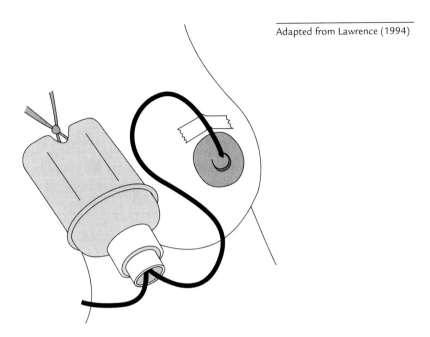

FIGURE 5.9
Nursing supplementer

saged for several minutes, or until the mother feels milk letdown. The mother should then wash her hands thoroughly with soap and water. A clean, wide-necked container to catch the milk should be held in one hand, while the other hand pumps the breast.

The thumb and forefinger are placed above and below the nipple, just behind the areolar margin. The finger and thumb are then gently pressed into the breast toward the chest wall so they grasp the milk sinuses behind the areola, and the breast is gently squeezed and pulled outward in a pumping motion. This motion is done repeatedly, and the grasp on the breast should be shifted occasionally to empty all of the sinuses.

Breast Pumps

Many women, particularly those that are away from their infants for prolonged periods (including mothers with infants in the hospital and those working outside the home) prefer to use a breast pump to collect milk. Pumps are often more efficient and more comfortable than manual expression. A variety of well-designed mechanical and electrical breast pumps are available. All of them simulate suckling by alternating negative and positive pressure on the breast to extract milk. Only as much suction as needed to maintain milk flow should be used, and pumping should stop as soon as milk output slows.

In a mother whose baby is preterm or ill and cannot be directly breast-fed, pumping should be done at least eight times each day. This will avoid pooling of milk in the glands, which can reduce milk secretion.

Milk Storage

One of the advantages of breast-feeding over formula feeding is the relative freedom from bacterial contamination of human milk. However, human milk that is extracted and stored poorly can easily become contaminated. Milk that is not refrigerated is generally only safe from bacterial contamination for about six hours after collection. Milk that is to be used within 48 hours of collection can be safely refrigerated, but milk that is stored longer should be frozen (Auerbach 1993). Freezing preserves most of the immune properties of the milk (it maintains SIgA levels) and all of its nutritional value.

Other techniques for milk storage, including heat treatment and lyophilization, are much more detrimental, as they reduce the digestibility and levels of protective factors in the milk (Schmidt 1982). Microwaving milk significantly reduces levels of immune factors, as shown in Table 5.9. Milk should be frozen in plastic (polypropylene) containers and gently thawed before use. It should not be refrozen after it thaws. Table 5.10 gives guidelines for safe storage of milk.

	No.	Control	Low Microwave	High Microwave
Lysozyme activity, µg/ml	22	23.7 ± 4.0	19.2 ± 3.4	0.9 ± 0.72
Total IgA, mg/dl	22	73.3 ± 16.1	48.9 ± 15.8	1.55 ± 1.54

Low power setting = 20–25 degrees C; high setting = 70–98 degrees C.

Adapted from R. Quan, et al., "Effects of microwave radiation on anti–infective factors in human milk," *Pediatrics* 89 (1992):667.

TABLE 5.9
Microwaving and anti-infective factors in milk

- Use a clean container.

- Label each container with date and time.

- Store milk in the approximate quantities that the baby is likely to need for one feeding.

- If refrigerated, use within two to five days.

- If frozen in a refrigerator freezer section, use within one month.

- If frozen in a deep-freezer, use within six months.

- Discard any remaining milk that is not used at the feeding for which it was thawed and warmed.

TABLE 5.10
Guidelines for storage of breast milk

MATERNAL CONCERNS DURING BREAST-FEEDING

INSUFFICIENT MILK AND FAILURE TO THRIVE

During the first few weeks postpartum, one of the main concerns of many breast-feeding mothers is having sufficient milk. It is one of the most common reasons given by mothers for early weaning and supplementation. However, nearly all women have adequate milk supplies, and if the milk letdown reflex is functioning well, milk production matches infant demand.

Recently, Swedish researchers studied 50 breast-feeding mothers. Although over half of the mothers experienced transient periods of thinking that their milk was insufficient, the infant's milk intake and growth during times of perceived insufficiency were no different from other periods during lactation (Hillervik-Lindquist et al. 1991). First-time mothers often need to be reassured that infants are restless, fussy, cry, and sleep poorly for a variety of reasons that may be unrelated to hunger. Also, nursing patterns are frequently irregular in the first month—mothers should not be overly concerned if feedings appear too short or too long, or if their babies occasionally fall asleep while feeding.

In the first four to six weeks postpartum, the mother can be reassured that her milk supply is adequate by observing the infant's feeding and stooling patterns, and by weekly weighing of the infant. Signs that a breast-feeding infant is receiving sufficient nourishment are (AAP 1993):

- feedings occur at least six times a day,

- the infant produces at least six wet diapers a day and stools several times each day,

- the infant is gaining sufficient weight (usually about 1/3 to 1/2 lb per week).

In a healthy infant, poor weight gain or **failure to thrive** may indicate that nutritional needs are not being met. There are many potential causes of failure to thrive (discussed in Chapter 6), including insufficient milk intake by the breast-feeding infant. There are several possible reasons for insufficient milk intake (see Figure 5.10). All of these should be considered before resorting to supplemental feeding:

- *The pattern of breast-feeding.* The infant may not be feeding frequently enough. Some mothers hold to a schedule of feedings every four hours, or allow their infants to sleep through the night without feeding. These patterns may provide insufficient milk for normal growth. Also, if the infant feeds too long at one breast, or does not suckle both breasts at each feeding, the stimulus for milk production in the underused breast may be reduced, and milk output may be affected.

failure to thrive: unsatisfactory growth; inability to maintain an established growth curve

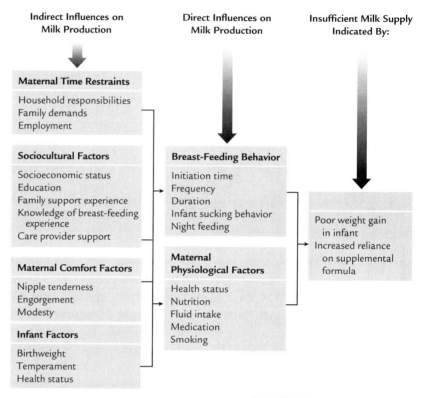

Indirect Influences on Milk Production

Maternal Time Restraints

Household responsibilities
Family demands
Employment

Sociocultural Factors

Socioeconomic status
Education
Family support experience
Knowledge of breast-feeding
  experience
Care provider support

Maternal Comfort Factors

Nipple tenderness
Engorgement
Modesty

Infant Factors

Birthweight
Temperament
Health status

Direct Influences on Milk Production

Breast-Feeding Behavior

Initiation time
Frequency
Duration
Infant sucking behavior
Night feeding

Maternal Physiological Factors

Health status
Nutrition
Fluid intake
Medication
Smoking

Insufficient Milk Supply Indicated By:

Poor weight gain
  in infant
Increased reliance
  on supplemental
  formula

FIGURE 5.10
*Potential reasons for
insufficient milk*

Adapted from P.D. Hill, and S. S. Humenick "Insufficient Milk Supply," *Image* 21:145–48.

■ *Inadequate infant suckling.* Poor positioning may interfere with the infant's ability to suckle and swallow effectively. If the nipple and areola are not properly grasped by the feeding infant, even vigorous suckling will not provide sufficient milk (Shrago and Bocar 1990).

■ *Sore nipples.* Nipple tenderness may cause the mother to limit the number and duration of feedings. Also, pain from nipple and breast soreness can inhibit the milk letdown reflex and reduce milk output. Causes of breast tenderness are discussed later in this chapter.

■ *Maternal stress and fatigue.* A new baby adds significantly to maternal responsibilities for family care and work. Insufficient rest and inattention to diet and fluid intake may impair milk letdown and diminish the mother's ability to provide sufficient milk (Ruvalcaba 1987).

■ *Early introduction of bottle feedings.* Many families choose to supplement breast-feeding with formula in early infancy, particularly if the mother works outside the home. However, the infant should not be fed from a bottle for the first two weeks after birth. This allows lactation and suckling patterns at the breast to become firmly established. Newborns may be confused by a rubber nipple, which requires slightly different tongue and mouth movements than the mother's nipple (AAP 1993).

- *Medications, alcohol, and smoking.* As discussed in chapter 4, medications can potentially interfere with milk production. Also, excessive alcohol use or cigarette smoking can impair milk output (Woodward and Hand 1988).

TOO MUCH MILK

Milk production in some women may exceed their infant's needs. The breasts may become uncomfortably full and leak milk between feedings. Also, milk output during infant suckling, particularly at the beginning of suckling during milk letdown, may be so fast and abundant that feeding may be uncomfortable for the infant. Proper positioning may ease the baby's struggle to keep up with the milk flow. When necessary, extra milk can be manually expressed from the breast and stored frozen for later use.

BREAST PROBLEMS

Engorgement

About two to four days after birth, most breast-feeding women experience a gradual build-up of fluid in the breasts, called breast fullness—a normal and welcome sign of increasing milk production. The breasts feel heavy and full for about 24 hours, remain compressible and soft, are not tender, and there is no discomfort while breast-feeding. The mother should nurse frequently during this period to reduce the risk of breast fullness developing into breast engorgement.

As lactation begins, blood flow to the breasts increases tenfold. In a few women, the increase in blood flow results in **venous stasis** and **edema** and, when added to the increased volume of milk in the glands, produces a painful condition called *breast engorgement.* Engorgement usually occurs during the first week postpartum and is often precipitated by a delay or reduction in breast-feeding during the period of breast fullness (Moon and Humenick 1989). The breasts become warm, firm, and tender. If the breasts become engorged, the mother should offer the breasts to her infant at least every three hours. Increasing the frequency and duration of breast-feeding may relieve the condition.

During engorgement, wearing a sturdy nursing brassiere can provide support for the breasts. Also, warm towels applied to the breasts or a warm bath may reduce discomfort (AAP 1993). Some women report that gentle massage is helpful in softening the breasts.

If engorgement becomes severe, the normally compressible tissue under the areola becomes firm and tight, and it is difficult for the infant

venous stasis: stagnated flow of blood in veins

edema: an abnormal accumulation of fluid in the intercellular spaces of the body

to grasp the nipple properly during feeding. The infant is less effective at suckling milk from the breast, and maternal discomfort increases as milk stasis occurs.

When breast-feeding, the areola should be lightly compressed between two fingers to allow the infant to grasp the nipple and areola. To relieve the pressure of accumulating milk, manual extraction of milk using a breast pump may be necessary. Breast engorgement is transient (usually lasting only a day or two), and as it subsides, normal patterns of breast-feeding can begin again.

Leaking

Many women experience leaking of milk from their breasts between feedings. This usually occurs in the first month postpartum, when feeding frequency and duration are most variable. Leaking can be caused by breast fullness because of a missed feeding or milk letdown occurring in response to stimuli other than infant suckling. The letdown reflex is easily conditioned, and it can occur if the mother sees or hears a baby, thinks about her baby, or even by the scent of her baby's clothes.

Usually the amount of milk leaked is small, and many mothers are reassured, perceiving leaking as a sign that they have ample milk. Light pressure for a short time on leaking breasts may be helpful in reducing the flow. For example, when a mother feels her milk "let down", she can cross her arms over her chest and gently press down for five to ten seconds. This maneuver, which can be done in public, will often reduce leakage.

Nipple Soreness

It is common for mothers to experience discomfort of the nipples and areola during feedings, particularly in the early postpartum period. Up to 95% of women experience nipple soreness during the first week (Ziemer et al. 1990). The soreness usually peaks around days three to six and then subsides. If discomfort persists beyond the first week, or is accompanied by redness or bruising of the areola, several factors could be responsible:

- Many mothers note mild aching in the breasts when the infant begins to suckle, which is relieved as the feeding goes on. This is caused by stimulation of the breast by oxytocin during milk letdown.

- Improper infant grasp of the nipple can produce nipple soreness. The infant's grasp should include some of the areola and should form a tight seal (see Figure 5.7). Nipple soreness can often be avoided if the position of the infant at the breast varies with each feeding, allowing the infant to suckle on slightly different portions of the areola.

- The mother should try to feed the infant on demand. A baby that must wait for a feeding may be hungrier and suckle more vigorously, increasing soreness. Feedings should be limited to about 7–10 minutes per side if the nipples are tender. Feeding should be stopped as soon as the breast is empty, and the nipple should be gently removed from the infant's mouth.

Nipple soreness or redness should be treated promptly, as the condition is easier to treat in the early stages, before bruising or cracking occurs. Ordinarily, nipples should be kept dry (AAP 1993). Air drying the nipples for a few minutes after each feeding before replacing the brassiere can reduce nipple soreness.

Allowing a small amount of breast milk to dry on the nipples may aid healing and prevent infection. Soaps, alcohol, and petroleum-based ointments can increase irritation and should be avoided. Specially formulated breast ointments are available, but because substances applied to the nipple may be ingested by the infant during feedings, nursing mothers should consult their health-care provider before using breast ointments.

Plugged Milk Ducts

Plugged milk ducts are a nagging problem for many women. They usually occur when feedings do not empty the milk ducts completely, and milk and cellular debris accumulate in the duct. As milk builds up behind the obstruction, the breast becomes tender, and if the duct is near the surface, a well-defined, firm lump may be felt under the skin. Occasionally, a small white plug is visible at the opening of a duct on the nipple.

Plugged ducts appear more often in women with ample milk supplies who do not sufficiently empty each breast during feedings. Plugged ducts are not harmful to the infant, but if untreated, a plugged duct may develop into mastitis. Care for a plugged duct should begin promptly.

- Apply damp heat to the breast, particularly before a feeding. The breasts can be soaked in warm water and gently massaged.

- Feed the infant often, beginning each feeding with the affected breast, and allow the infant to completely empty the breast. Change the position of the infant at the breast, ensuring that suckling drains all of the ducts. The breast can be massaged during feedings to stimulate flow.

When the plug is released, it may appear in the milk as tiny, white bits of "hardened" milk, or it may be brownish and stringy. Once the plug is loosened, relief comes quickly—symptoms often disappear in 12 to 24 hours.

Mastitis and Breast Abscesses

Mastitis, an infection of the breast, occurs in about 6% of breast-feeding women (Riordan and Auerbach 1993). The usual symptoms of mastitis are fatigue, headache, and fever (typical flulike symptoms), and the appearance of a reddened, hot, tender area on the breast. The infection is usually confined to one area on a breast, but can involve the entire breast, and rarely, occur bilaterally.

Mastitis can occur anytime during lactation, but most often appears in the first few weeks postpartum. There appear to be a number of predisposing factors, including maternal stress and fatigue, cracked nipples, breast engorgement, plugged milk ducts, and a change in the number of feedings. The infection is most often caused by staphylococci bacteria (Lawrence 1994).

Although mothers often mistakenly think they should stop breast-feeding when mastitis occurs, stopping feeding causes increased milk stasis and buildup in the glands, and increases discomfort. Therefore, in general, breast-feeding should continue during mastitis (Lawrence 1994). The treatment of mastitis involves application of warm heat to the affected area, continued breast-feeding, bed rest, and use of antibiotics.

Rarely, mastitis will develop into a *breast abscess.* An abscess is a sharply defined collection of pus and bacteria that produces a painful swelling in the breast. Abscesses require urgent treatment: they must be drained using a fine needle or a small surgical incision. Feedings from the affected breast may need to be discontinued while the abscess heals; the other breast can be used to feed the infant.

BREAST-FEEDING IN SPECIAL SITUATIONS

CESAREAN BIRTH

Almost all women who have undergone a *cesarean delivery* (an operative delivery, or C-section) are able to breast-feed. Depending on the type and amount of anesthesia used during the operation, the mother and baby (anesthetics cross the placenta) may be drowsy and lethargic for a time after birth. When the mother is fully alert and able to hold her baby, she may begin nursing; however, the newborn may suckle less vigorously until the anesthetic effects have worn off. Although the first feeding may be delayed, successful feedings can often begin within 12 hours of delivery (AAP 1993).

Finding a comfortable breast-feeding position is important, as postoperative pain after a cesarean delivery may inhibit the letdown reflex. Because the mother's lower abdomen may be tender—particularly if the infant rests on it while nursing—an alternative to the usual sitting up and cradling position can be used.

A comfortable position after a cesarean delivery is one in which the infant lies along the mother's side, while the infant's head is held in the palm of the hand at the breast, and the mother's forearm supports the baby's torso (see Figure 5.6).

After a cesarean delivery, most mothers need to take pain medication. In order to reduce the amount in the breast milk, mother should take pain medication after feedings (AAP 1993). This allows time for the level of the drug in the maternal blood to peak and fall before the next feeding, and the dose to the infant will be reduced.

Although most women who choose to breast-feed after a cesarean delivery are successful, several studies have found that cesarean births are associated with reduced rates of initiation of breast-feeding and early failure (Pietz 1989). Support and guidance from health-care providers and family are particularly important for successful breast-feeding after a cesarean delivery.

BREAST-FEEDING TWINS

Nature has provided women with the ability to suckle two infants—even simultaneously—if they choose to. Mothers who breast-feed twins maintain high rates of milk output: studies have reported milk outputs of 1500 to 2500 ml per day (Saint et al. 1986). The nutritional needs of the mother who is nursing two infants are sharply increased, and special attention to diet is important. Mothers nursing twins should seek professional guidance in selecting a diet and follow the general guidelines discussed in the section "A Healthy Diet for the Breast-Feeding Mother" at the end of Chapter 4.

Although some mothers of twins begin by feeding both infants simultaneously, many mothers later change to feeding only one infant at a time. This allows them to focus attention on only one infant and can facilitate maternal bonding and attachment to each individual baby. If two babies are nursed simultaneously, the football hold is a good position, using plenty of pillows to support the mother's arms (see Figure 5.11).

BREAST-FEEDING AFTER BREAST SURGERY

Surgery to augment or reduce the size of the breasts is not uncommon in the United States, with over 70,000 breast augmentations performed each year (Riordan and Auerbach 1993). Most of these procedures are performed on women during their childbearing years, and may affect a woman's ability to breast-feed. Surgery may disrupt the nerve pathways

FIGURE 5.11
*Positions for
simultaneously
nursing twins*

around the nipple and breast, the milk ducts, or the blood supply to the glands. In general, breast-feeding is possible for most women after augmentation surgery. However, after a breast reduction, it is often not possible to breast-feed unless breast milk is supplemented by formula.

A recent study found that women who had previous breast surgery were three to five times less likely to produce sufficient milk during lactation than women who had not had surgery (Neifert et al. 1990). Insufficient milk production was especially common in women who had incisions around the areola as part of their surgery. Clearly, women who are considering breast surgery and wish to breast-feed later should carefully consult with their physicians about their ability to do so after the procedure.

INFANT DISORDERS

Respiratory Infections

Upper respiratory tract infections (URIs) are common during infancy and can interfere with feeding patterns. Usually the culprit is a virus, and the infant develops rhinitis (a runny nose) and cough.

Because the infant breathes through his nose while feeding at the breast, nasal congestion during a URI can impair the ability to breast-feed. Proper positioning can make it easier for the infant to breathe while feeding. The mother should breast-feed while sitting up and hold the infant in a more upright position. Also, nasal secretions should be cleared before beginning feeding. Normally, infants lose some appetite when they have an infection. The mother may need to express some milk manually until the infant's appetite returns, usually in a day or two.

Down's Syndrome

Down's syndrome, a common congenital disorder, is characterized by typical eye and facial features, heart defects, hypotonia (lack of proper muscle tone), and mental retardation. Infants with Down's syndrome can breast-feed, although it may take them longer to establish full suckling behavior. A study of 60 breast-fed infants with Down's syndrome found that, although half of the infants had no difficulty suckling immediately, about a third took more than one week to establish full suckling (Aumonier and Cunnigham 1983).

Cleft Lip and Palate

Cleft lip and *palate* are common congenital malformations caused by incomplete closure of the lip and upper jaw during embryonic development. This leaves a gap in the lip or palate (or both, in about half of cases) that makes it difficult for the infant to suckle. Although many of these infants are fed formula, breast-feeding is possible with some adjustments.

The mother can assist her infant with a cleft lip in forming a seal over the nipple by pressing the cleft against the breast and covering the gap with her finger or thumb. Cleft lips are ordinarily repaired at three weeks postpartum, and afterwards the infant can usually breast-feed normally (Weatherly-White et al. 1987). In an infant with a cleft palate, there is an opening through the roof of the mouth into the nose, and during feeding, milk can flow into the nose and choke the infant. Because cleft palates are not repaired until later in infancy, a plastic plug that closes this gap is custom-made for the infant, and improves the ability to suckle.

Breast Milk Jaundice

One of the metabolic adaptations to life outside the uterus is a complete changeover of the hemoglobin in red blood cells, during which the adult form of hemoglobin replaces fetal hemoglobin. Older red cells containing fetal hemoglobin are broken down during this changeover, and the metabolism of hemoglobin produces large amounts of **bilirubin.**

In adults, bilirubin is promptly excreted by the liver and kidney, but the newborn's immature liver and kidney have only a limited capacity for secretion of the substance. Consequently, bilirubin often builds up in the newborn, as the liver and kidney struggle to keep up with the bilirubin load.

bilirubin: a breakdown product of hemoglobin; it is excreted by the liver and kidney

Newborn **jaundice** is caused by the accumulation of bilirubin in tissues, and it is visible as a yellow pigment in the skin and the white of the eyes. Jaundice is common during the first week postpartum, as fetal hemoglobin is broken down. In almost all infants, the jaundice is mild, transient, does no harm, and requires no treatment. However, in some infants, bilirubin levels go very high, and high levels can harm the developing nervous system (Hansen et al. 1986). Although a number of factors can elevate bilirubin levels, in a few infants, breast-feeding appears to be the cause.

Breast-fed infants tend to have slightly higher bilirubin levels and longer duration of jaundice, compared to formula-fed infants (Schneider 1986). Although the mechanism is not clear, there appear to be factors in breast milk that slow the excretion of bilirubin. Usually, the small increase in bilirubin during breast-feeding is insignificant; however, about 1% of breast-fed infants develop prolonged jaundice and elevated levels of bilirubin, a condition termed *breast milk jaundice* .

If breast milk jaundice is severe, many pediatricians recommend temporarily replacing breast milk with formula, usually for 24 hours (Newmann and Maisels 1992). Often, this pause in breast-feeding produces a large drop in the bilirubin level. Breast-feeding can then resume, with little or no increase in bilirubin. During the pause in breast-feeding, the infant is fed formula and the mother extracts milk regularly to maintain lactation. When breast milk jaundice subsides, mothers should be actively encouraged to resume breast-feeding. (Kemper et al. 1989).

Metabolic Disorders

Galactosemia is a rare, inherited metabolic disease found in about 1 in 60,000 infants at birth. Normally, the galactose in milk (galactose is a component of lactose) is converted to glucose before it is used by cells. In an infant with galactosemia, one of the enzymes that metabolizes the galactose to glucose—galactose-1-phosphate uridyltransferase—is absent. Consequently, when the infant consumes milk, large amounts of galactose build up in the body and have severe adverse effects, including mental retardation. Therefore, in galactosemia, breast-feeding (or any milk feeding) is contraindicated (AAP 1993). Infants who are given special galactose-free formula develop normally and are healthy.

About 1 in 12,000 babies in the United States are born with *phenylketonuria* (PKU). Infants with PKU cannot convert phenylalanine to tyrosine, the first step in the major pathway of phenylalanine metabolism. If the diet is rich in phenylalanine, the amino acid accumulates to high levels in the body, impairing normal development of the central nervous system and producing mental retardation (Smith and Leonard 1991). Individuals with PKU require small amounts of dietary phenylalanine (an essential amino acid) for protein synthesis, but they must consume

jaundice: yellow pigmentation of the skin, conjuctiva of the eyes and mucus membranes caused by accumulation of bilirubin

| Month | Serum Phenylalanine (µmol/L) | |
	Breast-Fed	Formula-Fed
1	993	1084
2	478	454
3	472	478
4	617	630
5	599	593
6	557	684

There are no significant differences in serum phenylalanine levels between the two groups.

Adapted from L. McCabe et al., "The Management of Breast Feeding among Infants with Phenylketonuria," *J Inher Metah Dis* 12 (1989):467–74.

TABLE 5.11

Comparison of blood levels of phenylalanine in breast-fed infants and infants fed low-phenylalanine formula

diets low in phenylalanine to avoid adverse effects. Babies born with PKU are otherwise healthy, and as long as phenylalanine intake is controlled, they will grow and develop normally.

Almost all infants in the United States, and in many countries worldwide, are screened for PKU at birth. Traditionally, if an infant tests positive for the disease, it is weaned from the breast and placed on a special low-phenylalanine formula. However, human milk has relatively low levels of phenylalanine. Breast-fed infants, on average, consume similar amounts of phenylalanine during their first six months as infants fed with low-phenylalanine formula (McCabe et al. 1989). See Table 5.11. Thus, PKU is no longer considered a contraindication to breast-feeding. Infants with PKU can be fully breast-fed as long as their blood phenylalanine levels are periodically checked and remain low.

MATERNAL DISORDERS

Diabetes Mellitus

Not only can women with diabetes successfully breast-feed their infants, but lactation can also have beneficial effects on blood sugar control in many diabetic mothers (Ferns et al. 1988). Blood sugar levels are often lower during breast-feeding because glucose is steadily pulled from the maternal blood and converted to galactose and lactose during milk synthesis.

Researchers compared two groups of new mothers with insulin-dependent diabetes (Ferns et al. 1988). One group of women breast-fed their infants, while the other group fed formula. Even though the breast-feeding mothers consumed more calories, their blood sugar levels were significantly lower, compared with the women feeding formula. Another study found that diabetic mothers needed less insulin during lactation—insulin requirements typically fell by 27% (Davies et al. 1989).

The milk produced by diabetic women is similar to that of nondiabetic women, although the fat content of milk from diabetics may be slightly reduced, and there may be more glucose in the milk. Although diabetic mothers tend to produce slightly smaller volumes of milk postpartum, weight gain and growth of their infants is normal (Bitman et al. 1989).

Chronic Diseases

There are very few medical conditions that preclude breast-feeding. Depending on the individual, most mothers with heart, lung, liver, and kidney disease can breast-feed successfully. Several chronic diseases require long-term therapy with drugs that can pass into the milk and harm the infant. If no safe alternative drugs are available, formula-feeding may be necessary. In cases of maternal alcoholism and drug abuse, or debilitating illnesses and severe maternal malnutrition, formula-feeding is usually preferable to breast-feeding (Lawrence 1994).

Maternal Infections

During some maternal infections, viruses and bacteria may be passed to the newborn in breast milk. However, breast-feeding is not contraindicated during most maternal infections. Antibodies and immune cells produced by the mother in response to infections are passed to the infant in milk (Glass et al. 1983). These agents can provide the infant with some protection from the infection. For example, the rubella (measles) virus can be passed through breast milk to the infant (Ruff 1994), but along with the virus travel antibodies and other immune factors that provide passive immunity from the virus, so the baby does not become ill.

human immuno-deficiency virus (HIV): the virus that causes the acquired immune deficiency syndrome (AIDS); it can be transmitted by sexual intercourse, contaminated needles or blood products, or from mother to infant during pregnancy or lactation

BREAST-FEEDING AND THE HUMAN IMMUNODEFICIENCY VIRUS: POLICIES IN INDUSTRIALIZED AND DEVELOPING COUNTRIES

Mothers infected with the **human immunodeficiency virus (HIV)** can pass the virus to their infants during breast-feeding (Ruff 1994). HIV has been isolated from breast milk, and in mothers who develop HIV infection postnatally, the risk of transmission during breast-feeding is between 27 and 40%. In 1985, the U.S. Centers for Disease Control recommended that HIV-infected women be discouraged from breast-feeding, a policy widely adopted by the other industrialized countries (Centers for Disease Control 1985).

However, whether HIV-positive mothers in impoverished developing countries should breast-feed remains a dilemma. In these areas, infant health is critically reliant on breast-feeding. Because breast milk protects against infant deaths from other diseases and safe and affordable breast milk substitutes are often unavailable, the risk of HIV infection must be balanced against the benefits of breast-feeding. In 1992, the World Health Organization issued a statement recommending that in areas where infectious diseases and malnutrition are the main causes of infant death and the infant mortality rate is high, breast-feeding should be recommended even to women who are HIV-infected (WHO 1992).

Poor nutritional status during pregnancy and the postpartum period may increase the likelihood of maternal–infant viral transmission. A recent study in Africa found that HIV-positive mothers deficient in vitamin A are more likely to transmit the virus to their infant during the perinatal period, compared to mothers with adequate vitamin A status (Semba et al. 1994).

BREAST-FEEDING AND INFANT HEALTH: REVIEW AND SUMMARY

THE UNIQUE QUALITIES OF HUMAN MILK

Human breast milk, like other mammalian milks, is species specific: it evolved to meet the unique nutritional requirements of human newborns and ensure optimal growth and development. Human milk is the best source of nutrition for full-term infants during the first four to six months postpartum (AAP 1993). No manufactured formula can duplicate the biologically specific physical structure and nutrient composition of human milk, as discussed in detail in Chapter 4. Because many common infant formulas use a base of cow's milk, it is worthwhile to discuss some of the differences between human milk and cow's milk.

Protein in Human Milk

Mature human milk contains much less protein (0.8 g versus 3.5 g/100 ml) than cow's milk. A major difference between cow's milk and human milk is the concentrations of the whey and casein proteins. Eighty percent of the proteins in cow's milk are casein proteins, while only 40% of human milk protein is casein. During digestion in the infant's gastrointestinal tract, whey and casein proteins have distinct characteristics. The caseins precipitate into sizable curds in the newborn's stomach that are tougher and less readily digestible than the softer, more flocculent clumps formed when whey proteins precipitate. Whey proteins are consequently often better tolerated by the newborn.

The protein content of breast milk is considered ideal for the human infant. Human milk proteins are complete and balanced, supplying optimal amounts of the essential and nonessential amino acids (AAP 1993). Breast milk is particularly rich in cystine and taurine (taurine is found at levels 30 times that in cow's milk). Compared to cow's milk, human milk contains only moderate amounts of phenylalanine, tyrosine, and methionine. The enzyme systems that metabolize these amino acids mature late in fetal life (Rassin 1991), and particularly in preterm infants, overconsumption of these amino acids can increase blood levels to potentially detrimental levels.

Vitamins, Minerals, and Enzymes in Human Milk

The vitamin content of human milk is uniquely suited to the newborn's needs. For example, most of the vitamin D in human milk is present as 25-OH vitamin D. The activity of the hepatic hydroxylase enzymes that normally convert vitamin D to 25-OH vitamin D are low in the newborn (Lawrence 1994). Therefore, provision of most of the vitamin in the 25-OH form may be advantageous during early infancy. Also, folate in breast milk is bound to specific folate-binding proteins in the whey fraction; the association enhances infant absorption of the bound folate.

Mature breast milk contains all of the essential minerals necessary for infant growth and development. The mineral content of human milk is well balanced, and the bioavailability to the infant is high, making milk a high-quality source of these nutrients. For example, calcium in breast milk is highly bioavailable: infants absorb over two thirds of the calcium in breast milk, compared with 25% in cow's milk (Riordan 1993).

Similarly, the iron in human milk is highly bioavailable: 50–70% of the iron is absorbed by the infant, compared with less than 10% from cow's milk or formula (Dallman 1986). Lactoferrin, found only in human milk, is an iron-binding protein that may increase the availability of iron. The bioavailability of zinc from human milk is also high compared to zinc in cow's milk and formula.

A variety of enzymes that have digestive functions are secreted into human milk. These enzymes, including lipases, amylases and proteases, are important in that they help the immature gastrointestinal tract of the newborn digest and absorb nutrients in the milk (Hamosh 1989).

PROTECTION FROM INFECTIOUS DISEASES

Breast-feeding protects the infant against infectious disease in several ways:

- Human milk contains a complex variety of anti-infective substances and cells, including lymphocytes, immunoglobulins, enzymes, anti-inflammatory agents, and other protective factors. Particularly abundant in colostrum, they interact to protect the newborn from disease, particularly from infections of the gastrointestinal tract and infant diarrheal disease (Bu'Luck et al. 1990). Most of these substances are not found in infant formula (or are present in insignificant amounts).

The frequency of gastrointestinal infections is much lower in breast-fed infants than in formula-fed infants, and protection is particularly strong in impoverished, developing countries where sanitation is poor (see Table 5.12). In Bangladesh, the rate of **cholera** infection is much lower in infants who are breast-fed (Glass et al. 1983). Breast-fed babies in industrialized countries (such as the United States and England) also have a lower incidence of gastrointestinal infections compared with infants fed formula (Kovar et al. 1984), (Bauchner 1990). In addition, breast-fed infants mount a more vigorous immune response to certain respiratory viruses—respiratory illnesses tend to be milder and shorter than those in formula-fed infants.

- Food and water, particularly in areas with poor sanitation, are often contaminated with potential pathogens. The newborn's immune system is still developing, and during this vulnerable period, breast-feeding limits infant exposure to potential pathogens in other foods and fluids.

- In impoverished areas, breast milk substitutes, such as formula and weaning foods, may be of poor nutritional quality. Malnourished infants are more susceptible to infection. Adequate breast-feeding ensures the nutritional status of the infant, thereby reducing vulnerability to disease (Wray 1991).

- Because of its contraceptive effects, breast-feeding allows adequate child spacing and reduces fertility, thereby reducing the number of infectious contacts (siblings) in the household (Wray 1991). Also, smaller families allow more human resources and attention to be focused on fewer children.

cholera: an acute infectious disease caused by the bacteria *Vibrio cholerae*, marked by severe diarrhea, vomiting and dehydration; often epidemic

TABLE 5.12

Differences in infant deaths from diarrhea and measles in Nigeria: breast-fed vs bottle-fed infants

Feeding Method	Number of Deaths	Sample size	Infant Deaths/1000
Breast	0	65	0
Mixed	20	282	70.9
Bottle	9	67	134.3
Total	29	414	70.0

Adapted from M. M. Scott-Emuakpor and U. A. Okafor, "Comparative Study of Morbidity and Mortality of Breastfed and Bottlefed Nigerian Infants," *East Afr Med J* 63 (1986): 452–457.

PROTECTION FROM ALLERGIES

Food allergies are thought to occur when intact proteins or larger peptides are absorbed from the intestine and recognized as antigens by the body's immune system, triggering allergic reactions (discussed in chapter 7). Between 0.5 and 7.5% of children are affected by food allergies, many of them during infancy (Metcalfe 1984).

Infants are thought to be susceptible to food allergy because their developing intestine is more permeable to proteins and peptides in the diet (Cant 1991). Normally proteins are efficiently and completely digested in the small intestine, and tight junctions between intestinal epithelial cells and ample secretory IgA (SIgA) form an effective barrier. In the newborn's intestine, this intestinal barrier is incomplete.

Breast-feeding for four to six months defers the introduction of foreign food proteins until the infant's IgA system becomes fully functional and the intestinal barrier to food proteins matures. Breast milk contains ample SIgA, as well as growth factors and other substances that may promote earlier maturation of the intestinal barrier.

Atopic eczema, a skin disorder that often affects infants (60% of individuals who develop atopic eczema do so during the first year of life), may be caused by allergens in food. Atopic disease has a strong hereditary component and often clusters in families. In a Finnish study in families with a history of **atopy,** infants who were exclusively breast-fed for 6 months developed eczema and food allergies at much lower rates than infants started on solid foods at three months (Kojosaari and Saarinen 1983).

Although allergies can occur with most foods, proteins in cow's milk are probably the most common allergens in infancy (Cant 1985). Several proteins in cow's milk, including lactoglobulin, bovine serum albumin, and casein, can act as allergens. Although high temperatures used in manufacturing formulas reduce the allergic potential of these proteins, infants fed cow's milk-based formula can develop allergies to these proteins. Symptoms often appear early in infancy and can be localized to the intestine (causing vomiting and diarrhea), or they can show up as cough, rhinitis, or skin reactions (Kramer 1988).

Breast-Feeding, Maternal Diet, and Infant Allergy

There is also evidence that foods eaten by a mother during lactation can produce allergies in breast-feeding infants (Cant et al. 1985), (Chandra et al. 1986). See Table 5.13. Components of food absorbed from the maternal diet can pass into breast milk and function as allergens in breast-feeding infants. Cow's milk consumed by the nursing mother has often been implicated—antigens from cow's milk have been found in the breast milk of women (Machtinger and Moss 1986).

atopy: a hereditary disposition toward the development of allergic reactions; manifestations include hay fever, asthma, and atopic eczema (a skin disorder)

In families with a strong history of allergy, when suspected food allergens (such as milk, eggs, meat, and peanuts) are excluded from the diets of lactating women, rates of atopic dermatitis in their infants are reduced (Chandra et al. 1986), (Hattevig 1989).

A nursing mother who notices particular foods in her diet that seem to produce an allergic reaction in her infant should consult with her physician. Allergies should be confirmed by testing, and mothers should not avoid healthful foods (such as dairy products) unless an allergy is proven. During weaning, solid foods should be introduced to the infant gradually (see discussion in Chapter 6), and the infant should be observed for signs of allergy when a new food is introduced.

- Cow's milk

- Eggs, especially egg whites

- Wheat

- Nuts

- Fish or shellfish

- Corn

- Citrus fruits

- Tomatoes

Adapted from: J. Riordan and K. G. Auerbach, *Breastfeeding and Human Lactation* (Boston: Jones and Bartlett, 1993).

TABLE 5.13
Foods consumed by lactating mothers associated with allergic reactions in their infacts

BREAST-FEEDING AND CHRONIC DISEASES OF LATER LIFE

Many investigations have tried to determine whether methods of infant feeding may influence health in later life. Studies have focused on obesity, atherosclerosis, and several common and serious immunological disorders.

Obesity

No firm associations between the method of feeding during infancy and later development of obesity have been found (Hamosh 1988). Although formula-fed infants tend to have greater weight for length at 6–12 months of age, size and growth rate among children from 3 to 12 years old are similar. Formula-feeding does not appear to increase the risk of obesity in adulthood (Institute of Medicine 1991).

Atherosclerosis

The cholesterol content in human milk is significantly higher than in manufactured milk formulas. Many studies have tried to determine whether the higher levels of cholesterol in human milk influence cholesterol levels and atherosclerosis in later life. Overall, research in both animals and humans has found no consistent links between methods of infant feeding and levels of adult cholesterol or rates of atherosclerosis (Institute of Medicine 1991), (Hamosh 1988).

Immune Disorders

Because human milk contains a variety of factors that hasten the maturation of the newborn's immune system, breast-feeding may influence predisposition to immunological diseases. Breast-feeding appears to exert long-term protective effects against several diseases that have immunological or infectious causes: Type I diabetes mellitus (Mayer et al. 1988) (see Table 5.14), childhood **lymphoma** (Davies et al. 1988), and **Crohn's disease** (Koletzo et al. 1989). For example, individuals who are fed formula during infancy were three times more likely to develop Crohn's disease in later life, compared with those who are breast-fed (Koletzo et al. 1989). These findings are intriguing, but preliminary. More research is needed to clarify the links between infant feeding and the development of these diseases.

lymphoma: a general term for cancer arising from the lymph nodes or other lymph tissues

Crohn's disease: a chronic inflammatory disease affecting the intestine

Months of Breast-Feeding	Odds Ratio of Developing IDDM
None	1.00
1–3	0.68
3–6	0.74
6–12	0.67
>12	0.54

Compared with individuals who had never been breast-fed, the risk of developing IDDM decreased as duration of breast-feeding increased. After adjusting for factors like birth year, race, income, maternal education, and maternal age, individuals breast-fed for at least 12 months had only half the risk of those who were never breast-fed.

Adapted from E. J. Mayer et al., "Reduced Risk of IDDM Among Breastfed Children," *Diabetes* 37 (1988): 1625–32.

TABLE 5.14
Breast-feeding and insulin-dependent diabetes mellitus risks

BREAST-FEEDING AND INFANT MORTALITY

In the developing countries of the world, formula-fed infants have sharply higher mortality rates than breast-fed infants, particularly in the newborn period. Although infants provided both human milk and formula have lower death rates than those provided only formula, exclusive breast-feeding is associated with the lowest mortality rates. See Table 5.15. The protective effect of breast-feeding is particularly strong during the first year, but mortality rates are also reduced in infants up to age 3 years. Differences in mortality are primarily due to differences in rates of infection between formula-fed and breast-fed infants (Wray 1991).

As shown in Figure 5.12, breast-feeding was also associated with lower infant death rates in the industrialized countries before 1950. Today, in countries such as the United States, mortality rates are low for both breast-fed and formula-fed infants, and it is not clear whether breast-feeding still confers a protective effect (Wray 1991).

TABLE 5.15
Infant mortality in the developing world; breast-fed vs. bottle-fed infants in Brazil

	Relative Risk of Death, by Disease	
Method of Feeding	**Diarrhea**	**Respiratory Infections**
Breast only	1.0	1.0
Breast plus formula	4.5	2.1
Formula only	16.3	3.9

Adapted from C. G. Victoria et al., "Infant Feeding and Deaths Due to Diarrhea," *Am J Epidemiol* 129 (1989): 1032–41.

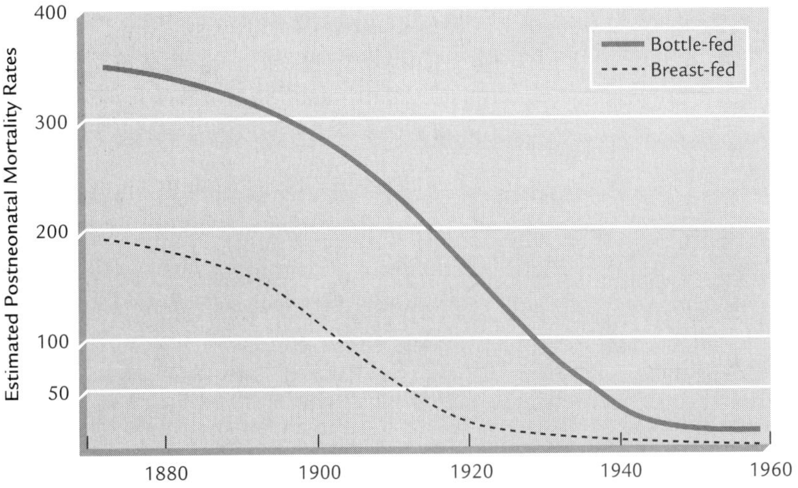

FIGURE 5.12
Infant mortality in industrialized countries: Breast-fed vs. bottle-fed (1870–1950)

Adapted from J.D. Wray, "Breastfeeding: An International and Historical Perspective," in *Infant and Child Nutrition*, ed., F. Falkner (Caldwell, NJ: Telford Press, 1990).

REFERENCES

Aberman, S., and Kichoff, K. T., "Infant-feeding Practices: Mothers' Decision–Making," *JOGN Nurs* 14 (1985):394–98.

American Academy of Pediatrics, Committee on Nutrition, and the Nutrition Committee of the Canadian Paediatric Society, "Breast-feeding: A Commentary in Celebration of the International Year of the Child," *Pediatrics* 62 (1978):591–601.

American Academy of Pediatrics, *Pediatric Nutrition Handbook,* (Elk Grove, Il: American Academy of Pediatrics, 1993).

Anderson, G. C., et al., "Development of Sucking in Term Infants from Birth to Four Hours Postbirth," *Res Nurs Health* 5 (1982):21–27.

Armstrong, H., "The International Code of Marketing of Breast-Milk Substitutes, Part Two of a Series," *J Hum Lact* 4 (1988):194–99.

Atkinson, L., "Prenatal Nipple Conditioning for Best Breastfeeding," *Nurs Res* 28 (1979):448–51.

Auerbach, K. G., and Avery, J. L., "Induced Lactation: A Study of Adoptive Nursing by 240 Women," *Am J Dis Child* 135 (1981):340.

Auerbach, K. G., "Maternal Employment and Breastfeeding," in *Breast-feeding and Human Lactation*, eds. J. Riordan and K. G. Auerbach (Boston: Jones and Bartlett, 1993).

Aumonier, M. E., and Cunningham, C., "Breastfeeding in Infants with Down's syndrome," *Child Care Health Develop* 9 (1983):247–55.

Bauchner, H., "Breastfeeding and Infections: Methodologic Issues and Approaches," in "Breastfeeding, Nutrition, Infection and Infant Growth in Developed and Emerging Countries," eds. S. A. Atkinson, L. A. Hanson and R. K. Chandra (St. John's,Canada: ARTS Biomedical, 1990).

Baumslag, N., "Breastfeeding: Cultural Practices and Variations," *Adv International Mat and Child Hlth* 7 (1987):36–50.

Biddulph J., "Promotion of Breast-feeding: Experience in Papua New Guinea," in *Advances in International Maternal and Child Health*, eds. D. B. Jelliffe and E. F. Jelliffe, (Oxford: Oxford University Press, 1981).

Bitman, J., et al., "Milk Composition and Volume during the Onset of Lactation in a Diabetic Mother," *Am J Clin Nutr* 50 (1989):1364–69.

Bocar, D. L., and Shrago, L., "Breastfeeding Education," in *Breastfeeding and Human Lactation*, eds. J. Riordan and K. G. Auerbach (Boston: Jones and Bartlett, 1993).

Bu'Lock, F., Woolridge, M. W., and Baum, J. D., "Development of Co-ordination of Sucking, Swallowing and Breathing: Ultrasound Study of Term and Preterm Infants," *Dev Med Child Neurol* 32 (1990):669–78.

Butte, N. F., et al., "Feeding Patterns of Exclusively Breast-fed Infants During the First Four Months of Life," *Early Hum Dev* 12 (1985):291.

Cant, A. J., "Food Allergy and Intolerance," in *Pediatric Nutrition*, 3d ed., edited by D. D. McLaren, et al., (Edinburgh: Churchill-Livingstone, 1991).

Cant, A. J., Marsden, R. A., and Kilshaw, P. J., "Egg and Cow's Milk Hypersensitivity in Exclusively Breastfed Infants with Eczema, and the Detection of Egg Protein in Breastmilk," *Br Med J* 291 (1985):932–35.

Casey, P. J., et al., "Nutrient Intake by Breastfed Infants during the First Five Days after Birth," *Am J Dis Child* 140 (1986): 933.

Centers for Disease Control, "Recommendations for Assisting in the Prevention of Perinatal Transmission of Human-T-Lymphotrophic Virus III/Lymphadenopathy-Associated Virus and Acquired Immunodeficiency Syndrome," *MMWR* 34 (1985):721–26.

Chandra, R. K., et al., "Influence of Maternal Food Antigen Avoidance During Pregnancy and Lactation on Incidence of Atopic Eczema in Infants," *Clin Allergy* 16 (1986):563–69.

Coates, M. M., "Tides in Breastfeeding Practice," in *Breastfeeding and Human Lactation*, eds. J. Riordan and K. G. Auerbach (Boston: Jones and Bartlett, 1993).

Copeland, C. A., Raebel, M. A., and Wagner, S. L., "Pesticide Residue in Lanolin," *JAMA* 261 (1989):242.

Dallman, P. R., "Iron Deficiency in the Weanling: A Nutritional Problem on the Way to Solution," *Acta Pediatr Scand Suppl* 323 (1986): 59–67.

Davies, H. A., et al., "Insulin Requirements of Diabetic Women Who Breastfeed," *Br Med J* 298 (1989):1357–58.

Davis, M. K., Savitz, D. A., and Graubard, B. I., "Infant Feeding and Childhood Cancer," *Lancet* 2 (1988):365–8.

Editorial, "Present Day Practice in Infant Feeding," *Lancet* 1 (1988):975–76.

Ferris, A. M., et al., "Lactation Outcome in Insulin-Dependent Diabetic Women," *J Am Diet Assoc* 88 (1988):317–22.

Fomon, S. J., "Trends in Infant Feeding Since 1950," in *Nutrition of Normal Infants* (St. Louis: Mosby, 1993).

Ford, C. S., *A Comparative Study of Human Reproduction, Publications in Anthropology No. 32* (New Haven: Yale University Press, 1945).

Gielen, A. C., et al., "Maternal Employment during the Early Postpartum Period: Effects on Initiation and Continuation of Breastfeeding," *Pediatrics* 87 (1991):298.

Glass, R. I., et al., "Protection Against Cholera in Breast Fed Children by Antibodies in Breast Milk," *N Engl J Med* 308 (1983):1389.

Grossman, L. K., et al., "The Infant Feeding Decision in Low and Upper Income Women," *Clin Pediatr* 29 (1990):30–37.

Guthrie, G. M., et al., "Early Termination of Breastfeeding Among Philippine Urban Poor," *Ecol Food Nutr* 12 (1983):195–202.

Hamosh, M., "Enzymes in Human Milk: Their Role in Nutrient Digestion, Gastrointestinal Function and Nutrient Delivery to the Newborn Infant," in *Textbook of Gastroenterology and Nutrition in Infancy*, 2d ed., edited by E. Lebenthal (New York: Raven Press, 1989).

Hamosh, M., "Does Infant Nutrition Affect Adiposity and Cholesterol Levels in the Adult?" *J Pediatr Gastrenterol* 7 (1988): 10–16.

Hansen, T. W. R., and Bratlid, D., "Bilirubin and Brain Toxicity," *Acta Paediatr Scand* 75 (1986):513.

Hattevig, G., "Effect of Maternal Avoidance of Eggs, Cow's Milk and Fish during Lactation upon Allergic Manifestations in Infants," *Clin Exp Allergy* 19 (1989):27–32.

Helsing, E., and King, F. S., *Breastfeeding in Practice* (Oxford: Oxford University Press, 1983).

Hillervik–Lindquist, C., Hofvander, Y., and Sjölin, S., "Studies on Perceived Milk Insufficiency III. Consequences for Breast Milk Consumption and Growth," *Acta Paediatr Scand* 80 (1991):297–303.

Institute of Medicine, *Nutrition during Lactation*, (Washington D. C.: National Academy Press, 1991).

Jimenez, M. H., and Newton, N., "Activity and Work during Pregnancy and the Postpartum Period: A Cross-Cultural Study of 202 Societies," *Am J Obstet Gynecol* 135 (1979):171–76

Kemper, K., Forsyth, B., and McCarthy, P., "Jaundice, Terminating Breast-Feeding, and the Vulnerable Child," *Pediatrics* 84 (1989):773–78.

Kojosaari, M., and Saarinen, U. M., "Prophylaxis of Atopic Disease by Six Month's Total Solid Food Elimination," *Acta Pediatr Scand* 72 (1983):411–14.

Koletzko, S., et al., "Role of Infant Feeding Practices in Development of Crohn's Disease in Childhood," *Br Med J* 298 (1989):1617–18.

Konner, M., and Worthman, C., "Nursing Frequency, Gonadal Function, and Birth Spacing Among ¡Kung Hunter-Gatherers," *Science* 207 (1980):788–91.

Kovar, M. G., et al., "Review of the Epidemiological Evidence for an Association Between Infant Feeding and Infant Health," *Pediatrics* 74 (1984): 615–38.

Kramer, M. S., "Does Breastfeeding Help Protect Against Atopic Disease? Biology, Methodology and a Golden Jubilee of Controversy," *J Pediatr* 112 (1988):181–90.

Lawrence, R. A., *Breastfeeding: A Guide for the Medical Profession*, 4th ed. (St. Louis: Mosby, 1994).

Machtinger, S., and Moss, R., "Cow's Milk Allergy in Breastfed Infants: The Role of Allergen and Maternal Secretory IgA Antibody," *J Allergy Clin Immunol* 77 (1986):341–47.

Mayer, E. J., et al., "Reduced Risk of IDDM Among Breastfed Children: The Colorado IDDM Registry," *Diabetes* 37 (1988):1625–32.

McCabe, L., et al., "The Management of Breastfeeding among Infants with Phenylketonuria," *J Inher Met Dis* 12 (1989):467–74.

Mennella, J. A., and Beauchamp, G. K., "Maternal Diet Alters the Sensory Qualities of Human Milk and Nurslings's Behavior," *Pediatrics* 88 (1991):737.

Metcalfe, D. D., "Food Hypersensitivity," *J Allergy Clin Immunol* 73 (1994):749-62.

Moon, J. L., and Humenick, S. S., "Breast Engorgement: Contributing Variables and Variables Amenable to Nursing Intervention," *JOGNN* 18 (1989):309–15.

Morse, J. M., and Harrison, M., "Social Coercion for Weaning," *J Nurs-Midwif* 32 (1987):205–10.

Mulford, C., "Subtle Signs and Symptoms of the Milk Ejection Reflex," *J Hum Lact* 6 (1990):177–78.

Neifert, M., et al., "The Influence of Breast Surgery, Breast Appearance, and Pregnancy-Induced Breast Changes on Lactation Sufficiency as Measured by Infant Weight Gain," *Birth* 17 (1990): 31-37.

Neifert, M. R., "Routine Management of Breastfeeding," in *Lactation; Physiology, Nutrition and Breastfeeding*, eds. M. C. Neville and M. R. Neifert (New York: Plenum Press, 1983).

Neifert, M. R., Seacat, J. M., and Jobe, W. E., "Lactation Failure Due to Insufficient Grandular Development of the Breast," *Pediatrics* 76 (1985):823-28.

Newmann, T. B., and Maisels, M. J., "Evaluation and Treatment of Jaundice in the Term Newborn: A Kinder, Gentler Approach," *Pediatrics* 89 (1992):809.

Pietz, C. L., "The Emotional Impact of Breastfeeding After a Cesarean," *Int J Child Educ* 4 (1989):20–21.

Post, R. D., and Singer, R., "Psychological Implications of Breastfeeding for the Mother," in *Lactation;Physiology, Nutrition and Breastfeeding*, eds. M. C. Neville and M. R. Neifert (New York: Plenum Press, 1983).

Rassin, D. K., "Amino Acid and Protein Metabolism in the Premature and Term Infant," in *Neonatal Nutrition and Metabolism*, ed. W. W. Hay (St Louis: Mosby-Year Book, 1991), 110–22.

Richards, M. P. M., "Breast Feeding and the Mother-Infant Relationship," *Acta Paediatr Scand Suppl* 299(1982):33.

Riordan, J., and Auerbach, K. G., "Breast-Related Problems," in *Breastfeeding and Human Lactation*, eds. J. Riordan and K. G. Auerbach (Boston: Jones and Bartlett, 1993).

Riordan, J., "The Biologic Specificity of Breastmilk," in *Breastfeeding and Human Lactation*, eds. J. Riordan and K. G. Auerbach (Boston: Jones and Bartlett, 1993).

Ruff, A. J., "Breastmilk, Breastfeeding, and Transmission of Viruses to the Neonate," *Sem Perinatol* 18 (1994):510–16.

Ruvalcaba, R. H. A., "Stress-induced Cessation of Lactation," *West J Med* 146 (1987):228–30.

Saint, L., Maggiore, P., and Hartmann, P. E., "Yield and Nutrient Content of Milk in Eight Women Breast-Feeding Twins and one Woman Breast-Feeding Triplets," *Br J Nutr* 56 (1986):49–58.

Sandiford, P., et al., "Why Do Child Mortality Rates Fall? An Analysis of the Nicaraguan Experience," *Am J Public Health* 81 (1991):30–37.

Schmidt, E., "Effects of Varying Degrees of Heat Treatment on Milk Protein and Its Nutritional Consequences," *Acta Paediatr Scand Suppl* 296 (1982):41.

Schneider, A. P., "Breast Milk Jaundice in the Newborn," *JAMA* 255 (1986):3270.

Semba, R. D., et al., "Maternal Vitamin A Deficiency and Mother-to-child Transmission of HIV-1," *Lancet* 343 (1994):1593–7.

Serdula, M. K., et al., "Correlates of Breast-feeding in a Low-Income Population of Whites, Blacks, and Southeast Asians," *J Am Diet Assoc* 91 (1991):41–45

Shrago, L., and Bocar, D., "The Infant's Contribution to Breastfeeding," *JOGNN* 19 (1990):209–15.

Smith, I., and Leonard, J. V., "Inborn Errors of Metabolism," in *Textbook of Pediatric Nutrition*, 3d ed., edited by D. D. Mclaren, et al., (Edinburgh: Churchill Livingstone, 1991).

Storr, G. B., "Prevention of Nipple Tenderness and Breast Engorgement in the Postpartal Period," *JOGNN* 17 (1988):203–8.

Switzky, L. T., Vitze, P., and Switzky, H. N., "Attitudinal and Demographic Predictors of Breast-Feeding and Bottle-Feeding Mothers," *Psychol. Rep.* 45 (1979):3–14.

U. S. Buruueau of Labor Statistics, "Facts on Working Women: 1989," no. 89-3 (Washington, D. C.: U. S. Government Printing Office, 1989).

UNICEF, Nutrition Cluster (H-8F) United Nations Children's Fund, "Innocenti Declaration on the Protection, Promotion and Support of Breastfeeding," (Florence, Italy, 1 August 1990).

Wade, N., "Bottle-Feeding: Adverse Effects of a Western Technology," *Science* 184 (1974):45–48.

Weatherley-White, R. C. A., et al., "Early Repair and Breast-Feeding for Infants with Cleft Lip," *Plast Reconstr Surg* 79 (1987):879–85.

Weaver, L. T., Ewing, G., and Taylor, L. C., "The Bowel Habit of Milk–fed Infants," *J Pediatr Gastroenterol Nutr* 7 (1988):568–71.

Wharton, B. A., "Weaning and Early Childhood," in *Textbook of Pediatric Nutrition*, 3d ed., edited by D. S. Mclaren et al. (Edinburgh: Churchill Livingstone, 1991).

WHO, "Consensus Statement From the WHO/UNICEF Consultation on HIV Transmission and Breastfeeding," *Wkly Epidemiol Rec* 67 (1992):177–84.

WHO, "The Prevalence and Duration of Breast-Feeding: Updated Information, 1980–1989. Part I," *Wkly Epidem Rec* 42 (1989):321–23.

WHO, "The Prevalence and Duration of Breast-Feeding: Updated Information, 1980–1989. Part II," *Wkly Epidem Rec* 43 (1989):331–34.

WHO, *Contemporary Patterns of Breast-Feeding* (Geneva: World Health Organization, 1981).

Winikoff, B., and Laukaran, V. H., "Breast Feeding and Bottle Feeding Controversies in the Developing World: Evidence From a Study in Four Countries," *Soc Sci Med* 29 (1989):859–868.

Winikoff, B., et al., "Dynamics of Infant Feeding: Mothers, Professionals, and the Institutional Context in a Large Urban Hospital," *Pediatrics* 77 (1986):357–65.

Woodward, A., and Hand, K., "Smoking and Reduced Duration of Breast-feeding," *Med J Aust* 148 (1988):477–78.

Woolridge, M. W., "The 'Anatomy' of Infant Sucking," *Midwifery* 2 (1986):164–71.

Woolridge, M. W., Ingram, J. C., and Baum, J. D., "Do Changes in Pattern of Breast Usage Alter the Baby's Nutrient Intake?" *Lancet* 336(8712) (1990):395–97.

Wray, J. D., "Breastfeeding: An International and Historical Review," in *Infant and Child Nutrition Worldwide: Issues and Perspectives*, ed. F. Falkner (Boca Raton: CRC Press, 1991).

Ziemer, M. M., et al., "Methods to Prevent and Manage Nipple Pain in Breastfeeding Women," *West J Nurs Res* 12 (1990):732–44.

Zuffoletto, J. M., "The Federal Family Medical Leave Act," *AORN J* 60 (1994):91–3.

infancy and nutrition

▼

NUTRITIONAL ADAPTATION TO EXTRAUTERINE LIFE

In utero, the nutritional needs of the fetus are provided in the form of basic substrates—amino acids, glucose, fatty acids, water, vitamins, and minerals—which are transported directly into the fetal blood through the placenta. Waste products of fetal metabolism are promptly transferred to the maternal circulation for excretion. The placenta also takes care of respiration, as oxygen continuously moves across to the fetal blood while carbon dioxide diffuses across to the maternal blood.

At birth, the fetus makes an abrupt transition to life outside the uterus. Adaptation to extrauterine life is one of the most important steps in the life cycle because, for many systems in the body, birth precipitates dramatic changes in function. The respiratory system undergoes major changes as the newborn's lungs fill with air during the first breath, and gas exchange across the pulmonary circulation abruptly replaces placental transfer.

Similarly, birth triggers a major transformation in nutritional pathways and metabolism of the newborn. At a time when nutritional needs are high (per unit body weight higher than at any other time after birth) the newborn must abruptly begin to swallow, digest, and absorb a variety of complex proteins, lipids, and carbohydrates. No longer are readily available simple substrates provided by the mother via the placenta.

The newborn begins to convert substrates-protein to glucose, glucose to fat, and must adapt to a discontinuous supply of nutrients. Glycogen stored during feedings must be converted to glucose to support energy needs between feedings. Adding further to the vulnerability of the newborn period, many of the digestive and absorptive functions of the newborn are incompletely developed (Schmitz 1991).

To ease the transition through this turbulent period, the newborn needs to be provided with easily digestible and bioavailable nutrition in the form of breast milk or infant formula. The infant depends on a diet of closely regulated composition and does not achieve nutritional independence until weaning.

This chapter focuses on infancy (the first two years of life). This period can be broadly divided into three stages: the newborn period, from birth to 1 month; early infancy, from 1 month to 1 year; and later infancy, the entire second year. At no other time in the life cycle are the connections between nutrition and development so clearly evident.

PHYSICAL GROWTH DURING INFANCY

HEIGHT AND WEIGHT

Physical growth is more rapid during the first few months postpartum than at any other time after birth. In the first three months, a healthy infant's birth length increases by 20% and, by the end of the first year, by 50%. By the end of infancy (2 years old) the average child has already achieved half the ultimate adult height. These changes are illustrated in Figure 6.1, which shows a cumulative curve, or the total amount of growth, for length from conception to 50 weeks postpartum.

Figure 6.2 shows a growth velocity curve, which measures the rate of growth, or amount of growth per unit time, during the same period. Although the rate of growth is remarkably high during infancy, it is actually decelerating from the extremely high rates achieved during growth in utero. A healthy infant gains about 25 cm during the first year, but growth rate slows during the second year. From age 1 to 2 years the average height increase is 12–13 cm.

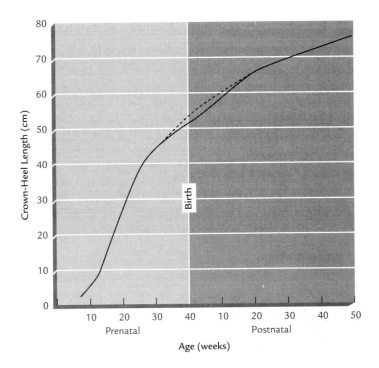

FIGURE 6.1

Cumulative growth curve for neonatal period

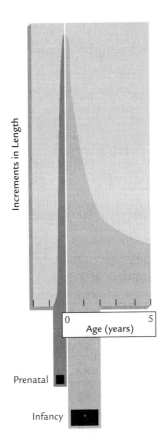

FIGURE 6.2

Growth velocity curve during neonatal period and infancy

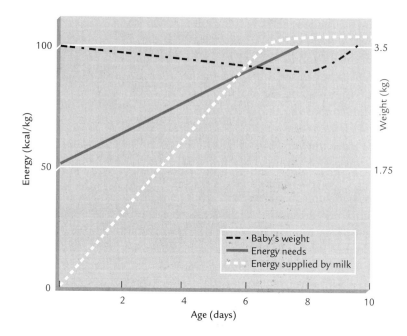

FIGURE 6.3

Weight loss and energy balance during the first ten days of life

As the newborn adjusts to life outside the uterus, changes in physiology produce losses of body water and small amounts of body tissues. Also, in the first few days after birth, the energy supplied by breastmilk does not yet meet the needs of the newborn. As a result, the infant loses about 6% of birthweight in the first few days after birth. By the end of the first week the infant begins to gain weight rapidly and, by about the tenth day, has regained birthweight (see Figure 6.3). By age 4 months, most infants have doubled their birthweights, and by the end of the first year, birthweight has tripled. Although the rate of weight gain during infancy is high, it decelerates from a peak rate obtained in utero. From birth to age 4 months the infant gains 20–25 g each day, and from 4 months to 1 year, weight gain slows to 15 g per day.

Growth Charts

Steady growth is a predictable characteristic of normal infancy. The National Center for Health Statistics (NCHS), has constructed *standard growth charts,* using data from cross-sectional national surveys of large numbers of infants in the United States (National Center for Health Statistics 1976). Standard curves show the percentile values for weight and length for age for both sexes. One set of charts is used for birth through 36 months, and a second set is used during later childhood. For infants under 24 months, length is measured while recumbent. Figure 6.4 shows the NCHS standards for girls and boys from birth to 36 months.

Healthy infants can be expected to maintain growth channels when serial measurements over time are recorded; normal variation in the size of

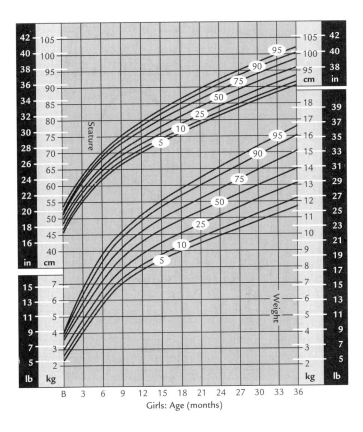

Girls: Age (months)

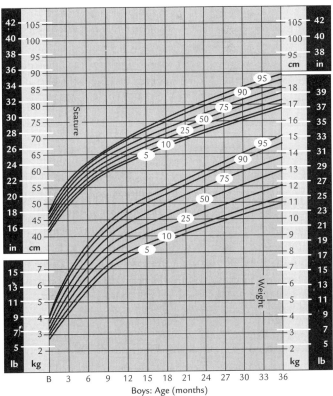

Boys: Age (months)

FIGURE 6.4

NCHS growth charts for birth through 36 months

individual infants will be apparent in the percentile growth channel followed by each growing infant. For example, in Figure 6.5, two 18-month-old infants are growing normally but following growth channels along different percentiles. Because the growth rate is sensitive to changes in nutrition, growth patterns are often used to evaluate nutritional status during infancy and childhood. A decrease in the growth rate during infancy is often the earliest indication of inadequate nutrition.

Figure 6.6 shows how growth during infancy can be impaired by poor nutrition. Because of insufficient energy and protein in the diet, the infant's weight has fallen from the 75th to the 10th percentile.

ETHNIC AND GENDER DIFFERENCES

In the United States, although Black newborns tend to have lower birthweights than Whites, Black infants grow more rapidly during infancy (Centers for Disease Control and Prevention 1992). Because of these dif-

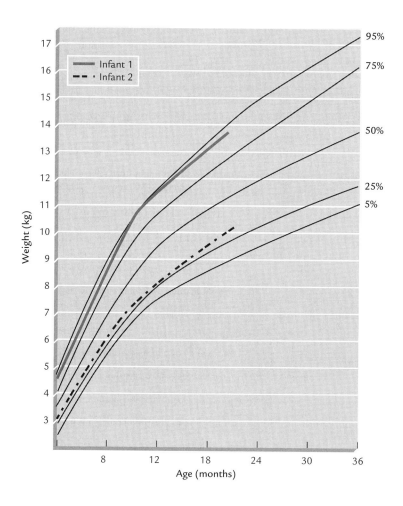

FIGURE 6.5

Two infants growing normally along different weight percentiles

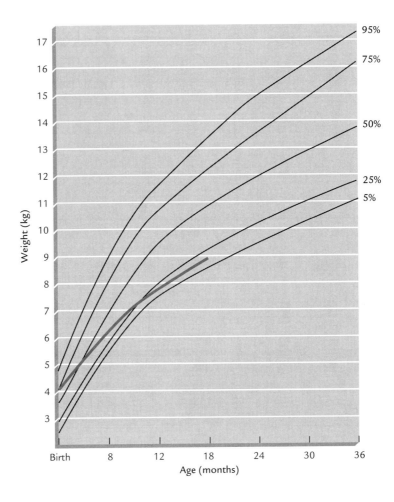

FIGURE 6.6

Growth patterns for three infants

ferences in infant growth, from age 2 through adolescence, Black children are taller than White children at the same age (Eveleth and Tanner 1990).

Male infants gain weight and length more rapidly than female infants in the United States (see Table 6.1). The gender difference in growth rate is greatest during the first four months after birth and, by the end of the first year, males weigh—on average—about 10% more than females and are slightly longer.

INFLUENCE OF NUTRITION ON GROWTH POTENTIAL

Intrinsic Factors

Intrinsic factors are conditions or events occurring before birth that influence an individual's potential for growth and development. Genetic endowment is a major intrinsic factor that determines growth potential, exemplified by the fact that infants and children of tall parents are more

Age (mo)	Length (cm)	Weight (g)
Males		
Birth	51.6	3,545
1	54.8	4,452
2	58.2	5,509
3	61.5	6,435
4	63.9	7,060
5	65.9	7,575
6	67.6	8,030
9	72.3	9,180
12	76.1	10,150
18	82.4	11,470
Females		
Birth	50.5	3,325
1	53.4	4,131
2	56.7	4,989
3	59.6	5,743
4	61.9	6,300
5	63.9	6,800
6	65.8	7,250
9	70.4	8,270
12	74.3	9,180
18	80.2	10,780

TABLE 6.1
Gender differences in weight and length during infancy

Adapted from S. J.Fomon, et al., "Body Composition of Reference Children from Birth to Age 10 Years," *Am J Clin Nutr* 35 (1982):1169–75.

likely to be taller than the offspring of shorter parents (Tanner 1990). Other intrinsic factors include congenital conditions such as maternal nutrition or disease that modify development in utero. For example, maternal malnutrition during pregnancy can limit fetal growth and produce a small-for-gestational-age (SGA) infant, as discussed in Chapter 2. Many SGA infants, even with adequate nutrition after birth, will be shorter than average for the rest of their lives, and they may show long-term impairments in intellect and mental development (Tanner 1990).

Extrinsic Factors

The extent to which an individual achieves his or her potential for growth after birth is also determined by *extrinsic factors* such as illness, climate, altitude, pollution, sanitation, and the amount of love and support. Nutrition is a major extrinsic factor during infancy and childhood, and the links between diet and development are the focus of this chapter and the following chapters on childhood and adolescence.

BODY PROPORTIONS

Changes in body proportions accompany the marked increases in height and weight during infancy. The parts of the body grow at different rates during different periods of development. In utero, the head is the fastest growing portion of the body, attaining about 50–60% of its final adult size by birth—when it accounts for about a quarter of total body weight. During the first year of life, the torso is the fastest growing portion, accounting for 60% of the increase in height during this period. During later infancy and childhood, the legs grow the fastest, accounting for about two thirds of the increase in stature.

BODY COMPOSITION

Dramatic changes in body composition occur during early infancy. As the infant grows and gains weight, all three major body compartments—water, fat, and lean tissue (mainly muscle and skeleton)—increase in size. However, proportions between the major compartments change. Total body water decreases from nearly 70% of total body weight at birth to 60% at 4 months, and thereafter remains stable. The 10% of body weight lost as water is replaced mainly by rapidly accumulating adipose tissue. Fat content of the body increases from about 15% at birth to 25% at 4 months. The percentage of body weight accounted for by lean tissue increases from about 11 to 15% during infancy, and rapid accretion of protein and minerals occurs as the skeleton, muscles, and teeth develop.

INFANT PHYSIOLOGY, DIGESTION, AND EXCRETION

THE STOMACH AND GASTROINTESTINAL MOTILITY

The newborn's stomach is small, with a capacity of only 10–15 ml, and is typically empty two to three hours after a feeding. Therefore, small, frequent feedings are needed during early infancy. Stomach capacity gradually increases to about 200 ml at the end of the first year.

Although the intricate nervous system that initiates peristaltic movements in the intestine is present as early as the 24th week in utero, the motility of the intestine and stomach may not be fully coordinated during the newborn period (Wershil 1991). In infants, this can cause less efficient mixing of digestive secretions with food, and slower movement of substances through the digestive tract.

DIGESTION AND ABSORPTION OF NUTRIENTS

The digestive and absorptive capabilities of the gastrointestinal tract, which begin to develop in utero, mature during the first year after birth.

Sugars

The enzymes in the **brush border** of the small intestine that metabolize sugars (sucrase, maltase, and isomaltase) appear as early as 28 weeks in utero, and they are well-developed at term (Riby 1994). However, the most important dietary carbohydrate during infancy is lactose, and the development of lactase activity in the intestine is somewhat delayed compared to other enzymes. At week 28 in utero, the level of lactase activity is still low, but by 34–38 weeks gestation it has reached maximal levels (Riby 1994). The full-term infant is readily able to absorb lactose and a number of other sugars—sucrose, glucose, corn syrup solids, and maltodextrins—included in infant formulas.

Starches

In adults, pancreatic amylase plays a central role in starch digestion. However, insignificant amounts of pancreatic amylase are produced in the first six months after birth, and other amylases compensate for this lack (see Table 6.2). Human milk contains an amylase that is resistant to stomach acid and aids in the breakdown of starches in breast-fed infants (discussed in Chapter 4). *Glucoamylase,* an enzyme produced by the brush border of the intestine, also contributes to starch digestion during the newborn period. Salivary amylase is present at low levels at birth and increases to about a third of adult levels at 3 months (AAP 1993).

Lipids

Lipids are important energy sources during the rapid growth of infancy. The ability to digest and absorb lipids develops late in utero and matures during early infancy (see Figure 6.7). An infant born preterm at 32 weeks can absorb only about 65–85% of dietary lipid, while a full-term infant can absorb about 80–90% (AAP 1993). Adult levels of fat absorption (greater than 95%) are not attained until about four to six months after birth. Several substances play key roles in lipid digestion in infancy:

brush border: the specialized surface of the epithelial cells that line the intestine, consisting of multiple microvilli that greatly increase the surface area available for absorption

- *Pancreatic lipase.* Levels of this enzyme at birth are only about 5% of adult levels. Activity increases fivefold in the first postpartum month and reaches adult levels at about 9 months.

Enzymes	(% of adult)
Protein	
H+	<30
Pepsin	<10
Trypsinogen	10–60
Chymotrypsinogen	10–60
Procarboxypeptidase	10–60
Peptidases	>100
Amino acid transport	>100
Fat	
Lingual lipase	>100
Pancreatic lipase	5–10
Bile acids	50
Medium-chain triglyceride uptake	100
Long-chain triglyceride uptake	10–90
Carbohydrate	
Alpha-amylases	
pancreatic	0
salivary	10
Lactase	>100
Sucrase-isomaltase	100
Glucoamylase	50–100
Monosaccharicle absorption	100

Adapted from E. Lebenthal and Y-K. Leung, "The Impact of Development of the Gut on Infant Nutrition," *Pediatr Ann* 16 (1987):215.

TABLE 6.2
Digestive enzymes: levels at birth as a percentage of adult levels

- *Lingual lipase.* The serous glands in the mouth of the newborn secrete a lipase that is fully active at birth. It is important in fat digestion only during the newborn period.

- *Human milk lipase.* Human milk contains several lipases that are stimulated by bile salts in the infant intestine (discussed in Chapter 4). They aid in the digestion of fat in breast milk.

- *Bile salts.* Fatty acids and monoglycerides produced by the hydrolysis of dietary triglycerides must be incorporated into micelles for efficient absorption. The newborn liver has a limited ability to produce bile acids necessary for micellar formation, and because of the small size of the newborn bile acid pool, the newborn capacity for fat absorption is limited (see Figure 6.8). Therefore, small, frequent feedings throughout the day result in more efficient fat absorption than larger feedings (AAP 1993).

The newborn's limited bile acid pool also influences the absorption of different types of fats during the newborn period. Because saturated fatty

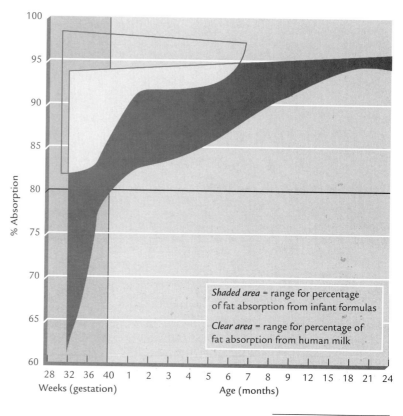

FIGURE 6.7

Development of fat absorption during infancy

Adapted from AAP (1993)

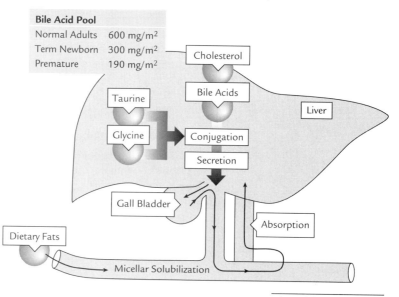

FIGURE 6.8

Bile acid metabolism and fat absorption

Adapted from AAP (1993)

acids require more efficient micellar formation for absorption than polyunsaturated fats, saturated fats (such as the butterfat in cow's milk) tend to be less well absorbed during early infancy (AAP 1993). **Medium chain triglycerides (MCTs)**, because they do not require micellar formation for absorption, are very well absorbed. Both polyunsaturated fats and MCTs are widely used in infant formulas, a subject discussed later in this chapter.

Protein

Proteolysis in the newborn stomach and intestine is limited. Levels of pepsin are only about 10% of adult levels during the first month (Wershil 1991). Also, the pH of the infant stomach during the first month of life may not be sufficiently low for optimal pepsin activity. In the intestine, levels of trypsin are usually adequate, but levels of chymotrypsin and carboxypeptidase are usually only 10 to 60% of adult levels (Lebenthal 1983).

Because of the developmental immaturity of these enzymes, protein digestion is incomplete. Incomplete digestion of proteins allows intact immunoglobulins in human milk to be absorbed and provide passive immunity during the newborn period, but they also may contribute to food allergies often seen in infancy. (These issues are discussed in Chapters 4 and 5.) Although protein digestion is limited in early infancy (estimates of capacity for protein absorption are 2 g/kg/day at birth, 3.75 g/kg/day at 4 months) (Lebenthal 1983), digestion and absorption are adequate in most infants to meet needs.

Micronutrients

The fat-soluble vitamins A, D, E, and K require solubilization and micellar formation for complete absorption. In early infancy (and particularly in preterm infants), absorption of these vitamins may be limited by low levels of pancreatic lipase and bile acids. Absorption of calcium and iron is more efficient during infancy and early childhood than it is in later life, and factors unique to human milk facilitate the absorption of zinc and iron during breast-feeding (AAP 1993). At the same time, absorption of lead is increased in infants, and they are at greater risk of lead toxicity (discussed in Chapter 7) (Ziegler et al. 1978).

KIDNEY FUNCTION

medium chain triglycerides (MCTs): triglycerides containing fatty acids with 6 to 10 carbons

The end products of protein metabolism, such as urea and creatinine, and excess electrolytes, such as sodium, chloride, and potassium, must be excreted by the kidney. The total amount of soluble waste products that must be filtered by the kidney is termed the *renal solute load*.

Diet can be a major source of soluble waste. For example, if protein intake is higher than requirements for growth and maintenance, more nitrogenous wastes will be produced. Also, if the diet is rich in sodium, extra sodium not used by the body must be excreted by the kidney. At the same time, water must be excreted to carry the soluble wastes out in the urine, and a high renal solute load may perturb water balance, particularly in the newborn.

The Newborn Kidney and Dehydration

Because it can produce a concentrated urine, the adult kidney can readily excrete large amounts of soluble wastes without excreting excess water. However, the newborn infant's immature kidney cannot concentrate urine efficiently. Also, secretion of **antidiuretic hormone** by the pituitary is limited in the newborn period, further limiting the infant's ability to concentrate urine and conserve body water (Foman, 1993). This compromises the infant's ability to maintain water balance in the face of fluid or electrolyte stress and increases the likelihood of **dehydration** during the newborn period.

The Newborn Kidney and Human Milk, Formula, and Cow's Milk

Human milk and commercial infant formulas are lower in protein and electrolytes than cow's milk or evaporated milk-based formulas. Thus they present less of a potential renal solute load (PRSL) to the newborn kidney (see Table 6.3). When water intake is reduced or water losses

antidiuretic hormone: a hormone produced by the pituitary gland that stimulates the kidney to reabsorb more water and thereby to excrete less; produced in response to dehydration or a high concentration of sodium in the blood

dehydration: the condition that results from undue loss of water from a body or tissue

Product	Solute Load (mosm/100 kcal)
Human milk	12
Whole cow's milk	33
Cow's milk (2% fat)	40
Skim cow's milk	68
Infant formula*	16
Strained foods	
Pears	5
Applesauce	5
Chicken with vegetables	30
Vegetable and beef	18

* Representative of the major commercial infant formulas

Adapted from AAP (1993)

TABLE 6.3

Potential renal solute load of milks, formula, and foods

	Daily Solute Load (mosmoles)	Water Balance (ml/day)	Days to Reach Dehydration and 10% Weight Loss
Human milk	70	–63	14
Whole cow's milk	230	–231	4

The hypothetical 1 month-old infant weighs 9kg, consumes only 750 ml milk each day, and has extrarenal losses of 700 ml/day. Renal concentrating ability is 1100 mosm/L.

Adapted from S. J. Fomon, *Nutrition of Normal Infants* (St. Louis: Mosby-Year Book, 1993).

TABLE 6.4

Water balance in an ill infant by type of milk

increase (during periods of illness, diarrhea, diminished fluid intake, or fever), newborn feedings with a high PRSL lead more rapidly to dehydration than feedings with lower PRSL (see Table 6.4). Because of these concerns, the U.S. Food and Drug Administration has set upper limits on the protein and electrolyte content of infant formulas (U.S. Food and Drug Administration 1985).

NUTRITIONAL REQUIREMENTS DURING INFANCY

ENERGY NEEDS

Table 6.5 shows the energy needs of a healthy newborn: resting energy expenditure is 40–60 kcal/kg/day, growth requirements are about 20 kcal/kg/day, and the thermic effect of food (energy expended for the digestion, absorption, transport, and storage of nutrients) constitutes about 8–10% of energy intake. When the energy needs of activity and loss of unabsorbed energy in the feces are added in, the total comes to about 90–120 kcal/kg/day (AAP 1993).

Category	Kilocalories (per kg per day)
Resting energy expendture	40–60
Activity	15–25
Thermic effect of food	10
Fecal loss of calories	5
Growth	20
Total	90–120

Adapted from AAP (1993)

TABLE 6.5

Energy requirements of a typical growing term infant

The actual energy requirement of individual infants will vary based on body size, level of activity, and rate of growth. Different infants have widely different activity patterns—some are restful and quiet while others spend more time crying, kicking, and moving. Energy spent in activity generally increases with age as the infant begins to move about, play, and explore.

To support the remarkable growth rate during early infancy, a substantial portion of energy intake goes toward deposition of new tissue. In contrast, during most stages of the life cycle, energy needs for new tissue deposition are small. For example, from age 3 years to puberty, only about 2% of energy is used for growth. In contrast, from birth to 4 months, up to 27% of energy is used for growth; from 4 to 6 months, about 11%; and from 6 to 12 months, about 5% of energy intake is used for deposition of new tissue (Fomon 1993). See Table 6.6.

Recommendations for Energy Intake

Total energy needs increase steadily throughout infancy as weight increases, while the energy needs per unit of body weight decline—reflecting a deceleration in the growth rate. Table 6.7 shows the daily need for energy in infants, based on the World Health Organization's estimations of energy intakes of healthy, normally growing infants from developed countries (WHO 1985). The first year is divided into two six-month periods: for the first half, recommended energy intake is 108 kcal/kg, and for the second, it is 98 kcal/kg. For the second year, recommended energy intake is 102 kcal/kg (NAS 1989).

Using these guidelines, the average 4-month-old infant who weighs 6 kg would need about 650 kcal/day, and a 14-month-old infant weighing 10 kg would need about 1000 kcal/day. Because energy needs can vary widely among individual infants, it is important to measure and record linear growth and weight gain during infancy. A normal pattern of growth is a good indication that energy needs are being met.

TABLE 6.6

Body increments in protein and fat, and energy used for growth during first year

| Age Interval (mo) | Body Increment (g/d) | | Energy for Growth | |
	Protein	Fat	kcal/d	$kcal \times kg^{-1} \times d^{-1}$
0 to 4	3.1	10.3	153	28
4 to 6	2.0	4.8	78	10
6 to 12	1.9	1.4	40	4

Adapted from S. J. Fomon, *Nutrition of Normal Infants* (St. Louis Mosby-Year Book, 1993).

Category	Age (years)	Weight		Height		REE	Average Energy Allowance (kcal)	
		(kg)	(lb)	(cm)	(in)	(kcal/day)	Per kg	Per day
Infants	0.0–0.5	6	13	60	24	320	108	650
	0.5–1.0	9	20	71	28	500	98	850
	1–3	13	29	90	35	740	102	1300

From National Academy of Sciences, Food and Nutrition Board of the National Research Council, *Recommended Dietary Allowances*, 10th ed. (Washington D.C.: National Academy Press, 1989).

TABLE 6.7

Recommended energy intake during infancy

The Importance of High Energy-Density Diets

Because of the small stomach capacity of newborn infants, they need small, frequent feedings with foods of high energy density to meet high energy needs. The consequences of feeding infants a diet of low energy density were shown in a study in which two groups of 3-month-old infants were fed different diets. One group received standard infant formula (with 67 kcal/100 ml) for two months; the other group received skim milk-based formula (36 kcal/100 ml). Although the skim-milk fed babies consumed larger amounts of food, they received about 30% less energy per day. The skim-milk fed babies continued to grow: they gained length at the same rates as the formula fed group, but they gained about 25% less weight. The skim-milk fed babies mobilized body fat to provide part of their energy needs, and skinfold thickness sharply decreased (Fomon et al. 1977). See Figure 6.9.

It appears that infants are at least partially able to compensate for a diet of greater or lesser energy density by adjusting the quantity of food

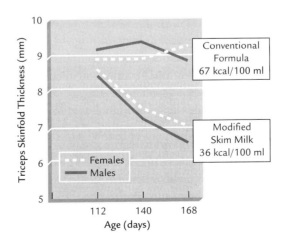

Adapted from Fomon et al. (1977)

FIGURE 6.9

Decrease in body fat in infants in a diet of low energy-density

consumed. In the study just cited, the infants fed skim milk consumed about 25% more milk than the infants receiving the higher energy formula. In another study, two groups of 1-month-old infants were fed formula containing either 100 kcal/100 ml or 54 kcal/100 ml for two months. Weight gain and daily energy intakes were not significantly different between the two groups (Fomon et al. 1975).

PROTEIN NEEDS

Protein requirements per kilogram of body weight are higher during infancy than at any other time in life. Protein provides amino acids and energy for the formation of new muscle, connective tissue, and bone, as well as for synthesis of a large number of enzymes, hormones, and plasma proteins. Protein accretion from birth to 4 months is more than 3 g/day, and from 4 to 12 months it is about 2 g/day.

The RDA for protein during infancy is shown in Table 6.8. The reference protein used for early infancy is human milk protein, and the protein requirement for the first few months after birth is based on the intake of protein by healthy breast-fed infants who are growing normally (NAS 1989). Human milk contains 0.8–0.9 g/100 ml of protein and 0.2–0.3 g/100 ml of nonprotein nitrogen. A 4-kg infant receiving 800 ml of breast milk each day consumes about 7.2 g of protein and 1.6 g of nonprotein nitrogen compounds, or about 1.8 g/kg and 0.4 g/kg per day of protein and nonprotein nitrogen, respectively.

Therefore, protein intakes of breast-fed infants average about 2 g/kg per day during the first three months after birth and 1.73 g/kg per day from 3 to 6 months. For later infancy, the protein allowances are derived from WHO's estimations of protein needs using a factorial method (WHO 1985), with the reference proteins being high-quality protein such as that in cow's milk or eggs. See Table 6.9.

Category	Age (years)	Weight (kg)	Recommended Dietary Allowance (g/kg)	(g/day)
Both sexes	0–0.5	6	2.2	13
	0.5–1	9	1.6	14
	1–3	13	1.2	16

From National Academy of Sciences, Food and Nutrition Board of the National Research Council, *Recommended Dietary Allowances,* 10th ed. (Washington D.C.: National Academy Press, 1989).

TABLE 6.8
The RDAs for protein during infancy

| Age | Growth | | | | | Allowance of Reference Protein (g/kg per day)[e] | |
	Nitrogen Increment (mg/kg per day)[a]	Nitrogen Increment x 1.5 (mg/kg per day)[b]	Nitrogen Increment x 1.5 plus Correction for Efficiency at 70% (mg/kg per day)[c]	Nitrogen Maintenance Level (mg/kg per day)[d]	Total Nitrogen (mg/kg per day)	Mean	+ 2 SD[f]
Both sexes							
Months							
3–5.9	47	70	100	120	220	1.38	1.73
6–11.9	34	51	73	120	193	1.21	1.51
Years							
1	16	25	36	119	155	0.97	1.21

[a] Increment for growth.

[b] 50% additional nitrogen increment to allow for daily variation in growth rate and inability to store amino acids to be available when maximum growth occurs.

[c] Assuming a 70% efficiency of dietary protein utilization for growth.

[d] Data from WHO (1985).

[e] High-quality, highly digestible protein such as egg or milk. Protein is total nitrogen x 6.25.

[f] Individual availability. The coefficient of variation for both maintenance and growth was assumed to be 12.5%.

From National Academy of Sciences, Food and Nutrition Board of the National Research Council, *Recommended Dietary Allowances*, 10th ed. (Washington D.C.: National Academy Press, 1989).

TABLE 6.9

Factorial derivation of protein allowances for infancy

Amino Acid Requirements

The nine amino acids that are essential for adults are also essential for infants. WHO has calculated the pattern of amino acid requirements for infants by dividing each essential amino acid requirement by the recommended allowance of the reference protein for infancy (human milk protein) (WHO 1985). The amino acid requirement pattern for infancy is shown in Table 6.10 and, for comparison, the amino acid composition of human and cow's milk. About 45% of the amino acids in human milk are essential amino acids.

In the newborn period, several additional amino acids are considered *conditionally essential* (see Table 6.11). In older children and adults, these amino acids are synthesized from precursors, but in the newborn the synthetic pathways and enzymes are not fully developed. Requirements must be at least partially met by dietary sources. For example, cysteine is an essential amino acid during early infancy, because the synthetic enzymes that convert methionine to cysteine are not fully developed.

Similarly, endogenous synthesis of taurine, arginine, and carnitine is inadequate to meet the needs of preterm and young term infants

	Requirement (mg/g protein)	Composition (mg/g protein)	
Amino Acid	**Infants 3–4 months**	**Human Milk**	**Cow's Milk**
Histidine	16	26	27
Isoleucine	40	46	47
Leucine	93	93	95
Lysine	60	66	78
Methionine plus cystine	33	42	33
Phenylalanine plus tyrosine	72	72	102
Threonine	50	43	44
Tryptophan	10	17	14
Valine	54	55	64
Total without histidine	412	434	477

Adapted from National Academy of Sciences, Food and Nutrition Board of the National Research Council, *Recommended Dietary Allowances*, 10th ed. (Washington D.C.: National Academy Press, 1989).

TABLE 6.10

Amino acid requirements during infancy compared to composition of milks

Essential Amino Acids	Nonessential Amino Acids	Conditionally Essential Amino Acids
Threonine	Glutamic acid	Cysteine
Leucine	Glycine	Arginine
Isoleucine	Aspartic acid	Carnitine
Valine	Proline	Taurine
Methionine	Tyrosine	
Phenylalanine	Glutamine	
Tryptophan	Alanine	
Lysine	Serine	
Histidine	Asparagine	

TABLE 6.11

Essential, nonessential, and conditionally essential amino acids in infancy

(Olson et al. 1989). Carnitine, an amino acid that plays a central role in the metabolism of fat, is supplied in ample amounts in human milk and cow's milk-based infant formulas but is lacking in soy-based formulas. Therefore, carnitine is added to soy-based infant formulas in the United States.

Taurine plays a central role in the production of bile acids, and an adequate supply is important for fat absorption from the intestine. The enzyme systems that normally synthesize taurine are still developing in the newborn, and endogenous synthesis is limited (Gaull 1989). Although taurine is abundant in human milk, it must be added to commercially prepared soy-based and cow's milk-based infant formulas. Supplementation of the diet of preterm infants with taurine increases fat absorption (Galeano et al. 1987).

Effects of Too Little or Too Much Protein

Protein deficiency in infants and children, a major nutritional problem worldwide, is discussed in the next chapter in the section on protein-energy malnutrition (PEM). Excessively high intakes of protein in early infancy can also harm the infant. Because of the high potential renal solute load, excess protein intake during infancy can trigger dehydration. Also, in some infants, large amounts of dietary protein can produce a **metabolic acidosis**.

FAT REQUIREMENTS

Infants require calorically dense foods to meet the high energy needs of growth. Human milk contains 50–55% of energy as fat, and most formulas contain 45–50% of energy as fat. Infants thrive and grow normally when fed diets with 30–60% of calories as fat; less energy-dense diets may result in inadequate energy intake.

Essential Fatty Acids

The fatty acids linoleic acid (18:2n6) and linolenic acid (18:3n3) cannot be synthesized by humans and are essential dietary components. Linoleic acid is important in maintaining the structure and function of cellular membranes, and both linoleic and linolenic acid are precursors of eicosanoids, a family of compounds with broad physiological functions (discussed in Chapter 9).

Infants fed formulas deficient in linoleic acid for as short a time as one week can develop a dry, eczemalike, flaky skin rash. Other symptoms include diarrhea, hair loss, and impaired wound healing. Long-term fatty acid deficiency may impair platelet function and lower resistance to infection. Because of poor fat absorption and low fat stores, preterm infants are particularly sensitive to essential fatty acid deficiency, and they quickly develop signs of deficiency if fat intake is delayed after birth.

Intake of linoleic acid at levels of 1–2% of total calories will prevent signs of deficiency— for infants consuming about 100 kcal/kg per day, this is equivalent to a daily intake of about 0.2 g/kg. The American Academy of Pediatrics and WHO (Joint FAO/WHO 1984) recommend that infant formulas provide at least 300 mg of linoleic acid per 100 kcal. This amount is readily available in human milk and commercially prepared infant formulas.

metabolic acidosis: abnormally high acidity in the body fluids caused by a disturbance in metabolism

The daily requirement for linolenic acid during infancy is less clear. Although the current RDAs do not include recommendations on essential fatty acid intake, the Canadian Recommended Dietary Intakes are for 0.5 g of n3-polyunsaturated fatty acids per day for infants during the first year (Scientific Review Committee 1990).

WATER REQUIREMENTS

Increased Risk of Dehydration

Infant needs for water per unit of body weight are significantly higher than in adults, and the risk of dehydration is sharply increased during the newborn period. There are several reasons for this:

- Because infants have a larger surface area per unit of body weight, **insensible losses** from the skin and lungs are increased.

- The newborn is unable to express thirst.

- Newborns have a higher percentage of body water, and water turnover is rapid.

- The capacity of the newborn to concentrate urine is limited in the face of high potential solute loads.

Water Balance during Infancy

insensible losses: losses that occur without being perceived, such as water loss from the skin and lungs

The bulk of water lost by the healthy newborn is insensible losses. Urinary and fecal losses also occur, and water is "lost" in growth, as it is incorporated into growing cells and tissues Water requirements of infants under usual conditions is shown in Table 6.12.

Age	Amount of Water (ml/kg/day)
3 days	80–100
1 week	120–150
3 mo	140–160
6 mo	130–155
9 mo	125–145
1 yr	120–135
2 yr	115–125

Adapted from R. E. Berkman, V. C. Vaughan III, and W. E. Nelson, eds., *Nelson Textbook of Pediatrics*, 11th ed. (Philadelphia: W. B. Saunders, 1987).

TABLE 6.12
Water requirements during infancy

The RDA for water during infancy is 1.5 ml/kcal of energy expenditure (NAS 1989). This is the water-to-energy ratio in human milk and commercial infant formulas. Unless the weather is very hot and evaporative losses are high, healthy infants who are exclusively breast-feeding or formula feeding require no supplemental water.

Later in infancy, as solid foods are introduced, extra water is usually needed to excrete the high renal solute load of foods such as strained meats, cow's milk, and egg yolks. To avoid dehydration, fluid intake in later infancy should be encouraged when losses are increased, such as during illness, diarrhea, and fever.

VITAMIN AND MINERAL NEEDS

Table 6.13 gives the RDAs and estimated safe and adequate intakes of vitamins and minerals for infants.

VITAMIN REQUIREMENTS

Fat-Soluble Vitamins

Vitamin A Vitamin A plays key roles in growth, cellular differentiation, and immune system integrity during infancy. Yet, many infants are born with low stores of vitamin A. Researchers measured liver stores in infants under 3 months of age: nearly three quarters had liver concentrations less than 20 μg/g (Olsen et al. 1984) (greater than 20 μg/g are considered adequate).

Breast milk from well-nourished mothers contains 40–70 μg/100 ml of preformed vitamin A and 20–40 μg/100 ml of beta-carotene. The intake of vitamin A by a breast-fed infant is about 300 μg/day, and this daily intake, adjusted for variability, is used to calculate the RDA (375 μg for ages 0–1 year, 400 μg for 1–2 years) (NAS 1989). These recommendations are aimed at supplying the needs of the rapidly growing infant and increasing body stores to adequate levels during the first year. Vitamin A is added to infant formulas as the retinyl ester, typically at levels of 300-400 IU/100 kcal. (The U.S. Food and Drug Administration requires levels in formula to be 250 IU/100 kcal.) (U.S. Food and Drug Administration 1985).

Although too much vitamin A during infancy can be harmful and small children are particularly susceptible to vitamin A toxicity if given large amounts in supplements or formula (Persson et al. 1965), the major nutritional problem of vitamin A in infancy and childhood worldwide is deficiency. Vitamin A deficiency affects millions of children in the developing countries and is the leading cause of blindness in children. Vitamin A deficiency during infancy and early childhood also increases mortality from infectious diseases, such as measles and

Recommended Dietary Allowances for Infancy

Fat-Soluble Vitamins

Age (years)	Weight (kg)	(lb)	Height (cm)	(in)	Protein (g)	Vitamin A (µg RE)	Vitamin D (µg)	Vitamin E (mg aαTE)	Vitamin K (µg)
0.0–0.5	6	13	60	24	13	375	7.5	3	5
0.5–1.0	9	20	71	28	14	375	10	4	10
1–3	13	29	90	35	16	400	10	6	15

Water-Soluble Vitamins

Age (years)	Vitamin C (mg)	Thiamin (mg)	Riboflavin (mg)	Niacin (mg NE)	Vitamin B$_6$ (mg)	Folate (µg)	Vitamin B$_{12}$ (µg)
0.0–0.5	30	0.3	0.4	5	0.3	25	0.3
0.5–1.0	35	0.4	0.5	6	0.6	35	0.5
1–3	40	0.7	0.8	9	1.0	50	0.7

Minerals

Age (years)	Calcium (mg)	Phosphorus (mg)	Magnesium (mg)	Iron (mg)	Zinc (mg)	Iodine (µg)	Selenium (µg)
0.0–0.5	400	300	40	6	5	40	10
0.5–1.0	600	500	60	10	5	50	15
1–3	800	800	80	10	10	70	20

Estimated Safe and Adequate Daily Dietary Intakes of Selected Vitamins and Minerals

Age (years)	Biotin (µg)	Pantothenic Acid (mg)	Copper (mg)	Manganese (mg)	Fluoride (mg)	Chromium (µg)	Molybdenum (µg)
0–0.5	10	2	0.4–0.6	0.3–0.6	0.1–0.5	10–40	15–30
0.5–1	15	3	0.6–0.7	0.6–1.0	0.2–1.0	20–60	20–40
1–3	20	3	0.7–1.0	1.0–1.5	0.5–1.5	20–80	25–50

From National Academy of Sciences, Food and Nutrition Board of the National Research Council, *Recommended Dietary Allowances*, 10th ed. (Washington D.C.: National Academy Press, 1989).

TABLE 6.13
RDAs for infancy

diarrheal illness. A thorough discussion of this topic can be found in Chapter 7.

Vitamin D Among its important roles throughout the body, vitamin D participates in the growth and differentiation of several tissues, including the skeleton and the bone marrow. Human milk contains only small amounts of vitamin D. In northern climates, during the winter months when maternal sunlight exposure is minimal, levels of vitamin D in breast milk may provide only about 30 IU to the infant daily. This level of vitamin D may not be sufficient to prevent rickets, if the infant is not exposed to adequate sunlight (Tsang 1983). In breast-feeding infants with abundant ultraviolet light exposure, additional vitamin D produced by the skin from synthesis from 7-dehydrocholesterol will maintain adequate vitamin D status.

Studies have found vitamin D status of breast-fed infants to be poorer than that of formula-fed infants (Tsang 1983), (Rothberg et al. 1982). Commercial infant formula contains higher amounts of vitamin D than breast milk, usually 60 IU/100 ml (AAP 1993). Infants who are exclusively breast-fed, particularly those who receive little sunlight exposure during darker winter months, can develop very low levels of 25-OH vitamin D (Specker and Tsang 1987). See Figure 6.10. Researchers have found lower bone mineral content in infants fed human milk without supplemental vitamin D, compared with a group fed human milk plus a 10-µg (400-IU) daily supplement of the vitamin (Greer and Marshall 1989).

Because of these concerns, breast-fed infants who are not exposed to sunlight regularly should receive a supplement of 5–7.5 µg vitamin D daily (NAS 1989). Infants fed formula and those who receive supplementary foods containing adequate vitamin D do not need supplementation. The RDA for vitamin D during the first six months after birth is set at 7.5 µg (300 IU); for the remainder of infancy and childhood, because of increasing body weight, it is increased to 10 µg (400 IU) (NAS 1989).

Toxicity can occur if infants are given high doses of vitamin D. In Europe in the 1940s, overuse of vitamin D (intakes of 70–100 µg per day) was associated with infantile hypercalcemia (British Pediatric Association 1956).

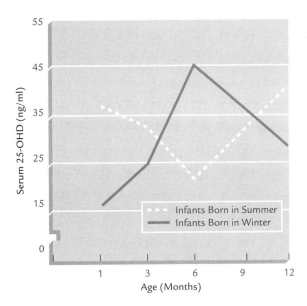

Adapted from Specker and Tsang (1987)

FIGURE 6.10

Seasonal changes in serum 25-OH vitamin D in breast-fed infants

Vitamin E A healthy term infant needs about 0.5 mg of tocopherol per 100 kcal (NAS 1989). The RDA for vitamin E during the first six months after birth is derived from the tocopherol intake of exclusively breast-fed infants consuming 750 ml of human milk/day (about 2.4 mg of tocopherols). The requirement for vitamin E increases as dietary intake of polyunsaturated fatty acids (PUFAs) increases, and the PUFA content of human milk is relatively high—about 6% of calories. The RDA for 0–6 months of age is set at 3 mg of tocopherol. The RDA increases to 4 and 6 mg for the second six months and second year, respectively, based on increasing body weight. These RDAs provide about 0.5 mg of tocopherol per kilogram of body weight (NAS 1989).

Premature infants have low body stores of vitamin E (NAS 1989). In addition, because of their rapid growth rates and reduced absorption of lipids, it is difficult for preterm infants to obtain sufficient vitamin E. During the 1960s and 1970s, preterm infants were often fed formulas high in PUFAs and with low tocopherol/PUFA ratios. In some infants, feeding of high PUFA formulas was associated with a **hemolytic anemia** thought to be due to a deficiency of vitamin E (Bell 1992). Formulas for preterm infants have been modified and now contain less PUFAs and more vitamin E. Also, to compensate for poor intestinal absorption, experts recommend that preterm infants receive daily supplementation with 5 to 25 mg of vitamin E (AAP 1993).

Vitamin K Adequate vitamin K during the newborn period ensures production of vitamin K-dependent blood clotting proteins and normal blood coagulation. Newborn infants with poor vitamin K status may develop **hemorrhagic disease of the newborn**, and they can die from bleeding in the brain and other organs (Gleason and Kerr 1989). In later infancy and childhood, vitamin K synthesized by intestinal bacteria contributes significantly to vitamin K requirements; but the newborn intestine does not contain an adequate population of microflora, and all vitamin K must be obtained from dietary sources.

Compared with human milk, cow's milk and commercial infant formulas contain much higher amounts of vitamin K. While infant formulas provide at least 4 µg of vitamin K/100 kcal, levels of vitamin K in human milk are low—only about 0.2 µg/100 ml. Because the RDA for vitamin K is set at 5 µg per day for the first six months after birth (NAS 1989), the recommended intake of vitamin K cannot be met by usual levels in breast milk. Therefore, to prevent bleeding problems and supply the newborn with adequate body stores, all infants in the United States routinely receive a single intramuscular injection of vitamin K soon after birth—the usual dose is 0.5–1.0 mg (AAP 1993).

Water-Soluble Vitamins

Vitamin C Infants exclusively breast-feeding typically receive about 25–40 mg of vitamin C per day. Vitamin C intakes as low as 7 mg per day

hemolytic anemia: anemia caused by lysis of red blood cells

hemorrhagic disease of the newborn: a generalized bleeding disorder in infants caused by a deficiency of vitamin K

from either breast milk or formula can protect infants against scurvy (Rajalakshmi et al. 1965). The RDA is based on the usual intake from breast milk and is set at 30 mg/day during the first six months of infancy. Thereafter, the RDAs are gradually increased to the adult level, and recommendations are for 35 mg and 40 mg for the second six months and second year, respectively (NAS 1989).

Infant formulas sold in the United States must contain at least 8 mg of vitamin C per 100 kcal (AAP 1993). In later infancy, supplementary foods can provide significant additional vitamin C—particularly citrus fruits, green vegetables, tomatoes, and potatoes. Families should carefully select supplementary foods rich in vitamin C—about one in five infants from 6 to 12 months old receive less than two thirds of the RDA for vitamin C (Fomon 1993). See Table 6.14.

Vitamin B$_6$ As discussed earlier in this chapter, protein requirements are higher per unit of body weight during early infancy than at any other time in life. Ample vitamin B$_6$ is therefore important for optimal protein metabolism during infant growth. Studies have shown that metabolic requirements for vitamin B$_6$ during infancy can be met by intakes of 0.015 mg/g protein, or about 0.04 mg/100 kcal (NAS 1989).

Infants who are breast-fed by mothers with low vitamin B$_6$ intakes (less than 2.0 mg/day) and who receive less than 0.1 mg/day of vitamin B$_6$ in milk, can show signs of deficiency (irritability, digestive problems, and seizures) (McCoy et al. 1985). The RDA is set at 0.3 mg/day for the first six months of infancy and 0.6 mg/day for ages 6 to 12 months. The recommendation for the second year increases to 1.0 mg/day, based on increasing body weight and protein intake (NAS 1989).

Food	Household Measure	Vitamin C (mg)
Human milk	8 oz	16
Cow's milk	8 oz	2
Broccoli	1 spear	22
Brussels sprouts	¼ cup	31
Cabbage, cooked, wedge	¼ cup	10
Cantaloupe, diced	½ cup	27
Grapefruit	½	37
Orange juice	½ cup	62
Potato, boiled (2½ inch)	½	11
Strawberries, whole	1 cup	88
Tomato, raw (2⅜ inch)	1	42
Tomato juice	½ cup	20

Adapted from C. F. Adams "Nutritive Value of American Foods in Common Units," *Agriculture Handbook* No. 456, Agriculture Research Service (Washington D.C.: U.S. Department of Agriculture, 1975).

TABLE 6.14

Selected infant foods and their vitamin C content

Folate Rapidly dividing cells in developing tissues require ample folate and vitamin B_{12} for protein and nucleic acid synthesis. Body stores of folate at birth are small (particularly in preterm infants) and can be quickly depleted by the high requirements of growth.

The RDAs for folate call for 3.6 μg/kg body weight for the first year, or about 25–35 μg of folate/day (NAS 1989). Healthy breast-fed infants who are growing normally receive about 36 μg of folate per day from milk. Cow's milk contains about 5 μg of folate per 100 ml, but boiling destroys about 50% of it (Ghitis 1966). Therefore, infants receiving boiled cow's milk or boiled evaporated milk often need supplemental folate.

Vitamin B_{12} The RDAs for vitamin B_{12} are derived from intakes of healthy infants who are breast-feeding from well-nourished mothers. The recommended intake for infants from 0–2 years is 0.05 μg/kg body weight (NAS 1989).

Infants born to mothers with adequate vitamin B_{12} status have accumulated significant stores of vitamin B_{12} in utero. However, in infants born to and breast-fed by mothers who are complete vegetarians, body stores and serum levels of vitamin B_{12} are often very low (Specker et al. 1988). Infants of complete vegetarians who are exclusively breast-fed can develop deficiency signs (anemia and neurological problems) as early as 4 months of age (Sklar 1986). In women who are complete vegetarians, maternal supplementation with vitamin B_{12} raises levels in breast milk and prevents signs of deficiency in their infants. Thus, it is recommended that pregnant and lactating women who are complete vegetarians take a vitamin B_{12} supplement at the RDA level (2.2 μg/day for lactating women and 2.6 μg/day for vegetarians) (NAS 1989).

Thiamin, Riboflavin, Niacin, Biotin, and Pantothenic Acid The level of these vitamins contained in human milk from well-nourished mothers is thought to be adequate to meet the needs of healthy infants who are growing normally. Therefore, the RDAs for these vitamins are based on the amounts obtained by infants from breast milk (NAS 1989). Most commercially prepared infant cereals and cereal–fruit combinations are enriched with thiamin, riboflavin and niacin, and they are rich sources of these vitamins during later infancy. Intakes of the B vitamins from several commercially prepared infant foods are shown in Table 6.15.

MINERAL REQUIREMENTS

Major Minerals

Calcium and Phosphorus The total body calcium of a newborn infant is approximately 30 grams, 99% of which is present in the skeleton. During the first year, an infant accumulates about 15g of additional calcium, or 1.2–1.3 g/month. The RDA for calcium during infancy is derived from the

Vitamin	10-g Dry Cereal	Wet-Pack Mixed with Applesauce and Bananas	Vegetable and Beef Dinner	Beef and Vegetable Dinner
Thiamin	160	200	15	20
Riboflavin	190	250	20	40
Niacin	1400	2800	360	880
Vitamin B$_6$	40	100	35	60
Folate	—	2.6	3	4
Vitamin B$_{12}$	—	—	0.2	0.3

TABLE 6.15

Intake of B-vitamins (in µg) from a 70 g serving of four commercially prepared infant foods

Values from J. A. T. Pennington, *Bowes and Church's Food Values of Portions Commonly Used*, 15th ed (New York: Harper and Row, 1987) and from tabular data of *Gerber Products, Gerber Nutrient Values* (Fremont, MI: Gerber Products, 1991).

Infant Age	Feeding	Absorption %
0–4 Months	Human milk	58
	Milk-based formulas	38
	Soy-based formulas	34
	Whole cow's milk	40
4–12 Months	Milk-based-formulas	45
	Soy-based formulas	43
	Whole cow's milk	30

TABLE 6.16

Percentage of calcium absorption from human milk vs. cow's milk vs. formula

Adapted from S. J. Fomon, *Nutrition of Normal Infants* (St. Louis: Mosby-Year Book, 1993).

calcium intakes of healthy breast-fed infants, which average about 240 mg/day during the first six months after birth. Allowing for differences in absorption between human milk and cow's milk (infants absorb nearly 60% of the calcium in human milk but only 25–40% the calcium in cow's milk), the RDA for formula-fed infants is set at 400 mg/day for the first six months, 600 mg for the second half of the first year and 800 mg for the second year (NAS 1989). Table 6.16 lists the percentage of calcium absorption from different types of milk and formula. The RDAs for phosphorus during infancy are shown in Table 6.13.

It is important that infants receive supplementary foods rich in calcium and other minerals during the later half of the first year. A recent study reported on the calcium intakes of two groups of infants during months 6–12: one group was fed a macrobiotic vegetarian diet and the other a balanced diet containing dairy products and meat (Dagnelie et al. 1989). The calcium intakes of infants being fed the macrobiotic diets were less than half the intakes of infants on nonvegetarian diets. Also, the vegetarian diet tended to be higher in fiber and phytates—substances that inhibit calcium absorption. Rickets can develop in infants who are fed supplementary foods low in calcium, minerals, and vitamin D (Dagnelie et al. 1990). Table 6.17 lists good sources of calcium for infants.

Feeding infants unmodified cow's milk can have adverse effects on calcium status. Cow's milk has a much higher amount of phosphorus than human milk (70 mg/100 ml versus 14 mg/100 ml, respectively), and the ratio of calcium to phosphorus is 1.2–1.4 :1 in cow's milk, while it is 2.3:1 in human milk. From the 1930s through late 1960s infants in the United States and United Kingdom were often fed formulas made from fresh, evaporated, or dried cow's milk.

Many newborns who were fed cow's milk-based formulas developed newborn hypocalcemia (low serum calcium) near the end of the first week of life, and if the hypocalcemia was severe, seizures could occur, (Mizraki et al. 1968). Newborn hypocalcemia is primarily caused by excessive phosphorus in cow's milk (the excess phosphorus is deposited into the skeleton, pulling calcium with it, and lowering blood levels of calcium). Cow's milk-based formulas are currently modified to reduce the phosphorus content to about 35 mg/100 ml, and the incidence of hypocalcemia in formula-fed infants has dropped dramatically.

Magnesium The amount of magnesium in human milk from well-nourished mothers is thought to be adequate to meet the needs of healthy infants who are growing normally. Therefore, the RDA for magnesium, shown in Table 6.13, is derived from the amounts obtained by infants from breast milk (NAS 1989).

Electrolytes

Sodium From birth to age 3 months, infants require approximately 0.5 mEq sodium/kg body weight daily to support rapid cell growth (NAS

Food	Household Measure	Calcium(mg)
Human milk	4 oz	40
Cow's milk	4 oz	144
Powdered milk, nonfat, instant dry	⅓ cup	293
Cheddar cheese	1 oz	213
Yogurt	½ cup	147
Custard, baked	½ cup	148
Ice cream	½	97

From C. F. Adams "Nutritive Value of American Foods in Common Units," *Agriculture Handbook* no. 456, Agriculture Research Service, (Washington D.C.: U.S. Department of Agriculture, 1975).

TABLE 6.17

Calcium content in milk and dairy products consumed during infancy

1989). By 6 months, as growth decelerates, sodium requirements decrease to 0.2 mEq/kg/day. Daily losses of sodium from the skin during infancy are estimated at about 0.4–0.7 mEq/kg. The estimated minimum daily requirement for sodium during infancy has been set at 1 mEq/kg, or about 120 mg sodium for 4.5 kg infant (NAS 1989). For most infants, breast milk easily covers sodium needs—human milk provides about 0.7 mEq/100 ml, or approximately 120 mg sodium/day during the first six months (based on intake of 750 ml of milk). Based on consumption of about 750 ml/day, infant formulas provide a minimum of 100 mg and a maximum of 300 mg of sodium/day (AAP 1993).

Potassium and Chloride As the major intracellular cation, potassium must be incorporated into new cells as they grow and divide. Due to the large increase in lean body mass during infant growth, potassium requirements are relatively high—about 60–80 mEq are needed per kilogram of weight gained. Using standard growth rates for infancy, the potassium requirement for growth is estimated to be 65 mg/day during the first year (NAS 1989). Potassium intake must also balance obligatory daily losses from the skin and in the urine and feces. The estimated minimum requirements for potassium, chloride, and sodium during infancy are shown in Table 6.18. For all of the electrolytes, minimum requirements are based on normal ambient and body temperatures; in cases of prolonged sweating or fever, losses will be greater, and requirements will increase.

Trace Minerals

Iron The rapidly growing infant requires large amounts of iron for the synthesis of hemoglobin, myoglobin, and a number of heme-containing enzymes (see Table 6.19). A healthy, full-term infant has accumulated ample iron stores at birth—total body iron at birth in a 3.5 kg infant is about 270 mg (Dahro et al. 1983). For four to six weeks after birth, production of new red blood cells is low—an adaptation by the newborn to the new, oxygen-rich environment. Storage iron increases during the first

Age	Infant Weight (kg)	Sodium (mg)	Chloride (mg)	Potassium (mg)
Months				
0–5	4.5	120	180	500
6–11	8.9	200	300	700
12–24	11	225	350	1000

From National Academy of Sciences, *Recommended Dietary Allowances,* 10th ed. (Washington D.C.: National Academy Press, 1989).

TABLE 6.18

Estimated minimum requirements of electrolytes for infancy

For an infant who is:	Requirement for Absorbed Iron during First Year of Life (mg)	
	3.5 kg at birth, 10.5 kg at 1 year	
Increment in total body iron		109
Body iron at 1 year	377	
Hemoglobin	270	
Myoglobin and enzymes	54	
Storage	53	
Body iron at birth	268	
Losses of iron		91
Gastrointestinal	62	
Dermal	29	
Total		**200** (0.55 mg/d)

Adapted from S. J. Fomon, *Nutrition of Normal Infants* (St. Louis: Mosby-Year Book, 1993).

TABLE 6.19

Estimated requirement for iron over the first year

weeks after birth as the breakdown of older red blood cells containing fetal hemoglobin releases large amounts of iron. But by the second month, the rapid synthesis of new red cells containing mature hemoglobin and the accumulation of myoglobin in growing muscle cells begin to draw heavily on iron stores. In most infants, iron stores are exhausted over the first four to six months after birth by the demands of growth, and needs must be met by dietary sources.

Estimates of the requirement for absorbed iron during the first year of infancy range from 0.55–0.75 mg/day, and for many infants, dietary iron is not sufficient to cover these needs. Iron-fortified infant formula and cereals are rich sources of iron for infants. Infants receiving iron-fortified formula (800–850 ml of formula/day containing 1.2–1.4 mg iron/100 ml) consume about 11 mg iron/day, of which 4–5% is absorbed (Fomon, 1993). Absorbed iron will be 0.45–0.55 mg/day, which is enough to cover needs.

Only small amounts of iron are present in human milk. A growing infant who is exclusively breast-fed will receive only about 0.26 mg of iron/day. Although the bioavailability of iron from human milk is high (about 50%) the amount of absorbed iron will only be about 0.13 mg/day, substantially below the estimated requirement of 0.55 mg/day. See Table 6.20.

In the later half of the first year, breast-fed infants not receiving supplemental iron are at sharply higher risk for iron deficiency and anemia, compared with infants fed iron-fortified formula (Pizarro et al. 1991). One study of low-income families in Chile found that by 9 months of age, over 27% of exclusively breast-fed infants had low hemoglobin values, compared with only 2–4% of formula-fed infants (Hertrampf et al. 1986).

Type of Feeding	Approximate mg Fe/liter	Absorption (%)	Iron Absorbed per Liter (mg)
Breast milk	0.5–1.0	49	0.25–0.5
Iron-fortified infant formula	12	4	0.5
Non-iron fortified infant formula	2.0	10	0.2

Adapted from P. R. Dallman, and A. Stekel, "Iron Deficiency in Infancy and Childhood," *Am J Clin Nutr* 33 (1980): 86.

TABLE 6.20

Iron content and percentage of absorption in infant formula and breast milk

	Breast-fed (%)	Fed Iron-Fortified Cow's Milk* (%)	Fed Soy-Based Formula** (%)
Anemia (Hemoglobin < 11.0 g/dl)	27.3	2.2	4.3
Transferrin saturation < 9%	37	17	22
Serum ferritin < 10 ng/ml	44	21	25

Adapted from E. Hertrampf, et al., "Bioavailability of Iron in Soy-based Formula and Its Effect on Iron Nutriture in Infancy," *Pediatrics* 78 (1986).

* Full fat powdered milk fortified with 15 mg of iron (from ferrous sulfate) and 100 mg of ascorbic acid per 100 g of powder.

**Isolated soy protein-based formula providing 2.5 mg of naturally present iron, 12 mg of added iron (from ferrous sulfate), and 54 mg of ascorbic acid per liter.

TABLE 6.21

Iron deficiency and anemia in 9-month-old infants in relation to type of feeding

Compared with formula-fed infants, twice as many breast-fed infants developed low serum ferritin values (a measure of iron stores) by nine months after birth. See Table 6.21.

Iron deficiency is very common worldwide, affecting an estimated 25% of all infants (WHO 1991). (See the discussion of iron deficiency in Chapter 7.) The incidence of iron-deficiency anemia among infants in the United States has fallen dramatically in the past 30 years. In the 1960s, iron-deficiency anemia was common among infants from lower socioeconomic groups (particularly among the urban poor). In the late 1960s, surveys of 18-month-old infants from urban areas in the U. S. found that 60–76% were anemic (Lanzkowsky 1970; Gutelius 1969). Among infants from higher socioeconomic groups, iron deficiency was less common but still affected about 6% of children (Yip 1989).

By 1985, prevalence of anemia among infants and young children had fallen to 3–9% nationwide (Filer 1995). Several factors contributed to this dramatic change: the feeding of cow's milk and evaporated milk to infants had decreased (feeding unmodified cow's milk contributes to iron deficiency during infancy); more women were breast-feeding, and iron was added to many infant formulas and cereals.

The Special Supplemental Food Program for Women, Infants, and Children (WIC) played an important role in reducing iron deficiency among the urban poor (Filer 1995). WIC programs distribute nutritious infant foods (including iron-fortified formula and cereals), encourage breast-feeding, and educate mothers about the importance of iron-rich foods during infancy. A detailed discussion of the aims and methods of the WIC program can be found in Chapter 2.

Iron deficiency can seriously harm a growing infant. Infants deficient in iron are more likely to suffer from infections, have poorer appetites, and grow more slowly than their healthy counterparts (Owen 1989). They are often irritable, inattentive, and withdrawn. Iron deficiency in childhood also impairs intellectual development (Sheard 1994). Iron-deficient anemic children do more poorly on tests of motor and mental development than their iron-sufficient counterparts (Idjradinata 1993). Children who have iron-deficiency anemia in infancy are at risk for long-lasting developmental impairment, even after the iron deficiency has been treated (Lozoff et al. 1991) (discussed in detail in Chapter 7).

Most expert groups, including the American Academy of Pediatrics (AAP) and the National Academy of Sciences (NAS), recommend iron supplementation for full-term, breast-fed infants beginning between 4 and 6 months of age. Rich sources of iron are iron-fortified infant formula or, for breast-feeding infants, two or more servings of iron-fortified infant cereal/day. Iron intake should be about 1 mg/kg body weight per day; the RDAs for iron during infancy are shown in Table 6.13. Preterm infants, who have smaller stores of iron and higher growth requirements compared with full-term infants, should be given supplemental iron no later than 2 months of age.

Besides iron-fortified infant cereals, nonfortified infant foods rich in iron include pureed green leafy vegetables and strained meats. Table 6.22 lists the iron content of selected infant foods. The AAP recommends against feeding unprocessed cow's milk to infants during the first year of infancy. Cow's milk contains little iron (0.15 mg of iron/100 kcal), and excessive ingestion can irritate the newborn digestive tract, causing trace amounts of bleeding that contributes to iron deficiency (AAP 1993).

Zinc The high rates of protein synthesis required for growth are dependent on a number of zinc-containing enzymes, and zinc deficiency during late infancy and childhood has been associated with impaired growth (Walravens et al. 1989), (Dirren et al. 1994) (discussed in Chapter 7).

Serum concentrations of zinc tend to be greater in breast-fed infants than in formula-fed infants, and zinc is more readily absorbed from

Foods	1 mg Iron
Meats and meat alternates	
Beans, dry (cooked)	3 tbsps (37.2 g)
Beef round steak	1.0 oz (28 g)
Egg yolk, medium	1¼ egg yolk (18.8 g)
Lentils, dry (cooked)	3.5 tbsps (47.6 g)
Liver	½ oz (11.6 g)
Peas, dry (cooked)	5 tbsps (58 g)
Breads and cereals	
Cream of wheat	½ cup
Enriched bread	1½ slices (40 g)
Infant cereal (dry)	1½ tsps
Iron-rich formula	⅓ cup
Oatmeal (cooked)	¾ cup
Ready-to-eat cereal iron-enriched (100% U.S.RDA)	2½ tsps
Whole wheat bread	1¼ slices (31.25 g)
Fruits and vegetables	
Oranges, small	2 (360 g) small
Raisins	3 tbsps (30 g)
Spinach (cooked)	¼ cup (45 g)

Data From C. F. Adams, "Nutritive Value of American Foods in Common Use," *Agriculture Handbook* no. 456, Agriculture Research Service (Washington, D.C.: U.S. Department of Agriculture, 1975).

TABLE 6.22
Iron content of selected foods for infants

human milk than from cow's milk-based formula (41% versus 31%, respectively) (Fomon 1993). The differences in absorption between human and cow's milk may result from different zinc binding—zinc in human milk is mainly bound to citrate, while most of the zinc in cow's milk is bound to casein (a less available form).

Information on zinc requirements during infancy are limited. Needs are provided by both zinc stores accumulated in utero and dietary sources. During early infancy, breast-fed infants consume about 2 mg of zinc/day, and healthy infants receiving only breast milk for the first four to six months of life show no signs of zinc depletion (NAS 1989).

Studies have shown that infants receiving formula with 0.58 mg zinc/100 ml grow better than infants consuming formula with only 0.18 mg/100 ml (Walravens and Hambidge 1976). The dietary requirements of formula-fed infants are significantly higher than the intakes of breast-fed infants because of lower zinc absorption from formula (Lönnerdal et al. 1984). The RDAs for zinc for the entire first year are set at 5 mg/day; for the second year, recommendations are for 10 mg/day (NAS 1989).

Infant formula sold in the United States must provide 0.5 mg of zinc/100 kcal, and most formulas provide about 0.75 mg/100 kcal (AAP

1993). Infant foods prepared from meats and legumes are good sources of zinc for older infants.

Selenium The RDA for selenium for infants has been extrapolated from adult requirements based on body weight. It is set at 5 µg/day and 10 µg/day for the first and second six months after birth, respectively (NAS 1989). In Finland and New Zealand, where selenium in the soil and water is low, breast-fed infants receive about 4–6 µg/day, and show no signs of deficiency. Intake of selenium by breast-fed infants in the U. S. (where selenium available from food sources is usually high) is about 13 µg/day from birth to six months (NAS 1989). Supplementary foods fed to infants contain significant selenium—the selenium content of selected infant foods in the U. S. is shown in Table 6.23.

Iodine The RDAs for iodine call for 40 µg/day for the first six months of infancy, 50 µg/day for the next six months, and 70 µg/day for the second year (NAS 1989). Because maternal iodine intakes in many industrialized countries are high, and maternal intake influences breast milk content, levels of iodine provided in breast milk in countries such as the United States and Canada are typically much greater than the RDA (Gushurst 1984). However, in many areas worldwide, iodine deficiency is common, and iodine deficiency in the neonatal period can have serious adverse effects on infant and child health (discussed in Chapter 7).

Manganese Human milk contains trace amounts of manganese, and the average intake of manganese by breast-feeding infants is only about 2 µg/day (Casey et al. 1985). Breast-feeding infants are often in negative balance—tissue manganese levels usually decrease during the first month after birth. However, there have been no reports of manganese deficiency during infancy (Lönnerdal et al. 1983). Stores accumulated during gestation may support manganese status during the newborn period before supplemental feedings are begun. As supplemental foods are introduced,

TABLE 6.23

Selenium content of selected commercially prepared infant foods

Food	Selenium Concentration (µg/kg)
Beef	40–116
Chicken	106–150
Egg	179–390
Cereal	
Oatmeal	30–109
Rice	21–39
Applesauce	1–2
Orange	1–15
Peach	1–3
Carrots	2–3
Green beans	2–5

Adapted from S. J. Fomon, *Nutrition of Normal Infants* (St. Louis: Mosby-Year Book, 1993).

manganese intake usually increases to 70–80 µg/kg body weight, or about 0.4 mg/day and 0.7 mg/day for typical infants in the first and second six months, respectively (Gibson and DeWolfe 1980). The estimated safe and adequate daily intake of manganese for infants is based on these levels of intake; they are shown in Table 6.13.

Copper The estimated safe and adequate daily intake of copper for infants is 75 µg/kg body weight, or 0.4–0.6 mg/day from 0–6 months of age and 0.6–0.7 mg/day from 6–12 months old (NAS 1989). The daily intake of copper by breast-feeding infants during the first four months after birth is below the recommended intake, averaging only 40 µg/kg (Butte et al. 1987). However, copper bioavailability from human milk is high, and newborns can draw on substantial hepatic stores of copper accumulated in utero.

Because copper levels in breast milk typically fall to low levels (20 µg/100 ml) by the sixth month (Shaw 1992), infants exclusively breast-fed in later infancy will not achieve the recommended intake for copper. Supplementary foods introduced at 4 to 6 months old increase copper intake and usually enable the infant to meet the recommended intake. The AAP recommends that infant formula provide 60 µg of copper per 100 kcal (AAP 1993). An infant consuming 700-750 ml of formula from birth to 6 months would thereby receive about 0.4 mg of copper/day.

Copper deficiency during infancy causes anemia, loss of pigment from the skin and hair, and psychomotor retardation. Preterm infants, infants who are malnourished or have chronic diarrhea, and newborns fed unmodified cow's milk are at higher risk of copper deficiency, compared with healthy term infants who are breast-fed (Shaw 1992).

Fluoride Fluoride is incorporated into bones and teeth as they mineralize during infant development. The recommended range of safe and adequate daily intake of fluoride is 0.1 to 0.5 mg during the first six months and 0.2–1.0 mg and 0.5–1.5 mg during the second six months and second year, respectively (NAS 1989). Fluoride intakes from all sources during infancy should not exceed 2.5 mg/day to avoid **mottling** of tooth enamel.

The unerupted primary teeth are mineralizing in early infancy, as shown in Table 6.24, and incorporation of fluoride into developing tooth enamel reduces the susceptibility to dental caries. Because only trace amounts of fluoride are found in breast milk, experts advocate supplementation for infants who are exclusively breast-feeding (AAP 1993). Fluoride supplementation during breast-feeding is controversial, however, because there is no firm evidence that supplementation during the first four to six months after birth reduces the level of dental caries in the permanent dentition.

Current recommendations from the AAP are to begin fluoride supplements in exclusively breast-fed infants and infants receiving ready-to-use

mottling: spotting and discoloration of tooth enamel

Primary Tooth	Hard Tissue Formation Begins (weeks in utero)	Enamel Completed Months after Birth	Eruption Age (mo)
Maxillary (upper jaws)			
Central incisor	14	1½	10
Lateral incisor	16	2½	11
Canine	17	9	19
First molar	15½	6	16
Second molar	19	11	29
Mandibular (lower jaw)			
Central incisor	14	2½	8
Lateral incisor	16	3	13
Canine	17	9	20
First molar	15½	5½	16
Second molar	18	10	27

Adapted from R. C. Lunt and D. B. Law, "A Review of the Chronology of Calcification of Deciduous Teeth," *J Am Dent Assn* 89 (1974):599.

TABLE 6.24
Calcification of the primary teeth

Food	Fluoride Concentration (µg/L)
Human milk	5–10
Cow's milk	30–60
Formula	
Concentrated liquid	
Milk-based	100–300
Isolated soy protein-based	100–400
Powdered	
Milk-based	400–1000
Isolated soy protein-based	1000–1600
Fruit juices	
Produced with nonfluoridated water	10–200
Produced with fluoridated water	100–1700
Dry cereals	
Produced with nonfluoridated water	90–200
Produced with fluoridated water	4000–6000

Concentration ranges have been rounded off; most reported values fall within the ranges listed in the table.

Adapted from Foman (1993)

TABLE 6.25
Fluoride concentration in infant foods

formula without fluoride soon after birth. A daily supplement of 0.25 mg of fluoride should be provided until the infant begins to consume fluoridated water (AAP 1993). The fluoride concentrations of human milk and other infant foods are shown in Table 6.25.

Molybdenum and Chromium Very little is known about the specific requirements for these trace elements during the newborn period. The estimated safe and adequate dietary intakes for infancy are extrapolated from adult requirements based on body weight (for molybdenum) and food intake (for chromium) (NAS 1989). They are shown in Table 6.13.

INFANT FORMULAS

DEVELOPMENT OF INFANT FORMULAS

Breast milk substitutes have undergone a remarkable evolution over the past 50 years. Until the 1950s, about half of the infants in the United States were fed formula prepared at home by mixing evaporated milk or fresh cow's milk with water and carbohydrate (corn syrup or sucrose). However, by 1970, commercially prepared infant formula was given to over three quarters of infants—and very few infants were receiving evaporated milk or fresh cow's milk formulas (Fomon 1993).

This changeover was spurred by the introduction in the early 1950s of convenient, concentrated infant formulas that required only added water before feeding. Many pediatricians and child-care workers accepted and endorsed these new products and they were widely promoted by the formula industry. Formulas based on evaporated milk were increasingly recognized to contribute to iron and vitamin C deficiencies and water imbalance (due to their high potential renal solute load).

Despite widespread recommendations for breast-feeding, in the 1990s formula is the sole source of nourishment for many infants in the United States. About half of new mothers choose to feed their infants formula at birth, and by age 6 months over 85% of infants in the United States are receiving infant formula—either as a supplement to breast-feeding, or as their major nutritional source (Fomon 1993).

Although no formula can completely duplicate the immunological benefits or the unique digestibility and nutritional bioavailability of human milk, properly prepared commercial formulas can meet all the energy and nutrient needs of healthy, growing term infants for the first six months after birth (AAP 1993).

USES OF INFANT FORMULAS

Many women, because of lifestyle, employment, or personal preference, choose to feed formula to their infants. Certain maternal or infant conditions require substitution of formula for breast milk. Maternal contra-indications to breast-feeding include certain infections (such as hepatitis, tuberculosis, and HIV) that can be transmitted to the infant during breast-feeding (Goldfarb 1993). Also, several maternal medical problems are treated with medications that pass into breast milk and can harm the infant. Certain rare inherited metabolic disorders can cause newborn intolerance to components of breast milk. These topics are discussed in the Chapter 5.

COMPOSITION OF INFANT FORMULAS

Human milk is the model for infant formulas. Most formulas for healthy term infants attempt to simulate the nutritional composition of breast milk. For example, the energy content of most formulas is similar to human milk—about 66–67 kcal/100 ml, with about 50% of calories from fat. Infant formulas also supply high-quality protein, carbohydrate, fat (including the essential fatty acids), vitamins and minerals, taurine, inositol, choline and stabilizers or emulsifiers (such as mono- and diglycerides, soy lecithin, or carageenan).

In general, formulas contain higher concentrations of many nutrients than human milk, to compensate for the lower bioavailability of nutrients from formula. The AAP (AAP 1993) and WHO (Joint FAO/WHO 1984) recommend a range of nutrient composition for infant formula (see Table 6.26 for the AAP's recommendations). In the United States, regulations cover many aspects of the production and composition of infant formulas. The Infant Formula Act of 1980 (revised in 1986) established the minimum levels of 29 nutrients in formula (and the maximum levels for 9 nutrients) (U.S. Food and Drug Administration 1985). It also requires manufacturers to assure by analysis the level of each nutrient in each batch of formula.

FORMS OF INFANT FORMULA

Infant formula comes in three general forms: modified cow's milk-based formulas, soy protein-based formulas, and protein hydrolysate formulas. Table 6.27 gives the nutrient composition of each form.

Nutrient	Range	
	Lowest Adequate	**Not to Exceed**
Protein (g)	1.8	4.5
Fat (g)	3.3 (30% of calories)	6 (54% of calories)
Including essential		
fatty acid (linoleate) (mg)	300 (2.7% of calories)	
Vitamins		
A (IU)	250	750
D (IU)	40	100
K (μg)	4	
E (IU)	0.7 (0.5 mg) at least	
	0.71U (0.5 mg)/g	
	linoleic acid	
C (mg)	8	
B$_1$ (μg)	40	
B$_2$ (μg)	60	
B$_6$ (μg)	35 (15 μg/g of protein)	
B$_{12}$ (μg)	0.15	
Niacin (μg)	250 (or 0.8 mg	
	niacin equivalents)	
Folic acid (μg)	4	
Pantothenic acid (μg)	300	
Biotin (μg)	1.4	
Choline (mg)	7	
Inositol (mg)	4	
Minerals		
Calcium (mg)	60	
Phosphorus (mg)	30	
Magnesium (mg)	6	
Iron (mg)	0.15	2.5
Iodine (μg)	5	25
Zinc (mg)	0.5	
Copper (μg)	60	
Manganese (μg)	5	
Sodium (mg)	20	60
Potassium (mg)	80	200
Chloride (mg)	55	150
Selenium (μg)	3	

TABLE 6.26

1992 AAP recommended levels for nutrients in infant formula (per 100 kcal)

*Where no upper limit is given, toxicity is not well defined; massive excesses may have adverse consequences.

Adapted from American Academy of Pediatrics, Committee on Nutrition, *Pediatric Nutrition Handbook* (Elk Grove, IL: AAP, 1993).

	Milk-based			Protein Hydrolysate-based	
	Casein Predominant	Whey Predominant	Isolated Soy Protein-based	Casein	Whey
Protein or protein equivalent (g)	14.5	15–15.2	16.5–21	18.6–19	16
Fat (g)	36	36–38	36–36.9	27–37.5	34
Fatty acids (g)					
Polyunsaturated	13	4.9–11	4.7–14	12.8–15.8	4.4
Saturated	16	15–19.1	14.9–18.1	3.5–18.2	14.5
Monounsaturated	6	5.4–14	5.1–14.2	2.6–7	15.1
Linoleic	8.8	3.3–8.8	3.3–8.8	10.8–13.6	4.3
Carbohydrate (g)	72	69–72	68–69	68.9–91	74
Minerals					
Calcium (mg)	492	420–470	600–710	640–710	430
Phosphorus (mg)	380	280–320	420–510	430–510	240
Magnesium (mg)	41	45–53	51–74	51–74	45
Iron (mg)	12#	12–12.8	11.5–12.8	12–12.8	10
Zinc (mg)	5.1	5–5.3	5–5.3	5.1–5.3	—
Manganese (µg)	34	100–106	170–200	200–210	—
Copper (µg)	610	470–640	470–640	510–640	—
Iodine (µg)	94.6	60–69	60–100	48–100	—
Selenium (µg)	15	12	7–15.6	15.6–19	—
Sodium (mg)	183	150–184	200–300	300–320	160
Potassium (mg)	708	560–730	700–830	730–740	660
Chloride (mg)	433	375–430	375–560	540–580	390
Vitamins					
Vitamin A (IU)	2030	2000–2100	2000–2100	2030–2100	3000
Vitamin D (IU)	410	400–430	400–430	305–430	600
Vitamin E (IU)	20	9.5–21	9.5–21	20–21	12
Vitamin K (µg)	54	55–58	100–106	100–106	82
Thiamin (µg)	680	530–670	410–670	410–530	600
Riboflavin (µg)	1010	100–1060	610–1000	610–640	1350
Niacin (µg)	7100	5000–8500	5000–9130	8500–9130	7500
Vitamin B_6 (µg)	410	420–430	410–430	410–530	750
Folate (µg)	100	50–106	100–106	100–106	90
Vitamin B_{12} (µg)	1.7	1.3–1.6	2–2.1	2.1–3	2.2
Pantothenic acid (µg)	3040	2100–3200	3000–3170	3200–5070	4500
Biotin (µg)	30	15–15.6	35–64	30–53	22
Vitamin C (mg)	60	55	55–81	55–60	80

Adapted from Foman (1993)

TABLE 6.27

Composition of typical milk-based, soy-based, and protein hydrolysate formulas in the United States

TABLE 6.27 *Continued*

	Milk-based			Protein Hydrolysate-based	
	Casein Predominant	Whey Predominant	Isolated Soy Protein-based	Casein	Whey
Other nutrients					
Taurine (mg)	45	40	40–45	40–45	—
Choline (mg)	108	100–106	81–85	54–90	120
Inositol (mg)	32	32	27–68	32–34	61
Potential renal solute load (mosm)	133	127–136	163–181	171–172	134

Data apply to formulas marketed in 1992. Values are units per liter at standard dilution (661 kcal).

Modified Cow's Milk-based Formulas

Milk-based formula is recommended as the formula of choice for healthy infants (AAP 1993). The protein source is fat-free cow's milk, to which additional lactose is added. The butterfat is removed from the milk and replaced by vegetable oils. This is done because the PUFAs from vegetable sources are better absorbed by infants than the saturated fatty acids in butterfat, and PUFAs provide essential fatty acids.

Some manufacturers use unaltered cow's milk protein (18% whey and 82% casein), while others modify the protein to make it resemble human milk more closely (60% whey and 40% casein). Infants grow equally well on both protein compositions. Some of the major minerals are derived from cow's milk, and many of the trace minerals and vitamins are added to bring concentrations to standard levels.

Although both low-iron (0.15–0.2 mg/100 ml) and iron-fortified (1.2 mg/100ml) formulas are available, the AAP strongly recommends iron-fortified formula as the preferred choice for infants. The low-iron formulas exist because, traditionally, the iron in the iron-fortified formulas was thought to cause gastrointestinal problems—such as constipation, colic, and spitting up—in some infants. However, several studies have shown no differences in gastrointestinal problems when infants on the low-iron and iron-fortified formulas were compared (AAP 1993).

Soy-based Formulas

Soy milks account for about 20% of all formula sold in the United States (AAP 1993). The composition of typical soy-based formula is shown in Table 6.27. The protein is derived from soybean flakes, and methionine is added to compensate for the low levels found in soy protein. The fats are similar to those added to milk-based formula, but the carbohydrate content is very different—soy milks contain cornstarch and sucrose in place of lactose. Because the phytate content of soy can interfere with mineral absorption, levels of minerals in soy formulas are often higher than those in milk-based formulas.

Although milk-based formulas are satisfactory for most infants, some develop intolerance or allergy to the proteins in cow's milk, and must be switched to soy milk. Soy milks are sometimes used for infants with gastrointestinal infections and diarrhea. The activity of lactase in the intestine may be reduced during and after diarrheal illnesses, and during these periods the carbohydrates in soy formulas may be better tolerated than lactose (AAP 1993). Infants with galactosemia should be fed soy-based formula (AAP 1993). Many mothers who are complete vegetarians and who are not breast-feeding choose soy formulas.

Growth of infants fed soy-based formulas is equivalent to infants who are breast-fed or fed with cow's milk-based formula (Fomon and Ziegler 1979). However, some experts think bone mineralization is suboptimal in infants fed soy-based formulas (Steichen and Tsang 1987), particularly preterm infants.

Protein Hydrolysate Formulas

Protein hydrolysate formulas have been developed for infants who cannot readily digest intact proteins or who are allergic to intact proteins. To prepare these formulas, milk proteins are extensively broken down by heat and enzyme treatment, and the resulting hydrolysate is a mixture of short peptides and amino acids. The allergenicity of the formula is reduced because intact proteins are no longer present to produce an immunological response (AAP 1989).

Manufacturers then add mixtures of sucrose, corn syrup, and starches (these formulas are usually lactose free), and mixtures of fats that often contain medium chain triglycerides (MCTs). MCTs are better absorbed by infants who have problems with fat malabsorption, such as preterm infants and those with cystic fibrosis (AAP 1993). Use of protein hydrolysate formulas is limited by their high cost and unappealing taste.

UNMODIFIED COW'S MILK IN INFANCY

The AAP recommends against feeding cow's milk (whether whole, low-fat, skim, or evaporated) to infants during the first twelve months after birth. Table 6.28 shows how specific nutrients in cow's milk compare with human milk. Only when at least two thirds of an infant's energy needs are met by supplemental foods should cow's milk be introduced. The reasons for this recommendation are the following:

	Human Milk	Cow's Milk
Energy (kcal/100 mL)	68	70
Protein (g/100 mL)	1.1	3.5
Proteins (% of total protein)		
Casein	40	82
Whey proteins	60	18
Fat (g/100 mL)	3.8	3.7
Minerals per liter		
Calcium (mg)	340	1170
Phosphorus (mg)	140	960
Sodium (meq)	7	22
Potassium (meq)	13	35
Chloride (meq)	11	29
Magnesium (mg)	40	120
Sulfur (mg)	140	300
Iron (mg)	0.5	0.5
Vitamins per liter		
Vitamin A (IU)	1898	1025
Thiamin (µg)	160	440
Riboflavin (µg)	360	1750

Adapted from AAP (1993)

TABLE 6.28

Selected nutrients in human vs. cow's milk

- The potential renal solute load of cow's milk is very high, placing infants at risk of dehydration.

- Cow's milk has little bioavailable iron and can cause trace amounts of bleeding in young infants because it irritates the newborn intestinal tract. Studies in infants who were iron deficient compared **occult bleeding** after ingestion of cow's milk and infant formula. Half of the infants fed cow's milk had significant losses of blood (equivalent to 0.5–0.6 mg iron/day) (Wilson et al. 1974). Because of these factors, many infants fed cow's milk for long periods become iron deficient and anemic.

- Incidence of milk allergy is higher with unmodified cow's milk than with commercial formulas (AAP 1993).

- Cow's milk contains little vitamin C, bioavailable zinc, and essential fatty acids, and it may produce deficiencies of these nutrients in infants fed mostly cow's milk.

occult bleeding: small, concealed amounts of bleeding; for example, bleeding into the stool that is not easily visible

FORMULA FOR OLDER INFANTS

Several manufacturers market special formulas for older infants that are higher in protein and lower in fat than standard formulas. They are meant to accompany the feeding of supplementary solid foods in the later half of the first year. However, older infants who are fed standard formula designed for younger infants while supplementary foods are introduced have satisfactory intakes of protein and energy. Most experts think use of these "older infant" formulas is unjustified—standard formulas are adequate for the entire first year after birth (AAP 1993).

PREPARATION OF FORMULA

Infant formulas come in several forms. Both powdered formulas and liquid concentrates are prepared by mixing a prescribed amount of powder or liquid in a specified volume of water. Ready-to-feed formulas are premixed and require no preparation at home.

Terminal Sterilization

Commercial infant formulas are sterilized before packaging. However, if not properly handled during preparation, they can become contaminated. In the past, groups of experts recommended *terminal sterilization*—boiling the prepared formula at home before feeding (Committee on Fetus and Newborn 1961). Newer recommendations state that single bottles of formula, if safely prepared with clean technique, can be fed without terminal sterilization.

Several studies have found no differences in infant illness or problems from contaminated formula when terminal sterilization and clean technique have been compared (Fomon 1993). Although often less convenient, preparing bottles of formula one at a time, just before a feeding, is preferable. If bottles are prepared as a batch—for example, all bottles for a 24-hour period—they should be heat treated before being refrigerated.

Clean Technique

During the preparation of formula, the bottle, nipple, measuring spoon, and top of the container of formula should be scrupulously cleaned with soap and water and allowed to dry. Also, the preparer's hands and the preparation area must all be thoroughly clean. If the local community water supply is not monitored, water used to dilute formula should be boiled. Prepared formula should be fed promptly or refrigerated, and liquid concentrate that is opened should be kept in a refrigerator (if refrigeration is not available, powdered formula is preferable).

Overdiluted and Underdiluted Formula

While preparing formula, careful attention to measurements is important—underdilution or overdilution of formula can have serious adverse effects on the infant. In impoverished areas of the developing world and in low-income families in the industrialized countries, many infants are fed overdiluted formula in an effort to make formula last longer (McJunkin et al. 1987), (Jelliffe and Jelliffe 1978). Feeding of overdiluted formula provides insufficient energy and essential nutrients and can result in undernutrition and growth failure.

Less commonly, feeding of underdiluted formula (because of preparation error or a conscious effort by a family to provide more dense and nutritious formula) can seriously harm an infant. When concentrated formula is fed, the potential solute load from the large amount of protein and electrolytes predisposes the infant to dehydration and metabolic acidosis, particularly during fever, illness, or high environmental temperatures. Clearly, families who choose to feed formula need to be carefully advised about proper preparation methods and the risks of under- or overdilution.

INFANT FEEDING BEHAVIORS AND PATTERNS

Over the first two years, an infant's eating pattern changes dramatically. Starting with an inborn set of instinctual reflexes to suckle and swallow, and a nutritious yet monotonous diet of breast milk or formula, the infant steadily progresses toward adult-type eating habits and food preferences. Infant feeding patterns evolve through three overlapping stages:

- The *nursing period*, during which breast milk or infant formula provides complete nutrition for the infant

- A *transitional period*, during which specially prepared semisolid foods are introduced and consumption of human milk or formula continues

- A *modified adult period*, during which most or all of the infant's food comes from adult-type table foods

There is no rigid timetable of infant feeding patterns. Different infants, depending on how quickly they develop and mature, move through these stages at different ages. See Table 6.29.

THE NURSING PERIOD

Development

For most infants, the nursing period lasts four to six months after birth. Born with the instinctive reflexes, the newborn's oral structure is ideal for

Age	Oral and Neuromuscular Development	Feeding Behavior
Birth	Rooting reflex	Stimulation to oral area causes lips, tongue, and head to turn toward stimulus
	Suckle-swallow	A stimulus introduced into the mouth elicits vigorous sucking followed by a swallow if liquid is present
	Bite reflex	Stimulation to gums elicits a rhythmical bite and release pattern
	Gag reflex	Stimulation to the posterior tongue or soft palate causes constriction of the posterior oral musculature to bring stimulating substance forward
1–10 weeks		Recognizes feeding position and begins sucking when placed in position
		Swallowing pattern changes and tongue ceases to come forward. There is a more definite jaw movement, a sign of beginning readiness for solid food
3–6 months	Beginning coordination between eyes and	Explores world with eyes, fingers, hands, and body movements mouth
	Brings hand to mouth Reaches for objects but overshoots	Finger sucking
	Voluntary grasp	Palmar grasp
6–8 months	Sucking pattern	A true sucking pattern with less jaw movement
	Lateral jaw movements	Enables early chewing. Can approximate lips to cup and cup-drinking begins. However, liquid dribbles from corners because tongue is projected before swallowing
	Sits erect with support Eyes and hands working together	
	Reaches for and grasps object on sight	Reaches for food and feeding utensils
	Sits alone without support	Freer to reach and grasp. Grasp has increasing finger-thumb opposition
9–12 months		Holds onto bottle and finger feeds
12 months		Uses cup independently with two hands. Beginning of independence with spoon. Chews well (but not yet mature chew pattern)

Adapted from A. Gesell and F. Ilg, *Feeding Behavior of Infants* (Philadelphia: J. B. Lippincott, 1937).

TABLE 6.29
Development of feeding skills during the first year

suckling: an arched palate, large tongue, slightly averted lips, and prominent fat pads in the cheeks for support. The strong extrusion reflex is present at birth, as solid foods or small objects placed in the mouth will elicit a rhythmic motion of the tongue and lips that pushes the material out of the mouth (Hervada and Newman 1992). This is a protective reflex that, in conjunction with a strong gag reflex, protects the newborn from choking, and usually precludes successful early introduction of solid foods. These reflexes typically disappear at about 3 to 4 months of age.

During the newborn period, digestive functions are maturing—gastrointestinal enzyme systems are developing, and levels of pancreatic amylase, pepsin, and gastric acid are low (Lebenthal 1983). The mucosal barrier of the digestive tract is developing and, if solid foods are introduced during the nursing period, absorbed intact proteins may provoke an allergic response. The infant's ability to support and steady his head is poor for the first several months after birth; when placed upright the infant's head will flex forward, making it difficult or impossible to swallow semisolid or solid food.

Feeding Patterns

Early introduction of supplemental foods should generally be avoided during the nursing period. Human milk or an appropriate formula should be the optimal and sole nutrient source (AAP 1993). The newborn infant typically feeds six to eight times a day, and the intake of formula or breast milk should be determined by infant demand. Although intake will vary from feeding to feeding, families should be reassured that nearly all infants satisfy energy requirements for growth and will not overeat. See Table 6.30.

The usual intake of formula or human milk is about 150–200 ml/kg body weight/day—providing 100–135 kcal/kg/day (a 5 kg infant would consume about 26–32 oz of formula/day), and weight gain should be 20–30 g/day (AAP 1993). For breast-feeding infants, supplemental bottle-feeding should be avoided during the first two weeks. This allows lactation to become firmly established and does not confuse the infant (the rubber nipple on a bottle requires a different tongue and jaw motion).

TABLE 6.30

Average number and volume of feedings during early infancy

Age	Approximate Feed Volume (single feed) (ml)	Number of Feedings per Day
1–2 weeks	50–70	7–8
2–6 weeks	75–110	6–7
3 months	170–220	5
6 months	220–240	4

Adapted from L. S. Taitz and B. L. Wardley, *Handbook of Child Nutrition* (Oxford: Oxford University Press, 1989).

A small amount of vomiting and spitting up are common for the first few months. For most formula-fed infants, colic, loose stools, spitting up, and other minor problems are unrelated to formula type. However, some problems may be due to cow's milk intolerance, and after talking to the infant's dietitian or pediatrician, a change to soy-based formula may be beneficial (AAP 1993). Families should avoid giving supplemental water during early infancy; after two to three months, water can be offered one to two times a day, particularly during hot weather.

Stooling Patterns

The breast-feeding infant's stools are a mustard yellow liquid with a yeasty odor; in contrast, stools from formula-fed infants are darker and more formed. While the breast-fed infant stools frequently (often with every nursing), the formula-fed infant usually stools about twice a day (Weaver et al. 1988).

THE TRANSITIONAL PERIOD

Development

During the transitional period (from 4 to 6 months to about 10 months old), the infant's neuromotor development is rapid. From 4 to 6 months, the typical infant develops a firm palmar grasp and learns to put hands to mouth to bite and suck objects. Chewing begins and the lips can be applied more accurately to the rim of a cup. The infant is able to indicate a desire for food, by opening his mouth and leaning toward the food, and show satiety or disinterest by pulling away. At 7 to 9 months, the infant begins to sit upright, hold a bottle alone, eat well from a spoon, and use a cup. Rotary chewing and well-coordinated swallowing of semisolid and solid foods are being established. See Table 6.29.

As supplementary foods are introduced, the oral structures begin to mature. The orofacial muscles enlarge and grow stronger, allowing the infant to suck and chew foods with good coordination for longer periods. The extrusion reflex is lost, the gag reflex is less strong, and the primary incisors erupt at 8 to 10 months. The tongue is more mobile, propelling food into the pharynx during swallowing, and can be protruded to receive foods or a spoon. The ability to digest and absorb fat, carbohydrate, and protein continues to mature, and the kidney can readily concentrate urine and excrete solutes without excessive water loss (Fomon, "Water and Renal Solute Load").

Introducing Solid Foods

The Committee on Nutrition of the American Academy of Pediatrics recommends supplementary foods be started in healthy infants between

4 and 6 months of age (AAP 1993). There is no nutritional advantage to introducing supplemental foods earlier than 4 months, and the potential risks of food sensitivity, allergy, and dehydration are greater if solid foods are begun early. However, many families around the world begin supplemental foods in the first few months (Raphael 1982), some believing it is a sign of maturity or precociousness, others because of cultural influences or traditions. Table 6.31 shows the age of introduction and types of supplementary foods in different regions of the world. Over two thirds of infants in the United States receive supplemental foods regularly by the third month, and many are given supplemental foods in the first month postpartum (Fomon 1993).

Supplemental feeding should not be delayed beyond six months in healthy infants who are developmentally ready. For infants who are exclusively breast-fed, intake of breast milk alone may not be sufficient to provide the total nutrient needs of the baby in the second half of the first year. Also, delaying supplemental foods much past 6 months may make their acceptance by the infant more difficult and can interfere with the normal pattern of development of feeding behaviors in later infancy (AAP 1993).

Choosing Supplementary Foods

Supplementary foods should be introduced one at a time at weekly intervals. Spacing new foods allows for the identification of food intolerance or allergy. Vitamin and mineral-enriched infant cereals are good first

Chinese	Supplementary foods are generally introduced between 3 and 8 months. Rice and rice products (soups, pastes, gruel), minced fish, pork, and eggs are given; fruit and vegetable purees generally are not given.
Egypt (village)	Between 1 week and 3 months, water with sugar, herbal tea is introduced. At 3 months, biscuits, breads, and egg yolks are given.
Jamaica (urban poor)	At about 2 weeks, milk, tea, water with sugar, and juice are introduced. From 2 weeks to 2 months, cornmeal with condensed milk, banana porridge, and strained oats are given.
Nigeria	From 1 to 3 months, water with sugar and milk are given. After 3 months, maize porridge, ground nuts, eggs, rice, beans, fruit, and yams are introduced.
Sardinia (village)	At about 6 weeks, infants are given sugar water and meat and vegetable broths. At about 3 months, semolina cereal in broth, and mashed spaghetti or potatoes are introduced.

Adapted from D. Raphael, *Lactation Rev* 6 (1982):1

TABLE 6.31

Ages of introduction and types of supplementary weaning foods in various regions of the world

choices, providing iron and energy. Three tablespoons of dry iron-fortified cereal provide about 5 mg of iron, which is half of the daily iron
requirement during months 6 through 12.

Rice is a particularly good first choice. Rice cereals are precooked and
partially hydrolyzed, and infants accept them and can digest them easily.
The AAP recommends against adding cereal routinely to bottles. Infants
should be fed the cereal by spoon so they can observe and begin to learn
to feed themselves. Some families, in an effort to increase satiety so the
baby will sleep through the night, give formula with added cereal in a
bedtime bottle. However, studies have found the addition of cereal does
not affect the need for nighttime feedings (Macknin et al. 1989).

Although the order of introduction is not critical, often vegetables and
fruits will be introduced after cereal, followed by meats, poultry, and fish
(see Table 6.32). Introduction of foods commonly implicated in infant
food allergies—such as eggs and tomatoes—is usually delayed until late
in the first year. Once infant tolerance for individual foods has been
established, mixtures of fruit and cereal, pureed mixed vegetables (good
sources of vitamins A and C), and strained meats (rich in highly available
heme iron and protein) are healthy choices.

Introducing infants to a wide variety of foods is the best way to ensure
a nutritionally balanced diet. A mixture of proteins (particularly when
cereals and legumes are the main protein sources) is important to provide all of the essential amino acids. Most infants continue to consume
generous amounts of breast milk or formula during the transitional
period. The high fat content of milk and formula is balanced by the
higher amounts of carbohydrate in cereals, vegetables and fruits. Most
infant diets will provide about 40% of energy as fat, and 45–50% as carbohydrate (Fomon 1993). Table 6.33 gives the nutrient content of many
supplemental foods, and Table 6.34 shows the popularity of different categories of foods.

Food texture is important during the transition period. Neuromuscular development allows most infants to swallow pureed and strained
semisolid food by 4 to 6 months of age, and by 8 to 10 months, many
infants can consume finely chopped solid foods without choking. However, infants should always be carefully watched when eating supplementary foods. Risk of choking and aspiration are highest in late infancy and
early childhood, and unless prepared properly, foods such as hot dogs,
nuts, carrots, and grapes (any food with firm chunks) should be avoided
during this period. See Table 6.35 for additional guidelines on prevention
of choking.

Fruits and Fruit Juices Certain fruits and fruit juices, particularly apple
and pear juice, should be fed to infants in moderation (Smith et al.
1995). These fruits can cause loose stools, bloating, and diarrhea in many
infants. One study compared the absorption of various fruit juices dur-

| Food | Age(mo) | | | | | |
	0–2	2–4	4–6	6–8	8–10	10–12
Breast milk or iron-fortified formula	5–9 feedings (16–32 oz)	4–7 feedings (20–36 oz)	4–6 feedings (24–40 oz)	3–4 feedings (24–32 oz)	3–4 feedings (16–32 oz)	3–4 feedings (16–32 oz)
Cereals			Infant cereal, single grain (rice, barley oatmeal)	Infant cereal, all varieties	Infant cereals, plain hot cereal (adult type), plain crackers, toast	Infant cereal, cooked cereal, bread products, rice, pasta
Fruit			Fruit juice (from cup)	Strained fruit, applesauce mashed bananas, fruit juice (from cup)	Soft peeled fruit (banana, pear, apple, peach), fruit juice (from cup)	Fresh fruits (peeled), canned fruits (packed in own juice), fruit juice
Vegetables				Strained or mashed vegetable (plain)	Cooked, mashed table vegetables	Cooked vegetable pieces, raw vegetables (finger foods)
Protein foods				Plain yogurt, strained meats, egg yolk, tofu, cottage cheese	Strained meats, egg yolk, tofu, cottage cheese, legumes (sieved)	Small pieces of chicken or meat (tender), whole egg, cooked, dry beans (mashed), peanut butter

TABLE 6.32
Recommended supplementary food introductions during the first year

ing infancy and reported the gastrointestinal symptoms produced by juice consumption. Apple or pear juice was poorly absorbed, compared with grape juice, and caused loose stools and other gastrointestinal problems in about 40% of the infants in the study (Hyams et al. 1988).

The high content of fructose and sorbitol (a sugar alcohol) was thought to be the reason apple and pear juice caused these symptoms. Sorbitol absorption is limited in many people, and the capacity to absorb fructose is low unless it is consumed with glucose (as sucrose—a disaccharide of glucose and fructose) (Fomon 1993). Both apple and pear juice have a high proportion of fructose to glucose, as well as large amounts of sorbitol. Table 6.36 shows the breakdown of fructose, glucose, sucrose, and sorbitol in common fruits.

Once the teeth have erupted in the later half of the first year, feeding of fruit juices or other sweetened liquids for prolonged periods—such as putting the infant to bed with a bottle—can increase the risk of dental caries.

Food	Portion	Energy (kcal)	Protein (g)	Iron (mg)	Vitamin A (IU)	Vitamin C (mg)
Cooked cereal (farina)	¼ cup	26	0.8	Dependent on level of fortification		
Mashed potato	¼ cup	34	1.1	0.2	10	5
Cheese	¼ oz	28	1.78		92	
Spaghetti	2 tbsp	19	0.6	0.2		
Macaroni and cheese	2 tbsp	54	2.1	0.2	107	
Hamburger	½ oz	41	3.4	0.5	5	
Eggs	1 medium	72	5.7	1.0	520	
Cottage cheese	1 tbsp	5	1.9		23.7	
Green beans	1 tbsp	3	0.15	0.2	43	1
Cooked carrots	1 tbsp	2.81			952	
Banana	½ small	40	0.5	0.35	90	5
Pudding	¼ cup	70	2.2	Trace	102	

From C. F. Adams "Nutritive Value in American Foods in Common Units," *Agriculture Handbook* no 456 (Washington D.D.: U.S. Department of Agriculture, 1975).

TABLE 6.33

Nutrient content of selected foods fed to infants

Preparing Foods at Home

Some mothers choose to feed their infants commercially prepared foods, while others prefer to prepare foods at home—and many families do both. For mothers who prepare most foods at home, experts recommend including some commercially prepared cereals for their iron content (AAP 1993). Healthy infant foods can be prepared from high-quality fresh, frozen, or canned fruits, vegetables, or meats. Many canned foods contain large amounts of added salt and sugar and are less desirable sources for infant foods.

The preparation area, food, and utensils should all be thoroughly cleaned when preparing infant foods at home. To preserve nutrients, foods should be cooked until tender in a small amount of water, or in a microwave. After cooking and pureeing or straining, foods should be

TABLE 6.34

Popularity of different commercially prepared infant food items in the United States

Item	% of Sales
Wet-pack cereals, juices, fruits	43
Meats, high-meat dinners, soups	30
Vegetables	13
Dry cereals	7
Desserts	7

Data from 1992 Supermarket Business.

The following steps should be taken to prevent choking episodes:

- Be sure that infants are attended to at all times while eating.

- Offer foods of a texture and size appropriate to the developmental level of the infant.

- Don't allow the mouth to be overstuffed.

- Encourage adequate chewing before food is swallowed.

- Be alert when infants are eating because they may not be able to make sounds when the airway is blocked.

TABLE 6.35
Prevention of choking during infancy

	Fructose	Glucose	Sucrose	Sorbitol
Apple	6.0	2.4	2.5	0.5
Pear	6.6	1.7	3.7	2.2
Plum	2.0	3.4	3.4	1.4
Sweet cherry	7.1	7.8	0.2	1.4
Peach	1.1	1.0	6.0	0.9
Grapes	6.5	6.7	0.6	—

Data from R. E. Wrolstad and R. S. Shallenberg, "Free Sugars and Sobitol in Fruits—A Compilation from the Literature," *J Assoc Off Anal Chem* 64 (1981):91–103.

TABLE 6.36
Concentrations of fructose, glucose, sucrose, and sorbitol in selected fruits (in g/100g of food)

divided into individual servings and refrigerated or frozen (an ice-cube tray is a convenient way to divide and store portions). Families should be careful to prevent spoilage of both home-prepared and commercial infant foods.

Certain foods should not be included in home preparations for infants. Honey should not be fed to infants during the first year, because it may contain heat-resistant spores of *Clostridium botulinum,* the bacteria that causes **botulism** (AAP 1993). Also, home-prepared carrots, spinach, beets, and collard and turnip greens should not be fed in early infancy (before 4 to 5 months of age). These vegetables may contain large amounts of nitrites that can impair oxygen binding by hemoglobin and cause **methemoglobinemia** (Fomon 1993).

THE MODIFIED ADULT PERIOD

Development

By the onset of the modified adult period (about age 10 months for most infants), the infant's digestive and absorptive capabilities have matured to adult levels, and flavor discrimination and food preferences

botulism: a paralytic disease caused by a toxin produced by *Clostridia* bacteria

methemoglobinemia: a disorder caused by abnormal oxidation of hemoglobin to methemoglobin (the ferric form); methemoglobin is unable to carry oxygen

are becoming established. The infant is moving toward self-feeding: by 10 to 12 months the infant has developed a refined pincer grasp and can finger-feed small cut-up foods and crackers. A spoonful of food can be held and licked by the infant, and by 11 to 12 months, the infant can usually drink from a cup. By the beginning of the second year, between six and eight primary teeth are present, well-defined chewing is possible, and the tongue and lips work in a coordinated way to accept and manipulate food.

Meeting Nutritional Needs

During weaning in the modified adult period, human milk or formula contribute a decreasing proportion of nutrients, as solid foods become more important to the infant. At age 9 months, supplemental foods usually contribute only about a third of caloric intake, whereas by the beginning of the second year, they provide over half of the energy in the infant's diet and up to two thirds of the infant's intake of iron and protein (Bronner and Page 1992). Although growth rate is much slower than in early infancy, energy needs continue to be high as the infant's activity level increases (the average infant needs over 1000 kcal/day by the tenth month) (NAS 1989). See Figure 6.11.

The water needs of early infancy are met by the water in formula or breast milk. But as the intake of solid foods increases, the renal solute load increases, particularly from high-protein foods such as strained meat and eggs. Additional water should be provided along •with formula or breast milk (AAP 1993). This will allow older infants to meet their water needs, particularly during hot weather or illness, without obligatory intake of calories.

Development of Food Preferences and Eating Habits

During the later half of the first year, infants begin to appreciate the variation in color and flavor of foods. Although infants eagerly try many new foods, most have an innate preference for sweet-tasting foods and an innate aversion to bitter taste (Beauchamp et al. 1986). Food preferences develop gradually during infancy and early childhood and are influenced by the types of food offered and the observations by the infant of the attitudes and reactions of parents and older siblings to foods.

A wide variety of foods should be introduced in a positive, nonjudgmental environment. How the infant is fed is as important as what is fed—meals should be pleasant and stress free, and the infant should decide when he or she has had enough to eat (Satter 1990). Recommendations for successfully feeding older infants are shown in Table 6.37.

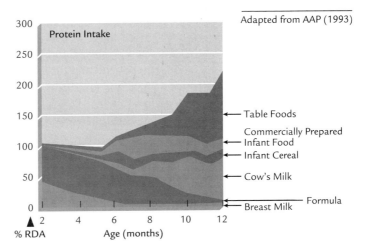

Adapted from AAP (1993)

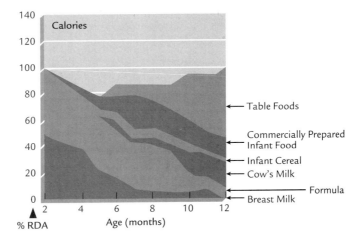

FIGURE 6.11

The increasing importance of supplementary foods as calorie and protein sources over the first year

- Talk in a quiet and encouraging manner.
- Seat infant so that he/she is straight up and facing forward.
- Wait for infant to pay attention and open up before feeding.
- Let infant touch food and feed him/herself with fingers.
- Let the infant decide how fast to eat.
- Respect the infant's food preferences.
- Respect the infant's caution; all children eventually learn to eat.
- Let the infant eat as much as desired.
- Stop when the infant indicates that he/she is full.

Adapted from E. Satter, "The Feeding Relationship: Problems and Intervention," *J Pediatr* 117, suppl. (1990):181.

TABLE 6.37

Successful feeding during later infancy

During feedings, infants are learning about the taste and texture of foods. They are curious, and as part of normal exploratory behavior, they play with food, often putting it into their mouths and taking it back out. Families should be patient, and the infant should be given positive reinforcement when eating is successful. Infants learn good eating habits by observation—when possible, they should be placed in a high chair at family meals where they can watch the rest of the family eat.

The Second Year

Early in the second year, infants begin to feed themselves. Their increased dexterity and coordination allowing them to nimbly finger-feed, drink from a cup, and use a spoon. They take more initiative in choosing foods and often have begun to establish definite food preferences. During this period, children should be offered three major meals each day and a number of nutritious snacks in between.

Although food intake may vary sharply from one meal to the next, most children meet their nutritional needs on a daily basis. Food preferences often change unpredictably, and today's favorite food may be unacceptable tomorrow. Because food intake slows compared to the first year (due to decelerating growth), some families erroneously perceive this as a sign of poor appetite. Exposure to a wide variety of foods will prevent overdependence on a single favorite food or drink (such as milk or juice), which may lead to nutritional imbalance. All of these issues are central themes in the feeding patterns of preschool children, which are discussed in detail in Chapter 7.

Guidelines on Dietary Fat and Cholesterol during Late Infancy

The American Heart Association and the Committee on Nutrition of the AAP recommend that fat and cholesterol not be restricted during the first two years after birth (AAP 1993). Although low-fat and low-cholesterol diets are recommended during later life (discussed in Chapters 7 and 9), low-fat diets may not meet energy requirements during rapid infant growth.

Cow's milk becomes an important part of many children's diets after the first year. Although some families, because of concerns about obesity and atherosclerosis in later life, provide skim or low-fat milk to their infants, the AAP recommends whole milk for children during the second year (AAP 1993). Low-fat and skim milks have low calorie densities and are relatively high in electrolytes and protein.

GROWTH AND DEVELOPMENT OF INFANTS FED BREAST MILK AND FORMULA

THE INFANT'S FIRST YEAR

In developing countries, breast-fed babies grow faster from birth to age 6 months, compared with babies fed formula (Wray 1991). This difference is probably due to a reduced incidence of infection among the breast-fed infants. In industrialized countries, the growth of breast-fed and formula-fed infants is similar up to age 2 to 3 months. For the rest of the first year, however, formula-fed infants typically grow at faster rates, particularly after the sixth month (Shepherd et al. 1988), (Nelson et al. 1989). Weight for length is significantly higher in formula-fed infants, while differences in length (height) tend to be small (Fomon et al. 1978). Because of this, differences in weight gain are thought to represent mainly differences in adiposity (Dewey et al. 1993; Dewey et al. 1992). See Table 6.38.

There is no evidence of any functional advantage to the more rapid growth of formula-fed infants. In one study, breast-fed infants, although they grew at slower rates, had less morbidity from infectious illness and normal or accelerated development (Dewey et al. 1991).

The reported energy intake of formula-fed and breast-fed babies is consistent with the differences in weight gain: formula-fed infants consume significantly more calories and more protein. The difference in energy intake is maintained throughout the first year, even when infants begin receiving supplemental foods at 4 to 6 months of age (Heinig et al. 1993). See Figure 6.12.

Are growth charts applicable for breast-fed infants? The commonly used NCHS charts (shown earlier in the chapter) are based on data collected between 1930 and 1975, mainly in infants who were fed formula. Moreover, the formulas consumed by most of the infants surveyed were less similar to human milk than modern formulas (they were higher in calories, protein, and fat). Also, many of the infants were given solid foods before 4 months of age. Because of these variables, many scientists and pediatricians think the NCHS charts are inappropriate for breast-fed infants (Heinig et al. 1993), (Hitchcock et al. 1985).

For example, when growth of healthy, active breast-fed infants in the United States is compared with the NCHS charts, nearly a third fall below the 5th percentile for weight gain during the first six months, and more than half fall below the 5th percentile during the second six months (Dewey 1989). Because growth below the 5th percentile is often used as a cut-off to indicate infants at risk of malnutrition—and most breast-fed babies are thriving—these charts are of little value in assessing the growth of breast-feeding babies.

	Breast-feeding Infants	Formula-feeding Infants
0–3 mo		
Weight gain (g/d)	28 ± 6.7	29 ± 6.7
Length gain (cm/d)	0.11 ± 0.02	0.12 ± 0.02
LBM gain (g/d)	16 ± 3.7	17 ± 3.4
Fat mass gain (g/d)	11 ± 4.1	12 ± 3.9
3–6 mo		
Weight gain (g/d)	16 ± 4.5	19 ± 4.0[‡]
Length gain (cm/d)	0.06 ± 0.01	0.06 ± 0.01
LBM gain (g/d)	11 ± 2.2	12 ± 2.1[‡]
Fat mass gain (g/d)	5 ± 2.4	7 ± 2.2[‡]
6–9 mo		
Weight gain (g/d)	10 ± 3.7	12 ± 3.0[‡]
Length gain (cm/d)	0.04 ± 0.01	0.05 ± 0.01
LBM gain (g/d)	8 ± 1.8	9 ± 1.4[‡]
Fat mass gain (g/d)	2 ± 2.21	3 ± 1.8[‡]
9–12 mo		
Weight gain (g/d)	8 ± 3.6	9 ± 4.1
Length gain (cm/d)	0.04 ± 0.01	0.04 ± 0.01
LBM gain (g/d)	7 ± 1.7	8 ± 2.1
Fat mass gain (g/d)	1 ± 2.1	1 ± 2.3
12–15 mo		
Weight gain (g/d)	8 ± 3.6	8 ± 4.2
Length gain (cm/d)	0.04 ± 0.01	0.04 ± 0.01
LBM gain (g/d)	7.3 ± 1.8	7.3 ± 1.8
Fat mass gain (g/d)	0.8 ± 2.1	0.6 ± 2.6

TABLE 6.38
Differences in growth during infancy for breast-fed vs. formula-fed infants

The values marked with (‡) are significantly different from the corresponding breast-feeding values. Values are ± SD.

Adapted from M. J. Heining, et al., "Energy and Protein Intakes of Breastfed and Formula-Fed Infants," *Am J Clin Nutr* 58 (1993):152-61.

LONG-TERM GROWTH

Different methods of infant feeding do not appear to affect long-term growth: the differences in growth between breast-fed and formula-fed infants do not persist into later childhood. Several studies have found no difference in size and growth rates among children from 3 to 12 years old, when those who were breast-fed as infants were compared with those who were formula-fed (Birkbeck et al. 1985), (Pomerance 1987). No firm associations between method of feeding during infancy and later development of obesity have been found (Institute of Medicine 1991).

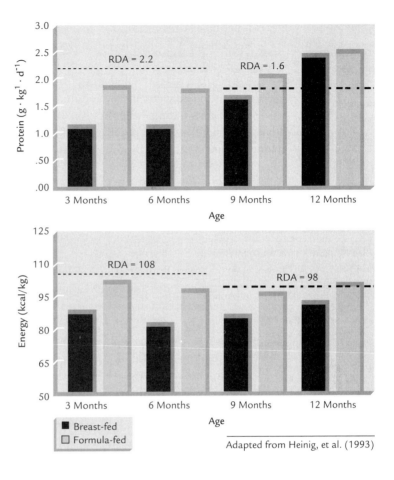

FIGURE 6.12

Energy and protein intake of breast-fed vs. formula-fed infants

Adapted from Heinig, et al. (1993)

COGNITIVE DEVELOPMENT

Studies have measured cognitive development in children up to age 15 years and compared the cognitive development of children who were breast-fed as infants with children who were fed formula. Even when potential confounding variables (such as socioeconomic status, maternal intelligence, education, and child-rearing attitudes) are controlled for, small but statistically significant differences have been consistently found—children who were breast-fed performed slightly better on tests of cognitive development than children who were fed formula (Morrow-Tlucak et al. 1988), (Taylor and Wadsworth 1984).

NUTRITIONAL CONCERNS DURING INFANCY

BREAST-FEEDING VERSUS FORMULA-FEEDING

Each mother's decision whether to breast-feed or formula-feed is a personal one, determined by a number of individual, family, and sociocultural factors (discussed in Chapter 5). Families can make an informed choice if they are aware of the differences in nutrition and overall health between feeding formula and feeding breast milk.

The nutritional goals of infant feeding—to supply the growing infant with sufficient energy and nutrients and to establish healthy eating habits—can be adequately met by both breast-feeding and formula-feeding. Nutrient bioavailability from breast milk is superior, but formulas are fortified with sufficient nutrients to compensate for the differences in availability.

Breast-feeding may have advantages during the immediate postpartum period. For example, colostrum is uniquely appropriate for the newborn infant, because it contains a variety of digestive enzymes that facilitate digestion and absorption. Nevertheless, infants can grow normally and thrive on either form of feeding.

Breast-feeding also provides benefits beyond optimal nutrition. Human milk contains a complex variety of anti-infective substances and cells that reduce infections of the gastrointestinal tract and infant diarrheal disease. Also, breast-feeding appears to be protective against food allergies. These topics are discussed in Chapters 4 and 5.

Many investigations have tried to determine whether methods of infant feeding may influence health in later life. Studies have focused on obesity, atherosclerosis, and several common and serious immunological disorders. Overall, research has found no consistent links between methods of infant feeding and levels of adult obesity or rates of atherosclerosis (Institute of Medicine 1991). However, as discussed in Chapter 5, breast-feeding appears to exert long-term protective effects against several diseases that may have immunological or infectious causes, including Type I diabetes mellitus (Mayer et al. 1988) and Crohn's disease (Koletzko 1989).

Therefore, breast-feeding has subtle but distinct benefits for infant health when compared with formula-feeding (AAP 1993). All expectant mothers should be encouraged to breast-feed and a decision to breast-feed should be strongly supported—through education, prenatal classes, and throughout the health-care system. At the same time, it should be appreciated that the decision on infant feeding is an individual and personal one, and no woman should be coerced into breast-feeding or made to feel guilty because she chooses to feed formula.

IS THE INFANT GETTING ENOUGH TO EAT?

During the first few weeks postpartum, one of the main concerns of many new mothers is providing their infants with sufficient breast milk or formula. Although nearly all women who choose to breast-feed have adequate milk supplies, a perception of insufficient milk is one of the most common reasons given by mothers for early weaning and supplementation (discussed in Chapter 5). First-time mothers need to understand that infants are restless, fussy, and they cry and sleep poorly for a variety of reasons that may be unrelated to hunger. Also, feeding patterns are frequently irregular in the first month, and the mother should not be overly concerned if feedings appear too short or too long or if her baby occasionally falls asleep while feeding.

Formula-fed infants should be provided as much formula as they want—the usual intake will be 150–200 ml/kg body weight/day, or 100–135 kcal/kg/day (AAP 1993). Women who are breast-feeding can be reassured that their milk supply is adequate by observing infant feeding and stooling patterns, and by weekly weighing of the infant (discussed in Chapter 5). During the first six weeks, the typical infant is receiving adequate nourishment if feedings occur at least six times a day. Another sign of sufficient intake is at least six wet diapers and several stools each day.

Probably the best indication that an infant is getting enough breast milk or formula is adequate weight gain. Weight gain should be measured regularly in all infants; gain of about 30 g/day (1/3 to 1/2 lb per week) after the first week typically indicates that the infant is receiving sufficient energy and nutrients during early infancy (AAP 1993). Infants who continue to lose weight after the first week or who have not regained their birthweight the third week should be evaluated by their pediatrician for failure to thrive (see the next section).

FAILURE TO THRIVE

Failure to thrive is a term used to describe an infant or child who fails to meet normal developmental standards for age. Although the term is occasionally used for children who show cognitive delays, it is most commonly applied to infants who are not growing or gaining weight normally. Infants who fail to thrive have values for length and weight below the 5th percentile on standard growth charts. There are many potential causes of failure to thrive, including nutritional, physical, and psychosocial factors (see Figure 6.13) (Lawrence 1994).

Infants who fail to thrive while being breast-fed may have an ineffective suck, and although they spend adequate time at the breast, may not be obtaining sufficient milk. Also, infants who are not fed frequently—for example, if they are allowed to sleep through the night and miss feedings—can have delayed growth. Maternal causes of failure to thrive include poor milk letdown (due to stress, anxiety, smoking, drug, or alcohol use) or, less often, poor milk production because of fatigue, a poor diet, illness, or drug use (discussed in Chapter 5).

Infant malnutrition is a major cause of failure to thrive in children worldwide. Deficiencies of energy and protein during childhood are common in the developing world and among disadvantaged groups in

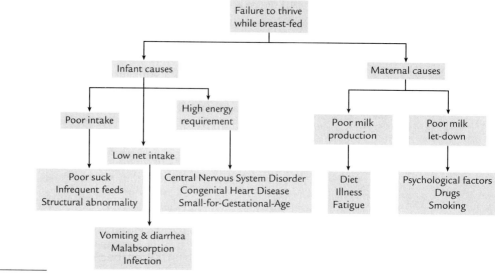

FIGURE 6.13
Failure to thrive

the industrialized countries. Micronutrient deficiencies, such as iron and zinc deficiency, can also produce failure to thrive.

For many infants, the cause of failure to thrive is their psychosocial environment—poverty, overcrowding, family instability, and inadequate mothering result in a non-nurturing environment. These infants may be irritable, temperamental, and difficult to care for. Both emotional deprivation and undernutrition play a role in psychosocial failure to thrive, and treatment must include both adequate nutrition and a provision of a caring and supportive family surrounding.

INFANTILE COLIC

A common but ill-defined condition, *infantile colic* typically occurs from 3 weeks to 3 months of age. It is characterized by repeated episodes of fussiness and inconsolable crying, associated with drawing up the legs, abdominal gas and distension—symptoms suggestive of spasmodic abdominal pain. Besides being distressing to the infant, episodes of colic often cause considerable anxiety in the parents.

Although a number of dietary and psychosocial factors have been proposed as causes of colic, in most cases the cause is unknown. Studies have shown that colic is equally common in breast-fed and formula-fed infants, and in infants fed cow's milk-based formula and soy-based formula. Careful studies have found that some cases of colic in formula-fed infants were caused by food intolerance or allergy to a component of formula, often the whey protein in cow's milk (Forsyth 1989), (Lothe and Lindberg 1989).

Also, although findings have been inconsistent, some studies have suggested that colic in breast-feeding infants may be linked to consumption of cow's milk by the mother (Jakobsson 1983). If nursing mothers choose to reduce their intake of milk in an attempt to prevent colic, they should strive to obtain calcium from other sources—calcium needs during lactation are high.

It is important that parents understand that infantile colic is not serious and that it will resolve by itself, usually by the third or fourth month. Mothers should not stop breast-feeding an infant because of colic. Changes in formula may help in a minority of cases, but often the cause is never found.

SPITTING UP AND VOMITING

Particularly during the first few weeks, many infants regularly spit up a small portion of breast milk or formula after feedings. The cause is *gastroesophageal reflux*—the regurgitation of material from the stomach back up into the esophagus and mouth—a common condition during infancy (Miller 1990).

A muscular ring, or sphincter, at the bottom of the esophagus normally closes tightly to prevent food from moving upward into the esophagus from the stomach. During early infancy, the gastroesophageal sphincter does not contract tightly. Also, because peristalsis in the esophagus is not fully developed, swallowed liquids can move up the esophagus to the mouth.

More than half of all infants regularly spit up or vomit during the first week (Keitel and Ziegra 1961). Although often a concern to parents, the percentage of a feeding lost is usually small, and most infants obtain adequate nutrition and grow normally. By the second month, the gastroesophageal sphincter has matured and peristalsis is well coordinated. In almost all infants, spitting up becomes less frequent, and the problem resolves itself.

NURSING BOTTLE CARIES

Nursing bottle caries is a characteristic pattern of dental caries found in infants who are given a bottle of juice, milk, or formula at naps or bedtime. It occurs less commonly in breast-fed infants who sleep for long periods at the breast. During sleep, salivary flow is low and the juice or milk, rich in fermentable carbohydrate, pools around the teeth. **Cariogenic** bacteria ferment the sugar to acids that produce severe and rapid decay. Usually all of the upper teeth and the lower back teeth are affected—the lower front teeth are usually covered and protected by the tongue. Infants should not be put to sleep with a bottle. It is preferable

cariogenic: capable of producing dental caries

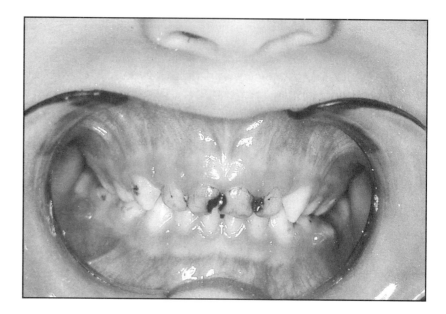

PHOTO 6.1
Nursing-bottle caries in a 25-month-old bottle-fed child. (Photo: Custom Medical Stock Photo)

to feed and burp them and then put them to sleep. Chapter 7 contains a detailed discussion of dietary sugar and dental caries.

THE PRETERM INFANT

An important aspect of the hospital care of preterm infants is optimal nutrition. Specially equipped modern nurseries can care for infants born as early as 26–28 weeks gestation. The digestive and absorptive capabilities of preterm infants are limited, yet they have very high growth rates and nutritional needs—energy requirements are 105–130 kcal/kg/day, and protein requirements are 3.5–4.0 g protein/kg/day (AAP 1993). Because well-coordinated suckling and swallowing does not develop until about 32–34 weeks gestation, preterm infants who are born younger than 32 weeks often need to be fed through a thin plastic tube that delivers formula to the stomach, or through intravenous routes.

Special infant formulas have been developed to meet the unique needs of preterm infants (see Table 6.39). Human milk does not contain sufficient amounts of protein and certain micronutrients (including calcium, sodium, zinc, and copper) to meet the high demands for growth in preterm infants (Forbes 1989).

However, the nutritional inadequacies of human milk can be corrected with the addition of commercial fortification mixtures containing protein, carbohydrate, and micronutrients. In general, preterm infants fed breast milk that has been collected from their mothers and fortified grow as well as those fed commercial formulas (Chessex 1983).

	Preterm Formula	Fortifier	Human Milk
Energy			
Kilocalories	80	3.5	62–68
Protein			
Content (g)	2.0	0.18	1.2–1.7
% of energy	9.5	20	7.1–10.6
Whey: casein ratio	60:40	60:40	60:40
Lipid			
Content (g)	4.1	0.05	3–4
% of energy	46	3	44–56
Composition			
Saturated (%)	62	63	43
Monosaturated (%)	13	32	42
Polyunsaturated (%)	25	2	15
Carbohydrate			
Content (g)	8.9	0.68	6–7
% of energy	45	77	38–44
Lactose (%)	40	26	100
Glucose polymers	60	74	—
Minerals and Trace Elements			
Sodium (mg)	27	1.75	25–29
Potassium (mg)	67	3.9	60–61
Chloride (mg)	58	4.4	46–51
Calcium (mg)	110	22.5	26–28
Phosphorus (mg)	55	11.25	15–17
Magnesium (mg)	5.0	0	2.6–3.0
Iron (mg)	0.17	0	0.03–0.08
Zinc (mg)	0.67	0.18	0.12–0.34
Copper (µg)	80	0.02	25–57
Manganese (µg)	8.3	2	0.6
Iodine (µg)	5.0	0	11
Vitamins			
Fat soluble			
Vitamin A (IU)	300	195	104–223
Vitamin D (IU)	46	65	0.5–2.2
Vitamin E (IU)	3.1	0.85	0.3–0.5
Vitamin K (µg)	8.7	2.3	0.2–2.0
Water soluble			
Vitamin B_1 (µg)	170	47	2–21
Vitamin B_2 (µg)	420	62	10–70
Vitamin B_6 (µg)	170	48	10–80
Vitamin B_{12} (µg)	0.38	0.05	0.01–0.06
Vitamin C (µg)	25	6.0	4.4–5.2
Biotin (µg)	2.5	0.2	0.01–0.80
Folic acid (µg)	25	5.8	5–6
Niacin (mg)	3.3	0.78	0.15–0.17
Pantothenate (mg)	1.25	0.2	0.18

TABLE 6.39

Formulas and breast-milk fortification supplements for preterm infants, compared with human milk

TABLE 16.39 *Continued*

	Preterm Formula	Fortifier	Human Milk
Other			
Carnitine (mg)	0.7	—	0.7
Choline (mg)	10.6	—	9.0
Inositol (mg)	20	—	14.9–56
Taurine (mg)	4.0	—	4.0

Neonatal and Perinatal Medicine: Diseases of the Infant and Fetus, 5 ed. eds. A. Fanaroff, R. J. Martin (St Louis: Mosby–Year Book 1992).

Values per 100 ml.

Many mothers can maintain lactation by manual pumping and collection of breast milk while the infant is being cared for in the hospital. Subsequently, when the infant comes home, breast-feeding can begin.

A HEALTHY DIET DURING INFANCY: REVIEW AND SUMMARY

VITAMIN AND MINERAL SUPPLEMENTATION

In the United States, vitamin and mineral supplements for infants are available as liquid drop preparations. All supplements for children under 4 years old are subject to FDA regulations designed to minimize misuse of these preparations. The regulations require that supplements contain appropriate combinations of vitamins and minerals, and they set limits for individual nutrients (the lower limits are set at about 25–50% of the RDA, the upper limits at 100–150% of the RDA).

To prevent hemorrhagic disease of the newborn and supply adequate body stores of vitamin K, all newborn infants should routinely receive vitamin K supplementation (AAP 1993). Vitamin K can be given as a single intramuscular injection of 0.5–1.0 mg or an oral dose of 1.0–2.0 mg. Recommendations on vitamin and mineral supplementation for breast-fed infants are the following (AAP 1993):

- *Vitamin D.* Human milk contains only small amounts of vitamin D. In northern climates, during the winter months when maternal exposure to sunlight is minimal, levels of vitamin D in breast milk may provide only about 30 IU to the infant daily—a level that may not be sufficient to prevent rickets. Therefore, breast-fed infants who are not exposed to sunlight, or who are deeply pigmented, should receive a supplement of 400 IU vitamin D daily. When the breast-fed infant begins to consume sufficient solid foods that are good sources of vitamin D, supplementation can be stopped.

Supplemental Fluoride Dosage Schedule (mg/day)			
	Concentration of Fluoride in Drinking Water (ppm)		
Age	<0.3	0.3–0.7	>0.7
2 weeks to 2 years	0.25	0	0

From American Academy of Pediatrics, Committee on Nutrition, *Pediatric Nutrition Handbook,* 3d ed. (Elk Grove, IL: AAP, 1993).

TABLE 6.40
Fluoride supplements

■ *Iron.* Breast-fed infants not receiving supplemental iron in the second half of the first year are at risk for iron deficiency and anemia. Iron supplementation for full-term, breast-fed infants should begin between 4 and 6 months of age. For infants beginning to take solid foods, two or more servings of iron-fortified infant cereal/day will provide adequate iron. For those who continue to exclusively breast-feed, a daily iron supplement in the form of ferrous sulfate (containing 15 mg of elemental iron) should be given.

■ *Fluoride.* Incorporation of fluoride into developing tooth enamel reduces the susceptibility to dental caries. Only trace amounts of fluoride are found in breast milk, and supplementation of infants who are exclusively breast-feeding is recommended. A daily supplement of 0.25 mg of fluoride should be provided until the infant begins to consume fluoridated water. See Table 6.40.

Because the level of certain nutrients in breast milk varies according to maternal status, malnourished mothers who are breast-feeding should be given a multivitamin/multimineral supplement, and a concerted effort should be made to improve their diets.

For formula-fed infants, no supplementation is needed for most full-term infants fed iron-fortified commercial infant formula. Those infants fed ready-to-use formulas (most are made from water low in fluoride) should be given a fluoride supplement, as should those infants in areas where the local water supply contains less than 0.3 ppm of fluoride.

FOOD CHOICES DURING INFANCY

The two basic goals of infant feeding are to provide sufficient nutrition for optimal growth and development, and to begin to establish food habits and eating patterns that lay the foundation for lifelong health. Breast milk is the preferred food during early infancy, although current commercially prepared infant formulas are adequate alternatives for mothers who choose not to breast-feed. For most infants, human milk or an appropriate formula should be the sole nutrient source for the first six months.

Supplementary foods should be started in most healthy infants at about 6 months of age. Earlier introduction of supplemental foods provides no nutritional advantage. Appropriate first foods are enriched infant cereals and strained fruits and vegetables. Subsequently, strained meats, poultry and fish, eggs, and fruit juices can be introduced, although certain fruits and fruit juices—particularly apple and pear juice—should be fed to infants in moderation. Both commercially prepared and home-prepared foods can supply adequate nourishment for infants. Families should be careful to prevent spoilage of any infant foods.

During the later half of the first year, progress from liquid to pureed to semisolid foods should parallel developmental readiness in the infant. Infants should be carefully watched when eating supplementary foods, as risk of choking and aspiration are highest in late infancy. A wide variety of foods should be introduced, to ensure nutritional balance. Although infants should continue to consume generous amounts of breast milk or formula during the weaning period, cow's milk should generally not be given during the first year.

Low-fat diets may not meet the high energy requirements of infancy. A balanced, energy-dense diet will ensure optimal growth and development, and in general, fat and cholesterol should not be restricted during infancy. Refer to Table 6.41 for food guidelines for the first year, including recommendations on healthy food choices and caloric intake.

Food preferences begin to develop during infancy and are influenced both by the types of food offered and the observations by the infant of the attitudes and reactions of family members to foods. Careful attention to food choices and appropriate reinforcement of positive mealtime behavior can lay the foundation for the development of healthy eating habits in later childhood and adolescence.

	0–2 Weeks	2 Weeks– 2 Months	2 Months	3 Months
Breast milk/formula				
Volume per feeding	2–3 oz	3–5 oz	5 oz	6–6.5 oz
Number of feedings (per day)	6–8	5–6	5–6	4–5
Average total	20 oz	28 oz	30 oz	32–34 oz
Recommended calories (115 kcal/kg)	403	483–598	598	656
Total calories	440	560	600	660–680

	4–5 Months	5–6 Months
Breast milk/formula		
Per feeding	7–8 oz	7–8 oz
Average total	28–32 oz	28–32 oz
Number of feedings	4–5	4–5
Food additions		
Apple juice		3–4 oz
Baby cereal, enriched	2–2.5 T, B & D	3 T, B & D
Strained fruits	1.5–2 T, B, L & D	1.5–2 T, B, L & D
Strained vegetables	1–2 T, L	2–3 T, L
Strained meat		1–2 T, L
Egg yolk		1/2 medium
Teething biscuit		1/2–1
Recommended calories (115 kcal/kg)	681–770	770–863
Total calories	729–780	780–860

	6–7 Months	7–8 Months	8–9 Months
Breast milk/formula			
Per feeding	8 oz	8 oz	8 oz
Number of feedings	3–4	3–4	3
Average total	28 oz	28 oz	24 oz
Food items			
Orange juice	4 oz	4 oz	4 oz
Fortified cereal	¼ C, B	⅓ C, B	½ C, B
Fruit, canned or fresh	4 t, B, L & D	4 t, B, L & D	2 T, B, L & D
Vegetables	1½ T, L & D	2 T, L & D	2 T, L & D
Meat, fish. poultry	1 T, L & D	2 T, L & D	2 T, L & D
Egg yolk or baby egg yolk	½ medium yolk, or 1 T	1 medium yolk, or 2 T	1 medium yolk, or 2 T

TABLE 6.41

Feeding guidelines from birth through 12 months

TABLE 6.41 *Continued*

	6–7 Months	7–8 Months	8–9 Months
Food items			
Teething biscuit or bread	1 biscuit	1 biscuit	½ slice bread
Butter			1 T
Recommended calories (105 kcal/kg)	788–840	840–893	893–945
Total calories	819	876	937

	9–10 Months	10–11 Months	11–12 Months
Breast milk/formula**			
Per feeding	8 oz	8 oz	8 oz
Number of feedings	3	3	3 (nutritious snacks are recommended)
Average total	24 oz	24 oz	24 oz
Food items			
Orange juice	4 oz	4 oz	4 oz
Fortified cereal	½ C, B	½ C, B	½ C, B
Fruit, canned or fresh	2 T, B, L & D	3 T, B, L & D	3 T, B, L & D
Vegetables	2 T, L & D	2 T, L & D	3 T, L & D
Meat, fish, poultry	2 T, L & D	2½ T, L & D	2½ T, L & D
Egg yolk or baby egg yolk	1 medium yolk, or 2 T	1 medium yolk, or 2 T	1 whole egg
Teething biscuit or bread	½ slice bread	½ slice bread	½ slice bread-
Starch—potato, rice, macaroni	2 T, D	2 T. D	2 T, D
Butter	1 t	1 t	1 t
Recommended calories (105 kcal/kg)	945–987	987–1019	1019–1053
Total calories	944	1007	1039

**Offer small amounts (2–4 oz) when milk is given from the cup.

C, cup: B, breakfast: T, tablespoon: L, lunch: D, dinner: t, teaspoon.

Adapted from The American Dietetic Association, *Manual of Clinical Dietetics,* developed by the Chicago Dietetic Association and the South Suburban Dietetic Association (Chicago: The American Dietetic Association, 1988).

REFERENCES

American Academy of Pediatrics, *Pediatric Nutrition Handbook,* (Elk Grove, IL: AAP, 1993).

American Academy of Pediatrics, Committee on Nutrition, "Hypoallergenic Infant Formulas," *Pediatrics* 83 (1989):1068–69.

Beauchamp, G. K., Cowart, B. J., and Moran, M., "Developmental Changes in Salt Acceptability in Human Infants," *Dev Psychobiol* 19 (1986):17–25.

Bell, E. F., "Vitamin E and Iron Deficiency in Preterm Infants," in *Nutritional Anemias,* eds. S. J. Fomon and S. Zlotkin, (New York: Raven Press,1992), 137–46.

Birkbeck, J. A., Buckfield, P. M., and Silva, P. A., "Lack of Long-Term Effect of the Method of Infant Feeding on Growth," *Hum Nutr Clin Nutr* 39C (1985):39–44.

British Paediatric Association, "Hypercalcaemia in Infants and Vitamin D," *BMJ* 2 (1956):149.

Bronner, Y.L., and Paige, D. M., "Current Concepts in Infant Nutrition," *J Nurse Midwifery* 37 (1992):43S–58S.

Butte, N. F., et al., "Macro- and Trace Mineral Intakes of Exclusively Breast-Fed Infants," *Am J Clin Nutr* 45 (1987):42–48.

Casey, C. E., Hambidge, K. M., and Neville, M. C., "Studies in Human Lactation: Zinc, Copper, Manganese, and Chromium in Human Milk in the First Month of Lactation," *Am J Clin Nutr* 41(1985):1193–1200.

Centers for Disease Control and Prevention, "Pediatric Nutrition Surveillance System— United States 1980–1991," in *CDC Surveillance Summaries, MMWR* 41 no. SS-7 (1992):1–25.

Chessex, P., et al., "Quality of Growth in Premature Infants Fed Their Own Mothers' Milk," *J Pediatr* 102 (1983):107–112.

Committee on Fetus and Newborn, "Sterilization of Milk-Mixtures for Infants," *Pediatrics* 28 (1961):674–75.

Dagnelie, P. C., et al., "High Prevalence of Rickets in Infants on Macrobiotic Diets," *Am J Clin Nutr* 51 (1990):202–08.

Dagnelie, P. C., et al., "Nutritional Status of Infants Aged 4 to 18 Months on Macrobiotic Diets and Matched Omnivorous Control Infants: A Population-Based Mixed-Longitudinal Study I. Weaning Pattern, Energy and Nutrient Intake," *Eur J Clin Nutr* 43 (1989):311–23.

Dahro, M., Gunning, D., and Olson, J. A., "Variations in Liver Concentrations of Iron and Vitamin A as a Function of Age in Young American Children Dying of the Sudden Infant Death Syndrome as Well as of Other Causes," *Int J Vit Nutr Res* 53 (1983):13–18.

Dewey, K. G., et al., "Adequacy of Energy Intake among Breastfed Infants in the DARLING Study: Relationships to Growth Velocity, Morbidity, and Activity Levels," *J Pediatr* 119 (1991):538.

Dewey, K. G., et al., "Breastfed Infants are Leaner than Formula-Fed Infants at One Year of Age: The DARLING Study," *Am J Clin Nutr* 57 (1993):140.

Dewey, K. G., et al., "Growth of Breastfed and Formula-Fed Infants from 0 to 18 Months: The DARLING Study," *Pediatrics* 89 (1992):1035.

Dewey, K. G., et al., "Infant Growth and Breast-Feeding," *Am J Clin Nutr* 50 (1989):1116–17.

Dirren, H., et al., "Zinc Supplementation and Child Growth in Ecuador," in *Nutrient Regulation during Pregnancy, Lactation and Infant Growth*, eds. L. Allen, J. C. King and B. Lonnerdal (New York: Plenum Press, 1994).

Eveleth, P. B., and Tanner, J. M., *Worldwide Variation in Human Growth*, 2d ed. (Cambridge: University Press, 1990).

Filer, L. J., " Iron Deficiency," in *Childhood Nutrition*, ed. F. Lifshitz, (Boca Raton: CRC Press, 1995).

Forbes, G. B., "Nutritional Adequacy of Human Breast Milk for Prematurely Born Infants," in *Textbook of Gastroenterology and Nutrition in Infancy*, ed. E. Lebenthal (New York: Raven Press, 1989):27–34.

Fomon, J., et al., "Influence of Formula Concentration on Caloric Intake and Growth of Normal Infants," *Acta Paediatr Scand* 64 (1975):172–81.

Fomon, S. J., and Ziegler, E. E., "Soy Protein Isolates in Infant Feeding," in *Soy Protein and Human Nutrition*, eds. H. L. Wilcke, D. T. Hopkins and D. H. Waggle, (New York: Academic Press, 1979):79–96.

Fomon, S. J.,   (St. Louis: Mosby-Year Book, 1993).

Fomon, S. J., et al., "Growth and Serum Chemical Values of Normal Breast Fed Infants," *Acta Paediatr Scand* 273, suppl. (1978):1.

Fomon, S. J., et al., "Skim Milk in Infant Feeding," *Acta Paediatr Scand* 66 (1977):17–30.

Forsyth, B. W., "Colic and the Effect of Changing Formula: A Double-Blind, Cross Over Study," *J Pediatr* 115 (1989):521–6.

Galeano, N. F., et al., "Taurine Supplementation of a Premature Formula Improves Fat Absorption in Preterm Infants," *Pediatr Res* 22 (1987):67–71.

Gaull, G. E., "Taurine in Pediatric Nutrition: Review and Update," *Pediatrics* 83 (1989):433–442.

Ghitis, J., "The Labile Folate of Milk," *Am. J. Clin. Nutr.* 18 (1966):452–57.

Gibson, R. S., and De Wolfe, M. S., "The Dietary Trace Metal Intake of Some Canadian Full-Term and Low Birthweigth Infants during the First Twelve Months of Infancy," *J Can Diet Assoc* 41 (1980):206–15.

Gleason, W.A., and Kerr, G. R., Jr, "Questions about Quinones in Infant Nutrition," *J Pediatr Gastroenterol Nutr* 8 (1989):285–287.

Goldfarb, J., "Breastfeeding: AIDS and Other Infectious Diseases," *Clin Perinatol* 20 (1993):225–44.

Greer, F. R., and Marshall, S., "Bone Mineral Content, Serum Vitamin D Metabolite Concentrations, and Ultraviolet B Light Exposure in Infants Fed Human Milk with and without Vitamin D_2 Supplements," *J Pediatr* 114 (1989):204–12.

Gushurst, C. A., et al., "Breast Milk Iodide: Reassessment in the 1980s," *Pediatrics* 731984.:354-357.

Gutelius, M. F., "The Problem of Iron Deficiency Anemia in Preschool Negro Children," *Am J Public Health* 59 (1969):290.

Heinig, J. M., et al., "Energy and Protein Intakes of Breastfed and Formula-Fed Infants during the First Year of Life and Their Association with Growth Velocity: The DARLING Study," *Am J Clin Nutr* 58 (1993):152.

Hertrampf, E., et al., "Bioavailability of Iron in Soy-Based Formula and its Effects on Iron Nurture in Infancy," *Pediatrics* 78 (1986):640–45.

Hervada, A. R., and Newman, D. R., "Weaning: Historical Perspectives, Practical Recommendations and Current Controversies," *Curr Prob Pediatrics* May/June (1992):223–40.

Hitchcock, N. E., Gracey, M., and Gilmour, A. I., "The Growth of Breast Fed and Artificially Fed Infants from Birth to Twelve Months," *Acta Paediatr Scand* 74 (1985):240–245.

Hyams, J. S., et al., "Carbohydrate Malabsorption Follwing Fruit Juice Ingestion in Young Children," *Pediatrics* 82 (1988):64–8.

Idjradinata, P., and Pollitt, E., "Reversal of Developmental Delays in Iron-Deficient Anemic Infants Treated with Iron," *Lancet* 341 (1993):1–4.

Institute of Medicine, Subcommitte on Nutritional Status and Weight Gain during Pregnancy, *Nutrition during Pregnancy* (Washington D.C.: National Academy Press, 1990).

Institute of Medicine, Subcommittee on Nutrition during Lactation, *Nutrition during Lactation* (Washington D.C.: National Academy Press, 1991).

Jakobsson, I., and Lindberg, T., "Cow's Milk Proteins Cause Infantile Colic in Breast-Fed Infants: A Double-Blind Cross-Over Study," *Pediatrics* 71 (1983):268–71.

Jelliffe, D. B., and Jelliffe, E. F. P., "World Consequences of Early Weaning," in *Human Milk in the Modern World: Psychosocial, Nutritional, and Economic Significance* (Oxford: Oxford University Press, 1978): 241–99.

Joint FAO/WHO Codex Alimentarius Commission, "Codex Standards for Foods for Special Dietary Uses Including Foods for Infants and Children and Related Code of Hygienic Practice," *CAC*/Vol 9, ed. 1, (Rome: Food and Agriculture Organization of the United Nations/World Health Oraganization, 1984).

Keitel, H., and Ziegra, S. R., "Regurgitation in the Full-Term Infant: A Controlled Clinical Study," *Am J Dis Child* 102 (1961):750.

Koletzko, S., et al., "Role of Infant Feeding Practices in Development of Crohn's Disease in Childhood," *Br Med J* 298 (1989):1617–18.

Lanzkowsky, P., "Iron Deficiency Anemia," *Pediatr Ann* 3 (1970):6.

Lawrence, R. A., "Normal Growth, Failure to Thrive and Obesity in the Breastfed Infant," in *Breastfeeding: A Guide for the Medical Profession*, 4th ed. (St. Louis: Mosby-Year Book, 1994).

Lebenthal, E., "Impact of Development of the Gastrointrestinal Tract on Infant Feeding," *J Pediatr* 102 (1983):1.

Lönnerdal, B., "Iron in Human Milk and Cow's Milk—Effects of Binding Ligands on Bioavailability," in *Iron Metabolism in Infants*, ed. B. Lönnerdal (Boca Raton, Fl: CRC Press, 1990), 87–107.

Lönnerdal, B., et al., "Iron, Zinc, Copper, and Manganese in Infant Formulas," *Am J Dis Child* 137 (1983):433–37.

Lönnerdal, B., et al., "The Effect of Individual Components of Soy Formula and Cow's Milk Formula on Zinc Biovailability," *Am J Clin Nutr* 40 (1984):1064–1070.

Lothe, L., and Lindberg, T., "Cow's Milk Whey Protein Elicits Symptoms of Infantile Colic in Colicky Formula-Fed Infants: A Double-Blind, Cross-Over Study," *Pediatrics* 83 (1989):262–6.

Lozoff, B., Jimenez, E., and Wolf, A. W., "Long-Term Developmental Outcome of Infants with Iron Deficiency," *New Engl J Med* 325 (1991):687–94.

Macknin, M. L., Vanderbrug-Mendendorp, S. V., and Maier, M. C., "Infant Sleep and Bedtime Cereal," *Am J Dis Child* 143 (1989):1066–68.

Mayer, E. J., et al., "Reduced Risk of IDDM Among Breast-Fed Children: The Colorado IDDM Registry," *Diabetes* 37 (1988):1625–32.

McCoy, E., Strynadka, K., and Brunet, K., "Vitamin B-6 Intake and Whole Blood Levels of Breast and Formula Fed Infants. Serial Whole Blood Vitamin B-6 Levels in Premature Infants," in *Vitamin B-6: Its Role in Health and Disease*, eds. R. D. Reynold and J. E. Leklem (New York: Alan R. Liss, 1985), 79–96.

McJunkin, J. E., Bithoney, W.G., and McCormick, M. C., "Errors in Formula Concentration in an Outpatient Population," *J Pediatr* 111 (1987):848–50.

Milla, P. J., "Reflux Vomiting," *Arch Dis Child* 65 (1990):996–99.

Mizraki, A., London, R. D., and Gribetz, D., "Neonatal Hypocalcemia: Its Causes and Treatment," *N Engl J Med* 278 (1968):1163–65.

Morrow-Tlucak, M., Haude, R. H., and Ernhart, C. B., "Breastfeeding and Cognitive Development in the First 2 Years of Life," *Soc Sci Med* 26 (1988):635–39.

National Academy of Sciences, Food and Nutrition Board, of the National Research Council, *Recommended Dietary Allowances*, 10th ed. (Washington D.C.: National Academy Press, 1989).

National Center for Health Statistics, "NCHS Growth Charts, 1976, *Monthly Vital Statistics Report*, vol. 25, no. 3, suppl. (HRA) 76-1120, (Rockville, MD: Health Resources Administration, 1976).

Nelson, S. E,. et al., "Gain In Weight and Length during Early Infancy," *Early Hum Dev* 19 (1989):223–39.

Olson, A. L., Nelson, S. E., and Rebouche, C. J., "Low Carnitine Intake and Altered Lipid Metabolism in Infants," *Am J Clin Nutr* 49 (1989):624–28.

Olson, J. A., Gunning, D. B., and Tilton, R. A., "Liver Concentrations of Vitamin A and Carotenoids, as a Function of Age and Other Parameters, of American Children who Died of Various Causes," *Am J Clin Nutr* 39 (1984):903–910.

Owen, G. M., "Iron Nutrition, Growth in Infancy," in *Dietary Iron: Birth to Two Years*, L. J. Filer, Jr. (New York: Raven Press, 1989) 103–13.

Persson, B., Tunell, R., and Ekengren, K., "Chronic Vitamin A Intoxication During the First Half Year of Life: Description of 5 Cases," *Acta Paediatr Scand* 54 (1965):49–60.

Pizarro, F., et al., "Iron Status with Different Infant Feeding Regiments: Relevance to Screening and Prevention of Iron Deficiency," *J Pediatr* 118 (1991):687–920.

Pomerance, H. H., "Growth in Breast-Fed Children," *Hum Biol* 59 (1987):687–93.

Rajalakshmi, R., Doedhar, A. D., and Ramarkrishnan, C. V., "Vitamin C Secretion during Lactation," *Acta Paediatr Scand.* 54 (1965):375–82.

Raphael, D., "Weaning is Forever," *Lactation Rev* 6 (1982):1.

Riby, J., "Perinatal Deveopment of Digestive Enzymes," in *Infant Nutrition*, eds. A. F. Walker and B. A. Rolls (London: Chapman and Hall, 1994).

Rothberg, A. D., et al., "Maternal-Infant Vitamin D Relationships during Breast-Feeding," *J Pediatr* 101 (1982):500–503.

Satter, E., "The Feeding Relationship: Problems and Interventions," *J Pediatr* 117 (1990):S181.

Schmitz, J., "Development of the Gastrointestinal Tract and Accessory Organs: Digestive and Absorptive Function," in *Pediatric Gastrointestinal Disease*, eds. W.A. Walker et al. (Philadelphia: BC Decker, 1991).

Scientific Review Committee, "Nutrition Recommendations: The Report of the Scientific Review Committee, 1990" (Ottawa: Canadian Government Publishing Centre, Supply and Services Canada, 1990).

Shaw, J. C. L., "Copper Deficiency in Term and Preterm Infants," in *Nutritional Anemias*, eds. S. J. Fomon and S. Zlotkin (New York: Raven Press, 1992):105–17.

Sheard, N. F., "Iron Deficiency and Infant Development," *Nutr Rev* 52 (1994):137–46.

Shepherd, R. W., et al., "Longitudinal Study of the Body Composition of Weight Gain in Exclusively Breast-Fed and Intake Measured Whey-Based Formula-Fed Infants to Age 3 Months," *J Pediatr Gastroenterol Nutr* 7 (1988):732–39.

Sklar, R., "Nutritional Vitamin B_{12} Deficiency in a Breast-Fed Infant of a Vegan-Diet Mother," *Clin Pediatr* 25 (1986):219–21.

Smith, M., et al., "Carbohydrate Absorption from Fruit Juice in Young Children," *Pediatrics* 95 (1995):340–4.

Specker, B. L., et al., "Cyclical Serum 25-Hydroxyvitamin D Concentrations Paralleling Sunshine Exposure in Exclusively Breast-Fed Infants," *J Pediatr* 110 (1987):744.

Specker, B. L., et al., "Increased Urinary Methylmalonic Acid Excretion in Breast-Fed Infants of Vegetarian Mothers and Identification of an Acceptable Dietary Source of Vitamin B_{12}," *Am. J Clin Nutr* 47 (1988):89–92.

Steichen, J. J., and Tsang, R. C., "Bone Mineralization and Growth in Term Infants Fed Soy-Based or Cow Milk-Based Formula," *J Pediatr* 110 (1987):687–92.

Tanner, J. M., *Fetus into Man: Physical Growth from Conception to Maturity* (Cambridge: Harvard University Press, 1990).

Taylor, B., and Wadsworth, J., "Breast-feeding and Child Development at Five Years," *Dev Med Child Neurol* 26 (1984):73–80.

Tsang, R. C., "The Quandary of Vitamin D in the Newborn Infant," *Lancet* 1 (1983):1370–72.

U.S. Food and Drug Administration, "Rules and Regulations: Nutrient Requirements for Infant Formulas," *Fed Reg* 50 (1985):45106–5108.

Walravens, P. A., and Hambidge, K. M., "Growth of Infants Fed a Zinc Supplemented Formula," *Am J Clin Nutr* 29 (1976):1114–1121.

Walravens, P. A., Hambidge, K. M., and Koepfer, D. M., "Zinc Supplementation in Infants with a Nutritional Pattern of Failure to Thrive: A Double-Blind, Controlled Study," *Pediatrics* 83: (1989):532.

Weaver, L. T., Ewing, G., and Taylor, L. C., "The Bowel Habits of Milk-Fed Infants," *J Pediatr Gastroenterol Nutr* 7 (1988):568–71.

Wershil, B. K., "Development of the Gastrointestinal Tract and Accessory Organs: Gastric Function," in *Pediatric Gastrointestinal Disease*, eds. W.A. Walker, et al.(Philadelphia: BC Decker, 1991).

Wilson, J. F., Lahey, M. E., and Heiner, D. C., "Studies on Iron Metabolism V. Further Observations on Cow's Milk-Induced Gastrointestinal Bleeding in Infants with Iron-Deficiency Anemia," *J Pediatr* 84 (1974):335–44.

World Health Organization, "Energy and Protein Requirements: Report of a Joint FAO/WHO/UNU Expert Consultation," *Technical Report Series* (Geneva: WHO, 1985).

World Health Organization, *National Strategies for Overcoming Micronutrient Malnutrition* (Geneva: WHO, 1991), EB 89/27.

Wray, J. D., "Breastfeeding: An International and Historical Review," in *Infant and Child Nutrition Worldwide: Issues and Perspectives*, ed. F. Falkner (Boca Raton: CRC Press, 1991).

Yip, R., The Changing Characteristics of Childhood Iron Nutritional Status in the United States," in *Dietary Iron: Birth to Two Years*, ed. L. J. Filer, Jr. (New York: Raven Press, 1989) 37–56.

Ziegler, E. E., et al., "Absorption and Retention of Lead by Infants," *Pediatr Res.* 12 (1978):29–34.

$$7$$

childhood and
nutrition

▼

WORLD HUNGER AND CHILDHOOD HEALTH

In 1992 the Food and Agriculture Organization of the United Nations, using new data and methodology, revised estimates of the number of food-poor households in the world. The estimates represent "the number of people who, on average during the course of the year researched, did not consume enough food to maintain body weight and support light activity". The data indicate that the proportion and absolute number of underfed in the world has declined since 1975. Despite the addition of about 1.1 billion people to the population, fewer are undernourished now than two decades ago. See Table 7.1.

This positive global trend conceals very different regional realities. In sub-Saharan Africa, South Asia, and South America the absolute number of chronically underfed has increased substantially.

385

	Sub-Saharan Africa	Near East & North Africa	Middle America	South America	South Asia	East Asia	China	All
Proportion (in percentages)								
1970	35	23	24	17	34	35	46	36
1975	37	17	20	15	34	32	40	33
1980	36	10	15	12	30	22	22	26
1990	37	5	14	13	24	17	16	20
Absolute Numbers (in millions)								
1970	94	32	21	32	255	101	406	942
1975	112	26	21	32	289	101	395	976
1980	128	15	18	29	285	78	290	846
1990	175	12	20	38	277	74	189	786

From ACC/SCN, Second Report on the World Nutrition Situation; *Volume 1 Global and Regional Results* (Geneva: ACC/SCN, 1992).

TABLE 7.1

The proportion and number of chronically underfed people worldwide

In contrast, China, East Asia, and the Near East have seen striking improvements. Changing political structures and reforms in economic policy are thought to be responsible for these dramatic shifts (Uvin 1994).

In spite of these improvements, the sad fact remains that one in five people worldwide are still chronically undernourished—even though the world produces enough food to feed its population an adequate, mostly vegetarian diet (see Table 7.2). In certain regions, such as South America and the Near East, there is abundant food to supply the entire population with a full and balanced diet, yet millions go hungry (Uvin 1994).

Young children are particularly vulnerable to the devastating effects of hunger. Thirty-five to forty-five percent of the world's children under 6 years old suffer from varying degrees of undernutrition and its consequences: growth retardation, decreased resistance to infection and disease, impaired learning ability, and increased mortality (Allen 1993).

	Numbers of People Potentially Supported by 1992 Global Food Supply with Different Diets
Basic diet (100% vegetarian)	6.3 billion (115% of world population)
Improved diet (85% vegetarian, 15% animal-derived)	4.2 billion (77% of world population)
Full-but-healthy diet (75% vegetarian, 25% animal-derived)	3.2 billion (59% of world population)

Adapted from P. Uvin, "The State of World Hunger," *Nutr Rev* 52, no.5 (1994):151–61.

TABLE 7.2

World food supply with different diets

In many developing countries, undernutrition and its sequela are the major causes of death in childhood (Pelletier 1994). Although the majority of malnourished children in the world live in the developing countries, child hunger and undernutrition persist in the United States. This is particularly true among the poor and certain minority groups.

NUTRITIONAL STATUS OF CHILDREN IN THE UNITED STATES

Studies examining the adequacy of diets of children in the United States have found that, in general, intake of most nutrients meets or exceeds recommended levels . However, several nutrients are often consumed in low amounts by children. Nearly half of all children consume less than 70% of the RDA for iron and zinc. The calcium, vitamin B_6, and folate content of many children's diets is also inadequate (see Table 7.3) (U.S. Dept. of Agriculture). Surveys examining the nutritional status of American children have revealed no signs or symptoms of frank dietary deficiencies in the majority of children, with the exception of iron (LSRO 1985) (see later discussion in this chapter).

POVERTY AND THE UNDERNOURISHED

Although most children in the United States are well nourished, a large number of young children among low-income and certain minority groups continue to suffer from undernutrition.

Socioeconomic status is an important determinant of nutritional adequacy in children, with children from families living below the

Nutrient	Percentage of Children with Low Intake
Vitamins	
Vitamin C	11
Folic acid	15
Vitamin B_6	15
Minerals	
Calcium	22
Iron	53
Zinc	46
Magnesium	14
Dietary fiber	50

Adapted from Nationwide Food Consumption Survey, Nutrition Monitoring Service, NFCS, CSF11 (Washington, D.C: U.S. Department of Agriculture, 1988).

TABLE 7.3
Percentage of children in the US with intakes below 70% of RDAs of selected nutrients

poverty line more likely to be undernourished (Miller and Korenman 1994). Over 20% of children in the United States live in poverty with Black and Hispanic children three times more likely to live in poverty than White children (U.S. Dept. of Health 1989). Inadequate intakes of iron, zinc, and vitamins C, B_6, and folic acid are more common in poor children, and iron-deficiency anemia is much more common in children from lower socioeconomic levels (particularly Black children). Although the overall iron status of children in the United States has improved in the past 15 years, the prevalence of iron-deficiency anemia in poor children in many parts of the country continues to be greater than 20% (Centers for Disease Control 1992).

Poverty is associated with impaired growth in children, and the impact can be severe—particularly in minority groups at high nutritional risk. Studies of low-income populations have found a significant number of children underweight or **stunted** (Alvarez et al. 1984). See Figure 7.1.

A recent nationwide survey by the Centers for Disease Control (CDC) found that low socioeconomic status did not significantly increase risk for underweight, thus suggesting that acute or severe malnutrition is rare. However, children from low-income families did have high rates of stunting thought to be caused by nutritional inadequacy and an increased frequency of infections. For children from 2 to 5 years, those from poor Asian families have the highest rates of stunting—over 12% (Centers for Disease Control 1992). See Table 7.4.

Homelessness

In recent years, homelessness has become a major problem in the United States, particularly in cities. The fastest growing segment within the homeless are families with children, making up approximately 34%

stunting: an abnormal pattern of growth characterized by very low weight and height for age, but with weight appropriate for height.

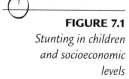

FIGURE 7.1

Stunting in children and socioeconomic levels

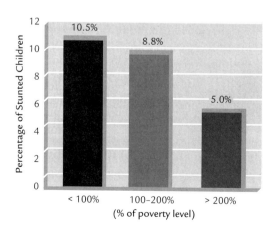

Source: Adapted from Massachusetts Department of Public Health, The Massachusetts Nutrition Survey, 1983.

Race/Ethnicity	Prevalence (%)
White	7.7
Black	5.3
Hispanic	7.6
Native American	5.7
Asian	12.2

From Centers for Disease Control Prevention, "Pediatric Nutrition Surveillance System-US, 1980–1991," *MMWR* 41, no.SS-7 (1992):1–24.

TABLE 7.4
The prevalence of stunting among children 2 to 5 years of age from low-income households, by ethnicity

of the homeless population (U.S Conference of Mayors 1990). It is estimated there are over 100,000 children living on the streets or in shelters, and nearly half of them are less than 6 years old (U.S. General Accounting Office 1989).

Homeless children are at high risk for nutritional problems. They tend to have low intakes of iron and zinc and high rates of iron-deficiency anemia (Taylor and Koblinsky 1993), (Drake 1992). Moreover, homelessness appears to increase nutritional risk beyond that of poverty alone. A recent study from New York City found that homeless children had higher rates of growth stunting when compared to domiciled children at comparable poverty levels (Fierman 1991). Even after controlling for other potential influences on growth, homeless children tended to have lower height for age with preservation of normal weight for height—a pattern consistent with exposure to mild to moderate undernutrition. Greater stunting was found in children from single-parent families and those with large numbers of children (Fierman 1991).

GROWTH AND DEVELOPMENT DURING CHILDHOOD

Although an individual's maximum potential size is genetically determined, nutrition during the growth years has a major influence on whether this potential is achieved

HEIGHT AND WEIGHT

pubertal growth spurt: the rapid increase in height and weight characteristic of puberty

Rapid growth in the first 12 months of life triples the birthweight and increases the infant's height by nearly 50%. The growth rate slows markedly during the second year, with the average height increasing only 12 to 13 cm, and the average weight increasing about 25%. After the second year the growth rate levels off even more. Childhood is a period of slow and steady growth between the explosive growth of infancy and the acceleration of the **pubertal growth spurt**. See Figure 7.2.

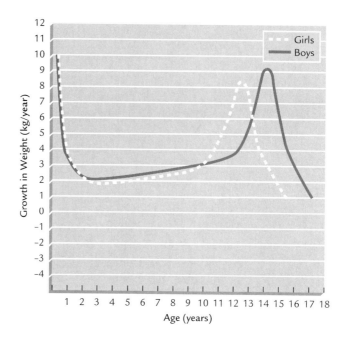

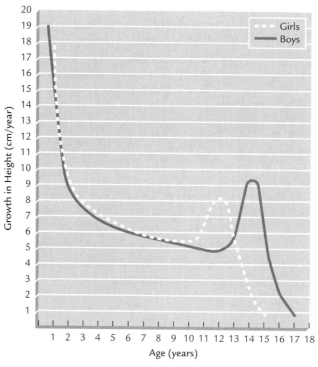

FIGURE 7.2
Growth velocity during childhood and adolescence

During the preschool years (ages 2 to 5) the average weight gain is 5 to 6 lb per year; it increases slightly to 7 lb per year from ages 5 to 10. Height increases at 2½ to 3 inches per year from age 2 to 10 years, with a doubling of birth length at about age 4.

Gender Differences

At around age 6, gender differences in body size gradually appear, with 6-year-old males being slightly taller and heavier than females at the same age. Females generally begin pubertal growth earlier than males, so by age 9 females are as tall as males and generally heavier (Tanner 1990).

Ethnic Differences

In the United States, Black infants have smaller birthweights than White infants, but they grow more rapidly until the end of the second year. From age 2 through adolescence Black children are taller than White children at the same age (Owen and Lubin 1973). The tall stature of school-aged Black children relative to comparable White children is due to differences in the timing and duration of development (Garn et al. 1973). For example, children of African ancestry tend to have more rapid skeletal maturation (Garn and Clark 1975). Asian children in the United States are shorter and lighter than their White and Black counterparts.

Growth Charts

Because growth is a predictable characteristic of normal children, and the rate of growth is sensitive to changes in nutrition, growth patterns during childhood can be used to evaluate nutritional status.

The National Center for Health Statistics (NCHS) has constructed standard growth charts using data from cross-sectional national surveys of large numbers of children in the United States (NCHS 1976). Standard curves show the 5th, 10th, 25th, 50th, 75th, 90th, and 100th **percentile** values for weight and height for age for both sexes. One set of charts is used for birth through 3 years, another from 2 through 18 years. See Figures 7.3 and 7.4.

The charts are used to separate normal from abnormal growth patterns and to draw attention to unusual body size. A single height or weight above or below a certain percentile should not be used to evaluate nutritional status. Useful information on growth rates can be obtained by examining **serial measurements.**

The three basic growth charts—height for age, weight for age, and weight for stature—reveal different things about a child's nutritional status. Because steady linear growth is a good measure of the long-term adequacy of a child's diet, height for age is a good indicator of chronic nutritional status. Weight for height is a better indicator of recent nutrient intake than of long-term changes, because factors slowing or increasing growth affect weight earlier than stature.

Healthy children can be expected to maintain **growth channels** when serial measurements over time are recorded. Normal variation in the size

percentile: the rank position of an individual in a serial array of data, stated in terms of what percentage of the group is equaled or exceeded

serial measurements: measurements taken in a systematic way over a period of time

growth channel: the progressive growth pattern of a child as that is genetically determined but can be influenced by the child's health and nutritional status

Adapted from AAP (1993).

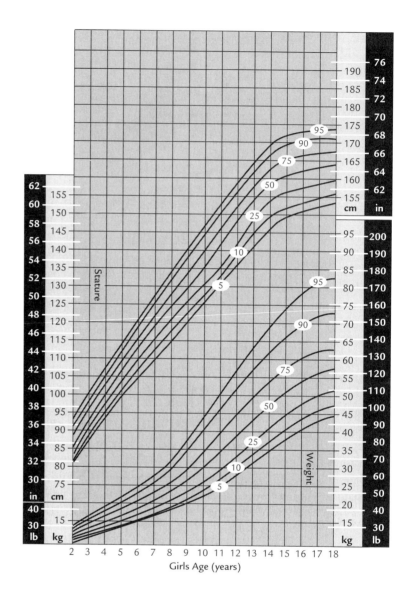

FIGURE 7.3
NCHS Standard Growth Charts: percentiles for weight and stature for age, 2–18 years

of individual children will be apparent in the percentile growth channel that each child follows. For example, a small child making steady progress along the 5th percentile growth channel may be growing normally.

There are no clear-cut guidelines to interpret shifts on growth charts. Short-term changes are less likely to be significant if they occur between the 25th and 75th percentiles. A distinct falloff in linear growth should raise suspicion that a growth disturbance, possibly related to undernutrition, may be present. A child whose weight increases from the 50th to the 90th percentile over a short time is at risk of becoming obese. See Fig-ure 7.5.

Weight for height is a standard criterion for determining obesity, but in some children weight for height may not predict the level of body fat. A British study found that among 4-year-old boys, less than a fifth of those in

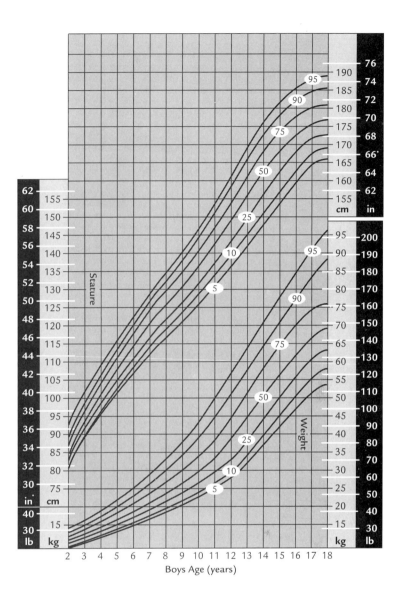

FIGURE 7.3
continued

Boys Age (years)

the top 10% of weight for height were also in the upper decile of fatness, as measured by skinfold thickness (Griffiths et al. 1985). Skinfold measurements or other measures of body fat can be used to distinguish children who are broad and muscular from those who are fat. Measurement of a triceps skinfold is the single most accessible and useful measurement—thickness above the 85th percentile is generally indicative of excess body fat.

Wasting and *stunting* are terms used to describe two patterns of poor growth in children. The child who has very low weight for height is termed wasted. Wasting usually indicates a child who has grown adequately in stature, but has recently been markedly undernourished, resulting in a slowdown in weight gain. A child whose weight and height are both low for age, but whose weight is appropriate for height, is termed stunted.

Adapted from AAP (1993)

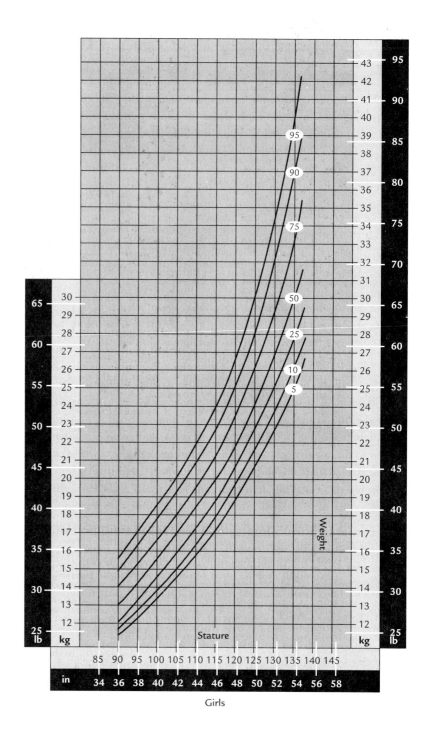

FIGURE 7.4

NCHS Standard Growth Charts: percentiles for weight for stature, prepubescent children

Stunting can be a growth adaptation to long-term undernutrition during childhood. Stunting due to chronic inadequate intake of a generally well-balanced diet is termed *hypocaloric dwarfism*. However, stunting can also occur for other reasons. The child may be genetically short, have insufficient growth hormone, or suffer from a number of other diseases unrelated to nutrition.

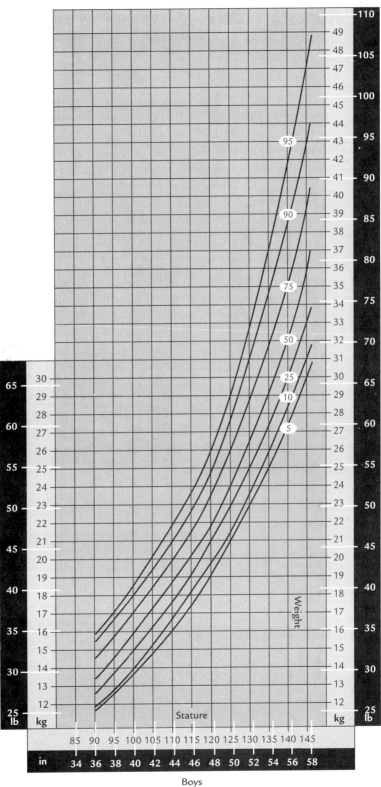

FIGURE 7.4
continued

Boys

Adapted from AAP (1993)

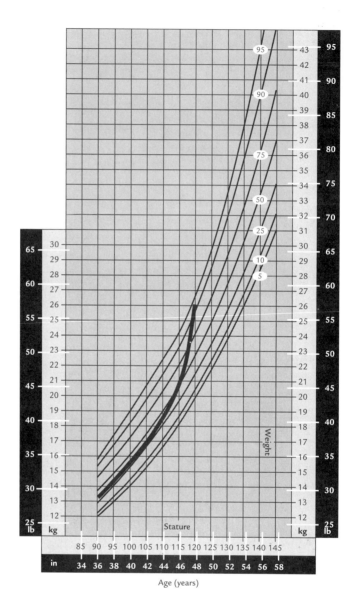

FIGURE 7.5
*Growth chart
showing weight gain
leading to obesity*

A characteristic of childhood growth is the ability to return to the predetermined growth channel after falling out of it because of undernutrition or disease. If a child's diet is inadequate, growth will slow down. But when the child begins to receive adequate food again, growth velocity increases to above normal. Unless the undernutrition has been prolonged or occurred very early in life, the child returns to the original growth curve and maintains it. This rapid growth following a period of growth restriction is called *catch-up growth*. See Figure 7.6.

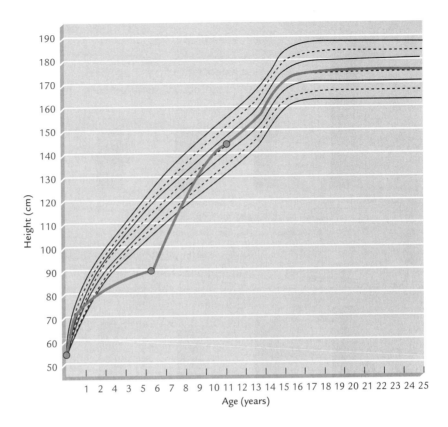

FIGURE 7.6

Catch-up growth during childhood. During the first year, the well-nourished child is growing along the 50th percentile. Growth retardation occurs from 1 to 5 years because of protein-energy mal-nutrition; and nutritional rehabilitation begins at 5. Complete catch-up growth occurs from 5 to 11 as growth accelerates to return to the original percentile and is subsequently maintained.

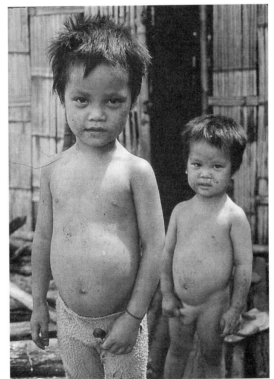

Two children of similar ages; the one on the right shows evidence of marked growth retardation, but weight/height ratio is normal.(Source: J. Lilianfeld/ Custom Medical Stock Photo)

BODY COMPOSITION

In infancy the percentage of body fat is greatest at age 9 to 12 months, when fat makes up about 25% of body mass (AAP 1993). From age 12 months to about 8 years children become leaner as they grow older. Nearly half of all new tissue formed during this period is skeletal muscle, and muscle mass accounts for an increasingly greater proportion of total body weight. At about 8 years, body fat percentage begins to increase (the prepubertal "fat spurt"). See Table 7.5.

ORGAN SYSTEMS

Many organ systems (skeletal muscle, bone, and liver) develop in a general pattern similar to that for height and weight. They develop rapidly during infancy and then grow at slower—but steady—rates until the adolescent growth spurt. However, growth of several organ systems differs from this pattern. Growth of neural tissue is rapid during infancy and is nearly complete before adolescence—brain growth is 75–80% complete by age 2, and by age 10 the brain has achieved 95% of its adult weight. In contrast, genital tissues do not begin rapid development until adolescence. Lymphoid tissues (tonsils, thymus, spleen) develop rapidly through preadolescence and begin to involute during adolescence (Tanner 1990).

	Male			Females		
Age	LBM (kg)	Fat (kg)	% Fat	LBM (kg)	Fat (kg)	% Fat
Birth	3.06	0.49	14	2.83	0.49	15
6 mo	6.0	2.0	25	5.3	1.9	26
12 mo	7.9	2.3	22	7.0	2.2	24
2 yr	10.1	2.5	20	9.5	2.4	20
4 yr	14.0	2.7	16	13.2	2.8	18
6 yr	17.9	2.8	14	16.3	3.2	16
8 yr	22.0	3.3	13	20.5	4.3	17
10 yr	27.1	4.3	14	26.2	6.4	20
12 yr	34	8	19	32	10	24
14 yr	45	10	18	38	13	25
16 yr	57	9	14	42	13	24
18 yr	61	9	13	43	13	23

TABLE 7.5
Lean body mass and body fat of children and adolescents

From Committee on Nutrition, American Academy of Pediatrics, *Pediatric Nutrition Handbook*, 3rd ed. (Elk Grove, IL: AAP, 1993).

The Digestive System

The digestive system develops rapidly in early childhood, allowing the growing child to meet increasing nutritional needs. The salivary glands are fully functional by age 2 years. The capacity of the stomach increases throughout childhood, from about 250–300 cc at 1 year old, to 500 cc at 2, to 900 cc toward the end of the first decade. The small intestine—about 3 meters long at birth—doubles in length by the beginning of adolescence.

Pancreatic and intestinal enzyme systems are only partially developed at birth, but they reach adultlike function early in childhood. Pepsin, trypsin, and amylase secretion are fully developed by 2 years old. Development of lipase secretion occurs more slowly, but is complete at 3 to 4 years old (Lebenthal 1983), (Riby 1994). The liver continues to develop the ability to store glycogen and provide glucose between meals.

As the large intestine matures, the ability to absorb water and secretions improves, and the feces become more solid. Regularity in timing and character of defecation is usually established by 2 years old. Further neuromuscular development enables the preschooler to control defecation and become toilet trained. By the end of the preschool period, a child's digestive system is functionally mature.

The Dentition

Calcification of the **primary teeth** begins during the third trimester of pregnancy and eruption begins at 5 to 6 months of age (AAP 1993). By age 3, the 20 primary teeth have erupted. Establishment of the primary dentition and growth of the jaws in early childhood enables the preschool child to masticate and swallow an increasing variety of solid foods . Calcification of permanent teeth begins shortly after birth and continues until the late teens; eruption begins at 6 to 7 years.

The Urinary System

primary teeth: nonpermanent teeth that are shed during childhood and adolescence; also called deciduous teeth

The urinary system matures in early childhood. By age 2 to 3 years the developing kidney is fully able to concentrate and dilute urine to maintain water balance. Dehydration can occur less readily in preschool children than infants because of their greater ability to control water exchange (AAP 1993).

NUTRITIONAL REQUIREMENTS DURING CHILDHOOD

RECOMMENDED DIETARY ALLOWANCES FOR CHILDREN

Table 7.6 gives the RDAs for children (NAS 1989). Preadolescent children are currently divided into three age groups: ages 1–3, 4–6, and 7–10. Note that gender differences in size and body composition are minimal before the onset of pubertal growth, so dietary allowances are identical for both sexes before age 10. It is also important to note that

Recommended Energy Intakes for Children

Age (years)	Weight (kg)	Height (cm)	Energy Allowances/Day (kcal)	Energy Allowances/kg/Day (kcal)
1-3	13	90	1300	102
4-6	20	112	1800	90
7-10	28	132	2000	70

Recommended Dietary Allowances of Protein

Age (years)	Protein(g/kg)	Protein (g/day)
1-3	1.2	16
4-6	1.1	24
7-10	1.0	28

Recommended Dietary Allowances of Vitamins

Vitamin	1-3	4-6	7-10
Fat-Soluble			
Vitamin A (µg RE)	400	500	700
Vitamin D (µg)	10	10	10
Vitamin E (mg α-TE)	6	7	7
Vitamin K (µg)	15	20	30
Water-Soluble			
Thiamin (mg)	0.7	0.9	1.0
Riboflavin (mg)	0.8	1.1	1.2
Niacin (mg)	9	12	13
Vitamin B_6 (mg)	1.0	1.1	1.4
Folate (µg)	50	75	100
Vitamin B_{12} (µg)	0.7	1.0	1.4
Ascorbic acid (mg)	40	45	45

Age (years)

TABLE 7.6
The RDAs for Children

these allowances are calculated to meet the needs of children of average height, weight, and activity. Since individual children of the same age often vary significantly in body size, growth rate, and activity level, there can be a wide range of adequate daily intakes for many nutrients. Recommended allowances for most nutrients increase through this period, reflecting the steady growth in body size and activity.

While the total need for most nutrients increases through childhood, requirements actually decline per unit of body weight (see Figure 7.7). This occurs because the lean body mass of children contains a greater proportion of metabolically active tissue (such as liver and brain) than in adults. In adults the major component of lean body mass is skeletal muscle, which has a lower rate of resting metabolism. Also, nutrient needs per unit body weight fall as the growth rate decelerates through childhood.

TABLE 7.6 *Continued*

Recommended Daily Dietary Allowances for Minerals

Age (years)	Calcium (mg)	Phosphorus (mg)	Iodine (µg)	Iron (mg)	Magnesium (mg)	Zinc (mg)	Selenium (µg)
1-3	800	800	70	10	80	10	20
4-6	800	800	90	10	120	10	20
7-10	800	800	120	10	170	10	30

Estimated Safe and Adequate Daily Dietary Intakes of Selected Trace Minerals for Children

Age (years)	Copper (mg)	Manganese (mg)	Fluoride (mg)	Chromium (µg)	Molybdenum (µg)
1-3	0.7-1.0	1.0-1.5	0.5-1.5	20-80	25-50
4-6	1.0-1.5	1.5-2.0	1.0-2.5	30-120	30-75
7-10	1.0-2.0	2.0-3.0	1.5-2.5	50-200	50-150

Estimated Minimum Requirements of Electrolytes

Age (years)	Sodium (mg)	Potassium (mg)	Chloride (mg)
2-5	300	1400	500
6-9	400	1600	600
10-18	500	2000	750

From: *Recommended Dietary Allowances;* 10 ed., 1989 National Academy of Sciences. Published by National Academy Press, Washington, D. C..

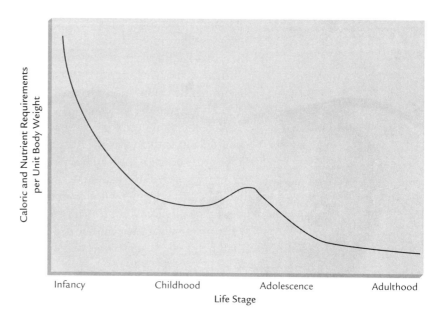

FIGURE 7.7

*Changes in energy
and nutrient needs from
infancy to adulthood, per
unit body weight*

ENERGY NEEDS

Adequate energy is of prime importance during childhood growth. The U.S. recommended energy allowances during childhood, shown in Table 7.6, are estimated from reported energy intakes during childhood associated with satisfactory growth (NAS 1989). Recommendations are based on estimates of average energy intake related to average body weights for the population.

A child's need for energy is determined by resting energy expenditure (REE), level of activity, and needs for growth. REE varies primarily with the amount of lean body mass. Because gender differences in body composition are minimal until children reach about age 10, the REE for males and females during childhood is similar. See Table 7.7. Activity levels vary considerably among children and in individual children from day to day. If not constrained by illness or the environment, children are usually very active, and their energy expenditure of activity is typically 1.7 to 2.0 times the REE (NAS 1989).

During growth, energy is required for the synthesis of new tissue, and the average energy cost of new tissue is about 5 kcal/gram (Roberts and Young 1988). In contrast to early infancy (when the energy required for rapid growth is nearly a third of the total energy requirement), during childhood, energy for growth is only a small component of the total energy requirement (about 1–2%) (AAP 1993). See Figure 7.8. Because growth needs, activity levels, and the REE of male and female children are similar, no distinction in the energy allowance is made between sexes before age 10. Gender differences in energy needs only become significant in adolescence.

		Resting Energy Expenditure	
Weight (kg)	Male	Kilocalories/24 hr Male and Female	Female
3		140	
5		270	
7		400	
9		500	
11		600	
13		650	
15		710	
17		780	
19		830	
21		880	
25	1020		960
29	1120		1040
33	1210		1120
37	1300		1190
41	1350		1260
45	1410		1320
49	1470		1380

TABLE 7.7

Resting energy expenditure during childhood and adolescence

Adapted from R. E. Behrman nad V.C. Vaughn, eds., *Nelson Textbook of Pediatrics,* 14th ed. (Philadelphia: W. B. Saunders, 1992).

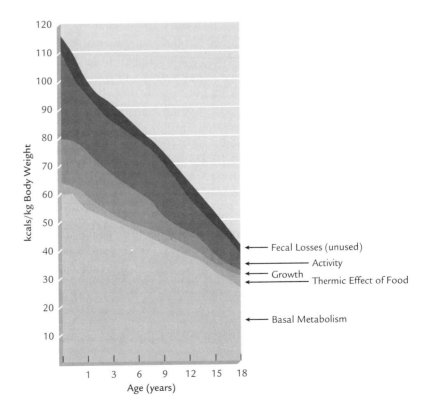

FIGURE 7.8

Apportionment of daily caloric expenditure (birth to 18 years)

Both the 1977–78 Nationwide Consumption Survey (USDA 1984) and the 1976–80 National Health and Nutrition Examination Survey (NCHS 1979) showed that the energy intakes of most children in the United States are adequate to meet their needs. A child's appetite is usually a good measure of appropriate energy intake to meet nutritional demands, but to ensure that energy needs are met, growth should be monitored in all school-age children. Adequacy of intake should be based on satisfactory growth as determined from growth charts and, if indicated, measurement of body fat.

PROTEIN NEEDS

Protein synthesis and deposition into new tissue are characteristics of childhood growth. During the explosive growth of infancy, nitrogen retention is high, approaching 200 mg/kg/day during the first few months of life (Fomon, 1993). As growth slows in the preschool years, nitrogen retention falls, and by age 4 years old it is approximately 11 mg/kg/day (WHO 1985). Estimates of protein needs for growth are 1–4 g of protein/kg of new tissue.

Protein needs for children are calculated based on maintenance requirements, changes in body size and composition, and growth rates. Childhood allowances are then estimated assuming that variability in protein requirements among individual children is similar to that of other age groups. Although the overall need for protein increases during childhood as body mass increases, there is a slow decrease in protein needs relative to weight during childhood, from 1.2 g/kg at 2 years old to 1.0 g/kg at 10 years (NAS 1989). Recommended intakes for protein during childhood are listed in Table 7.6.

Most children in the United States have no trouble obtaining adequate amounts of daily protein, and the typical childhood diet is rich in protein (USDA 1984). Three cups of milk a day alone provide about 24 grams of protein (the RDA for protein during childhood is 24–28 grams).

Evaluation of protein intake during childhood should consider not only the quantity but also the quality of the dietary protein. Appropriate adjustments should be made for dietary proteins of lower quality than reference proteins. Judgments on the adequacy of protein intake in children should be based on satisfactory growth rate as determined from growth charts.

FAT REQUIREMENTS

A major source of energy during childhood, fat provides 36–38% of total energy in the diet. Because children have small appetites, fats are important as concentrated sources of calories for growth. In response to concerns about dietary fat as a risk factor for heart disease, the American Academy of Pediatrics (AAP) has made recommendations on fat intake during childhood. For children older than 2 years dietary fat should provide 30% of total calories, and saturated fat less than 10%. Dietary cholesterol should be less than 300 mg per day (AAP 1993). See discussion later in this chapter.

The AAP recommends that approximately 3% of total energy come from the essential fatty acids, linoleic acid and alpha-linolenic acid (AAP 1993).

CARBOHYDRATE REQUIREMENTS

The relative requirements of children for carbohydrate do not differ from those of adults: generally about 40–60% of total calories should come from carbohydrates (AAP 1993). Carbohydrates provide a readily available source of energy, and children should be provided ample complex carbohydrates from a varied selection of whole-grain cereals, legumes, and fruits and vegetables.

Many children have a preference for sweet, carbohydrate-rich foods, but overconsumption of foods high in sugar increases the risk for dental caries (Gustafsson et al. 1954). Decreasing refined sugar intake during childhood can be difficult. Sugar is often added to foods popular with children.

Although use of artificial sweeteners has increased significantly in children's diets in the past decade, sugar consumption by children has not decreased. Children tend to eat slightly more added sugar than adults (14% vs. 11% of dietary energy, respectively) (American Dietetic Assoc. 1993). Families trying to reduce sugar consumption should set moderate goals, as rigorous elimination of sugar-containing foods from a child's diet without adequate energy substitution may lead to hypocaloric intake and poor growth.

FIBER REQUIREMENTS

The intake of dietary fiber is low in many children in the United States: surveys show that as many as 75% of children eat fewer fruits and vegetables than is currently recommended (AAP 1993). Some groups of experts have recommended an increase in fiber consumption for children, citing studies showing that including adequate dietary fiber in a child's diet may lessen constipation and lower chances of becoming obese (AAP 1993). Other groups have raised concerns about increasing fiber intake in children. Children have small stomachs, and foods high in fiber are bulky and often low in calories, so children on high-fiber diets may be unable to consume adequate calories for normal growth. Increasing fiber intake in the school-age child may produce abdominal pain, bloating, and flatulence. Also, high amounts of dietary fiber may interfere with the absorption of nutrients such as zinc and magnesium (LSRO 1987).

The AAP currently recommends that children's diets should provide modest amounts of fiber by including whole-grain cereals and breads, fruit, and vegetables. An unbalanced diet that emphasizes high-fiber, low-calorie foods should be avoided during childhood.

VITAMIN REQUIREMENTS

Fat-Soluble Vitamins

Vitamin A plays a central role in cellular growth and differentiation during development, particularly in epithelial tissues (Sporn et al. 1994). Adequate intake is important during childhood growth. The RDAs for vitamin A for children are extrapolated from the adult level based on body weight. The allowance increases from 400 RE at age 2 to 700 RE at age 10 for both sexes (NAS 1989).

Normal skeletal growth during childhood requires adequate vitamin D. The RDA for vitamin D for children older than 6 months is set at 10 µg (400 IU) (NAS 1989). In the United States, because of the fortification of foods with Vitamin D, most children achieve intakes at or above the RDA.

Requirements for vitamin E increase with increasing body weight during childhood growth. Allowances increase from 6 mg at age 2 to 7 mg at age 10. There are no specific data regarding the vitamin K requirement during childhood. The RDA for vitamin K is set at approximately 1 µg/kg body weight (NAS 1989).

Water-Soluble Vitamins

Thiamin, riboflavin, and niacin are important in energy metabolism. Few studies have been conducted to determine whether children have specific requirements for these vitamins. Available information suggests that intakes of thiamin, riboflavin, and niacin based on energy intake (similar to those used for adults) are adequate during childhood. Because energy intake steadily increases during childhood growth, the RDAs for thiamin (0.5 mg/1000 kcal), riboflavin (0.6 mg/1000 kcal), and niacin (6.6 NE/1000 kcal) rise proportionally (NAS 1989).

Because of its central role in protein utilization and synthesis, pyridoxine is required in increasing amounts during childhood growth. The RDA for vitamin B_6 is based on protein intake (0.02 mg/g protein). Allowances increase during childhood from 1.0 mg at age 2 to 1.4 mg at age 10.

Synthesis of large amounts of new blood cells in a growing vascular system requires ample dietary folate and vitamin B_{12}. Because there are few specific data on requirements for these vitamins in children, the RDAs for folate and vitamin B_{12} are interpolated from the allowances for adults (NAS 1989). Recommended intakes increase steadily through childhood based on increasing body weight.

Because of its central role in collagen synthesis, adequate vitamin C is necessary for optimal growth and development of supporting tissues, including cartilage, bone, and the connective tissue in skin and blood vessels. There is little data on ascorbic acid requirements in children. Intakes of as little as 10 mg per day protect infants from scurvy; the RDA beyond 6 months of age is gradually increased to the adult level. For children aged 1 to 3 years, the RDA is 40 mg; for ages 3 through 10 it increases to 45 mg (NAS 1989).

MINERAL REQUIREMENTS

Major Minerals

Calcium and Phosphorus Although skeletal growth during childhood requires a strong positive balance of calcium and phosphorus, specific data on calcium and phosphorus needs during childhood are lacking. The RDA for calcium during childhood has been estimated to be 800 mg/day for ages 1 through 10. The RDA subcommittee has estimated that a 1:1 ratio of phosphorus to calcium will provide sufficient phosphorus during childhood; therefore, the phosphorus allowance is also set at 800 mg for ages 1 through 10 (NAS 1989).

Milk supplies most of the calcium and phosphorus consumed by children in the United States. However, certain children (many Blacks, Asian Americans, and American Indians) can drink little or no milk because of intestinal lactase deficiency (Riby 1994). Diets for these children should include other sources of calcium, such as dark green leafy vegetables and sesame seeds (see table 7.17). Fermented milk products, such as yogurt, often are better tolerated, and are rich sources of calcium.

Magnesium The RDA for magnesium during childhood is estimated to be 6 mg/kg per day. The allowance increases from 80 mg at age 2 to 170 mg at age 10 (NAS 1989).

Iron   Children require ample iron to synthesize hemoglobin during steady growth of the red cell mass and for myoglobin synthesis in developing skeletal muscle (Oski 1993). To build body iron stores to a target iron storage level of 300 mg by age 20, the RDA for iron during the years 2 through 10 is set at 10 mg/day for both sexes (NAS 1989).

The average diet in the United States provides about 6 mg iron per 1000 kcal (NAS 1989). But children's diets are often lower in iron because of limited intake of iron-rich foods such as meat and eggs. Iron deficiency is the most common nutritional deficiency in children in the United States, particularly those under 3 years old (AAP 1993).

Zinc   Ample dietary zinc is important in protein synthesis and normal growth, and the RDA is set at 10 mg/day. Severe zinc deficiency during childhood and adolescence causes marked stunting of growth and delayed sexual development (Halsted 1972). Even marginal zinc deficiency during childhood may slow growth.

A study in the United States found low plasma and hair zinc levels in children whose zinc intake was 5–6 mg/day. These children had heights below the 10th percentile for their age. Their rate of linear growth improved markedly when they received additional dietary zinc (approximately 10 mg/day) (Walravens et al. 1983). Children of low-income families are at an increased risk for suboptimal zinc nutrition because of low dietary intake of meat, fish, and whole grains.

Iodine   Because there are few studies examining iodine needs during childhood, relative energy requirements are used to set the iodine allowance for children. The RDA for iodine during childhood increases from 70 μg at age 2 to 120 μg at age 10 (NAS 1989).

Although the iodine intake of the large majority of children in the United States is considered to be adequate, worldwide, iodine deficiency has adverse effects on the growth and development of millions of children (MDIS 1993). See the discussion later in this chapter.

Selenium   Little is known about selenium needs during childhood. Allowances for selenium have been extrapolated from adult values on the basis of body weight, with an additional amount estimated for growth requirements. For ages 1 through 6, the RDA is set at 20 μg; for ages 7 to 10 it increases to 30 μg (NAS 1989).

Trace Minerals

Very little is known about childhood requirements for many of the trace minerals (Milner 1990). The RDA committee of the National Research Council has estimated ranges of safe and adequate dietary intakes for copper, manganese, fluoride, chromium, and molybdenum in childhood (NAS 1989). See Table 7.6. Since the toxic levels for many trace minerals, particularly in children, may be only several times usual intakes, the RDA

Age	Body Weight (kg)	Water Requirement (ml/kg)
10 days	3.2	100–150
1 yr	9.5	120–140
6 yr	20.0	90–100
12 yr	38.0	60–80
Adult	70.0	20–40

TABLE 7.8

Daily water requirements

committee has recommended that the upper levels of intake for these trace elements not be routinely exceeded.

WATER AND ELECTROLYTES

During infancy, daily turnover of water is rapid (about 15-20% of total body water is taken in from food and water and excreted each day). As children grow, this decreases steadily, and by early adolescence water exchange is similar to adult rates (about 5%).

As water exchange slows through childhood, water requirements per kilogram of body weight fall steadily, from 120–140 ml/kg at age 1 to 60–80 ml/kg at age 12. Because of increasing body size, total water requirements increase, on average, from about 1200–1400 ml/day at 1 year to about 2800 ml/day at 12 years. See Table 7.8.

Control of water balance is less precise in early childhood, when water exchange rates and water requirements per kilogram of body weight are higher. Consequently, the younger child is more prone to dehydration than the older child and adult. Parents need to pay special attention to the substantial water needs of infants and young children.

Along with the increase in total water requirement, electrolyte needs increase steadily during childhood growth (NAS 1989). Estimated minimum requirements during childhood for sodium, potassium, and chloride are shown in Table 7.6.

FEEDING SKILLS

Children progress from bottles and pureed baby food to eating from the regular family menu at differing rates, with some beginning as early as 10 months and some as late as 2 years. By age 1 most infants have developed a coordinated pincer grasp, and finger-feeding becomes common and easy. About midway through the second year, children begin to scoop food into a spoon, losing much of the food because they lack wrist control. Later in the second year, the coordination of the elbow and wrist allows smooth transfer of the spoon (and its contents) to the mouth.

2 Years	2½ Years
· Holds small glass in one hand and drinks, moderate spilling. · Inserts spoon in mouth without turning upside-down, moderate spilling. · Can use straw.	· Moderate spilling · Distinguishes between finger and spoon food. · Chews with mouth closed. · Uses fork, but held in fist.

3 Years	4 Years	5 Years
· Minimal spilling from cup, glass, and spoon. · Rotary action of jaws to chew food.	· Manages spoon with liquids and solids with little spilling. · Swallows food before speaking. · Eats with fork held in fingers.	· Chooses fork over spoon when appropriate.

6 Years	School-Age
· Spreads with knife.	· Cuts with knife: 7 years. · Can prepare simple food with help (e.g., eggs, popcorn, jello): 9–10 years.

TABLE 7.9

Development of feeding skills

Although most 2-year-olds can move a cup steadily without spilling the contents and can handle a spoon, food is generally transferred from the plate to the spoon by fingers. Around age 4, most children begin cutting their foods and, by age 5, most can handle a knife and fork as well as the spoon.

Rotary chewing movements begin around 12 months, as primary dentition is established. The ability to chew hard, brittle, or fibrous food increases as the permanent teeth develop during the school years. A summary of feeding skill development is shown in Table 7.9.

Food should be prepared with the aim of supporting the development of self-feeding skills, and until the child develops the dexterity and motor skills to manage utensils, food should be served in a way that enables them to learn feeding skills without great anxiety or frustration. Most foods served to 2- and 3-year-olds should be divided into bite-sized pieces and prepared so they can be eaten—if necessary—with the fingers. Younger children quickly lose patience with foods that are too small to eat easily (such as peas), or those requiring trimming or cutting. Child-sized utensils and a cup that is easy to grasp should be required equipment for meals.

PRESCHOOL YEARS: AGES TWO TO FIVE

CHANGING FAMILY STRUCTURES

Over the past 25 years, there have been striking changes in family structure and child care in the United States. Nearly one in four children now live in single-parent families. Almost all single-parent families consist of children living with their mothers, and nearly half of them are living in poverty (Allen 1994). In addition, many more mothers are working outside the home, with half of the mothers in the United States with preschool age children and nearly two thirds with school-age children in the labor force (Hofferth and Phillips 1987). Demands for child-care services have increased substantially: nearly 25 million children required day care in 1994.

The nutritional implications of these societal changes are wide-ranging. With more mothers working outside the home, there has been a shift in responsibility for feeding children to nonfamily child-care workers. Concerns have been expressed about the child-care providers' knowledge of nutrition, the quality and adequacy of food provided, and the need to establish nutrition standards for day-care programs. Federal food assistance programs now provide meals to nearly 2 million children in day care (American Dietetic Association 1994).

Changing family structures also influence the nutrition of children at home. Working parents may lack the time, interest, and skills to prepare meals for the family. More families are depending on take-out and convenience foods—meals that generally are nutritionally imbalanced and high in fat, salt, and sugar (Young et al. 1986).

EATING HABITS AND SCHEDULES

Children who have had vigorous appetites as infants develop a reduced appetite, usually soon after their first birthday. Appetite slows because nutritional demands fall as the growth rate decelerates during the second and third years of life. This normal pattern is a concern for many parents who think their child's appetite is poor, even though the child's diet is meeting all dietary requirements.

The average preschool-age child stops eating iron-fortified infant cereals, drinks less milk than during infancy, and often refuses vegetables. As a result, intakes of calcium, riboflavin, iron, and vitamin A usually fall. At 2 to 3 years of age, as appetite slows, families need to pay more attention to selecting varied and nutrient-dense foods so that total nutrient intake remains adequate.

Normal fluctuations in appetite among children are often a cause of unnecessary concern for families. Excitement, fatigue, and emotional state all affect appetite in the normal, healthy child, and this should be respected by parents. Children can recognize hunger and satiety, and a child's appetite is a good measure of appropriate energy intake to meet nutritional demands.

In a recent study in preschool children researchers looked at energy intake of children who self-selected foods and amounts of food from a menu. Children were offered a wide variety of foods—including snack foods such as cookies and potato chips—in addition to standard meal-time foods. Although there was considerable day-to-day variability in the type and quantity of food consumed, the children accurately self-selected appropriate caloric intake when all the study days were averaged (Birch et al. 1991). It appears that normal preschool children can regulate energy intake on their own— despite meal to meal variability—and maintain energy balance.

FOOD PREFERENCES

By 4 or 5 years of age, most children have established a wide range of food preferences and aversions (Hammer 1992). Food preferences and habits are shaped by what type of food the child is offered, how it is offered, and parental and peer attitudes toward foods (Birch 1987). Many children become "picky" eaters, and they will reject food not only for its unpleasant taste, but also because they perceive certain foods as offensive or disgusting, or potentially dangerous (Fallon et al. 1984). Preschoolers generally prefer carbohydrate-rich foods that are easy to chew and swallow—foods such as cereals, breads, and crackers. Most also like milk, fresh fruit, and fruit-flavored beverages. Easy-to-chew meats, such as ground meats, are popular, as well as cheese and yogurt. Vegetables are typically the least preferred food group.

INTRODUCING NEW FOODS AND ENCOURAGING HEALTHY EATING

Building good food habits and introducing new foods to children requires perseverance, as the appetites of young children are unpredictable, and their food likes and dislikes change quickly. However, preschool children will usually eat, enjoy, and develop preferences for what is served to them regularly. There are several ways to introduce new foods and encourage healthy eating in young children (Staneck et al. 1990), (Birch 1987):

■ Eat in pleasant surroundings, don't rush, be affectionate, and pay attention to the child while he or she is eating. Companionship and support at mealtime are important.

- Maintain a reasonably consistent eating schedule by providing meals and snacks at predictable times. A schedule helps the child to come to a meal hungry and makes it more likely that the food will be accepted. Providing food at three-to four-hour intervals is often effective.

- Don't force children to eat. Offer a variety of nutritious foods at each meal: milk; bread, potato, or rice; fruit or vegetable; and a main dish. Then allow children to choose what they like. If they appear uninterested, don't jump up and prepare something else that they predictably eat. If they climb down from the table without eating and then return soon after asking for a snack, be firm, and wait until the next scheduled snacktime. This will reinforce the value of eating at mealtimes and teach the child to take the meal more seriously.

- Avoid using rewards such as "Eat your vegetables and you can have dessert." Although this approach may achieve short-term results, children recognize that external pressure is being applied to get them to eat a less-preferred food. They will eat little of that food if the reward is removed.

- Problems occur when children learn to use food to manipulate their parents. This most often happens when food is offered regularly as a diversion or pacifier, or as a reward for other behavior. On the other hand, presenting an unfamiliar food as a reward, or with the reinforcement of adult attention, may increase the child's preference for the new food.

- When introducing new foods to a preschooler, try the "one-bite policy" (at least one bite of a new food should be tried). Introduce new foods at the beginning of meals, when the child is most hungry. Regularly repeat exposure to the new food and be patient, as most foods will eventually be accepted.

- Large portions discourage children. It is usually better to offer less than children usually eat, and offer seconds if they ask for them. The review and summary at the end of this chapter contains a table that lists appropriate portions for children of different ages.

- Role models are important in acceptance of a new food. If parents and older siblings are seen enjoying a new food, children are more likely to try it. Parental food preferences have a strong influence on preschoolers' food choices (Klesges et al. 1991).

- Allow the child to participate in food preparation. This may encourage interest in new foods.

Children are picky about food temperature, taste, texture, and appearance. Many children prefer their food at room temperature and dislike very hot or cold foods. Young children tend to like foods with mild flavors and, in general, like food only lightly salted. Many find crisp-textured foods fun to eat. So that children can develop chewing skills but not have too much to chew at one sitting, soft-textured foods should be provided along with chewy foods. Children may prefer meals that include colorful foods, such as carrots, green peppers, and fresh fruit.

If a child refuses to eat vegetables, try mixing them with familiar foods or serving them with flavorful, nutritious dips. At about age 4 to 5, children should be able to eat raw vegetables such as carrots, broccoli, and celery sticks. Preschool children generally prefer raw vegetables to cooked vegetables.

SNACKS AND FOOD JAGS

Young children have a small stomach capacity and cannot eat large meals. During periods of high nutritional need, it may be difficult for a child to get adequate calories from only three meals a day. Scheduling several small meals and snacks throughout the day is a good plan for most children. The important thing is not when a child eats, but what is eaten. Several nutritious options should be available at snacktime (see Table 7.10), and the child should be allowed to choose among them. Snacks are important sources of nutrients in children's diets.

In preschool children, snacks often supply a significant portion of daily energy, calcium, riboflavin, and vitamin C needs. Studies in 10-year-old children have shown that snacks are even more important contributors to overall nutrition—supplying a third of the daily energy and fat intake, 20% of the daily protein, and over 40% of the daily carbohydrate for many children (Farris et al. 1986).

Food jags are patterns of eating in which a very few food items are eaten to the exclusion of all others. They are very common in preschoolers. During a food jag, a child will ask repeatedly for the same food at

- Milk
- Fruit shake (milk blended with fresh fruit)
- Fresh raw vegetables, served with a dip of yogurt or cottage cheese
- Fresh fruit
- Frozen fruit cubes (freeze pureed fruit or apple sauce)
- Gelatin with added fruit, vegetables, or cottage cheese
- Whole-grain cereals and milk
- Graham crackers
- Sandwiches, tortillas, or bagels with fillings such as peanut butter, banana, lean meat, cottage cheese, shredded carrots, tuna fish

TABLE 7.10
*Nutritious snacks
and beverages*

- Potato, corn, and tortilla chips
- Popcorn
- Nuts and seeds
- Hot dogs
- Raw vegetables
- Small fruits such as grapes
- Dried fruit: raisins, dates, apricots
- Hard candy

TABLE 7.11
Inappropriate foods: choking prevention in preschoolers

meals and snacks. For example, a child may demand a banana with every snack and meal for several weeks and then completely reject bananas for the next few months. As long as the preferred food is nutritionally adequate, the temporary monotony of the diet should be of little concern. There is no evidence that food jags during childhood lead to limited food choices in later life.

PREVENTION OF CHOKING

Young children, particularly preschoolers, are at risk from choking on food. Choking and death by **asphyxiation** most often occur in children under 2 years old. Foods most often implicated are hard candy, grapes, nuts, chips, and hot dogs (AAP 1993). Certain foods are not appropriate for preschoolers (see Table 7.11). An adult should always be present when a young child is eating, and children should be encouraged to sit down when snacking. Choking is more likely to occur when a child is eating while running or playing.

SCHOOL YEARS: AGES FIVE TO TEN

NUTRITION AT SCHOOL

School provides an ordered routine, and meal and snack patterns become more predictable for older children. For many, school lunch and breakfast programs contribute significantly to nutrient intake. School-age children begin to make more of their own food choices, as many have spending money and are increasingly able to choose foods at vending machines and grocery stores without parental guidance. Sweetened beverages, cookies, cakes, and other foods of poor nutritional

asphyxiation: suffocation; deprivation of oxygen

value are commonly chosen. At home, many children become responsible for preparing their breakfasts, bag lunches, and after-school snacks.

School Lunch Programs

There are several federally funded food assistance programs for school-age children in the United States. The National School Lunch Program was established by the U.S. Department of Agriculture in 1946 to promote the health of children by providing them with nutritious meals at school. In 1966, the School Breakfast Program was begun in response to widespread concerns that children were coming to school hungry. Another active program is the Child Care Food Program, which provides food to organized child-care programs. Currently, over 25 million students and over 95% of U.S. public schools participate in the School Lunch Program, for which the U.S. government spent $4.5 billion in 1994 (Community Nutrition Institute 1994).

All of these programs provide cash reimbursement and supplemental foods to food service programs that comply with federal regulations. Meals must meet established nutritional guidelines; lunches should provide at least a third of the RDAs for all nutrients, and breakfasts must include milk, fruit or vegetable juice, and bread or cereal (USDA 1982). Table 7.12 gives the components and quantities for school lunches. Also, meals must be served at reduced prices or be free to children of families that cannot afford to buy them. Children from families of four whose household income is less than about $15,000 a year are eligible for free meals.

Along with nutritious food, schools are encouraged to provide education in nutrition so students can begin to make healthy food choices when they are on their own. Federal regulations require that schools involve students and parents in menu planning. By providing school lunches offering several food choices, programs can increase variety and allow students to participate in the selection of a healthy diet.

Children participating in the School Lunch Program tend to have higher nutrient intakes and eat a greater variety of foods at lunch, compared with children who bring lunches from home (Hanes et al. 1984). Research has also indicated that many children participating in school food programs show improvements in learning (Meyers et al. 1989), (Meyers 1989).

There have been concerns about the high total fat, saturated fat, and salt content of many school meals (Walker and Walker 1986), (Farris et al. 1992). Programs are now encouraged to keep salt and sugar content at moderate levels, and reduce fat and cholesterol levels to conform with current American Heart Association recommendations. In many schools, menus have been revised and more lower-fat food options are now being provided (Snyder et al. 1992).

Components		Minimum Quantities: Group I Age 1–2; Preschool	Group II Age 3–4; Preschool	Group III Age 5–8; K–3	Group IV Age 9 and Older; 4–12	Recommended Quantities for Group V 12 Years and Older; 7–12
Milk	Unflavored, fluid low-fat, skim, or buttermilk must be offered	¾ cup (6 fl oz)	¾ cup (6 fl oz)	½ pint (8 fl oz)	½ pint (8 fl oz)	½ pint (8 fl oz)
Meat or meat alternate	Lean meat, poultry, or fish Cheese Large egg Cooked dry beans or peas Peanut butter or an equivalent quantity of any combination of any of above	1 oz 1 oz ½ ¼ cup 2 tbsp	1½ oz 1½ oz ¾ ⅜ cup 3 tbsp	1½ oz ½ oz ¾ ⅜ cup 3 tbsp	2 oz 2 oz 1 ½ cup 4 tbsp	3 oz 3 oz 1½ ¾ cup 6 tbsp
Vegetable or fruit	2 or more servings of vegetable or fruit or both	½ cup	½ cup	½ cup	⅜ cup	⅜ cup
Bread or bread alternate (servings per week)	Must be enriched or whole grain, —at least 1/2 serving for group 1 or one serving for groups II-V must be served (Serving: 1 slice of bread; or 1/2 cup of cooked rice, macaroni, noodles, other pasta products, other cereal product)	5	8	8	8	10

Adapted from Food and Nutrition Service, Department of Agriculture, National School Lunch Program Regulations, Nov. 16, 1982, Federal Register.

TABLE 7.12
National School Lunch Program recommended quantities

EATING PATTERNS IN OLDER CHILDREN

As children move into their school years, many of the feeding problems of the preschooler disappear. Appetites and food preferences are more predictable, and increasing activity stimulates a steady appetite. Differences in intake between males and females become apparent about age 6 to 7— differences that become more marked as children move into adolescence. Males tend to eat more and consume more protein and micronutrients than females. In school-age children in the United States, fat intake is generally about 35–40% of total calories, while sugar contributes approximately 25% of calories (USDA 1987).

Breakfast

Eating breakfast rebuilds glycogen stores depleted during the night and provides energy for morning activities. Skipping breakfast can diminish a child's attention span and morning performance at school, play, or sports (Simeon and Grantham-McGregor 1989). Although most children eat breakfasts that contribute at least a quarter of their daily nutrient needs, about 10 to 30% of school-age children do not eat breakfast, particularly those in the middle and upper grades (Morgan et al. 1981).

The following reasons are often given by children who skip breakfast:

- They are not hungry.

- They have no time.

- They don't like foods served at breakfast.

- The foods they like aren't available at breakfast.

Children should be reminded that breakfast does not have to consist of traditional foods, such as cereal and eggs. They should be encouraged to find nutritious foods they like to eat in the morning (McIntyre 1993).

INFLUENCES ON NUTRITION IN CHILDHOOD

PARENTAL INFLUENCES

The evening meal is a time for family interaction and socializing, and school-age children should be encouraged to participate in meal planning and food preparation. The dinner table should be a place to relax and rest after the hectic school day, and parents should refrain from mealtime criticism of children. Mealtime criticism of children by their parents reduces food consumption and reduces intake of certain nutrients, particularly vitamins A and C (Lund and Burk 1969). Parents should continue to encourage healthy eating habits and food preferences. Eating habits formed during the school years often carry into later life, and they may influence long-term health.

THE IMPACT OF TELEVISION ON CHILDHOOD NUTRITION

Children in the United States spend more time watching television in a year than they spend in the schoolroom. They average over 26 hours a week and watch over 10,000 commercials each year (Strasburger 1989). Television has a significant impact on many children's attitudes toward food and food preferences (AAP 1993), as the hours on television directed toward children's programming feature commercials for children's foods. Foods most commonly advertised during these periods are presweetened breakfast cereals, potato chips, soft drinks, fruit-flavored beverages, candy, and fast foods (Cotugna 1988).

Food advertising on TV aimed at children is effective. Children's TV viewing patterns are closely correlated with their choice of advertised foods, the amount of snacking they do (Dietz and Gortmaker 1985), and the foods and snacks their parents bring home (Crawford et al. 1978).

Consequently, many teachers, families, and pediatricians believe that television advertising aimed at children promoting high-fat, high-sugar foods should be limited. Excessive television viewing encourages a sedentary lifestyle in children and may contribute to obesity.

In a study of children from 6 to 11 years old, the risk of obesity was strongly correlated with the amount of television watched each day (Dietz and Gortmaker 1985). Higher cholesterol levels have been found in children who watch over two hours of TV a day, when compared with their more active counterparts (Wong et al. 1992). Families and schools should encourage school-age children to develop active, imaginative alternatives to television.

CHILDHOOD DIET AND HEALTH

DIET, BEHAVIOR, AND HYPERACTIVITY

Behavior during childhood is often erratic; many children unpredictably become unruly, excitable, or inattentive. Sudden changes may be due to a lack of sleep, lack of physical activity, emotional state, desire for attention, anxiety, or other factors. Nutritional factors can also influence childhood behavior.

Two common disorders of childhood, iron deficiency and lead toxicity (discussed later in this chapter), have detrimental effects on childhood behavior. Timing of meals and snacks can also affect children's behavior and performance at school. Many children who skip breakfast are less able to concentrate at school and have shorter attention spans than children who eat breakfast (Meyers et al. 1989), (Simeon and Grantham-McGregror 1989).

There has been substantial research into whether diet may be involved in hyperactivity in childhood. Hyperactivity (also called attention deficit–hyperactivity disorder, or ADHD) is a childhood behavioral disorder characterized by severe and chronic impulsiveness, inattention, and restlessness. It usually develops before age 7 and affects 3 to 5% of school-age children in the United States. Children usually outgrow the condition during their teen years.

Although the cause of ADHD is not known, most scientists think it results from a combination of genetic and environmental factors. Many different foods and food additives have been blamed for aggravating the disorder, including refined sugar, various food additives, and artificial flavors and colors (McLoughlin and Nall 1989). However, nearly all studies have found that diet has little, if any, role in this baffling disorder (Prinz et al. 1980), (Rosen et al. 1988), (Conners 1991).

Sweeteners

Although parents often blame sugar as a cause of hyperactive behavior in their children, there is no convincing evidence that refined sugars or aspartame cause behavioral problems in children (Kanarek 1994). Both dietary challenge and dietary replacement studies have consistently shown that refined sugar has no measurable adverse effects on behavior. In a recent **double-blind study** (Wolraich et al. 1994), children reported by their parents to be sugar-sensitive were fed large amounts of sucrose, aspartame, or a placebo for three weeks. Neither sucrose nor aspartame affected the children's behavior or function.

Artificial Flavors and Colors

The idea that food additives are a major cause of hyperactivity was popularized in the 1970s in a book by Dr. Benjamin Feingold. Eliminating certain additives by following the Feingold Diet was reported to help most hyperactive children (Feingold 1975). Double-blind studies of diets eliminating artificial colors and flavors have generally found that the great majority of hyperactive children do not benefit. However, a small minority of children with ADHD may benefit from elimination of artificial colors and flavors from their diet (AAP 1993).

Some experts recommend a short-term elimination trial to determine whether the child will respond (Carter et al. 1993), (Kinsbourne 1994). Implementing the diet requires additional work by the family, but it is a nutritionally healthful and adequate diet. If no improvement is seen during the elimination period, the child should resume a normal diet.

Caffeine

Over three quarters of school-age children consume caffeine regularly in food and beverages, and the stimulant effects of caffeine may be more pronounced in children than adults (Arbeit 1988). Chocolate, ice cream, and carbonated beverages often contain caffeine, and some children become inattentive, restless, and have difficulty sleeping after ingesting these products. Also, irregular heartbeats have been reported in children

double-blind study: a study of the effects of a specific agent in which neither the administrator nor the recipient, during the study, knows whether the active or inert substance is given

Item	Caffeine (mg)
Colas (12 oz)	30–50
Iced tea (12 oz)	70
Cocoa beverages (5 oz)	2–20
Chocolate milk beverage (8 oz)	2–7
Milk chocolate candy (1 oz)	1–15
Dark chocolate, semisweet (1 oz)	5–35
Chocolate-flavored syrup (1 oz)	4

TABLE 7.13

Caffeine content of foods and beverages consumed by children

after ingesting caffeine-containing products. Caffeine consumption should be limited during childhood, particularly in those children who appear sensitive to its effects. Table 7.13 lists the caffeine content of some popular children's foods.

VITAMIN AND MINERAL SUPPLEMENTATION

Although use of vitamin–mineral supplements among school-age children is extensive (Kover 1985)—over a third of children under 18 consume supplements regularly—the AAP does not recommend routine supplementation for healthy children (AAP 1993). One exception is the use of fluoride supplements in areas where there is insufficient fluoride in the drinking water (discussed later in this chapter). A vitamin-mineral supplement may be indicated for children in special situations, as discussed in Table 7.14. Supplements should provide nutrients at approximately the RDA levels.

1. Children and adolescents from deprived families, or those who suffer from parental neglect or abuse.

2. Children and adolescents who have anorexia, or those who have poor and capricious appetites, or those who consume fad diets.

3. Children who are on dietary regimens to manage their obesity.

4. Pregnant teenagers.

5. Children and adolescents consuming vegetarian diets without adequate dairy products may need supplementation, particularly with vitamin B_{12}, which is absent from vegetable foods.

6. Children with chronic diseases such as cystic fibrosis, inflammatory bowel disease, liver disease.

Adapted from Committee on Nutrition, American Academy of Pediatrics, *Pediatric Nutrition Handbook*, 3rd ed. (Elk Grove, IL: AAP, 1993).

TABLE 7.14

Recommendations from the AAP: special situations in which vitamin-mineral supplementation may be indicated in children

VEGETARIAN DIETS

Plant-based diets supplemented with milk or with eggs and milk are generally very similar nutritionally to diets containing meat. Children raised on well-balanced vegetarian diets including milk and eggs have demonstrated excellent health (Tayter and Stanek 1989), and such diets may provide some nutritional benefits including rarity of obesity (Labin et al. 1986), and tendency toward lower blood pressure (Beilin et al. 1987). However, such diets may not provide adequate iron for the growing child, and children eating diets without meat are at increased risk of iron-deficiency anemia (AAP 1993). Dietary iron should be emphasized by encouraging the child to eat ample nuts, seeds, and legumes, in addition to milk and eggs.

Although vegetarian diets including milk and eggs can provide adequate nutrition for growing children (ADA 1988), (Committee on Diet and Health 1989), vegan diets—particularly during early childhood—may have significant drawbacks (see Table 7.16). Because vegans exclude all animal products from the diet, combinations of cereals and legumes are necessary to provide ample protein of high biological value (ADA 1988). Plant proteins must be properly combined to ensure that amino acid intake will be adequate to support growth. Table 7.15 lists some typical vegetarian foods and their protein content.

A nutritionally adequate diet can be devised through careful selection and preparation of foods, but the energy density of many vegan diets is low (ADA 1988). Because these diets are bulky, young children may not be able to consume an adequate volume of food for their needs. Inadequate energy or protein may result in failure to grow (Shull et al. 1977).

TABLE 7.15

Protein content in vegetarian food combinations for children

Food	Portion	Usable Protein (gm)
Cooked oatmeal with milk	¾ cup cooked oatmeal with 2 tsps soy grits, ⅓ cup milk	8
Scalloped potatoes	½ cup potatoes, ½ cup milk	4
Macaroni and cheese	½ cup cooked macaroni, 3 tbsp grated cheese	7
Rice pudding	½ cup cooked rice, ¼ cup milk	5
Peanut butter sandwich on whole-wheat bread	½ tbsp peanut butter, 1 slice bread	4
Tortillas and beans	2 tortillas, ¼ cup cooked beans	4
1 whole-wheat soy muffin	2 tbsp whole wheat, 1 tsp soy flour, 1 tbsp milk	3

Dietary Components that May Be Inadequate	Cause of Inadequacy
Energy	Bulkiness, indigestibility, low energy and fat content
Protein	Indigestibility, amino acid imbalance
Iron, calcium, zinc	Possibly lower intake, decreased bioavailability
Vitamin B_{12}	Not available from plant sources
Vitamin D	Possibly lower intake, insufficient exposure to sunlight during certain seasons

TABLE 7.16

Potential dietary inadequacies in vegan diets in childhood

Some studies, but not all, have reported that vegan children tend to be smaller than their peers consuming mixed diets (AAP 1993). In the Netherlands, vegan children have heights and weights at the 5th to 15th percentile more often than their counterparts who eat mixed diets including animal foods (Van Staveren 1985), (Dagnelie et al. 1994). Generous intake of vegetable oils and fortified soy milk can help provide necessary calories and protein to support growth.

Vegan diets lack the milk products and meats that are rich sources of bioavailable calcium, zinc, and iron for growing children. They are also high in phytates, which inhibit the absorption of these nutrients. Generous intake of unrefined cereals, legumes, seeds, and dark green leafy vegetables is important to avoid deficiencies of calcium, iron, and zinc (see Table 7.17).

Vitamin D obtained from exposure to sunlight may be inadequate in some climates and seasons, but suitably fortified margarines can provide additional vitamin D for a growing child on a vegan diet. Because of the difficulty in obtaining adequate micronutrients, the AAP recommends the inclusion of a multivitamin supplement with iron and vitamin B_{12} in a vegan diet for young children (AAP 1993).

DENTAL HEALTH

Nutritional disorders during childhood, such as protein–calorie malnutrition or rickets, can interfere with the calcification and eruption of the teeth. Proper diet during childhood supports formation of healthy teeth— ample protein, calcium, phosphate, and vitamins C and D are particularly important. Diet is also important in the prevention of dental caries— one of the most widespread diseases of childhood. Tooth decay occurs when a susceptible tooth is exposed to cariogenic bacteria and sugar. **Plaque** on the tooth surface contains several strains of bacteria that are able to break down dietary sugar; one of the products is lactic acid. Lactic acid dissolves

dental plaque: the noncalcified accumulation on the teeth of oral bacteria and their products: desquamated cells and salivary proteins

Calcium (mg per serving)			
100+	**150+**	**200+**	**250+**
Brazil nuts, 10	Ice cream, 1 cup	Beet greens, 1 cup	Almonds, 1 cup
Broccoli, 1 med. stalk	Rhubarb, 1 cup cooked	Cheddar cheese, 1oz	Cheese (Swiss or
Instant farina, 1 cup	Spinach, 1 cup cooked	Cottage cheese, 1 cup	Parmesan), 1 oz
Canned herring, 3 oz			Collards, 1 cup cooked
Cooked kale, 1 cup			Dandelion greens, 1 cup
			cooked
Blackstrap molasses, 1 tbsp			Turnip greens, 1 cup cooked
Light molasses (regular),			
3 tbsp			
Navy beans, 1 cup cooked			
Soybeans, 1 cup cooked			
Soybean curd (tofu), 3.5 oz			
Sunflower seeds, 3.5 oz			
Maple syrup, 5 tbsp			

Zinc		Iron	
		Good Sources	**Excellent Source**
1 to 3 mg per serving	**3+ mg per serving**	**(1–1.5 mg per serving)**	**(>1.5 mg per serving)**
Beans, common, ½ cup	Wheat germ, ⅓ cup	Egg, 1 medium	Navy beans, lima beans,
Bran flakes, 1 oz		Prunes, 4 medium	soybeans, lentils, split peas,
Cheese, 1 oz		Strawberries, 1 cup	½ cup
Chick peas, cow peas, lentils,		Apple juice, 1 cup	Prune juice, ½ cup
½ cup		V-8 juice, 1 cup	Broccoli, 1 stalk
Eggs, 2 medium		Collards, 1 cup	Tomato juice, 1 cup
Nuts (Brazil, pecan, cashews),		Green peas, ½ cup	Oatmeal, ¾ cup
1 oz.		Sweet potato,	Bran flakes with
Wheat flours, whole,		1 medium	raisins, 1 oz
80% extraction, 3½ oz			Cream of Wheat, ¾ cup
			Spinach, ½ cup
			Molasses, 2 tbsp
			Corn syrup, 2 tbsp
			Wheat germ, 2 tbsp

Adapted from Committee on Nutrition, American Academy of Pediatrics, *Pediatric Nutrition Handbook*, 3d ed. (Elk Grove, IL: AAP, 1993).

TABLE 7.17

Sources of zinc, iron, and calcium in vegetarian diets in childhood

cariogenic: promoting the development of dental caries

the enamel covering the tooth and leads to cavity formation. If allowed to progress, the cavity will deepen and allow bacteria to invade the dental pulp, causing infection, swelling, and pain (AAP 1993).

Sucrose is the most **cariogenic** sugar, followed by glucose, maltose, lactose, and fructose (Rugg-Gunn and Hackett 1993). The physical form of the sugar is important. Sticky, retentive forms of sugar—such as candy, cakes, and cookies—cling to the teeth longer. They are more cariogenic than sugars in liquid form, which can be rapidly cleared from the mouth.

The frequency of sugar ingestion also influences dental decay (Gustafsson et al. 1954). After being exposed to sugar, bacteria can produce acid for a limited period—only about 20 minutes to an hour (saliva eventually washes away the food particles and contains buffers that can neutralize the acid). Therefore, reducing the number of times sugar is introduced into the mouth will reduce the acid challenge to the enamel. Repeated exposure of the teeth to sugar by frequent snacking on high-carbohydrate food and drink substantially increases the risk of dental caries (Rugg-Gunn and Hackett 1993). It is not feasible or desirable to completely eliminate sugar from the diet, but foods of less cariogenicity should be eaten as snacks, and sugar intake should be restricted, as much as possible, to mealtimes. See Table 7.18.

Certain foods contain protective factors against formation of caries. Fats and protein cannot be metabolized by cariogenic bacteria to produce acid. Fats can coat the teeth, reducing the retention and cariogenicity of sugars in the plaque. Protein increases the buffering capacity of the saliva; thus foods rich in protein and fat eaten after carbohydrates may reduce the risk of caries (AAP 1993). For example, milk or cheese rather than sugary foods at the end of meals may be beneficial. Effective cleaning of the teeth after meals is also important.

Resistance to dental caries increases if the diet contains optimal amounts of fluoride. Fluoride is incorporated into the crystals that form the tooth enamel, making them more resistant to demineralization from

Food Group	Less Cariogenic	More Cariogenic
Dairy	Milk, cheese, plain yogurt	Ice cream, chocolate milk, milkshakes, yogurt with sweetened fruit
Meat and meat alternates	Lean meat, eggs, fish, nuts, peanut butter without added sugar	Meats with sugared glazes, nuts with sugar or candy coatings, peanut butter with added sugar
Vegetables and fruits	Fresh fruits and vegetables, unsweetened fruit and vegetable juices	Dried fruits, jams and jellies, fruit packed in sweetened syrup, fruit drinks with added sugar, candied sweet potatoes, sugar-glazed carrots
Grains	Popcorn, pretzels, nonsweetened crackers	Cookies, pies, pastries, pre-sweetened breakfast cereals
Other	Sugarless gum and soft drinks, coffee and tea without added sugar	Gum with sugar, candy, honey, syrups; coffee or tea sweetened with sugar

Adapted from A. Ehrlich, *Nutrition and Dental Health* (Albany: Delmar Publishers, 1987).

TABLE 7.18

Food choices to reduce the risk of dental caries

acid. Water fluoridation can markedly reduce the prevalence of dental caries among children.(Arnold et al. 1956). See Figure 7.9. In many areas, fluoridation of the water supply provides children with ample fluoride, and additional supplementation by other means, such as fluoride mouth-washes or tablets, is unnecessary. In areas where the fluoride content of the water is less than 0.3 parts per million, supplements are recommended (AAP 1993). See Table 7.19. The best time to give a fluoride supplement is at bedtime, after teeth cleaning.

The declining prevalence of dental caries in many developed countries is strongly correlated with widespread fluoridation of water supplies and decreased consumption of snacks high in sugar. However, in the developing world, increasing availability of sugary foods combined with a lack of fluoridated public water is contributing to a rising rate of dental caries (Rugg-Gunn and Hackett 1993).

DIETARY FAT AND CHOLESTEROL

In adults, increased blood cholesterol is a major risk factor for cardiovascular disease, and reduction of cholesterol levels reduces the risk (See discussion in Chapter 9). The significance of blood cholesterol levels in childhood has been the subject of much recent debate (Newman et al. 1990), (AAP 1992). Compelling evidence now suggests that cardiovascular disease, the leading cause of death in the United States, has its roots in childhood (NCEP 1991).

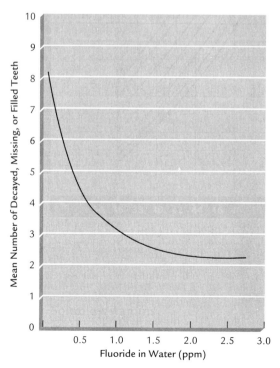

Source: Adapted from Rugg-Gunn and Hackett (1993)

FIGURE 7.9

Dental caries vs. fluoride in water supply

	Fluoride mg/Day		
	Concentration of Fluoride in Drinking Water (ppm)		
Age	**<0.3**	**0.3–0.7**	**>.7**
2–3 yr	0.50	0.25	0
3–16 yr	1.00	0.50	0

Adapted from Committee on Nutrition, American Academy of Pediatrics, *Pedatric Nutrition Handbook*, 3d ed. (Elk Grove, IL: AAP, 1993).

TABLE 7.19
AAP supplemental fluoride dosage schedule

Atherosclerosis appears to begin early in life and progress slowly into adulthood. Elevated cholesterol levels in childhood are thought to play a role in the initiation and development of atherosclerosis, and high total and LDL-cholesterol are correlated with the extent of early atherosclerosis in adolescents (NCEP 1991). Many experts think that eating patterns during childhood affect blood cholesterol and that reducing intake of fat and cholesterol will be beneficial (NCEP 1991), (AAP 1992).

Because of these concerns, experts have recommended a strategy that combines two complementary approaches (NCEP 1991):

- Populationwide reduction in fat and cholesterol in the diets of all children

- An individualized approach aimed at identifying and treating children who are at greatest risk of having high blood cholesterol and an increased risk of cardiovascular disease in later life

As part of the population approach, the following dietary recommendations have been widely endorsed. All healthy children and adolescents over the age of two years should follow the American Heart Association Step-One Diet:

- Total fat intake should be less than 30% of total calories.

- Saturated fat intake should be less than 10% of total calories.

- Daily cholesterol intake should be less than 300 mg per day.

- Nutritional adequacy should be achieved by eating a wide variety of foods.

- Adequate calories and other nutrients must be provided to support growth and development.

The AAP has also endorsed these recommendations with one minor change. Because the AAP believes that recommending "less than" 30% of calories from fat may lead to the inappropriate use of more restrictive

atherosclerosis: deposits of plaques containing cholesterol, smooth muscle cells, calcium, and macrophages (a type of white blood cell) formed in the inner wall of large and medium-sized arteries, often narrowing the arteries and decreasing blood flow

diets, the AAP recommends total fat intake be maintained at 30% of total energy (AAP 1992).

These recommendations are not intended for infants from birth to age 2 years, whose demand for energy during rapid growth requires a larger percentage of calories from fat. During the third year of life, as children make the transition to eating with the family, these guidelines may be safely implemented (AAP 1993).

Changing from whole-milk products to reduced-fat milk products can substantially reduce fat intake in preschool children. A survey of children aged 2 to 5 years found that over 80% exceeded the current recommendations for fat and cholesterol intake, with most of the fat coming from whole milk (Thompson and Dennison 1994). The researchers calculated that if all dairy products consumed were replaced by lower-fat alternatives, fat intake would be reduced to recommended levels (Thompson and Dennison 1994). Table 7.20 lists the amount of saturated fat in many popular children's foods.

Food Item	Grams of Saturated Fat/ Serving
Ice cream, 1 cup 10% fat	9
Hamburger, 3 oz, fast food	8
Hot dog, 2 oz	7
Butter, 1 tbsp	7
Pizza with cheese, 1 slice	6
Cheese, 1 oz, American or Swiss	6
Milk, 1 cup whole	5
Beef, pork, veal, 3 oz lean	5
Heavy cream, 1 tbsp	4
Milk, 2% low fat, 1 cup	3
Bacon, 2 slices crisp	3
Egg, 1 medium	3
French fries, 10 medium	3
Peanut butter, 2 tbsps	3
Cheese, part-skim mozzarella, 1 oz	3
Cream cheese, 1 tbsp	3
Mayonnaise, 1 tbsp	2
Brownie, frosted	2
Potato chips, 10	2
Chicken, light meat, 3 oz	1
Milk, 1% low fat, 1 cup	1
Baked potato, 1 medium	0
Milk, skim, 1 cup	0
Sorbet, 1 cup	0

Note: The average 10–year-old child on an 1800–calorie Step One diet should consume less than 20 g saturated fat per day.

Adapted from C. L. Williams, "Intervention in Childhood," *Prevention of Coronary Heart Disease* (Boston: Little Brown, 1992).

TABLE 7.20

Sources of saturated fat in children's diets

The guidelines for individual screening of children do not recommend that all children over 2 years old have their cholesterol checked—only those children who have a parent whose cholesterol level is greater than 240 mg/dl, or a family history of early (less than 55 years of age) heart disease (NCEP 1991). See Table 7.21.

Children who have an elevated LDL-cholesterol should be placed on a managed AHA Step-One Diet. If careful adherence to this diet for three months fails to achieve acceptable goals, then the AHA Step-Two Diet should be started under qualified supervision. Because the Step-Two Diet requires stringent reduction of saturated fat and cholesterol intake, careful planning is required to ensure adequate intake of all necessary nutrients (AAP 1992). The AHA diets are shown in Table 7.22.

There have been concerns about the safety of implementing a lower-fat diet during childhood. Poor growth in children on low-fat diets has been reported (Lifshitz and Moses 1989), (Kaplan and Toshima 1992), but this has occurred in situations where an important guideline of the AHA Step-One Diet was not followed. Inappropriate dietary restriction resulted in inadequate total caloric intake as well as lower fat intake (see Figure 7.10).

Concern has also been expressed about the potential vitamin and mineral inadequacy of low-fat diets, particularly of iron and calcium. However, there is no good evidence that reducing fat intake while maintaining normal calories and protein will lead to a deficiency of these or other micronutrients. Evaluation of sample AHA Step-One Diets for children have shown that they provide adequate amounts of micronutrients and essential fatty acids (Dobrin-Seckler and Dockelbaum 1991).

Another controversy surrounding cholesterol levels in childhood is whether high levels during childhood predict elevated levels in later life. Although most available evidence suggests that cholesterol levels track well from childhood through adolescence into adulthood (Laver et al. 1988), (Webber et al. 1991) not all studies agree. One recent study, for example, showed that only a quarter to a half of children with cholesterol levels higher than desirable during childhood continued to have elevated levels at 20 and 30 years of age (Laver and Clarke 1990).

	Total Cholesterol mg/dl	Low-Density Lipoprotein Cholesterol mg/dl
Acceptable	<170	<110
Borderline	170–199	110–129
High	≥200	≥130

From National Cholesterol Education Program, *Report of the Expert Panel on Blood Cholesterol Levels in Children and Adolescents* (Bethesda, MD: National Heart, Lung and Blood Institute, 1991).

TABLE 7.21

Recommended cholesterol levels in childhood

Nutrient	Recommended Intake	
	Step-One Diet	**Step-Two Diet**
Total fat	Average of no more than 30% of total calories	Same as Step-One Diet
Saturated fatty acids	Less than 10% of total calories	Less than 7% of total calories
Polyunsaturated fatty acids	Up to 10% of total calories	Same as Step-One Diet
Monounsaturated fatty acids	Remaining dietary fat calories	Same as Step-One Diet
Cholesterol	Less than 300 mg/d	Less than 200 mg/d
Carbohydrates	About 55% of total calories	Same as Step-One Diet
Protein	About 15% of total calories	Same as Step-One Diet
Calories	To promote growth and development	Same as Step-One Diet

From National Cholesterol Education Program "Report of the Expert Panel on Blood Cholesterol in Children and Adolescents," *Pediatrics*, 89 suppl (1992):525–84.

TABLE 7.22
American Heart Association's Step-One and Step-Two diets for children and adolescents

A growing consensus of experts recommends the AHA Step-One Diet for children over 2 years old and adolescents (AAP 1992), (NCEP 1991). Many think if this diet could be adopted during childhood and carried into adulthood, it would not only reduce the incidence of cardiovascular disease, but also help prevent a variety of other chronic conditions and diseases of later life.

FOOD ALLERGIES IN CHILDHOOD

Early childhood is a common age for food allergies. They occur in 2 to 15% of young children (Cant 1991), and those with a family history of allergy are much more likely to develop food allergies (Van Asperen et al. 1983). Environmental factors during infancy appear to influence the development of food allergies, and exposure to food antigens—such as cow's milk protein—in early infancy may increase the risk of developing food allergies (AAP 1993).

Most dietary proteins are broken down into single amino acids before they are absorbed. Food allergies occur when dietary proteins are incompletely broken down before absorption, allowing large molecules of protein to enter the body and interact with the immune system (Cant 1991). The immune cells identify the food molecule as an antigen and react by

FIGURE 7.10

Impairment of growth during late childhood due to overzealous fat restriction. Beginning at age 8, both linear growth and weight gain were slowed because energy intake was insufficient on a very low-fat diet. After nutritional intervention and provision of adequate calories, the child exhibited catch-up growth and full recovery.

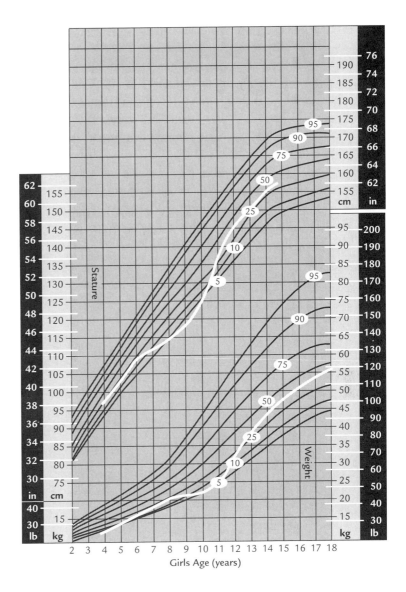

histamine: an amine, beta-imidazolylethyl-amine, that is a chemical mediator of allergic reactions, capillary dilation, and gastric acid secretion

anaphylactic reaction: an unusually strong and exaggerated allergic reaction, which, if severe, can produce a life-threatening fall in blood pressure and constriction of the bronchial tubes

producing antibodies, **histamine,** and other defensive compounds (Sampson et al. 1989).

Production of these agents can cause a variety of symptoms—the most common are listed in Table 7.23. Rarely, **anaphylactic reactions** to food occur, which can be fatal (Sampson et al. 1992). Food allergies occur more often in early childhood than in later life because the developing intestinal tract and immune system are immature and inexperienced in handling food antigens.

Allergy to single foods is much more common than to multiple foods. Table 7.24 shows foods that most often provoke allergies—the most common offenders are milk, fish, peanuts, and eggs (Bock and Atkins 1990). Allergic reactions to food can be immediate or delayed, depending on how

- Anaphylaxis

- Eczema

- Hives

- Angioedema (swelling of the lips and face)

- Asthma

- Rhinitis (stuffy, runny nose)

- Iron-deficiency anemia

- Inflammation of the intestine

- Migraine

Adapted from A.J. Cant, "Food Allergy and Intolerance" in, *Textbook of Pediatric Nutrition*, 3d ed., edited by D. D. McLaren, et al. (Edinburgh: Churchill Livingston, 1991).

TABLE 7.23
Potential manifestations of food allergies

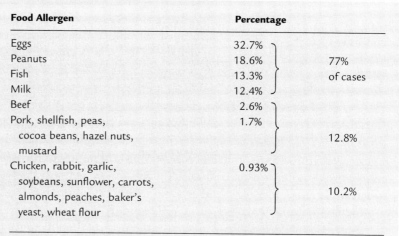

Food Allergen	Percentage	
Eggs	32.7%	
Peanuts	18.6%	77%
Fish	13.3%	of cases
Milk	12.4%	
Beef	2.6%	
Pork, shellfish, peas, cocoa beans, hazel nuts, mustard	1.7%	12.8%
Chicken, rabbit, garlic, soybeans, sunflower, carrots, almonds, peaches, baker's yeast, wheat flour	0.93%	10.2%

Adapted from D. A. M. Moneret-Vautrin, "Food Allergy and Intolerance in Infancy," in *Infant Nutrition*, eds. A .F. Walker and B. A. Rolls (London: Chapman and Hall, 1994).

TABLE 7.24
Common allergy-causing foods in children ages 5 months to 14 years

quickly the immune system reacts to the food antigen. Symptoms can be delayed up to 24 hours, making it difficult to identify the offending food.

The preferred way to test for food allergy is an elimination diet. In an elimination diet, the suspected foods are omitted from the diet for two to three weeks. If symptoms improve, the foods are reintroduced one at a time to see if a return of symptoms can identify the culprit. Extensive elimination diets in children should be supervised by a nutrition professional, and nutritional adequacy must be carefully monitored (AAP 1993).

Children often grow out of food allergies. In a recent study, researchers retested 34 children who had marked food allergies several years earlier. They found that over three fourths no longer reacted to the foods they had been allergic to previously (Pollack and Warner 1987).

LEAD POISONING

In the United States, elevated levels of lead in the blood are a major health risk for children. It is estimated that over 3 million children under 6 years old have blood lead levels (BLLs) high enough to decrease intellectual performance and produce other adverse health effects (U.S. Public Health Service 1988). The prevalence of elevated blood levels is estimated to be about 12% in Black children and 2% in Whites ("Lead Poisoning" 1989).

Lead is distributed throughout the environment, and it makes its way into food through contaminated soil and water. Levels of lead in foods today are 90% lower than 15 years ago, mostly due to the food industry's elimination of lead solder on food cans and the reduction in lead from automobile exhaust that settles onto crops and water (Foulke 1993).

However, food and drink continue to be sources of lead. Warm or hot tap water in homes with lead pipes is contaminated with lead. Dishes are potential sources: small amounts of lead can leach from the glazes and decorative paints on ceramic ware, lead crystal, and pewter and silver-plated hollowware. Acidic liquids, such as coffee, fruit juices, and tomato soup, have a greater tendency to cause leaching of lead (Foulke 1993).

A primary source of lead exposure, particularly in urban areas, is lead-based paint. Although residential use of leaded paint was banned in 1978, about 12 million children under 7 years old reside in older homes containing lead-based paint (Centers for Disease Control 1992). In areas containing lead-based paint (most often poor urban areas), over two thirds of children have elevated BLLs (U.S. Public Helath Service 1988). Interior paints used before 1940 contain up to 50% lead. Children can ingest lead by eating paint chips (which are often colorful and sweet tasting), or by ingestion of lead-contaminated dust and dirt during normal mouthing and exploratory behavior.

Children absorb more lead and are more sensitive to its effects than adults. Children absorb up to 50% of ingested lead compared with adults, who absorb about 10% ("Lead Poisoning" 1989). Our bodies cannot distinguish well between lead and calcium, so lead is absorbed and distributed much like calcium. Because young children absorb calcium avidly to meet their extra needs, they also efficiently absorb lead. Deficiencies of protein, iron, or calcium enhance the absorption of lead and may increase its toxic effects in children. Compared with adults, children are more sensitive to lead toxicity because less can be deposited into their smaller skeleton, leaving a higher percentage of the lead in soft tissues and blood—where it is more toxic.

Many children repeatedly ingest small doses of lead that slowly accumulate in the body. Lead is a biochemical poison that interferes with cellular enzymes and metabolism. It affects almost every organ system—the kidney, bone marrow, and brain are particularly sensitive. It can slow growth, damage hearing, and impair coordination and balance. A child with chronic lead intoxication may be listless and irritable, and low levels of lead exposure in childhood can impair neuropsychological development and classroom performance (Needleman and Bellinger 1991). A recent study showed that low-level exposure during early childhood had adverse and enduring effects on intellectual performance through the first seven years of life. Moderate elevations of blood lead levels were associated with a reduction in IQ of 4 to 5%. (Baghurst et al. 1992). See Figure 7.11.

The AAP now recommends that all children undergo blood lead screening when they are 9 to 12 months old and again at 2 years (Schaffer and Campbell 1994). A child's history of possible lead exposure should be assessed during regular checkups between 6 months and 6 years. Parents and children can guard against exposure to lead in food by observing the guidelines in Table 7.25.

IRON DEFICIENCY

Worldwide, iron-deficiency anemia is a public health problem of staggering dimensions—it is estimated that over 2 billion people are anemic because of iron deficiency (Uvin 1994). Iron deficiency is probably the most common nutrient deficiency in children worldwide. In many developing countries, a mostly cereal diet with little meat, fish, and ascorbic acid is low in bioavailable iron. Intestinal parasites are common—hookworm infection is a frequent cause of increased blood loss and iron defi-

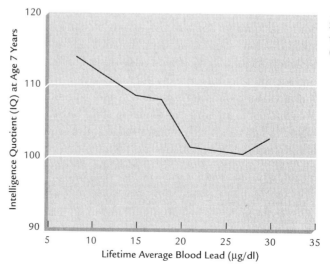

Source: Adapted from Baghurst (1992)

FIGURE 7.11

Blood concentrations of lead and IQ

Exposure to lead in food and beverages at home can be reduced by observing the following guidelines:

- Avoid the use of lead crystalware, and do not feed infants from lead crystal bottles.

- Do not store acidic foods such as fruit and vegetable juices and tomato sauce in ceramic containers.

- Do not store beverages in lead crystal containers.

- Limit the use of antique or collectible housewares.

- Avoid using items that show a dusty or chalky gray residue on their glaze after they are washed.

- Avoid use of imported foods from cans sealed with lead solder.

- Carefully follow label directions on ornamental products that have warnings such as: "Not for food use—for decorative purposes only."

- Ensure that children consume adequate amounts of calcium and iron; deficiencies of these nutrients increase lead absorption.

- If wine is sealed with a foil capsule, wipe off the rim of the bottle carefully before pouring.

- When pregnant, avoid the use of ceramic mugs when drinking hot beverages such as coffee and tea.

Adapted from *FDA Consumer*, April 1993.

TABLE 7.25

Guarding against exposure to lead in food and beverages

ciency. Iron deficiency may cause anemia, decreased performance, and impaired mental and motor development (Filer 1995).

In the past two decades, there has been a general improvement in the iron status of children in the United States and a decline in the prevalence of iron-deficiency anemia. In 1970, 8 to 10% in the United States of children aged 6 months to 5 years were anemic; by 1985, the figure had fallen to 3% (Filer 1995). The fortification of infant formulas and breakfast cereals, which began in the early 1970s, is thought to be a major factor in this heartening trend.

Iron-deficiency anemia is rare before 4 to 6 months of age, because the healthy infant has ample iron stores at birth. Iron deficiency develops between 6 months and 3 years if increased needs for rapid growth are not met by an adequate dietary supply (Oski 1989). Milk is a major source of calories at this age, but it has only about 1.5 mg of iron/1000 kcal. After age 3, iron nutrition usually improves and anemia is less likely because of the greater iron content of a typical mixed diet (about 6 mg/1000 kcal).

In parts of the world where intestinal parasites, such as hookworm, are a problem, iron-deficiency anemia remains common through childhood into adolescence. Poverty also increases the risk of childhood anemia. Children from families with lower socioeconomic status tend to have a higher incidence of iron deficiency—rates are as high as 20–30% in some poor urban areas (Yip 1989).

The symptoms of iron deficiency are easy to recognize when anemia becomes severe—children appear listless and develop pallor and easy fatigue (see Table 7.26). But anemia is only one manifestation of the widespread systemic disease produced by iron deficiency. Children who are deficient in iron have poor appetites, are more likely to develop infections, and grow more slowly than their healthy counterparts. They are often irritable, inattentive, and withdrawn (Owen 1989).

Iron deficiency in childhood also impairs intellectual development (Sheard 1994). A recent study in Indonesia confirmed that iron-deficient anemic children do more poorly on tests of motor and mental development than their iron-sufficient counterparts (Idjradinata and Pollitt 1993). See Table 7.27. Although supplemental iron reversed the developmental delays in the Indonesian children, another study in Costa Rica produced more troubling results. Researchers found that children who had iron-deficiency anemia in infancy were at risk for long-lasting developmental impairment, even after the iron deficiency was treated. Compared with their peers who had normal iron status during infancy,

- Pallor

- Weakness and fatigue

- Anorexia

- Anemia

- Fast heartbeat

- Poor growth

- Impaired immune system

- Cognitive disabilities

TABLE 7.26

*Symptoms and signs of
iron-deficiency anemia*

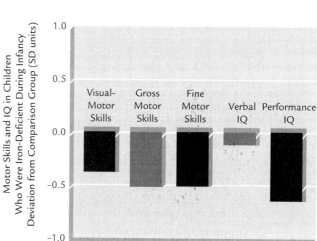

Source: Adapted
from Lozoff et al.
(1991)

FIGURE 7.12

*Development and
iron deficiency*

	Iron-Deficient Anemic		Iron-Deficient Nonanemic	
	Ferrous Sulfate	Placebo	Ferrous Sulfate	Placebo
Mental development score				
Before	88.8	92.4	102.4	101.8
After	108.1	92.9	107.7	109.3
Motor development score				
Before	88.5	92.4	102.9	103.5
After	112.0	97.5	107.8	106.6

Iron deficiency impairs mental and motor development in infants, and the reversibility of these effects continues to be debated by scientists. In a recent Indian study, iron-deficient (anemic and nonanemic) infants aged 12–18 months were given 4 months of ferrous sulfate (3 mg/kg/day) or placebo. The iron-treated children showed significant improvements in hemoglobin concentrations and mental and motor development compared with the placebo-treated group.

Adapted from P. Idjradinata and E. Pollitt, "Reversal of Developmental Delays in Iron-Deficient Anemic Infants Treated with Iron," *Lancet* 341 (1993):1–4.

TABLE 7.27

Motor and mental development scores and iron status

children who were iron deficient at 1 to 2 years had impairment in mental and motor functioning that persisted into the school-age years (Lozoff et al. 1991). See Figure 7.12.

What can parents do to ensure ample dietary iron for their children? In preschool-age children, most dietary iron is nonheme iron, while only a small proportion is present as heme iron. Heme iron, primarily from meat, is well absorbed regardless of the content of the rest of the meal. Nonheme iron is much more poorly absorbed, and its absorption is markedly influenced by other foods eaten at the same meal. For example, ascorbic acid is a potent enhancer of absorption, while milk, bran, and tea tend to inhibit absorption of nonheme iron (Oski 1993).

A child's choice of beverage with meals is important—orange juice doubles the absorption of nonheme iron from a meal, while tea decreases it by up to 75%. Heme iron itself promotes the absorption of nonheme iron: about four times as much nonheme iron is absorbed from a mixed meal when the principal protein is meat, fish, or chicken than when it is dairy products or eggs (Oski 1993).

In order to prevent iron-deficiency anemia in children, regular sources of iron, such as iron-fortified breakfast cereals, green leafy vegetables, and lean meat, should be provided. Because very young children often prefer foods that are softer, providing meat in a form that is easy to chew, such as ground beef, may be helpful. In children whose iron intake is limited, a supplement containing the RDA (10 mg) may be indicated (AAP 1993).

UNDERNUTRITION IN CHILDHOOD: A WORLDWIDE PERSPECTIVE

Millions of children worldwide suffer and die each year because they lack adequate nourishment. Although undernutrition is still occasionally found in developed countries, particularly among the poor and certain minority groups (Miller and Korenman 1994), the highest incidence is in the economically underdeveloped countries. As discussed in the introduction to this chapter, the problem is not inadequate supplies of food energy, but its global distribution (Uvin 1994).

Poverty, as we have seen, is a major determinant of undernutrition in children; but other economic, social, cultural, and educational factors contribute to the problem. Weaning an infant at an early age without a nutritious replacement for breast milk is perhaps the most important single cause of childhood undernutrition (McLaren 1991).

Infections—particularly diarrheal illnesses—are a major contributor to morbidity and mortality in the malnourished child. Diarrhea aggravates undernutrition by causing malabsorption of energy, protein, and micronutrients, and increasing water and electrolyte losses.

It is estimated that over 200 million of the world's children under 5 years old suffer from varying degrees of undernutrition (de Onis et al. 1993). The consequences of undernutrition include growth retardation, decreased resistance to infection and disease, impaired learning ability, and increased mortality. Undernutrition is a contributing cause in a third to a half of all child deaths worldwide (Pelletier 1994). See Figure 7.13.

PROTEIN–ENERGY MALNUTRITION

The term *protein–energy malnutrition (PEM)* encompasses a wide range of chronic undernutrition, from moderate growth failure to more severe

Source: Adapted from Pelletier et al. (1994)

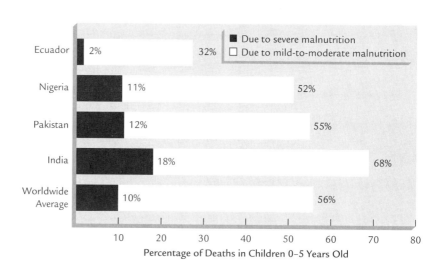

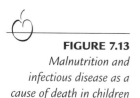

FIGURE 7.13

Malnutrition and infectious disease as a cause of death in children

All of the syndromes of PEM—growth failure, kwashiorkor, marasmic kwashiorkor, and maramus—have the following features in common:

- Protein and/or energy-deficient diet

- Retarded growth

- Loss of body protein and diminished muscle mass

- Frequent or persistent diarrhea, often with dehydration

- High morbidity and mortality from infections

- Mental changes: apathy and/or irritability

- Anemia

Adapted from J. L. Hansen and J. M. Pettifor, "Protein Energy Manutrition," in *Textbook of Pediatric Nutrition*, 3d ed., edited by D. S. McLaren, et al. (Edinburgh: Churchill Livingston, 1991).

TABLE 7.28
The clinical spectrum of PEM

	% Underweight	% Stunted	% Wasted
Africa	27.4 (31.6)	38.6 (44.6)	7.2 (8.3)
Asia	42.0 (154.1)	47.1 (172-8)	10.8 (39.6)
Latin America	11.9 (6.5)	22.2 (12.1)	2.7 (1.5)
Oceania	29.1 (0.3)	41.9 (0.4)	5.6 (0.1)
All developing countries	35.8 (192.5)	42.7 (229.9)	9.2 (49.5)

Figures in parentheses are millions of children.

From *WHO Bulletin* OMS 71(1993)

TABLE 7.29
Prevalence of underweight and stunted children in developing countries

conditions such as marasmus and kwashiorkor. Table 7.28 lists the central features of PEM.

Mild PEM

The mildest and most common form of PEM is growth failure alone (Hansen and Pettifor 1991). Growth impairment, most often seen in the post-weaning period from 9 months to 3 years of age, is common in children from developing countries. Current WHO estimates, shown in Table 7.29 and Figure 7.14, indicate the disturbing magnitude of the problem:

- In 1990, the prevalence of stunting in children less than 5 years old was estimated at about 43% worldwide (de Onis et al. 1993). Of the 230 million children affected, 80% are in the developing countries of southern Asia. Most stunted children in developing countries never regain their full height (Allen 1993). Stunting is associated with lower

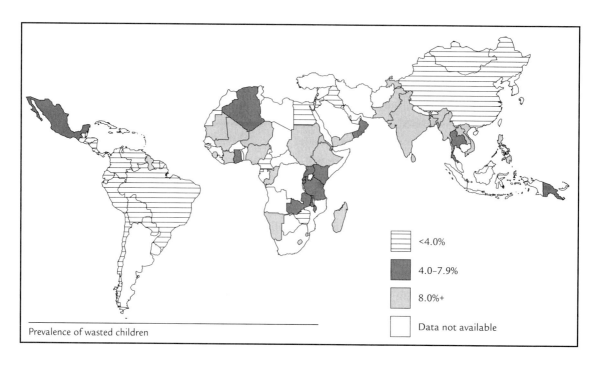

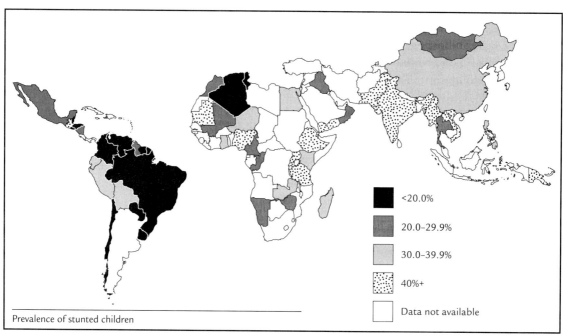

Prevalence of wasted children

Prevalence of stunted children

FIGURE 7.14

Prevalence of wasted and stunted children worldwide, 1980–1992

intelligence, impaired learning ability, and poor school achievement (Hansen and Pettifor), (Agarwal 1987).

- In 1990, 184 million children—or over a third of all children worldwide—were underweight (more than 2 SD below the expected weight for age) (de Onis et al. 1993). Although the proportion of underweight children has fallen since 1975, the absolute numbers have increased from 164 to 184 million (Uvin 1994).

- The number of children in 1990 who were wasted (more than 2 SD below expected weight for height) was nearly 50 million (de Onis et al 1993). Because wasting reflects severe short-term undernourishment, these numbers imply that nearly one in ten children in the developing world are acutely starving.

Mild PEM in children increases the risk of infections, particularly diarrheal illnesses and measles, and increases mortality (Hansen and Pettifor 1991). Anemia is very often found in children with mild PEM and growth failure. Growth retardation in early childhood is associated with functional impairment and diminished work capacity in adult life (WHO 1974).

Severe PEM

Kwashiorkor is a word derived from west African language meaning "the disease suffered by the displaced child." The syndrome occurs when a child is displaced from the breast by a younger sibling and is weaned onto a food that contains inadequate protein. Because in poor families in the developing world weaning may be delayed, and because it takes time for signs of deficiency to develop, kwashiorkor is most common in the second year of life (Hansen and Pettifor 1991).

Kwashiorkor occurs when a child consumes a diet with adequate energy but with a very low protein: energy ratio. Protein intake is inadequate and dietary protein is often of low quality. It is often seen in young children fed exclusively carbohydrate in the form of **cassava**, maize, or rice. Children with kwashiorkor will not grow, and they develop anorexia, diarrhea, and characteristic hair and skin changes. Electrolyte and protein losses cause fluid retention—edema is a hallmark of kwashiorkor. Subcutaneous fat may be preserved and, with edema, may mask the wasting of underlying tissue. Severe muscle wasting often results in the child's being unable to stand or walk.

Marasmus is a form of severe PEM caused by starvation—inadequate intake of energy, protein, and other nutrients results in a shrunken, wasted child. Body weight is less than 60% of expected weight for age. Although it can occur at any age, it is most common in early infancy associated with failure of breast feeding or severe diarrhea. Marasmic children are often severely anemic, suffer from chronic infections, and have a high mortality.

cassava: the large, tuberous starchy root of the shrub *Manihot esculenta*; also called manioc

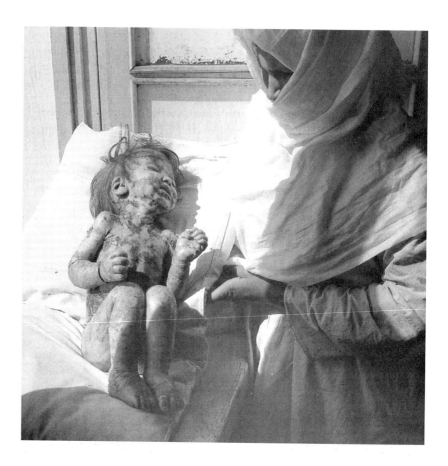

*Child with
kwashiorkor
(Source: F. Perabo)*

More often than either syndrome alone, severely malnourished children develop characteristics of both kwashiorkor and marasmus. The term *marasmic kwashiorkor* is used for children who are less than 60% of expected weight and have edema and other signs of kwashiorkor (Hansen and Pettifor 1991). See Table 7.30.

Associated deficiencies of vitamins and minerals contribute to the clinical picture of the child with severe PEM. Treatment for severe PEM consists of providing adequate amounts of both calories and protein and treating intercurrent infections. Although the mortality rate is high, the response of children who do recover is rapid.

MICRONUTRIENT DEFICIENCIES

Micronutrient deficiencies are common in children in the poorer regions of the world. They are often associated with protein–energy malnutrition and can have significant adverse effects on growth, learning ability, and the immune system. Studies have shown micronutrient deficiencies, particularly of vitamin A and iron, play major roles in childhood stunting in certain developing countries (Allen 1993).

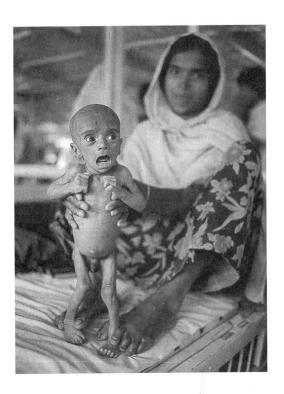

*Marasmic child
(Source: Lisa
Taylor/Panos Pictures)*

	Edema	
Body Weight (% of standard)	**Present**	**Absent**
80–60	Kwashiorkor	Underweight
< 60	Marasmic kwashiorkor	Marasmus

Adapted from Wellcome Trust Working Party, "Classification of Infantile Malnutrition," *Lancet* 2 (1970):320–3.

TABLE 7.30
Classification of PEM

In a recent long-term survey among the poor in Mexico, Kenya, and Egypt, researchers were surprised to find that stunting was explained to a large extent by inadequate micronutrient intake and not by inadequate calories or protein. Moreover, micronutrient deficiencies were associated with widespread retardation of cognitive, motor, and psychosocial development in children (NCRSP 1992).

WHO has identified three micronutrient deficiencies as particularly important public health concerns worldwide: vitamin A, iron, and iodine (McLaren 1991). Iron deficiency has been discussed earlier in this chapter; consequences of vitamin A and iodine deficiency are described here. See Table 7.31 and Figure 7.15 for WHO's estimation of micronutrient deficiencies worldwide.

Source: Adapted from McLaren (1991)

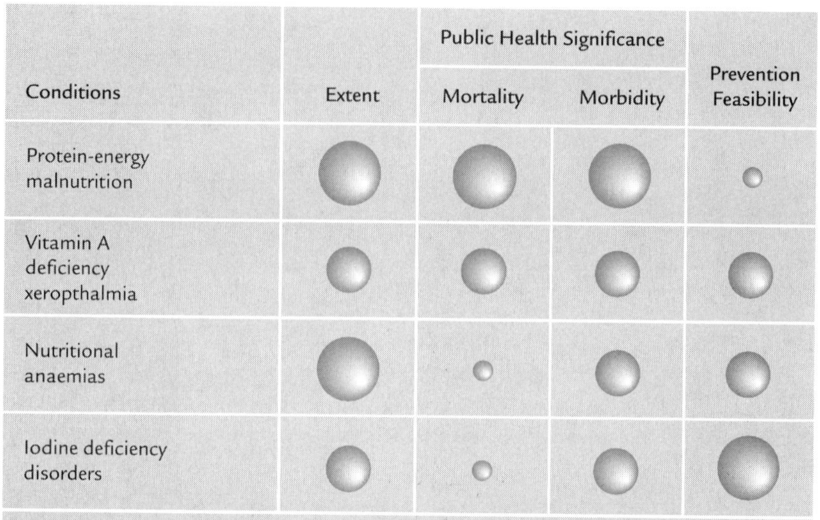

Conditions	Extent	Public Health Significance		Prevention Feasibility
		Mortality	Morbidity	
Protein-energy malnutrition	⬤	⬤	⬤	∘
Vitamin A deficiency xeropthalmia	⬤	⬤	⬤	⬤
Nutritional anaemias	⬤	∘	⬤	⬤
Iodine deficiency disorders	⬤	∘	⬤	⬤

The circle gives an indication of the magnitude in each case.

FIGURE 7.15

Magnitude of major nutritional problems worldwide

	Iodine Deficiency	Vitamin A Deficiency	Iron Deficiency
	Goiter	Xerophthalmia	Anemia
Africa	86	1.3	206
Asia & Oceania	317	11.4	1674
Americas	63	0.1	94
Europe	97	0	27
Eastern Mediterranean	93	1	149
World	655	13.8	2150

Note: Numbers are people in millions.

Adapted from P. Uvin, "The State of World Hunger," *Nutr Rev* 52, no.5 (1994):151–61.

TABLE 7.31

Number of people worldwide affected by micronutrient malnutrition

Vitamin A Deficiency

Millions of children worldwide do not obtain adequate vitamin A in their diets. Epithelial tissues, especially the eyes, are particularly sensitive to vitamin A deficiency. Damage to the eyes from lack of vitamin A, termed *xerophthalmia*, is the major cause of blindness in young children worldwide (McLaren and Thurnham 1991). Deficiency also impairs immunity. Young children with vitamin A deficiency often die of overwhelming infections—many from severe measles.

Preschool-age children (1 to 6 years old) are most susceptible to vitamin A deficiency. Milk and other supplementary foods consumed by young children are low in vitamin A, and rapid growth increases requirements. Even mild deficiency of vitamin A is associated with an increased susceptibility to infections and increased mortality (Sommer et al. 1986). The WHO estimates 120 million preschool children worldwide are at risk of vitamin A deficiency (MDIS 1993), and nearly 14 million have xerophthalmia. Each year, over 250,000 children go blind from xerophthalmia, and 50–80% of blinded children die (McLaren and Thurnham 1991).

Diets that are mainly vegetarian, with a staple such as rice or cassava, and that lack dark green vegetables and yellow fruits, will be deficient in vitamin A. Most cases in children are found in the rice-dependent areas of south and east Asia and in east Africa, where cassava is a major portion of the diet.

In many countries where vitamin A deficiency is endemic, public health programs have begun to promptly treat early xerophthalmia and control vitamin A deficiency. Strategies for preventing deficiency involve modification of diets to include more sources of vitamin A, food fortification (such as sugar in Central America and monosodium glutamate in southeast Asia), and distribution of high doses of the vitamin to young children every three to six months (WHO 1991) (see Table 7.32). Effective vitamin A supplementation reduces childhood mortality in certain regions by as much as 34% (Sommer et al. 1986).

Iodine Deficiency

Dietary deficiency of iodine occurs mainly in areas where the soil and water, and therefore the food produced on it, are low in iodine. Deficiency disorders are also found in regions where there is adequate iodine in the

Children over 1 year and under 6 years old	200,000 IU of vitamin A orally every 3-6 months
Infants 6-12 months old and any older children who weigh less than 8 kg	100,000 IU of vitamin A orally every 3-6 months Immunization against measles provides a good opportunity to give one of these doses*
Lactating mothers	200,000 IU of vitamin A orally once: at delivery or during the next 2 months This will raise the concentration of vitamin A in the breast milk and therefore help to protect the breast-fed infant

* When infants less than 6 months old are not being breast-fed, supplementation with 50,000 IU of vitamin A before they reach 6 months should be considered.

From WHO/UNICEF/IVAGG Task Force, 1988.

TABLE 7.32

WHO preventive schedule for vitamin A for preschool children in areas where deficiency is widespread

Goiter (Source: Sanjay Acharyu/Panos Pictures)

goiter: an enlargement of the thyroid gland, visible as a swelling at the front of the neck

cretinism: growth stunting and impaired mental development in children that results from inadequate maternal intake of iodine during pregnancy and poor iodine nutriture during infancy

diet, but the food supply contains substances that inhibit iodine absorption and metabolism. Such substances, termed *goitrogens*, include cassava, soy beans, and cabbages.

The two most commoncsyndromes of iodine deficiency are **goiter** and **cretinism;** they are endemic in many parts of the world. Currently over 1 billion people worldwide are estimated to be at risk of iodine deficiency; goiter is present in over 600 million. Prevalence of iodine deficiency is highest in central Africa and the mountainous areas of southeast Asia and South America (MDIS 1993).

Iodine-deficiency disorders are most common in young children and pregnant women, when iodine requirements are high. The effects of iodine deficiency are particularly severe for the fetus and growing child. A pregnant woman who is iodine deficient is at risk of delivering a baby with varying degrees of neurological damage. Iodine deficiency during early childhood growth is also associated with impaired mental function and retarded physical development. In areas where iodine deficiency is severe, such as the Himalayas, Andes, and central Africa, cretinism affects up to 5% of children (MDIS 1993). According to the World Health Organization, iodine deficiency is the leading cause of preventable mental retardation in the world today (WHO 1991).

Major public health programs have been successful in many parts of the world in reducing the prevalence of iodine deficiency. The most widely used and effective means of controlling iodine deficiency are fortification of salt with iodine and widespread oral or injected administration of iodinated oil (WHO 1991). In China, free supplies of iodinated salt and administration of iodinated oil has sharply reduced the prevalence of iodine deficiency in childhood.

Childhood malnutrition is one of our most daunting global problems. A number of complex barriers—political, economic, agricultural, and social—must be overcome. Although organizations such as the World Bank, WHO, and the FAO of the United Nations continue to be successful in addressing many of these issues, nutrition is a basic human need that remains unmet for vast numbers of children.

PLANNING A HEALTHY CHILDHOOD DIET: REVIEW AND SUMMARY

Diets for children should be high in nutrient density. A broad range of basic foods, including dairy products, whole-grain cereals, lean meat, eggs, fish, fruits and vegetables, should be encouraged. Foods rich in iron, zinc, and calcium should be emphasized—milk products are important sources of calcium, riboflavin, and protein, while meat is an excellent source of iron, zinc, and vitamin B_{12}.

Fat and cholesterol intake should be moderate, and sugar intake, particularly from foods that promote tooth decay, should be limited. Healthful

	Serving Size		
	2–3 Years	**4–6 Years**	**7 Years to Puberty**
Milk and Cheese			
4 servings daily in the amount recommended			
Milk, yogurt	½–¾ cup	¾ cup	1 cup
Cottage cheese	4–6 tbsp	6 tbsp	6–8 tbsp
Custard, milk pudding, and ice cream (served only after a meal)			
Cheese (1 oz = 1 slice or a 1" cube) (limit due to saturated fat)	⅔–1 oz	1 oz	1 oz
Meat, poultry, fish and beans			
4 servings daily in the amounts recommended			
Beef, pork, lamb, veal, and fish	1½ oz	4 tbsp or 2 oz	2 oz
Eggs	¾	1	1
Peanut butter	3 tbsp	3 tbsp	3 tbsp
Cooked legumes, dried beans, or peas	⅜ cup	½ cup	½–¾ cup
Fruits and vegetables			
4 servings daily in the amounts recommended			
Citrus fruits, berries, melons, tomatoes peppers, cabbage, cauliflower, broccoli	¼ cup	¼ cup	½ cup
Melons, peaches, apricots, carrots, spinach, broccoli, orange squash, pumpkin, sweet potatoes, peas, beans, and Brussels sprouts	3–4 tbsp	4–5 tbsp	5–6 tbsp
Breads and cereals			
4 servings daily in the amount recommended			
Whole grain or enriched breads	¾ slice	¾–1 slice	1–1½ slices
Cooked cereals, rice, and pasta	⅓ cup	½ cup	¾ cup
Whole-grain or fortified ready-to-eat cereals	¾ oz	1 oz	1–1½ oz
Fats/oils			
Margarine, oils, mayonnaise, and salad dressing	1 tsp	1 tsp	1 tsp
Other foods			
Jams, jellies, soft drinks, candy, sweet desserts, salty snacks, gravies, pickles, and catsup	use in moderation		use in moderation

Adapted from J. Endred and R. Rockwell, *Food and Nutrition and the Young Child* (St. Louis: C. V. Mosby, 1980) and M. E. Lowenberg, "The Development of Food Patterns in Young Children," in *Nutrition in Infancy and Childhood*, P. Pipes (St. Louis: Mosby College Publishing, 1985).

TABLE 7.33

Dietary guide for childhood

food preferences established during childhood can influence lifelong health. The recommendations in Table 7.33 can help parents provide a well-balanced and healthy diet for the active and growing child by ensuring ample intake of calories, protein, and micronutrients.

REFERENCES

Agarwal, D. K., et al., "Nutritional Status, Physical Work Capacity and Mental Function in School Children," *Scientific Report* no. 6 (New Delhi: Nutrition Foundation of India, 1987).

Allen, C. E., "Families in Poverty," *Nurs Clin North America* 29 (1994):377–94.

Allen, L., "The Nutrition CRSP: What Is Marginal Malnutrition, and Does It Affect Human Function?" *Nutr Rev* 51 (1993):255–67.

Alvarez, S. R., Herzog, L.W., and Dietz, W,. Jr., "Nutritional Status of Poor Black and Hispanic Children in an Urban Neighborhood Health Center," *Nutrl Res* 4 (1984):583–89.

American Academy of Pediatrics, *Pediatric Nutrition Handbook,* (Elk Grove, IL: AAP, 1993).

American Academy of Pediatrics, "Statement on Cholesterol," *Pediatrics* 90 (1992):469–73.

American Dietetic Association. "Nutrition Standards in Child Care Programs: Technical Support Paper," *J Am Diet Assoc* 94 (1994):324–28.

American Dietetic Association, "Position of the American Dietetic Association: Vegetarians Diets," *J Am Diet Assoc* 88 (1988): 351–55

American Dietetic Association, "Use of Nutritive and Nonnutritive Sweeteners," *J Am Diet Assoc* 93 (1993):816–21.

Arbeit, M. L., et al., "Caffeine Intakes of Children from a Biracial Population: The Bogalusa Heart Study," *J Amer Diet Assoc* 88 (1988):466–71.

Arnold, F. A., Jr, et al., "Effect of Fluoridated Public Water Supplies on Dental Caries Prevalence. Tenth Year of the Grand Rapids-Muskegon Study," *Pub Health Rep* 71 (1956):652, 1136.

Baghurst, P. A., et al., "Environmental Exposure to Lead and Children's Intelligence at Age of Seven Years," *N Engl J Med* 327 (1992):1279–84.

Beilin, L. J., et al., "Vegetarian Diet and Blood Pressure," *Nephron* 47, suppl 1 (1987):37–41.

Birch, L. L. et al., "The Variablility of Young Children's Energy Intake," *N Engl J Med* 324 (1991):232.

Birch, L. L., "The Acquisition of Food Acceptance Patterns in Children," in *Eating Habits: Food, Physiology and Learned Bahaviour,* eds. R. A.

Boakes, D. A. Popplewell, and M. J. Burton (New York: John Wiley & Sons, 1987).

Birch, L. L., et al., "The Influence of Social-Affective Context on the Formation of Children's Food Preferences," *J Nutr Educ* (1981):115.

Bock, S. A., and Atkins, F. M., "Patterns of Food Hypersensitivity During Sixteen Years of Double-Blind, Placebo-Controlled Food Challenges, *J Pediatrics* 117(1990): 561–67.

Cant, A. J., "Food Allergy and Intolerance," in *Textbook of Pediatric Nutrition.*, 3d ed. edited by D. S. McLaren et al. (Edinburgh: Churchill Livingstone, 1991).

Carter, C. M., et al., "Effect of a Few Food Diet in Attention Deficit Disorder," *Arch Dis Child* 69 (1993):564–68.

Centers for Disease Control, "Fatal Pediatric Poisoning from Leaded Paint," *JAMA* 265 (1991):2050–51.

Centers for Disease Control, "Pediatric Nutrition Surveillance System— United States 1980-1991. in CDC Surveillance Summaries, *MMWR* 41. no. ss-7 (1992);1–25.

Committee on Diet and Health, *Diet and Health: Implications for Reducing Chronic Disease Risk* (Washington, D. C.: National Academy Press, 1989):76–77.

Community Nutrition Institute, "Free School Lunches at Record Level; Program Backers Express Concern," *Nutrition Week* 24 no.(28)1994:1.

Conners, C. K., "Sugars and Hyperactivity," in *Sugars and Sweeteners*, eds. N. Kretchmer and C. B. Hollenbeck (Boston: CRC Press, 1991):115–29.

Cotugna, N., "TV Ads on Saturday Morning Children's Programming— What's New?" *J Nutr Ed* 20 (1988):125–27.

Crawford, P. B., et al., "Environmental Factors Associated with Preschool Obesity," *J Am Diet Assoc* 77 (1978):589.

Dagnelie, P. C., et al., "Effects of Macrobiotic Diets on Linear Growth in Infants and Children Until Ten Years of Age," *Eur J Clin Nutr* 48, Suppl. 1 (1994):S103–12.

de Onis, M., et al., "The Worldwide Magnitude of Protein-Energy Malnutrition: An Overview from the WHO Global Database on Child Growth," *Bull WHO* 72 (1993):703–12.

Dietz, W. H., and Gortmaker, S. L., "Do We Fatten our Children at the Television Set? Obesity and Television Viewing in Children and Adolescents," *Pediatrics* 75 (1985):807.

Dobrin-Seckler, B. E., and Deckelbaum, R. J., "Safety of the American Heart Association Step-One Diet in Childhood," *Ann NY Acad Sci* 623 (1991):263–68.

Drake, M. A., "The Nutritional Status and Dietary Adequacy of Single Homeless Women and Their Children in Shelters," *Pub Health Rep* 107 (1992):312–19.

Fallon, A. E., Raozin, P., and Plinder, P., "The Child's Conception of Food: The Development of Food Rejections, with Special Reference to Disgust and Contamination Sensitivity," *Child Dev* 55 (1984):566–75.

Farris, R. P,. et al., "Macronutrient Intake of Ten-Year-Old Children, 1973 to 1982," *J Am Diet Assoc* 86 (1986):765.

Farris, R. P., et al., "Nutrient Contribution of the School Lunch Program: Implications for Healthy People 2000," *J School Health* 62 (1992):180–4.

Feingold, B. F., *Why Your Child Is Hyperactive* (New York: Random House, 1975).

Fierman, A. H., et al., "Growth Delay in Homeless Children," *Pediatrics* 88 (1991):918–25.

Filer, L. J., "Iron Deficiency," in *Childhood Nutrition*, ed. F. Lifshitz, (Boca Raton: CRC Press, 1995).

Fomon, S. J., "Protein," in *Nutrition of Normal Infants* (St. Louis: Mosby-Year Book, 1993).

Foulke, J. E., "Lead Threat Lessens, but Mugs Pose Problem," *FDA Consumer*, April (1993):91–93.

Garn, S. M., and Clark, D. C., "Nutrition, Growth, Development and Maturation: Findings from the Ten-State Nutrition Survey of 1968–70. P*ediatrics*, 56 (1975):306–19.

Garn, S. M., Clark, D. C., and Trowbridge, F. L., "Tendency Toward Greater Stature in American Black Children," *Am J Dis child* ,126 (1973):164–66.

Griffiths, M., Rivers, J. P. W., and Hoinville, E. A., "Obesity in Boys: The Distinction Between Fatness and Heaviness," *Hum Nutr Clin Nutr*, 39C (1985):259–69.

Gustafsson, B. E., et al., "The Vipeholm Dental Caries Study: The Effect of Different Carbohydrate Intake on Caries Activity in 436 Individuals Observed for Five Years," *Acta Odontol Scand* 11 (1954):232–40.

Halsted, J. A., et al., "Zinc Deficiency in Man: The Shiraz Experiment," *Am J Med* 53 (1972):277–84.

Hammer, L. D., "The Development of Eating Behavior in Childhood," *Ped Clin NA* 39 (1992):379–94.

Hanes, A., Vermeersch, J., and Gale, S., "The National Evaluation of School Lunch Programs: Program Impact on Dietary Intake," *Am J Clin Nutr* 40 (1984):390–413.

Hansen, J. D. L., and Pettifor, J. M., "Protein Energy Malnutrition," in *Textbook of Pediatric Nutrition.* McLaren, D. S., Burman, D., Bellton, N. R., and Williams, A. F., eds. Third edition. (Edinburgh: Churchill Livingstone, 1991).

Hofferth, S. L., and Phillips, D. A., "Child Care in the United States 1970-1995," *Journal of Marriage and the Family* 49 (1987):559–71.

Idjradinata, P., and Pollitt, E., "Reversal of Developmental Delays in Iron-Deficient Anemic Infants Treated with Iron," *Lancet* 341 (1993):1–4.

Kanarek, R. B., "Does Sucrose or Aspartame Cause Hyperactivity in Children?" *Nutr Rev* 52 (1994):173–5.

Kaplan, R. M., and Toshima, M. T., "Does a Reduced-Fat Diet Cause Retardation in Child Growth?" *Prev Med* 21 (1992):33–52.

Kinsbourne, M., "Sugar and the Hyperactive Child," *N Engl J Med* 330 (1994):335–6.

Klesges, R. E., et al., "Parental Influence on Food Selection in Young Children and Its Relationships to Childhood Obesity, *Am J Clin Nutr* 53 (1991):859–64.

Kovar, M. G., "Use of Medications and Vitamin-Mineral Supplements by Children and Youths," *Public Health Report* 100 (1985):470–73.

Lauer, R. M., and Clarke, W. R., "Use of Cholesterol Measurements in Childhood for the Prediction of Adult Hypercholesterolemia: The Muscatine Study," *JAMA* 264 (1990):3034–38.

Lauer, R. M., Lee, J., and Clarke, W. R., "Factors Affecting the Relationship between Childhood and Adult Cholesterol Levels: The Muscatine Study," *Pediatrics* 82 (1988)309–18.

"Lead Poisoning," in *Pediatric Medicine*, eds. M. E. Avery and L.R. First (Baltimore: Williams & Wilkins, 1989).

Lebenthal. E., "Impact of Development of the Gastrointrestinal Tract on Infant Feeding," *J Pediatr* 102 (1983):1.

Lebin, N., Rattan, J., and Gilat, T., "Energy Intake and Body Weight in Ovo-Lacto Vegetarians," *J Clin Gastroenterol* 8 (1986):451–53.

Life Sciences Research Office, "Physiological Effects and Health Consequences of Dietary Fiber" (Bethesda, MD: Federation of American Societies for Experimental Biology, 1987).

Lifshitz, F., and Moses, N., "Growth Failure: A Complication of Dietary Treatment of Hypercholesterolemia," *Am J Dis Child* 143 (1989):537–42.

Lozoff, B., Jimenez, E., and Wolf, A. W., "Long-Term Developmental Outcome of Infants with Iron Deficiency," *New Engl J Med* 325 (1991):687–94.

LSRO, "Summary of a Report on Assessment of the Iron Nutritional Status of the U. S. Population," *Am J Clin Nutr* 42 (1985):1318–30.

Lund, L. A., and Burk, M. C., "A Multidisciplinary Analysis of Children's Food Consumption Behavior," *Technical Bull.* no. 265 (St Paul: University of Minnesota, Agricultural Experimental Station 1969).

McIntye, L., "A Survey of Breakfast Skipping and Inadequate Breakfast Eating Among Young School Children in Nova Scotia," *Can J Public Health* 84 (1993) :410-4.

McLaren, D. S., and Thurnham, D. I., "Vitamin Deficiency and Toxicity," in *Textbook of Pediatric Nutrition.* McLaren, D. S., Burmna, D., Bellton, N. R., and Williams, A. F., eds. Third edition. (Edinburgh: Churchill Livingstone, 1991).

McLaren, D. S., "The Nutritional State of the Third World's Children," in *Textbook of Pediatric Nutrition.* McLaren D. S., Burman, D., Bellton, N. R., and Williams, A. F., eds. Third edition. (Edinburgh: Churchill Livingstone, 1991).

McLoughlin, J. A,. and Nall, M., "Teacher's Opinion of the Role of Food Allergy on School Behavior and Achievment," *Ann Allergy* 61 (1988):89–91.

Meyers, "Child Nutrition Programs: Issues for the 101st Congress," *School Food Service Research Review* 13 (1989):6–8.

Meyers, A. F., et al., "School Breakfast Program and School Performance," *Am J Dis Child* 143 (1989):1234–39.

Micronutrient Deficiency Information System (MDIS), "Global Prevalence of Iodine Deficiency Disorders" (Geneva: WHO, 1993).

Miller, J. E., and Korenman, S., "Poverty and Children's Nutritional Status in the United States," *Am J Epidemiol* 140 (1994):233–43.

Milner, J. A., "Trace Elements in the Nutrition of Children," *J Pediatr* 117 (1990):S147–55.

Morgan, K. J., et al., "The Role of Breakfast in Nutrient Intake of 5- to 12-Year-Old Children," *Am J Clin Nutr* 34 (1981):1418.

National Academy of Sciences, Food and Nutrition Board of the National Research Council, *Recommended Dietary Allowances*, 10th ed. (Washington D. C.: National Academy Press, 1989).

National Center for Health Statistics, "Dietary Intake Source Data," *DHEW* publ. no. 79-1221 (Hyattsville, MD: NCHS, 1979).

National Center for Health Statistics, "NCHS Growth Charts, 1976," *Monthly Vital Statistics Report*, vol. 25, no. 3, suppl (HRA) 78-1120 (Rockville, MD: Health Resources Administration, June 1976).

National Cholesterol Education Program, "Report of the Expert Panel on Blood Cholesterol Levels in Children and Adolescents," *NIH Publication* no. 91-2732, (Bethesda MD: National Heart Lung and Blood Institute, 1991).

Needleman, H., and Bellinger, D. , "The Health Effects of Low Level Exposure to Lead," *Annu Rev Publ Health* 12 (1991):111–10.

Newman, T. B., Browner, W. S., and Hulley, S. B., "The Case Against Childhood Cholesterol Screening," *JAMA* 264 (1990):3039–43.

Nutrition Collaborative Research Support Program, "Functional Implications of Malnutrition. Final Report," Human Nutrition Collaborative Research Support Program, 1992.

Oski, F. A., "Iron Deficiency in Infancy and Childhood," *N Engl J Med* 329 (1993):190–93.

Oski, F. A., "The Causes of Iron Deficiency in Infancy," in *Dietary Iron: Birth to Two Years*, ed. L. J. Filer, Jr., (New York: Raven Press, 1989).

Owen, G. M,. and Lubin, A. H., "Anthropometric Differences Between Black and White Preschool Children," *Am J Dis Child* , 126 (1973):168–69.

Owen, G. M., "Iron Nutrition and Growth in Infancy," in *Dietary Iron: Birth to Two Years*, ed. L. J. Filer Jr., (New York: Raven Press, 1989).

Pelletier, D. L., "The Potentiating Effects of Malnutrition on Child Mortality: Epidemiologic Evidence and Policy Implications," *Nutr Rev* 52 (1994):409–15.

Pollock, I., and Warner, J., "A Follow-Up Study of Childhood Food Additive Intolerance," *J Royal College of Physicians London* 21 (1987): 248–50.

Prinz, R. J., Roberts, W. A., and Hantman, E., "Dietary Correlates of Hyperactive Behavior in Children," *J Consult Clin Psychol* 48 (1980):760–9.

Ray, J. W., and Klesges, R. C., "Influences on the Eating Behavior of Children," *Ann NY Acad Sci* 699 (1993):57–69.

Riby, J., "Perinatal Deveopment of Digestive Enzymes," in *Infant Nutrition*, eds. A. F. Walker and B. A. Rolls (London: Chapman and Hall,1994).

Roberts, S. B., and Young, V. R., "Energy Costs of Fat and Protein Deposition in the Human Infant," *Am J Clin Nutr* 48 (1988):951–55.

Rosen, L. E., et al., "Effects of Sugar (Sucrose) on Children's Behavior," *J Consult Clin Psychol* 56 (1988):583–9.

Rugg-Gunn, A. J., and Hackett, A. F, *Nutrition and Dental Health* (Oxford: Oxford University Press, 1993).

Sampson, H. A., Broadbent, K. R., and Bernhisel-Broadbent. J., "Spontaneous Release of Histamine from Basophils and Histamine-Realising Factor in Patients with Atopic Dermatitis and Food Hypersensitivity," *N Engl J Med* 321 (1989):228–32.

Sampson, H. A. , Mendelson, L., and Rosen, J. P., "Fatal and Near-Fatal Anaphylactic Reactions to Food in Children and Adolescents," *N Engl J Med* 327 (1992):380–4.

Schaffer, S. J,. and Campbell, J. R., "The New CDC and AAP Lead Poisoning Prevention Recommendations: Consensus vs. Controversy," *Ped Annals* 23 (1994):592–9.

Sheard, N. F., "Iron Deficiency and Infant Development," *Nutr Rev* 52 (1994):137–46.

Shull, M. W., et al., "Velocities of Growth in Vegetarian Preschool Children," *Pediatrics* 60(4) (1977):410–17.

Simeon, D. T., and Grantham-McGregor, S., "Effects of Missing Breakfast on the Cognitive Functions of School Children of Differing Nutritional Status," *Am J Clin Nutr* 49 (1989):646–53.

Snyder, M. P., Story, M., and Trenkner, L. I., "Reducing Fat and Sodium in School Lunch Programs: The LUNCHPOWER! Intervention Study," *J Am Diet Assoc* 92 (1992):1087–91.

Sommer, A., et al., "Impact of Vitamin A Supplementation on Childhood Mortality: A Randomized Controlled Community Trial," *Lancet* I (1986):1169–173.

Sporn, M. B., Roberts, A. B., and Goodman, D. S., eds., *The Retinoids*, 2d ed. (New York: Raven Press, 1994).

Stanek, K., et al., "Diet Quality and the Eating Environment of Preschool Children," *J Am Diet Assoc* 90 (1990):1582.

Strasburger, V. C., "Children, Adolescents, and Television: The Role of Pediatricians," *Pediatrics* 83 (1989):446.

Tanner, J. M., *Fetus into Man: Physical Growth from Conception to Maturity* (Cambridge: Harvard University Press, 1990).

Taylor, M. L., and Koblinsky, S. A., "Dietary Intake and Growth Status of Young Homeless Children," *J Am Diet Assoc* 93 (1993):464–66.

Tayter, M., and Stanek, K. L., "Anthropometric and Dietary Assessment of Omnivore and Lacto-Ovo-Vegetarian Children," *J Am Diet Assoc* 89 (1989):1661–63.

Thompson, F. E., and Dennison, B. A., "Dietary Sources of Fats and Cholesterol in U. S. Children aged Two-Five Years," *Am J Pub Health* 84 (1994):799–806.

U. S. General Accounting Office, "Children and Youths: About 68,000 Homeless and 186,000 in Shared Housing at Any Given Time," (Washington, D. C.: U. S. General Accounting Office, 1989).

U. S. Public Health Service, Agency for Toxic Substances and Disease Registry, "The Nature and Extent of Lead Poisoning in Children in the United States: A Report to Congress" (Atlanta: U. S. Department of Health and Human Services, 1988).

U. S. Conference of Mayors, *A Status Report on Hunger and Homelesssness in America's Cities: 1990. A 30-City Survey*, (Washington, D. C.: U. S. Conference of Mayors, 1990).

U. S. Department of Agriculture "Nationwide Food Consumption Survey. Nutrient Intakes: Individuals in 48 States, Year 1977–78," Report no. I-2 (Hyattsville, MD: Consumer Nutrition Division, Human Nutrition Information Service, U. S. Department of Agriculture, 1984).

U. S. Department of Agriculture, Food and Nutrition Service, "National School Lunch Program Regulations," *Federal Register*, 16 Nov. 1982.

U. S. Department of Agriculture, Human Nutrition Information Service, Nutrition Monitoring Division, "Nationwide Food Consumption Survey. Continuing Survey of Food Intakes by Individuals: Women 19–50 Years and Their Children 1–5 Years, 4 Days, 1985," Report no. 85-4. Nutrition Monitoring Division, Human Nutrition Information Service. (Hyattsville, MD: U. S. Department of Agriculture, 1987).

U. S. Dept. of Health and Human Services, Bureau of Maternal and Child Health and Resources Development, Child Health USA '89 (Washington, D. C.: U. S. Department. of Health and Human Services, 1989).

Uvin, P., "The State of World Hunger," *Nutr Rev* 52 (1994):151–61.

Van Asperen, P. P., Kemp, A. S., and Mellis, C. M., "Immediate Food Hypersensitivity: Reactions on First Known Exposure to Food," *Archives of Desease in Childhood* 58(1983):253–56.

Van Staveren, W. A., et al., "Food Consumption and Height/Weight Status of Dutch Preschool Children on Alternative Diets," *J Am Diet Assoc* 85 (1985):1579–84.

Walker, A. R. P., and Walker, B. F., "School Nutrition Programmes—Do They Fulfill Their Purpose?" *Hum Nutr: Apl Nutr* 40A (1986):125–35.

Walravens, P. A., Krebs, N. F., and Hambidge, K. M., "Linear Growth of Low-Income Preschool Children Receiving a Zinc Supplement," *Am J Clin Nutr* 38 (1983):195–201.

Webber, L. S., et al., "Tracking of Serum Lipids and Lipoproteins from Childhood to Adulthood: The Bogalusa Heart Study," *Am J Epidemiol* 133 (1991):884–99.

Wolraich, M. L., et al., "Effects of Diets High in Sucrose or Aspartame on the Behavior and Cognitive Performance of Children," *N Engl J Med* 330 (1994):301–7.

Wong, N. D., et al., "Television Viewing and Pediatric Hypercholesterolemia," *Pediatrics* 90 (1992):75-9.

World Bank, *World Development Report 1993* (New York: Oxford University Press, 1993).

World Health Organization, "Energy and Protein Requirements. Report of a Joint FAO/WHO/UNU Expert Consultation," *Technical Report Series* 724 (Geneva: WHO, 1985).

World Health Organization, "Malnutrition and Mental Development," *WHO Chronicle* 28 (1974):95–102.

World Health Organization, "National Strategies for Overcoming Micronutrient Malnutrition," EB 89/27 (Geneva: WHO, 1991).

Yip, R., "The Changing Characteristics of Childhood Iron Nutritional Status in the United States," in *Dietary Iron: Birth to Two Years*, ed. L. J. Filer, Jr., (New York: Raven Press, 1989).

Young, E. A., et al., "Fast Foods Update: Nutrient Analysis," *Dietetic Currents* 13, no. 6 (1986).

adolescence and nutrition

▼

ADOLESCENCE, SELF-IMAGE, AND DIET

Recently, as part of a nationwide survey, female students in U.S. high schools were asked the following questions:

- Do you think you are too thin, too fat, or about the right weight?

- Are you trying to lose or gain weight?

- During the past week, did you skip meals either to try to lose weight or keep from gaining weight?

- Have you taken diet pills or vomited on purpose either to try to lose weight or keep from gaining weight?

The results of the survey were striking. Over two thirds were dieting to lose weight or keep from gaining weight. Half reported skipping meals regularly in an attempt to lose or maintain weight. Twenty-one percent reported using diet pills, and 14% admitted using self-induced vomiting to try to lose or control weight (Centers for Disease Control 1990).

Diet, body weight, and self-image are major concerns in today's society and are clearly a central preoccupation of most teenagers. For most of this century, and particularly over the past 25 years, the U.S. cultural ideal of a woman's body has evolved toward a thinner shape. At the same time, the female population as a whole has steadily gained weight. Magazines are filled with advertisements for high-calorie foods on one page and diets for weight loss on the next.

Adolescents today, struggling to form a stable, positive self-image during pubertal development—and growing up in a food-centered society that prizes thinness—are caught in a bind. Images in the media and advertising, and sometimes peers and family, pressure young women to pursue unrealistic ideals for body image and weight. Yet nutritional needs are higher in adolescence than at any other time in life. Limiting food choices and strenuous dieting can have a profound impact on immediate and long-term health.

Adolescence is a time of rapid physical, physiological, social, and emotional changes that influence eating patterns and nutritional needs. About 15% of the U.S. population—over 35 million people—are teenagers (Joffe 1994). This chapter focuses on the unique nutritional needs of this age group and discusses the impact of diet on growth and development during puberty.

BIOLOGICAL CHANGES OF ADOLESCENCE

Adolescence is the period between childhood and adulthood, or, the life stage from the beginning of **puberty** until maturity. Generally, adolescence in females spans the years from 10 to 18, in males from 12 to 20. Adolescence is characterized by a predictable sequence of developmental changes—rapid increases in height and weight, changes in body composition, and the attainment of sexual maturity.

Throughout life, the brain produces small amounts of the follicle-stimulating hormone (FSH) and luteinizing hormone (LH). These hormones, secreted by the pituitary gland into the blood, send signals to the sex organs. LH and FSH control the production of the sex hormones testosterone (produced by the testes in the male) and estrogen and progesterone (produced by the ovaries of the female). In childhood, the brain produces only small amounts of FSH and LH, but two to three years before puberty, production increases markedly. This stimulates a sudden upsurge in the production of the sex hormones by the ovaries and testes.

The dramatic changes of puberty are caused by the sudden appearance of these sex hormones, but what is the trigger that produces this

puberty: the stage of adolescence in which a person becomes physiologically capable of sexual reproduction

surge in hormones that initiates puberty? It appears that nutritional factors play a significant role.

Many scientists think the trigger that signals the initiation of puberty is the attainment of a **critical body weight** or body composition. Generally, when a child reaches approximately 30 kg (66 lb), or a body composition of about 17% fat, a signal is given for the production of hormones that produce the adolescent growth spurt. Menarche in females also has been linked to attainment of a critical body weight and body composition: a weight of about 47 kg (104 lb), or about 17–22% body fat, appears to be necessary for the beginning of menstruation (Frisch 1991), (Frisch and McArthur 1974). Therefore, nutritional status may be an important determinant of growth and maturity.

THE GROWTH SPURT IN FEMALES VERSUS MALES

The height growth spurt in females begins at about 9 or 10 years of age, and the weight growth spurt follows about six months later. The period of rapid growth lasts for two and a half to three years, with the velocity of height growth peaking at about age 12. Sexual development occurs simultaneously with growth: the breast begins to develop at about age 9 to 10, coinciding with the onset of the height spurt; pubic and axillary hair then appear.

Menarche follows the growth spurt, occurring about 9 to 12 months after the peak in height growth velocity—about 12 and 13 years of age in most females—and represents the attainment of sexual maturity in women. Growth continues after menarche, but at a much lower rate (most females will grow only two to three more inches after their first menstrual period) (Tanner 1990). Females achieve their final adult stature about age 18 to 20.

Most males begin their growth spurt at age 12, two years later than females, and reach peak growth velocity at age 14. Sexual development coincides with the growth spurt—early changes (enlargement of the testes and penis) occur at age 11 to 12. Most of the **secondary sexual characteristics** and the achievement of sexual maturity are complete within two years after the onset of puberty. Later manifestations of puberty, such as the deepening of the voice and the appearance of axillary and facial hair, take place after peak growth. Muscle development, growth of body hair, and height gains may continue at a slower rate until age 19 to 25.

Although patterns of growth for both sexes are similar, as shown in Figures 8.1 and 8.2, males reach greater adult height because of their longer period of preadolescent growth and their greater peak growth velocity during the growth spurt. On average, adolescent males gain 8 inches of height during the growth spurt, and females gain 6 inches. Males also gain more weight, adding an average of 45 pounds, while females add about 35 pounds. (Tanner 1991).

critical body weight: the concept that the attainment of a minimal body weight during growth triggers pubertal changes

secondary sexual characteristics: the changes associated with pubertal development that include, in females, broadening of the hips, enlargement of the breast and areola, and development of pubic and body hair; in the male, development of the penis and testes, growth of body hair, and deepening of the voice

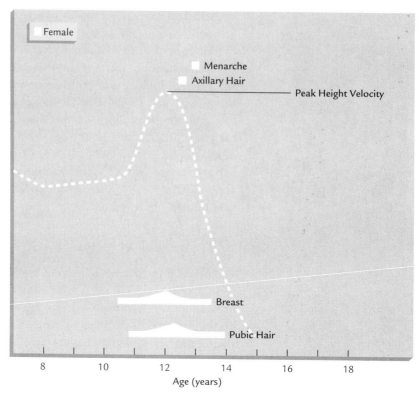

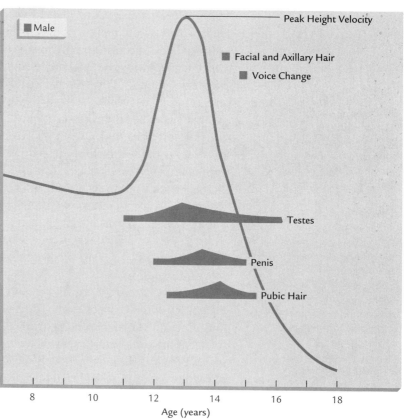

FIGURE 8.1

Sequence of growth and pubertal events in females and males

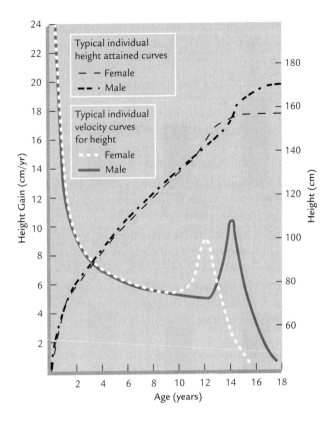

FIGURE 8.2

Velocity and cumulative curve for height in males and females

BODY COMPOSITION

Changes in body composition during adolescent growth show marked gender differences, as shown in Figure 8.3. Before puberty, the proportions of lean body tissue and fat in both sexes are similar—body fat is about 16 to 18%, and muscular strength is roughly equal. During adolescent growth, females gain proportionately more fat, and males gain proportionately more lean tissue (muscle and bone). By adulthood, women have twice as much body fat as men, while men have 50% more lean body mass than women. These normal adult gender differences in body fat—23% for females and 12% for males—are formed during puberty .

Gender differences in adolescent growth create differences in nutritional requirements. Because males grow more rapidly and gain more lean tissue, they generally require more protein, energy, calcium, and zinc than females.

THE SECULAR TREND

In the industrialized countries since the early 1800s, successive generations of adolescents have been generally growing taller and heavier, and attaining puberty at younger ages. Menarche has been occurring in

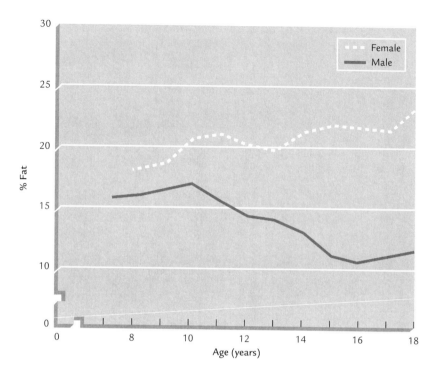

FIGURE 8.3

Gender and age differences in percentage of body fat

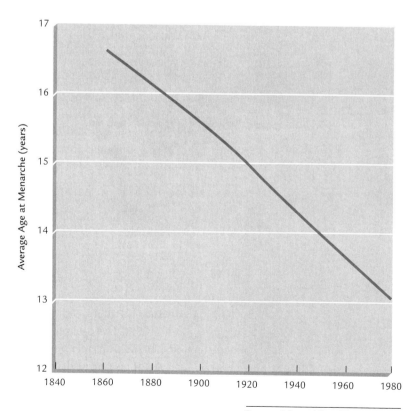

FIGURE 8.4

Declining age of menarche in developing countries

Source: Adapted from Tanner (1973)

females at younger ages—the grandmothers of today's teenagers began puberty about two years later than their granddaughters do today (see Figure 8.4). The decline in the age of menarche and, in males, the age of reproductive maturation, is generally attributed to better nutrition: improvements in diet and higher caloric and protein intake during childhood (Tanner 1990).

NUTRITIONAL REQUIREMENTS DURING ADOLESCENCE

The large height and weight gains of adolescence are the most visible changes, but a similar growth spurt is seen in tissues throughout the body. Muscle, skeletal, and red blood cell mass increase dramatically, with many organs doubling in size. Demands for energy, protein, certain vitamins, calcium, and iron are higher during adolescence than at any other stage in the life cycle. They are exceeded only by special requirements during pregnancy and lactation (NAS 1989).

The onset of puberty and the timing of the adolescent growth spurt can vary substantially between children. Earlier developing females may start the growth spurt as young as 7 or 8, while later developing males may not begin rapid growth until age 13 or 14. This variability makes it difficult to evaluate the nutritional needs of adolescents based solely on their chronological age.

For example, two 12 year-old males, one at the peak of growth velocity and one who has not yet entered the growth spurt, will have dramatically different nutritional needs. Adolescents at the peak of the growth spurt may need up to twice the protein, calcium, iron, and zinc compared with others in the same age group who are not in the growth spurt (Forbes 1981). See Table 8.1.

RECOMMENDED DIETARY ALLOWANCES DURING ADOLESCENCE

The RDAs for adolescents are divided into separate recommendations for ages 11 to 14 and 15 to 18 (NAS 1989). The RDAs are established by estimating the average physiological requirements for populations, not individuals. When considering individual teenagers, nutritional requirements should be based more on the individual's physiological age, rate of growth, and maturity (see Table 8.2).

ENERGY NEEDS

During adolescence, energy needs are greater than at any other time in life. Although the current RDAs suggest average energy intakes for

		Average for Period 10–20 Yr (mg)	At Peak of Growth Spurt (mg)
Calcium	M	210	400
	F	110	240
Iron	M	0.57	1.1
	F	0.23	0.9
Nitrogen	M	320	610
	F	160	360
Zinc	M	0.27	0.50
	F	0.18	0.31
Magnesium	M	4.4	8.4
	F	2.3	5.0

Adapted from G. B. Forbes, "Nutritional Requirements in Adolescence," in *Textbook of Pediatric Nutrition,* ed. R. M. Suskind (New York: Raven Press, 1981).

TABLE 8.1
Differences in daily increments in body content of selected nutrients during adolescence

				Fat-Soluble Vitamins		
Age (yr)	Energy (kcal)	Protein (g)	Vitamin A (µg RE)	Vitamin D (µg)	Vitamin E (mg α-TE)	Vitamin K (µg)
Males						
11–14	2500	45	1000	10	10	45
15–18	3000	59	1000	10	10	65
Females						
11–14	2200	46	800	10	8	45
15–18	2200	44	800	10	8	55

	Water-Soluble Vitamins					
Vitamin C (mg)	Thiamin (mg)	Riboflavin (mg)	Niacin (mg NE)	Vitamin B6 (mg)	Folate (µg)	Vitamin B$_{12}$ (µg)
Males						
11–14	50	1.3	1.5	17	1.7	150	2.0
15–18	60	1.5	1.8	20	2.0	200	2.0

TABLE 8.2
1989 U.S. RDAs for ages 11 through 18

TABLE 8.2 *Continued*

Water-Soluble Vitamins continued

	Vitamin C (mg)	Thiamin (mg)	Riboflavin (mg)	Niacin (mg NE)	Vitamin B6 (mg)	Folate (µg)	Vitamin B$_{12}$ (µg)
Females							
11–14	50	1.1	1.3	15	1.4	150	2.0
15–18	60	1.1	1.3	15	1.5	180	2.0

Minerals

	Calcium (mg)	Phosphorus (mg)	Magnesium (mg)	Iron (mg)	Zinc (mg)	Iodine (µg)	Selenium (µg)
Males							
11–14	1200	1200	270	12	15	150	40
15–18	1200	1200	400	12	15	150	50
Females							
11–14	1200	1200	280	15	12	150	45
15–18	1200	1200	300	15	12	150	50

Estimated Sodium Chloride and Potassium—Minimum Requirements of Healthy Persons

	Weight (kg)	Sodium (mg)	Chloride (mg)	Potassium (mg)
10–18	50.0	500	750	200

Estimated Safe and Adequate Daily Dietary Intakes (1989)

	Vitamins		Minerals				
Age (yrs)	Biotin (µg)	Pantothenic Acid (mg)	Copper (mg)	Manganese (mg)	Fluoride (mg)	Chromium (µg)	Molybdenum (µg)
Males							
11–18	30–100	4–7	1.5–2.5	2.0–5 0	1.5–2.5	50–200	75–250
Females							
11–18	20–100	4–7	1.5–2 5	2.0–5 0	1.5–2 5	50–200	75–250

adolescents at different ages (see Table 8.3), energy needs for individual teenagers vary greatly because of varying growth rates, gender, physical activity, and body size and composition (Lifshitz et al. 1993).

Dietary surveys indicate that average energy intakes are greatest in males at age 16 (3400 kcal/day) and in females at 12 to 13 (2600 kcal/day) (LSRO 1989). This is illustrated in Figure 8.5. Female energy needs peak at earlier ages and typically fall in later adolescence, particularly if physical activity is not maintained. Because individual needs vary considerably with growth rate and physical activity, the best indication that teenage energy intake is appropriate is maintenance of an acceptable rate of growth and a desirable weight.

Age (years)	Average Allowance (kcal/cm height)*	
Males		
11–14	15.9	
15–18	17.0	
Females		
11–14	14.0	
15–18	13.5	2

*Average energy allowance (kcals) and median height (cm) for the age group. Adapted from National Academy of Sciences, Food and Nutrition Board of the National Research Council. *Recommended Dietary Allowances*, 10th ed. (Washington, D.C.: National Academy Press, 1989).

TABLE 8.3

Recommended energy intakes for adolescents in the United States

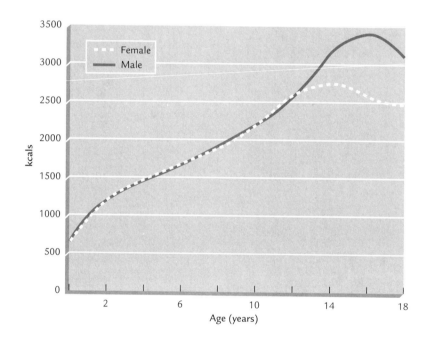

Source: Adapted from LSRO (1989)

FIGURE 8.5

Typical daily caloric intakes for males and females from birth to 18 years old

FAT REQUIREMENTS

Some teenagers consume fat-restricted diets, either by personal choice or because they are guided by parents seeking perceived health benefits. Although the average adolescent diet is high in fat (Kimm et al. 1990)— and lowering fat intake to recommended levels is desirable—overzealous restriction of dietary fat may result in dietary inadequacies and impaired growth (Lifshitz and Moses 1989), (Pugliese et al. 1983). Because fat is an important energy source for growing teenagers with high caloric needs, strenuous fat restriction may lead to inadequate energy consumption. Adequate dietary supply of the essential fatty acids is important for normal growth.

Also, avoiding fat by eating fewer animal products, such as eggs and red meat, may contribute to nutrient deficiencies common in adolescence, as these foods are important sources of iron and zinc (Lifshitz et al. 1993). Limiting portion sizes of high-fat items and substituting reduced-fat animal products can help meet healthy dietary guidelines without compromising energy and micronutrient intakes.

PROTEIN NEEDS

Adolescents need generous amounts of protein, particularly during the growth spurt. A substantial portion of the weight gain during puberty is new muscle and bone, and synthesis of these tissues requires large amounts of protein. At the peak of the growth spurt, the daily increment in body content of protein in males is nearly 4 g (610 mg of nitrogen) and in females, 2.2 g (360 mg nitrogen)—over twice the average increment for the teen decade (Forbes 1981). Protein requirements during adolescence are 0.9–1.0 g/kg, and the energy value of daily protein intake should be about 8% of total energy intake (NAS 1989).

Most teenagers in the United States consume adequate calories and protein (USDA 1984). But either due to choice (intentional dieting) or because of poverty, some teens during late childhood or early adolescence consume diets providing low or marginal kcalories and protein. Chronic, mild underconsumption of calories can have significant effects in this age group, diminishing the growth rate in late childhood and early adolescence, reducing skeletal growth, and delaying the onset of puberty (Lifshitz and Moses 1988). Severe and protracted undernutrition can cause permanent stunting of growth.

VITAMIN REQUIREMENTS

Fat-Soluble Vitamins

Vitamin D requirements for adolescents are higher than those for adults because of the rapid skeletal growth during the teenage years. Although bones reach their full size by the late teens, bone density continues to increase throughout the 20s (Recker et al. 1992). Thus, the RDA for vitamin D remains elevated in the 19 to 24-year-old age group to ensure deposition of calcium and other minerals into the maturing skeleton.

Vitamin A plays a central role in cellular differentiation and proliferation and is essential for normal growth (Sporn et al. 1994). Vitamin A requirements are calculated on the basis of both body weight and needs for growth. The requirement for vitamin A during adolescent growth is higher than amounts needed later to maintain adequate reserves in adults (NAS 1989). Therefore, the RDA remains constant through adolescence and adulthood as the growth rate falls but the body weight increases.

Adequate vitamin E is important during adolescence to maintain the structural and functional characteristics of the rapidly increasing numbers of new cells. The RDA for vitamin E increases to 10 mg for males and 8 mg for females, which is the same allowance as for adults (NAS 1989). Requirements for vitamin K are 1 μg/kg body weight (Sutie et al. 1988), so the RDA increases through adolescence as body size increases.

Water-Soluble Vitamins

Requirements for thiamin, riboflavin, and niacin peak during the teenage years. Demand for these B vitamins increases in proportion to energy needs—and caloric requirements are highest during adolescence (NAS 1989).

Vitamin C plays a central role in the formation of collagen, an important structural protein found throughout the body. During adolescence, rapid growth of connective tissues and the musculoskeletal system requires large amounts of new collagen. The RDA for vitamin C increases to 50 mg in the 11 to 14-year-old age group, and 60 mg in the 15 to 18 age group (NAS 1989). Adolescents who smoke cigarettes need nearly twice the amount of vitamin C of nonsmokers—at least 100 mg/day (NAS 1989).

Amino acid incorporation into new proteins is very high during adolescent growth. Synthesis of plasma proteins, muscle, and connective tissue requires ample vitamin B6. The requirement for vitamin B6 for adolescents is based on protein intake, and the RDA is calculated using the ratio of 0.02 mg of vitamin B6/g protein (NAS 1989).

Rapid cell division and DNA and RNA synthesis in growing tissues increase the demand for folate and vitamin B12. Although the 1989 RDA for folate for adolescent females is set at 150 mg for 11 to 14-year-olds, and 180 mg for 15 to 18-year-olds (NAS 1989), adolescent females who have begun their menstrual periods and are capable of becoming pregnant may need higher amounts. The U.S. Public Health Service recommends that all women of childbearing age who are capable of becoming pregnant consume at least 400 mg of folic acid per day (Centers for Disease Control 1992). Higher folate intake by women during the periconceptional period may reduce the risk of certain birth defects (Centers for Disease Control 1992). Links between dietary folate and neural tube defects are discussed in Chapter 2.

MINERAL REQUIREMENTS

During the teenage years, mineral needs are high. Calcium and phosphorus needs are higher during adolescence than at any other age because of the rapid growth of the skeleton. About half of bone growth occurs during adolescence—most of it during the adolescent growth spurt (Christansen et al. 1975). The RDA for calcium for males and females ages 11 to 24 years is 1200 mg (NAS 1989), but intake of most adolescents is under 800 mg/day (USDA 1984). The connections between calcium intake during adolescence and early adulthood and lifelong bone health are discussed later in this chapter.

About 60% of the body's magnesium is in the skeleton, and much of the remainder is incorporated into muscle (Aikawa 1981). Because the musculoskeletal system undergoes rapid growth during adolescence, magnesium needs are high. The daily requirement for magnesium in adolescence is 6.0 mg/kg body weight. As growth slows in the late teens, daily requirements fall to those of adulthood—about 4.5 mg/kg (NAS 1989).

Trace Elements

Adolescent growth increases the requirement for iron, which is needed for incorporation into hemoglobin in new red cells as blood volume expands (see Figure 8.6). Large amounts of iron are also used in the synthesis of myoglobin in growing muscle. The increase in muscle mass and the increase in blood volume, hemoglobin, and red blood cell mass are greater in males than females. The demand for iron in males peaks during the years 11 to 18, with requirements of 12 mg/day; requirements then fall to 10 mg/day in adulthood, as growth slows. Iron requirements

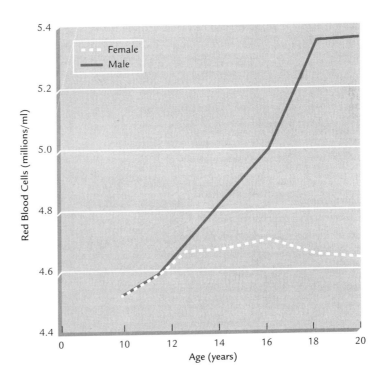

FIGURE 8.6

Increase in red blood cell mass during adolescence by gender

jump to 15 mg/day in adolescent females to replace iron lost in menstrual bleeding and supply needs for growth.

During adolescent growth, large amounts of zinc are deposited into developing muscle and bone. Particularly at the peak of the growth spurt, the daily demand for zinc is high. Almost twice the amount of zinc is deposited into developing tissues during the growth spurt than is deposited during the remaining teen years (Forbes 1981). Researchers studied a group of young Middle Eastern men whose diets were very low in bioavailable zinc (Presad 1982). In these zinc-deficient adolescents, growth was markedly impaired and sexual development delayed. Refer to Table 8.2 for the RDAs for the remaining trace elements.

WATER AND ELECTROLYTE REQUIREMENTS

The committee that establishes the RDAs has estimated minimum requirements for the electrolytes sodium, potassium, and chloride for adolescents, as shown in Table 8.2. Water requirements for adolescents are approximately 1.0–1.5 ml/kcal energy expenditure (NAS 1989). However, needs will vary greatly depending on ambient temperature and humidity, and they can be substantially increased by illness, fever, vomiting, or diarrhea.

NUTRITIONAL HEALTH OF ADOLESCENTS IN THE UNITED STATES

Because of rapid growth, nutritional demands are higher during adolescence than at any other stage in the life cycle, yet many teenagers do not eat well. In the U. S., dietary inadequacies are found more often among adolescents than in any other age group (NCHS 1979), (USDA 1984).

ENERGY INTAKE

National surveys show energy intakes below recommended levels for many females aged 12 to 17, Blacks of both sexes, and adolescents from lower socioeconomic groups (USDA 1984). Low energy intake in adolescence may reflect intentional dieting, economic limitations, or decreased needs due to low physical activity. Most adolescent males obtain recommended levels of caloric intake, while many older teenage females do not.

PROTEIN INTAKE

Almost all adolescents in the United States achieve adequate protein intake. Calories from protein represent 14% of the energy in many adolescent diets, which is well above recommended levels (LSRO 1989).

FAT INTAKE

Similar to adult diets in the United States, most adolescent diets contain more total and saturated fat, more sodium, and less fiber than is currently recommended. A recent survey found that 27% of high school females and 43% of high school males eat three or more servings a day of foods with high fat content (ASHA 1989). The typical adolescent diet contains 36% calories from fat, 13% of which are saturated fat (Kimm et al. 1990). Expert committees recommend that diets of adolescents should contain less then 30% of calories as fat—with less than 10% from saturated fat—and dietary cholesterol should be less than 300 mg/day (NCEP 1992).

Atherosclerosis and Adolescence

Atherosclerosis is usually considered a disease of older adults, but among young American soldiers killed in Vietnam, nearly half showed evidence of early atherosclerosis (McNamara 1971). Because of these and other findings, the U.S. National Cholesterol Education Program (NCEP) recommmends limits on dietary fat, similar to those in adults, for older children and adolescents (NCEP 1992). Dietary fat and cholesterol in childhood and adolescence are discussed in Chapter 7.

MICRONUTRIENT INTAKE

Diets of teenage males more consistently provide adequate amounts of most micronutrients than those of teenage females (LSRO 1989). Males from 14 to 18 years-old are much more likely than females at the same ages to consume the recommended five or more servings of fruits and vegetables a day (USDA 1984). In general, males eat more food than females, and the extra 800 to 1000 kcalories each day increase their chances of obtaining adequate amounts of micronutrients.

Because of their lower energy requirements, adolescent females must be careful to select nutrient-dense foods. Studies show that many teenage females, concerned about weight control, regularly consume only 1600 to 1800 calories per day (LSRO 1989). At this level of intake, unless foods are chosen wisely, obtaining adequate amounts of micronutrients is difficult—particularly nutrients that are present in low concentrations in most foods, such as iron, magnesium, and vitamin B6.

VITAMIN INTAKE

The vitamin A and C intakes of adolescents are often inadequate, with less than half of adolescents in the United States regularly obtaining the RDA for vitamin A. Less than two thirds of teenagers consume the RDA for vitamin C, reflecting the lack of fresh fruit and vegetables in the diets of many teenagers. Only about one in seven high school students in the United States consume the recommended five or more fruits and vegetables daily (LSRO 1989).

Vitamin B_6 intake is frequently low. About 50% of adolescent females obtain less than two thirds of the RDA for vitamin B_6; one study found evidence of marginal or deficient vitamin B_6 status in half of adolescent females, whose mean daily intake of vitamin B_6 was only 1.25 mg (Driskell et al. 1987). Other B vitamins, particularly folate, niacin, thiamin, and riboflavin, are often deficient in teenage diets. In a study of 400 U.S. adolescents, dietary intake of folate was generally poor, and 40% of those surveyed had low levels of folate in red blood cells (indicating poor folate status) (Bailey and Cerda 1988).

CALCIUM INTAKE

Many adolescents, particularly females, do not obtain enough calcium in their diets. Only about one in five teenage females obtain the RDA for calcium (1200 mg), and the median daily intake in adolescent females is less than 800 mg—only 66% of the U.S. RDA. Black adolescents tend to consume less calcium than their White counterparts, primarily because teenage Blacks eat fewer dairy products—perhaps due to the higher prevalence of lactose intolerance among Blacks. (LSRO 1989).

IRON INTAKE

Iron deficiency is the most common dietary deficiency in adolescents in the United States. It is most common in older adolescent females, females with heavy menstrual bleeding, teenagers of both sexes from lower socioeconomic groups, and those consuming vegetarian diets (Lifshitz et al. 1993).

The amount of iron in the average U.S. diet is approximately 6 mg/1000 kcal. Because the daily energy needs of many teenage females is only 2000–2200 kcal, they may have difficulty consuming 15 mg of iron every day. A national survey of iron status in teenagers found that 6% of females aged 11–14 and 14% of those aged 15–19 have low iron stores. In males, 12% of those 11–14 and 1% of those 15–19 have low stores. Anemia, a late stage of iron deficiency, was found in 5% of males aged 11–14, 2.5% of females 11–19, and less than 1% of older adolescent males (LSRO 1989).

Increasing evidence suggests that iron-deficiency anemia, even when asymptomatic, can have adverse effects on cognitive function in adolescents (Ballin et al. 1992). Also, iron deficiency, which is more common among adolescent athletes than nonathletes, decreases exercise capacity and endurance (Nickerson et al. 1989). For both males and females, diet during adolescence should emphasize iron-rich foods.

INFLUENCES ON THE ADOLESCENT DIET

Adolescents face dramatic change and upheaval in their lives, not only the physical changes of puberty and growth but also broad psychological and psychosocial changes. All of these factors affect food choices and nutrition during this developmental stage.

INFLUENCE OF FAMILY AND PEERS

For most teenagers, the family provides major structure and support during this turbulent time. Younger adolescents generally see their attitudes toward drugs, education, sex, and most other matters as being close to their parents'. Particularly in early adolescence, teenagers continue to eat most meals at home, and parental influences on food choices and diet remain strong.

However, as adolescents begin to develop self-identities and establish independence from their parents, peer acceptance becomes important. Teenagers do not want to differ substantially from their friends in rate of growth, body weight, or behavior. Social pressures influence what teenagers eat, as they learn new food preferences and abandon old eating habits. Peer attitudes also influence decisions about alcohol, tobacco, and drug use—decisions that have a major impact on nutritional health.

Because of the emphasis on appearance and physique, teenagers often experiment with diet and supplements, trying to improve the way they look or feel. Over a third of U.S. teenagers take vitamin and mineral supplements regularly, believing they can improve athletic performance or clear their complexions (Kovar 1985). Adolescents begin to make more decisions for themselves, including choosing what and where to eat.

Busy with school, after-school activities, social life, and often a job, adolescents develop irregular eating habits. They skip meals, and eat many meals away from home (ASHA 1989). In contrast to the family dinner table, meals outside of the home are often bought at vending machines, fast-food restaurants, and convenience or grocery stores. Many of these meals are of low nutritional value. Studies have shown that as the number of meals eaten away from home increases, adolescent diets are more likely to be inadequate in a number of micronutrients (Bull 1992). Energy-dense, but "micronutrient-empty" foods, such as soft drinks, baked goods, candy, french fries, and other fried foods contribute significantly to the energy intake of most teenagers (Bigler-Doughten and Jenkins 1987).

Skipping Breakfast

Skipping breakfast is widespread among adolescents, and it establishes an eating habit that may persist into adulthood. In a large recent study of dietary habits in adolescence, one in three eighth graders and nearly half of tenth-grade students surveyed reported eating breakfast less than three times a week (ASHA 1989). Breakfasts are important sources of calcium, vitamin C, and riboflavin—nutrients that are important and often deficient in teenage diets.

FAST FOODS AND SNACKS

Fast food is often criticized for its perceived contribution to the poor diets of adolescents. Many current items on fast food menus have more than 50% of calories from fat and sodium content is often high. Certain nutrients are usually lacking, particularly vitamins A, C, and folate (Young et al. 1986). For these reasons, fast foods should not be a major part of adolescent diets (Portnoy and Christenson 1989).

However, in moderation fast foods can be part of an acceptable diet for teenagers, as many fast foods have considerable nutrient content. For example, a meal of a broiled hamburger with lettuce and tomato, french fries, and a milkshake has a fairly good nutritional profile, at a calorie level growing teens can afford (Young et al. 1986). If breakfast and dinner compensate for the nutrients that are missed in the fast food meal, daily intake of most nutrients will be adequate. The addition of fruit juice and salad bars to the menus of many fast-food restaurants has improved their nutritional value significantly.

Snacking is a way of life for teenagers—over 90% of teenagers eat regularly between meals, and nearly 50% eat an average of three snacks per day (ASHA 1989). Many adolescents obtain up to a third of their daily energy intake from snacks (Bigler-Doughten and Jenkins 1987). Rapidly growing and physically active adolescents have very high energy demands, and snacks can be important calorie sources.

However, many teenagers make poor choices in snacking—60% of the snack foods consumed by teens are high in fat or low in nutritional quality, (Witischi et al. 1990). Snack foods commonly eaten by teenagers lack vitamins A and C, folate, calcium, and iron. Dietary surveys have found that the snacks most often consumed by adolescents include cookies and other baked goods, milk, soft drinks, chips and other salty snack foods, fruit, and frozen desserts (ASHA 1989). Teenage snacking is not necessarily harmful—snack foods, if chosen wisely, can contribute substantially to nutritional needs of adolescence. The focus should be not on trying to reduce snacking but on making healthier choices. Table 8.4 lists some nutritious fast foods and snacks.

- Cheeseburger with lettuce and tomato
- Lean roast beef on a bun
- Grilled chicken sandwich
- Peanut butter sandwich
- Pizza with green peppers or other vegetables
- Cheese sandwich with sliced tomato
- Chef's salad, tossed salad, coleslaw, carrot sticks
- Soft taco with chicken or beef and vegetables
- Bean burrito
- Chinese vegetables and meat with plain rice
- Hard-cooked egg
- Milk, low-fat milk, chocolate milk, milkshakes
- Flavored yogurt
- Fresh fruit
- Fruit juice
- Dried fruit and nuts

TABLE 8.4
*Nutritious fast foods
and snacks*

NUTRITIONAL RISK FACTORS IN ADOLESCENCE

A recent survey of U.S. teenagers indicated that certain factors regularly contribute to nutritional inadequacy in adolescents (ASHA 1989). Skipping meals regularly and poor choices of snack foods were common causes of inadequate intake of nutrients. Other behaviors contributing to less than optimal diets during adolescence are listed in Table 8.5. Diets of males generally improved as they went through adolescence, while females ate less nutritiously as they moved into their later teens.

ADOLESCENT HEALTH PROBLEMS AND NUTRITION

ACNE

Many teenagers are bothered by acne. Acne is chronic irritation and inflammation of the *sebaceous glands* of the skin—the glands that surround hair follicles and are found all over the body (many of them are on the face, chest, and back). They normally produce small amounts of oils and fats (called sebum) that help keep the skin and hair from drying.

TABLE 8.5
*Behavioral risk factors
for nutritional inadequacy
in adolescence*

1. Skipping meals regularly

2. Poor choices in snack foods

3. Strong fear of gaining weight

4. More meals eaten away from home without supervision

5. Lack of time for meals

6. Eating meals alone

7. Alcohol use

8. Consuming few dairy products

During adolescence, testosterone and other androgens stimulate increased production of sebum by the sebaceous glands. The sebum often accumulates inside the hair follicles, plugging them and causing inflammation and pressure, and producing a pimple. Acne is a normal characteristic of growth during adolescence—indeed almost all teenagers have some acne. Males tend to have greater amounts than females, because they produce more testosterone and other androgens. Changes in hormone levels, sun exposure, and other unknown factors cause acne to come and go, often unpredictably (Stern 1992).

Although dietary factors such as fatty foods, chocolate, and sugar have traditionally been blamed for acne, acne does not appear to be affected by diet. Studies have generally shown that there is no connection between ingestion of foods and appearance or severity of acne (Tolman 1985).

Teenagers often take vitamin and mineral supplements in an effort to treat or prevent acne, and some take large doses of vitamin A in hopes of clearing acne. Although several effective anti-acne drugs have been derived from vitamin A, they are chemically very different from the micronutrient. Swallowing vitamin A supplements will not affect acne and may produce dangerous overdoses of the vitamin (AAP 1993). Because one of the signs of severe zinc deficiency are skin blemishes that look like acne, some adolescents mistakenly try to clear their complexions with zinc supplements. However, most studies show that zinc supplements are ineffective for treatment of acne (Savin 1983). To minimize acne, teenagers should keep their skin clean with soap and water, eat a nutritious diet, avoid high doses of vitamin–mineral supplements, and seek medical help if acne becomes severe.

HYPERTENSION

High blood pressure, or *hypertension,* is a major risk factor for heart and blood vessel disease in adulthood. Over 50 million adults in the United States suffer from hypertension (Williams and Moore 1995) and for many, the disorder begins in adolescence. Transient elevations in blood pressure occur in up to 13% of adolescents, and about 1% of adolescents develop sustained high blood pressure (NHBLI 1987). The cause of most cases of hypertension in adolescents and adults is unknown.

Longitudinal studies have shown that adolescent hypertension often tracks into adulthood, particularly if there is a history of hypertension in the family (Shear et al. 1986), (Hoynarowska et al. 1985). Also, teens with blood pressure in the highest part of the normal range have a greater risk of hypertension in adulthood. For example, a 15-year-old boy with blood pressure at the top end of the normal range for his age has twice the risk of having high blood pressure at age 35, compared with a 15-year-old with average blood pressure (Rochinni et al. 1988).

Adolescent hypertension is often associated with obesity. Diet and lifestyle adjustments are the first line of therapy for adolescent hypertension (NHBLI 1987). Weight reduction in obese adolescents can have a beneficial effect on lowering blood pressure. In a study of the effects of diet and exercise in obese adolescents, average weight loss of 2.5 kg over 20 weeks resulted in a significant fall in blood pressure (Rochinni et al. 1988). Dietary restriction of sodium can also be valuable in lowering blood pressure in some adolescent hypertensives (NHBLI 1987). Hypertensive teens may benefit from avoiding alcohol and maintaining ample dietary intake of potassium.

Blood pressure in hypertensive adolescents can also be reduced by regular aerobic exercise. In a study of over 200 students aged 7–17, increased physical fitness was significantly associated with lower blood pressure in both males and females (Fraser et al. 1983). In another study, six months of exercise training in nonobese, hypertensive teens resulted in a significant lowering of blood pressure (Hagberg et al. 1983). This effect was seen even though there was no change in body weight or body fat.

Because hypertension is a "silent" disorder that can be present for years without symptoms, adolescents should have their blood pressure checked regularly. To reduce the risk of hypertension, adolescents (particularly those with a family history of hypertension) should be encouraged to maintain a normal weight, be physically active, eat foods low in sodium and high in potassium, and avoid alcohol.

TEENAGE PREGNANCY

Over 1 million adolescents become pregnant in the United States each year, and one in eight babies born in the United States are born to teenage mothers (Davies et al. 1989). We have seen how many adolescent females have difficulty meeting their own nutritional needs for growth, and the additional burden of a pregnancy at this age places both mother and fetus at high risk for nutritional deficiency. Teenage pregnancy is discussed in detail in Chapter 2.

CALCIUM INTAKE AND BONE HEALTH

Calcium needs are higher during adolescence than at any other age because of rapid skeletal growth. Bone growth is particularly rapid during the adolescent growth spurt, when about 40 to 45% of the total skeleton is formed (Christiansen et al. 1975). Although bones have nearly attained their adult size and shape by the end of the growth spurt, bone density continues to increase and minerals are deposited into the skeleton at a slow rate through the 20s (Davies et al. 1989). Peak bone mass is finally achieved around 30.

Whether a person develops osteoporosis in later life depends, in part, on how much bone mass has accumulated during growth (Matkovic et al. 1990). More calcium deposited into the skeleton during adolescence and young adulthood means a greater calcium "bank" to draw from during aging and the **menopause**.

Peak bone mass is determined by many factors—nutrition, heredity, hormones, and physical activity all play a role. One of the most important nutritional determinants of peak bone mass is calcium intake during adolescence and early adulthood (Matkovic et al. 1990), (Johnston et al. 1992), and insufficient dietary calcium during adolescence can have lasting consequences. In a retrospective study of women between 50 and 65 years old, researchers found that regular consumption of dairy products during adolescence was associated with lower amounts of postmenopausal bone loss (Sandler et al. 1985).

Unfortunately, the calcium intake of many adolescents, particularly females, begins to fall during puberty and late adolescence. The RDA for calcium for males and females from 11 to 24 years is 1200 mg (NAS 1989). Surveys have found that the average calcium intake of adolescent females in the United States is only about 66% of the RDA. Only about 40% of adolescent males and 15% of adolescent females obtain the RDA for calcium (LSRO 1989).

Poor intakes of calcium can compromise bone health and increase the incidence of bony fractures both during adolescence and later in life (Matkovic 1992). In a recent study of calcium intake in teenagers, 100 adolescent females were divided into two groups. For 18 months, one group consumed 900 mg of calcium a day (about 80% of the RDA), while the other group consumed 1400 mg a day (110% of the RDA). The group with the higher intake gained significantly more bone (Lloyd et al. 1993).

The poor eating habits of many adolescents increases the risk of not achieving peak bone mass. Milk and other dairy products are the primary source of calcium in the teenage diet, yet many adolescents regularly substitute soft drinks, coffee, or alcoholic beverages for milk. Diet soft drinks replace milk in the diets of many teenagers who are trying to lose weight. A recent study compared the calcium intakes of teenage girls who consumed soft drinks with those who did not. High consumers of diet soft drinks had calcium intakes of only 59% of the RDA, while non-

menopause: the time of life marked by the permanent and natural cessation of the menses, generally occurring between the ages of 45 and 55

consumers had intakes significantly higher—about 75% of the RDA (Guenther 1986).

Recommendations from different committees of experts on adequate calcium intake during adolescence vary considerably. For example, in the United States the adolescent RDA for calcium is set at 1200 mg (USDA 1984), while the Food and Agriculture Organization/ World Health Organization (WHO) recommends a minimum of 500 to 700 mg a day for teens (WHO 1962). A recent National Institutes of Health conference suggested optimal daily calcium intakes of 1200 to 1500 mg/day for teens (Levenson and Bockman 1994). Future research should provide us with a better understanding of calcium requirements during adolescent growth. Until then, teenagers and young adults, and particularly females, should strive to maintain calcium intakes at or near the RDA. Rich, low-calorie sources of calcium include nonfat or low-fat dairy products and leafy greens such as kale and collard greens.

SUBSTANCE ABUSE

Most teenagers in the United States experiment with alcohol, smoking, and marijuana. Many begin to smoke and drink regularly, and this will have a significant impact on their nutritional and overall health, both today and later in life. It is estimated that 70 to 80% of adolescents experiment with alcohol and drugs, about 10% abuse these substances, and about 1% are chemically dependent. Although teenage use of alcohol has fallen over the last decade, cigarette smoking, cocaine, and use of hallucinogens is rising (Centers for Disease Control 1990), (Moss et al. 1992).

Alcohol

In the United States, nearly 40% of adolescents 15 to 16 years old drink alcoholic beverages regularly. High school seniors often abuse alcohol: in one U.S. study 28% reported binge drinking, and nearly 4% reported daily use (Centers for Disease Control 1990). Alcohol can have far-reaching effects on nutritional health and Chapter 9 contains a thorough discussion of this topic.

Alcohol use plays a significant role in the three leading causes of death in adolescents—accidents (about two thirds of which are motor vehicle crashes), suicide, and homicide. Of all fatal motor vehicle accidents involving 16 to 19-year-old drivers, in about one third the driver had been consuming alcohol, and in a fifth the driver was legally intoxicated. Overall, about 20% of all deaths among adolescents aged 15 to 20 are from alcohol-related motor vehicle accidents (Centers for Disease Control and Prevention 1990). Alcohol and drug use is also involved in 50% of homicide deaths and 30% of suicides among adolescents (U.S. Public

Health Service 1990). Most adults who abuse alcohol began using it during adolescence (Kandel and Logan 1984).

Smoking

Since 1966, the total number of smokers in the United States has steadily declined, but slightly more than one in four adults (over 45 million people) continue to smoke (USDHHS 1989). The trends in adolescent smoking are troubling—after plateauing during the 1980s, since 1991 the number of young smokers is actually rising (Moss et al. 1992), (Allen et al. 1989). Recent studies have shown that 16% of adolescents smoke, and 22% of adolescent males use smokeless tobacco. Smoking is increasing most among white female adolescents.

Today, most people who become regular smokers begin at about age 17 to 19 (Escobedo et al. 1990). Smoking that begins during adolescence is strongly predictive of smoking in adulthood (Kandel and Logan 1984): over 90% of adult smokers began smoking as teenagers.

The medical costs and health consequences of smoking are enormous. Cigarette smoking is a major risk factor in nearly all of the leading causes of death in the United States, as well as in many of the common disabling diseases (USDHHS 1989). The long-term consequences of smoking include sharply increased rates of heart attack and stroke. About 35% of all cancer deaths among men and 18% of cancer deaths among women are caused by smoking (Loeb et al. 1984).

Figure 8.7 gives examples of tobacco-related diseases and death rates for smokers and nonsmokers. A third of all 35-year-olds who continue to smoke will die of diseases caused by their smoking (USDHHS 1989). In the United States, the costs to society—for patient care and lost work time and production—are a staggering $56 billion each year (USDHHS 1989).

Precursors of adult heart and lung disease are already apparent in teen smokers. Adolescents who smoke have significantly higher LDL-cholesterol and lower HDL-cholesterol levels in their blood than do nonsmoking adolescents (Craig et al. 1990). Teen smokers also show early signs of decreased lung function (Tager et al. 1984).

The use of smokeless tobacco also has adverse health effects and should not be considered a safe alternative to smoking. Health problems associated with smokeless tobacco include detrimental changes in blood lipid levels and a sharply increased risk of oral cancer (Elster 1994). Cigarette smoke contains many toxic compounds, including tars, carbon monoxide, arsenic, and dozens of known cancer-causing agents, **Nicotine** is the addictive compound in tobacco smoke.

Discouraging cigarette smoking is one of our biggest public health challenges. Studies have shown that school health programs designed to help adolescents resist social pressures to smoke can reduce the onset of smoking by up to 50% (Best et al. 1988), (Hirschman and Leventhal

nicotine: a poisonous alkaloid found in tobacco; the small amount inhaled when smoking causes constriction of blood vessels, raises blood pressure, and has other systemic effects

Source: Adapted from
USDHHS (1989)

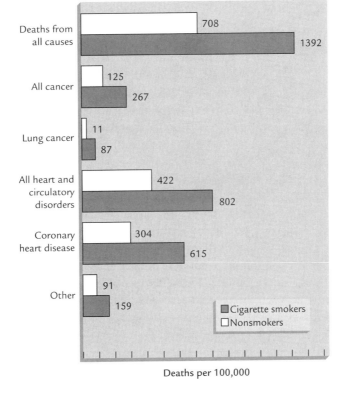

FIGURE 8.7

*Diseases associated
with smoking: death
rates of cigarette
smokers and nonsmokers,
by selected diseases
(men aged 45 to 64)*

1989).Quitting smoking or never starting can have substantial benefits on nutritional, and overall, health.

Nutritional Health of Smokers Smoking causes profound and widespread changes in the body's physiology and metabolism. It has direct effects on levels of macro- and micronutrients in the body and also influences energy expenditure and caloric requirements (Preston 1991).

Smokers generally weigh less than nonsmokers, despite caloric intakes similar or higher than that of nonsmokers. Although they may weigh less, the body fat of smokers is more often distributed in a central, or abdominal pattern—a pattern associated with greater risk of heart attack and diabetes (Shimokata et al. 1989).

Smokers have higher 24-hour energy expenditures (about 10% more than nonsmokers) (Hofstetler 1986), which may explain why smokers are thinner than nonsmokers. When people stop smoking, their metabolic rate falls, they often eat more, and many gain weight as fat. Major weight gain of over 25 pounds occurs in about 10% of people who quit (Williamson et al. 1991). People who return to smoking lose weight again, but they lose lean tissue. Therefore, repeated efforts at quitting followed by a return to smoking produce unfavorable changes in body composition, as fat replaces lean tissue (Preston 1991).

Fear of weight gain, particularly in women, may influence attempts to quit smoking (Williamson et al. 1991). In general, a person who quits will gain between 4 and 9 pounds, but half of those who quit will gain less (USDHHS 1990). The health benefits of quitting smoking (including cutting in half the risk of a heart attack) are much greater than the potential problems associated with such a small gain in weight. Smokers trying to quit should practice effective methods of weight control.

Smoking affects protein metabolism. Compared with nonsmokers, smokers are less efficient at retaining nitrogen from dietary sources (Preston 1991). Blood cholesterol levels are higher in smokers. Cholesterol rises approximately 8–10 mg/dl for each pack of cigarettes smoked each day. Regular cigarette smoking increases the level of LDL-cholesterol in the blood, which is a major risk factor for heart attack and stroke (Preston 1991).

Smoking influences the status of many vitamins in the body. Smokers metabolize vitamin C more rapidly and have lower blood levels of vitamin C than nonsmokers at equal intakes (see Figure 8.8). Smokers need more vitamin C in their diets than nonsmokers to maintain blood levels of the vitamin. Because of this, the National Research Council subcommittee that prepares the RDAs now recommends that regular cigarette smokers ingest at least 100 mg of vitamin C each day—compared with the normal adult RDA of 60 mg (NAS 1989). However, in the United States, the average daily vitamin C intake of smokers from dietary sources is only 45 to 55 mg (Preston 1991).

Smoking is also associated with lower blood levels of beta-carotene, vitamin D, and pyridoxal-5-phosphate (the active form of vitamin B_6) (Preston 1991). Smoking lowers levels of folate and vitamin B_{12} in both blood and tissues, and poor folate and vitamin B_{12} status in smokers may increase their already high risk for lung cancer (Heimburger et al. 1988).

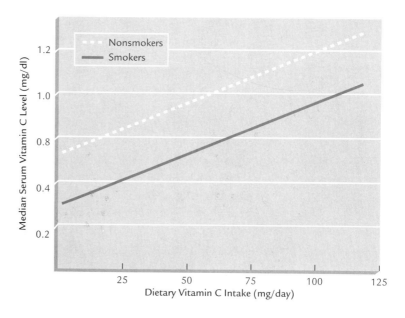

FIGURE 8.8

The effect of smoking on blood vitamin C

Mineral status can also be influenced by cigarette smoking: it has been shown to increase bone mineral losses and increase the risk of osteoporosis in older women.

Diet and Smoking Large surveys in the United States and Great Britain have shown that smokers' food and nutrient intakes differ significantly from those of nonsmokers (Subar et al. 1990), (Bolton-Smith et al. 1991). Smokers are less likely to consume fruits and vegetables, whole-grain breads, and low-fat milk products. A recent survey of over 1000 women in the United States revealed that smokers have significantly lower protein, fiber, and vitamin C intakes than nonsmokers, and higher intakes of cholesterol (Larkin et al. 1990). For both sexes, smokers are less likely to obtain adequate amounts of vitamins A and C, fiber, and beta-carotene in their diets. Smokers consume greater amounts of alcohol and fewer vitamin–mineral supplements than nonsmokers.

Illicit Drugs

Use of cocaine among teenagers has been rising—the number of adolescents regularly using cocaine has quadrupled in the past two decades. About 2 to 6% of adolescents in the United States report frequent cocaine use (Johnston et al. 1991). Cocaine abuse adversely affects the nutritional health of teenage users. Compared to nonusers, young cocaine users eat fewer meals, drink more alcohol, and consume more fatty foods (Castro et al. 1987). There is also a strongly elevated risk for bulimia and anorexia nervosa among regular cocaine users (Jonas et al. 1987).

In the United States, 14 to 15% of high school seniors use marijuana regularly (Centers for Disease Control 1991). In clinical studies of marijuana, use appears to increase appetite and food intake. However, regular marijuana use by adolescents is associated with a significant decrease in dietary quality accompanied by signs of nutrient deficiency. Compared to nonusers, regular marijuana users consumed fewer dairy products, fruits, and vegetables, and they consumed more snack foods and meat products. Regular users were noted to have more frequent indigestion, they complained more often of loss of appetite, and had lower plasma zinc concentrations (Mohs et al. 1990).

EATING DISORDERS

The most prevalent nutritional problem among today's adolescents are the eating disorders—obesity, anorexia nervosa, and bulimia nervosa. Alarming statistics show that the incidence of all three of these disorders is increasing in teenagers in the United States (Lucas et al. 1991), (Gortmaker et al. 1990):

- Over 20% of adolescents are obese.

- Over 2 million people in the United States, most of them teenage women, suffer from anorexia nervosa or bulimia.

- About 1 in 20 adolescent and college-aged females are bulimic, and 1 in 200 have anorexia nervosa (Lucas et al. 1988).

- Many more young people with symptoms of bulimia or anorexia nervosa are not identified and remain untreated.

An *eating disorder* is a disturbance in eating behavior that jeopardizes physical, emotional, or social health. Although most of the symptoms and signs of eating disorders are physical and nutritional, the underlying cause is psychological. Many people in U.S. society use food inappropriately as a response to the pressures, loneliness, or stress of daily life. Abnormal eating behaviors may be deeply confused efforts to avoid or solve what are perceived as unsolvable problems.

Teenagers are especially vulnerable to eating disorders because of the heightened awareness of physical appearance during adolescence. Teens grappling with a rapidly changing self-image during pubertal growth become sensitive about how they look to others and are susceptible to body-image problems. At the same time, they are struggling to become adults, and they strive to gain independence, autonomy, and control of their often turbulent lives. Eating disorders may be a maladaptive response to these challenges—the stress of trying to maintain control may find an outlet through rigid self-control of body weight.

U.S. society prizes leanness and puts enormous pressure on people to pursue a thin and athletic "ideal" body. The media and advertising surround us with images of ultra-trim, lean, and muscular bodies. Thinness is perceived as beauty and success. Many normal, healthy adolescents compare themselves to these images and begin to view themselves as thick and fat—without a realistic perception of what is a normal weight for age and height.

Societal pressure for thinness falls heaviest on females, and eating disorders occur predominantly in women. A recent survey of 326 adolescent females revealed that most were unhappy with their body weight and had made efforts to restrict their diets (Moses et al. 1989). Among females who were underweight, 51% were fearful of obesity and over a third were preoccupied with body fat. Among the females who dieted, many had inappropriate and exaggerated views of the amount of weight they wanted to lose.

In a recent study that looked at ethnicity and dieting in adolescence, researchers found that White and Hispanic adolescent females more frequently perceive themselves as overweight than do African American females (Centers for Disease Control 1991). Among adolescent females, chronic dieting was reported by 15% of Hispanics, 12.5% of Whites, 11% of Native Americans, 10.5% of Asian Americans, and 8% of African Americans (Story et al. 1991).

Males are most often concerned with physical and muscular development, but many adolescent males also regularly diet for weight loss. In one study, about 5% of male adolescents reported dieting four times or more during the previous year (ASHA 1989). Concerns about overweight are common even in prepubertal children of both sexes.

Strenuous dieting during adolescence can disrupt normal growth and development and produce nutritional deficiencies. Female athletes, for example, who strive to achieve a very thin body shape, may develop amenorrhea. Amenorrhea in young women may be due to undernutrition, low body weight, or very low fat stores (Baker 1981), (Biller 1989). Amenorrhea may increase loss of minerals from the skeleton, and it is associated with lower bone density, particularly in the bones of the spine (Drinkwater et al. 1984). Figure 8.9 shows the lower bone mineral density typically found in athletic females with amenorrhea. Lower bone density during adolescence and young adulthood increases the risk of osteoporosis in later years (Matkovic et al. 1990).

Repeated, strenuous dieting can also push a vulnerable individual into a chaotic eating pattern that may involve abnormal and unhealthy eating behaviors. Three common types of abnormal eating patterns are bingeing, purging, and restricting.

Binge eating is rapid consumption of large amounts of food accompanied by a feeling of loss of control over eating. Binge eating is usually done in secret, is often associated with feelings of anxiety and shame, and

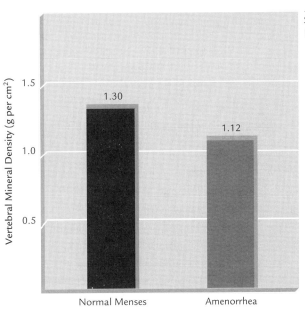

Source: Adapted from Drinkwater et al. (1984)

FIGURE 8.9

Bone mineral density in athletic females with amenorrhea compared to females with normal menstrual periods

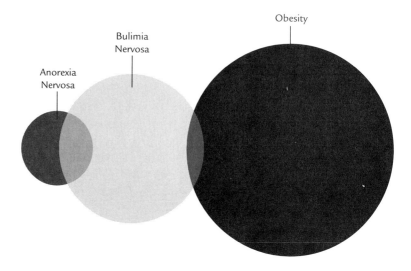

FIGURE 8.10

The spectrum of eating disorders

may be followed by purging. *Purging* is the use of self-induced vomiting, or laxative or diuretic abuse, in an attempt to prevent weight gain or cause weight loss. *Restricting* is severe, chronic dieting or fasting in an effort to lose weight.

Many adolescents adopt these abnormal eating patterns in an attempt to lose or control weight. For example, a study of eighth and tenth grade students found that 32.5% of females and 12% of males fasted, 11% of females and 3% of males used diet pills, 7.5% of females and 2.5% of males induced vomiting after eating, and 4% of females and 2.5% of males used laxatives to control weight (ASHA 1989).

Eating behavior in U.S. society spans everything from the extremes of overeating and obesity at one end, to the semistarvation of anorexia nervosa at the other (see Figure 8.10). Restricting, bingeing, and purging behavior can occur in normal-weight, underweight, or overweight people at all ages. For example, binge eating has been reported in a quarter to a half of overweight people seeking help with weight loss, and incidence of binge eating appears to increase with severity of obesity (deZwann et al. 1992), (Marcus et al. 1985).

OBESITY

In the United States, obesity is the most prevalent eating disorder at all ages, including adolescence. The National Health and Nutrition Examination (NHANES) III survey (1988–91) found that 1 out of 3 adults in the United States is obese(McGinnis and Lee 1995)—a sharp increase from 1980, when 26% of adults were obese.

Three generally accepted ways to classify people as overweight or obese are as follows:

- Comparison of body weight with published life insurance tables (Metropolitan Life Insurance Company 1983): 20% above the recommended weight is considered obese; 10% above is overweight.

- Measurement of skinfold thickness and comparison to published standards.

- Calculation of body mass index (BMI) (Kushner 1993). BMI is calculated as the weight in kg/height in m^2. The NHANES III defined overweight as a BMI equal to or greater than that at the 85th percentile for young adult men and women in the United States. Adult men were considered overweight if BMI was equal to or greater than 27.8; women if BMI was equal to or greater than 27.3. Because BMI normally increases steadily during adolescence, BMI cutoff points for obesity vary throughout adolescence, as shown in Figure 8.11.

Adolescent Obesity

National surveys suggest that among adolescents aged 12 to 17, approximately 26% of White females, 25% of African-American females, 20% of White males, and 13% of African-American males are obese (Gortmayer et al. 1989). Rates among adolescents of Native-American and Hispanic backgrounds are even higher. Similar to trends in adults, obesity rates in the teenage population are steadily increasing (Rising 1995).

Obesity in a child entering adolescence influences the events of puberty. Obese children often develop pubertal changes earlier than normal-weight children, gain stature earlier but gain less total height (because of decreased time for prepubertal growth), and have slightly accelerated skeletal maturation (Tanner 1990).

Obesity during adolescence can have both immediate and long-lasting effects on physical, social, and psychological health. Teenagers who are significantly overweight have increased risk of high blood pressure, diabetes, and elevated blood cholesterol (Rochinni et al. 1988).

Although not all overweight children and adolescents become obese adults, there is a strong tendency for obesity to track into adulthood. Of obese adolescents, 75 to 80% become obese adults, and they are generally more overweight than those who become obese during adulthood (Zack et al. 1979). Overweight adolescents who are poor, or are from certain minority groups (African Americans, Hispanics, and Native Americans), are more likely to maintain obesity as adults.

Moreover, independent of eventual adult weight, obesity in adolescence is a strong predictor of future risk of health problems in adulthood—particularly atherosclerosis and heart disease. Individuals who are

Source: Adapted from
Himes, J. H., Dietz, W. H.,
*Guidelines for Overweight in
Adolescents Preventive Services*
(Chicago: American Medical
Association, 1993).

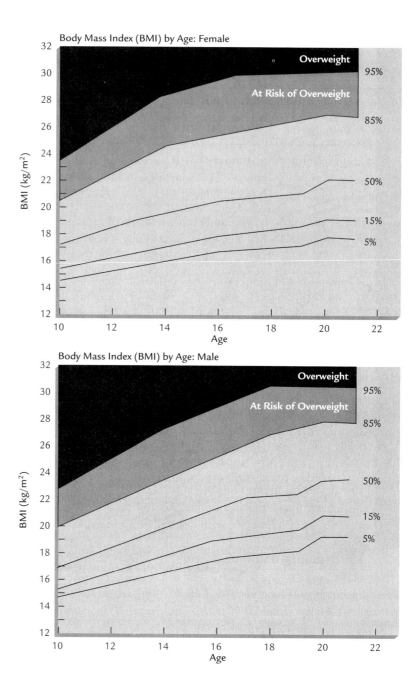

FIGURE 8.11
*Criteria for overweight
(BMI) in adolescence*

overweight at age 13 to 18 have an increased risk of mortality 50 to 60 years later, compared with individuals who have normal weights as teenagers (Must et al. 1992).Certain risk factors for future cardiovascular disease, including high blood pressure and high levels of blood cholesterol, may be established in adolescence. Prevention and treatment of adolescent obesity may help prevent adult morbidity and premature mortality from cardiovascular disease.

Obese teenagers develop behaviors that may reinforce their weight problem. Many react to the stigma of obesity by becoming passive and socially isolated: they may feel uncomfortable in social situations and spend increasing amounts of time watching television or reading. Some avoid physical activities and school gym classes because they draw attention to their weight, thereby perpetuating the problem by adoption of a sedentary lifestyle. Compared to normal-weight children, obese children are at higher risk of developing low self-esteem and depression (Kaplan and Wadden 1986), (Martin et al. 1988).

The psychological and social consequences of adolescent obesity may be lasting, and more severe for females than for males. In a recent study, females who had been overweight during adolescence were surveyed during later adulthood. Compared with a group that had not been overweight, the overweight women had completed fewer years of school, were less likely to be married, and had lower household incomes (Gortmaker 1993). In addition to the adverse physical effects, obesity can be a social and economic handicap as well.

Physical Activity in Adolescence   There is evidence that many adolescents diet regularly, and that energy intakes in adolescence are often lower than those recommended by experts. Yet teenagers in the United States appear to be getting fatter, and over 20% of adolescents in the United States are significantly overweight (Rising 1995).

Changes in activity patterns among adolescents may be a contributing factor. It appears that adolescents are becoming more sedentary, with current studies indicating that physical activity declines through adolescence—falling about 50% from age 12 to 18. In the United States, the number of children participating in physical education classes at school is 98% at age 10, but it decreases to about 50% at age 17 (Centers for Disease Control 1991).

Although studies have shown that over 80% of the typical teenager's physical activity occurs outside of school, a 1990 survey of high school students in the United States found that only a third exercised vigorously at least three times a week (Centers for Disease Control 1992). In contrast, the hours adolescents spend watching television is increasing in the United States, and there is an inverse relationship between watching TV and physical activity among adolescents (Gortmaker et al. 1990). Greater time spent in front of the TV is a risk factor for obesity during adolescence.

Current recommendations for exercise in adolescence include aerobic exercise at moderate intensity (at about 75% of maximum heart rate) for 20 to 30 minutes at least three times a week (Simons-Morton et al. 1988). However, even low-to-moderate levels of activity are associated with beneficial effects on health (Blair et al. 1992). Increased levels of physical activity in adolescence can reduce blood pressure and improve blood lipid levels (Sallis et al. 1988), as well as reduce depression and improve emotional well-being (Brown and Lawton 1986).

Treatment of Adolescent Obesity

Because of the adverse physical and psychological effects of adolescent obesity, treatment should begin early. The recommended approach is a program of diet, exercise, and behavior modification designed to meet the needs of both the individual and the family (Epstein et al. 1990), (Wadden et al. 1990), (Becque et al. 1988).

Careful attention to the growth pattern of the adolescent is important: even in an overweight teenager, strenuous dieting can interfere with normal growth. In adolescents who are still gaining height, the level of caloric restriction required for weight loss may impair growth and should usually be avoided. The goal should be to hold weight at a stable level or reduce the rate of weight gain while height gain continues, allowing them to grow out of their obesity. In teenagers who have attained full stature, the next step should be gradual, steady weight loss of one to two pounds a week. The emphasis should be on modifying the eating and exercising patterns that led to weight gain (Epstein et al. 1990).

Adolescents should be encouraged to exercise regularly, eat a balanced diet, and avoid overconsuming high-calorie foods. Changing poor lifestyle habits in adolescence can have lifelong benefits. Successful prevention and treatment of adolescent obesity may be an effective means of decreasing adult obesity and obesity-associated chronic diseases in adulthood.

Health Risks of Obesity

The relationship between weight and mortality is U-shaped—the weight associated with the least mortality is 5 to 15% underweight, and mortality increases as weight increases or decreases from there (Van Itallie 1979). Although some studies have not confirmed this relationship, studies that have followed people for longer periods nearly all agree—overweight increases mortality in both sexes, independently of the effect of obesity on other risk factors.

Obesity increases the risk of many of the chronic diseases that commonly affect the adult population (Pi-Sunyer 1991).

- *Hypertension.* Overweight adults are three times more likely to be hypertensive compared with normal-weight people. In young adults (20 to 45), obesity increases the risk nearly six times. On average, there is about a 2 to 3 mm increase in blood pressure for every 10 kg increase in body weight.

- *Blood lipids.* Obesity tends to increase total cholesterol, particularly in younger adults (ages 20 to 45). Also, high-density lipoprotein (HDL) concentrations are lower in obese individuals.

■ *Heart disease.* Overweight people have more heart attacks and die more often from them, compared with normal-weight people.

■ *Type II (noninsulin-dependent) diabetes.* Obese people have more than three times the risk of developing this form of diabetes compared with normal-weight individuals.

These health problems often cluster in overweight individuals, and all of them usually respond to weight loss. In general, at all ages and all levels of obesity, for each 10% loss of weight, blood pressure drops about 5 mm Hg, blood glucose falls 2–3 mg/dl, and cholesterol falls about 10 mg/dl. Other adverse effects of obesity are aggravation of osteoarthritis, increased risk for gallstones, and increased risk for cancer— colon and prostate cancer in men and cancer of the uterus in women (Pi-Sunyer 1991).

Obesity and Fat Distribution In obesity, not only total fat but the distribution of body fat appears to be important in determining health risks. Fat deposited around the abdomen (central, or android, pattern) appears to be more dangerous than fat distributed in the hips, thighs, or under the skin (peripheral, or gynoid, pattern). Figure 8.12 illustrates the two body types.

Compared with a peripheral distribution, fat located centrally increases the chances of developing heart disease, diabetes, and high blood pressure, and increases mortality (Pi-Sunyer 1991). Abdominal fat

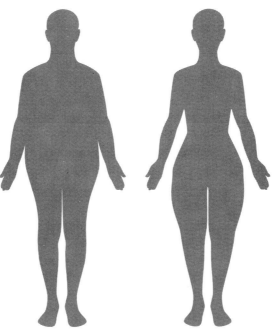

FIGURE 8.12

Fat distribution patterns

Central, or android obesity:
fat distributed on the trunk
and abdomen.

Peripheral, or gynoid obesity:
fat distributed in hips and
thighs.

cells appear to break down triglycerides more readily compared to the cells elsewhere in the body, and this may contribute to the health risk. Metabolism of fat released from the abdomen may produce a more **atherogenic** lipid profile in the blood (Leibel et al. 1989).

Reducing the Prevalence of Obesity

Reducing rates of obesity would be a powerful step toward improving the health of adults in the United States. The latest figures show that 34% of the adult population is obese, compared with 26% in 1980 (McGinnis and Lee 1993).

In trying to reduce obesity rates, prevention is critical. Once people are overweight, less than 5% who attempt weight loss actually lose weight and keep it off. The National Institutes of Health (NIH) has suggested three specific actions to reduce obesity in the U.S. population (Ganner 1993):

1. Education: Teach people that increasing physical activity while decreasing energy intake is the most effective way to lose weight.

2. Prevention: Ensure that individuals already at an acceptable body weight remain that way.

3. Treatment: Motivate obese individuals to adopt a plan to reduce energy intake and increase physical activity, so they can attain and maintain a desirable weight.

BULIMIA

Bulimia nervosa (often called just bulimia) is an eating disorder in which large amounts of food are eaten rapidly (the binge) and then eliminated from the body (the purge) by self-induced vomiting, laxative abuse, and other means. The term comes from the Latin words *bous*, meaning ox, and *limos*, meaning hunger. The ravenous appetite during a binge is compared to the "hunger of an ox." The distinct combination of characteristics that make up bulimia nervosa are listed in Table 8.6. Besides recurrent episodes of bingeing and purging, people with bulimia are overly concerned with their body weight and feel they lack control over their abnormal eating behavior.

Because bulimia is often difficult to detect—bingeing and purging are done secretly, and most people with the disorder are normal or slightly overweight—it is hard to know how common bulimia is. Using the strict criteria in Table 8.6, bulimia nervosa affects about 2% of adolescent and college-aged females (Lucas and Huse 1988). Milder bulimic-type behavior is much more common—up to 20% of college-aged females use self-induced vomiting as a way to control weight (Comerci et al. 1989).

atherogenic:: contributing to or causing atherosclerosis

- Recurrent episodes of binge eating (rapid consumption of a large amount of food in a discrete period of time).

- A feeling of lack of control over eating behavior during the eating binges.

- The person regularly engages in either self-induced vomiting, use of laxatives or diuretics, strict dieting or fasting, or vigorous exercise in order to prevent weight gain.

- A minimum average of two binge eating episodes a week for at least three months.

- Persistent overconcern with body shape and weight.

From American Psychiatric Association, *Diagnostic and Statistical Manual of Mental Disorders,* rev. ed. 3 (Washington D.C.: American Psychiatric Association, 1987).

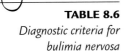

TABLE 8.6

Diagnostic criteria for bulimia nervosa

Bulimia is most frequently found in young females, with only one in ten cases in males. However, male bulimia rates are increasing, particularly among athletes such as wrestlers, who need to strictly control their weight.

Most people with bulimia have a history of fluctuating weight and have made multiple attempts at weight-reduction diets without success. Many are high achievers and perfectionists. They are often perceived by others as successful, but they usually have low self-esteem (Buckner 1991).

For many individuals with bulimia, binge eating is an abnormal reaction to anxiety and stress; it may also be triggered by loneliness, depression, or anger. The binge and purge cycle may occur several times a day, or several times a week (to fulfill the strict definition of the disorder in Table 8.6, it must occur at least twice a week for three months).

People with bulimia may set aside special times to binge, usually times when they are unlikely to be interrupted. Foods eaten during a binge are easily and quickly ingested and require little preparation or chewing—for example, a half-gallon of ice cream may be rapidly eaten without pause. The most common binge foods are high-fat, high-carbohydrate foods such as cookies, cakes, pastries, and ice cream. For some bulimic people, even individual pastries may be perceived as a binge, and must be purged. More commonly, enormous amounts of food are gulped down. Binge–purge cycles can last for several hours, as repeated self-induced vomiting is interspersed with binge eating.

Bulimia is characterized by a loss of control during binges—most people with the disorder do not taste or enjoy food during a binge. They may stop only if they are interrupted, run out of food, or if distention of the stomach becomes too great to continue. Individuals with bulimia often describe feelings of eager anticipation and anxiety to begin the binge–purge cycle; followed by a short respite of relaxation and relief; and finally, guilt, shame, disappointment, and renewed anxiety. Bulimic episodes may be repetitive, or they may be followed by a period of return to normal eating or self-imposed starvation.

Bingeing and purging has serious physical consequences. Repeated vomiting is dangerous and destructive. Vomit contains gastric acid, which demineralizes the teeth—causing erosions, decay, and tooth loss. Vomiting causes painless enlargement of the parotid glands (the salivary glands below the ears that enlarge in the mumps). Though rare, the stomach or esophagus may tear and rupture during vomiting, causing severe injury or death. (Comerci 1990).

Vomiting, particularly when combined with laxative or diuretic use, may cause dehydration and dangerous changes in blood chemistry. Blood potassium may fall, causing weakness, muscle cramping, and irregular heartbeats. A disturbance in the rhythm of the heart due to low blood potassium is a leading cause of death in people with bulimia (Comerci 1990). Overuse of laxatives during purges causes chronic constipation. Some individuals may use ipecac syrup, traditionally used in homes as emergency treatment for childhood poisoning, as an emetic. Ipecac can cause severe inflammation of the heart muscle and heart failure. People with severe bulimia nervosa have nearly ten times the mortality rate compared with the general population.

People with bulimia are acutely aware of their abnormal eating behavior and may desperately seek help. They may appeal to doctors, school counselors, ministers, family, or friends. Everyone should be aware of the seriousness of this disorder, and prompt professional help should be sought.

Treatment for Bulimia

Therapy for bulimia is best handled with a team of experienced doctors, dietitians, psychologists, and family counselors. Treatment consists of four components: correction of medical complications, normalization of dietary patterns, psychological counseling, and, occasionally, medications. The main focus of psychotherapy is to help the person acquire a more realistic body image and to develop alternative methods of coping with doubt, anxiety, and stress. Sometimes antidepressants are of benefit in treating bulimia, along with psychotherapy (Buckner 1991).

In order to break the starve–binge–purge cycle, individuals are placed on a structured routine of three regular meals a day. Content of meals and snacks are planned in advance, and meals should be eaten, when possible, with another person. Foods should contain ample complex carbohydrates to increase satiety (see Table 8.7).

Because binge eating often follows a period of strict dieting, the person must learn to avoid crash dieting, which might trigger bulimia. Individuals with bulimia need to be reassured that the treatment team understands the difficulty of overcoming the disorder and will help them to regain control of eating and not gain weight.

1. Plan regularly scheduled meals and snacks, and record meal content in a food diary.

2. Eat meals and snacks sitting down and using utensils. Avoid finger foods.

3. Increase meal satiety by eating warm foods rather than room-temperature or cold foods.

4. Eat a balanced diet and don't eat in a rush: prolong meal duration by including salads, vegetables, and fruits, as well as high-fiber breads and cereals.

5. Choose foods that are naturally divided into portions, such as potatoes (rather than pasta or rice), small (4 oz) containers of yogurt or cottage cheese, or packaged, frozen meals.

6. Include foods containing ample amounts of complex carbohydrates and adequate amounts of fat. This will help slow gastric emptying, increase satiety, and enhance the feeling of fullness after meals.

Adapted from C. L. Rock and J. Yager, "Nutrition and Eating Disorders: A Primer for Clinicians, *Internat J Eating Disorders* 6 (1987): 276.

TABLE 8.7

Dietary recommendations for bulimia nervosa

How successful are treatments for bulimia nervosa? Many treatment programs report that bulimic episodes can be reduced in about three quarters of individuals but only about half of the people with bulimia are able to stop bingeing and purging completely. Bulimia nervosa is a chronic disease, and controlling it often takes years of intense effort (Herzog et al. 1991).

ANOREXIA NERVOSA

Anorexia nervosa (AN) is the rarest, but the most dangerous and best known of the eating disorders. The term comes from the Greek word *anorektos*, meaning without appetite, and the Latin word *nervosa*, meaning emotional disorder. It is not a very accurate term because a person with AN does not lack an appetite. Rather, affected individuals refuse to eat, out of fear of gaining weight. They are, literally, dying to be thin.

AN is characterized by the relentless pursuit of a thinner body shape and an intense fear of weight gain or becoming fat. People with AN perceive their bodies as being fatter or larger than they actually are. Even though they are dangerously underweight, they are convinced they are overweight. They firmly believe there is nothing abnormal about their diet and body image (Comerci et al. 1989). Table 8.8 compares AN to bulimia nervosa, and Table 8.9 lists the primary characteristics of AN.

Anorexia nervosa is not a new disease. It was well described over a century ago, and scattered reports of the disorder date back to the 1600s.

Anorexia Nervosa	Bulimia Nervosa
Extreme weight loss, refusal to maintain minimal normal weight for age	Weight fluctuations
Intense fear of becoming obese	Possible fear of fatness but a greater fear of loss of control of eating
Distorted body image	Overcome with body shape/weight
Amenorrhea (in females)	Possible menstrual irregularities
Intense preoccupation with food	Intense preoccupation with food
Severe caloric restriction	Secretive binge-eating episodes Self-induced vomiting (laxatives, diuretics, emetics), vigorous exercise
Excessive physical activity	Other compulsive behaviors (alcohol/drug use)
Isolation, asexuality	Outgoing personality, sexual relationships
Denial of Illness	Distress, willingness to accept help

TABLE 8.8

Comparison of anorexia nervosa and bulimia nervosa

- Refusal to maintain body weight over a minimal normal weight for age and height, e.g., weight loss leading to maintenance of body weight 15% below that expected; or failure to make expected weight gain during period of growth, leading to body weight 15% below that expected.

- Intense fear of gaining weight or becoming fat, even though underweight.

- Disturbance in the way in which one's body weight, size, or shape is experienced, e.g., the person claims to "feel fat" even when emaciated; believes that one area of the body is "too fat" even when obviously underweight.

- In females, absence of at least three consecutive menstrual cycles when otherwise expected to occur.

From American Psychiatric Association, *Diagnostic and Statistical Manual of Mental Disorders* rev. ed. 3 (Washington D.C.: American Psychiatric Association. 1987).

TABLE 8.9

Diagnostic criteria for anorexia nervosa with subtypes

AN appears to be increasingly prevalent, particularly among young women. For women in the United States aged 15 to 24, the number of reported cases increased five times between 1935 and 1984 (Lucas et al. 1991). Using the strict definition in Table 8.9, AN occurs in less than 1% of females aged 15 to 40 (Lucas and Huse 1988). However, many more women have some of the characteristics of AN—as many as 3% of adolescent females may have milder forms of the disorder.

Although it is most common among White females from middle and upper socioeconomic levels, AN is well represented across the entire racial and economic spectrum. It is seen in rural as well as urban settings, and

Adapted from Lucas et al. (1992)

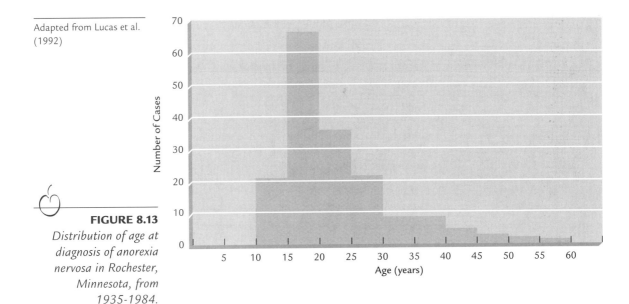

FIGURE 8.13

Distribution of age at diagnosis of anorexia nervosa in Rochester, Minnesota, from 1935-1984.

certain athletes (gymnasts, endurance runners) and performers (dancers, figure skaters) are at increased risk. While it occurs most often in adolescence—three quarters of all cases occur in people younger than 25—the age of onset can range from prepubertal childhood to middle age (see Figure 8.13). Only about 1 in 20 cases occur in males (Garner 1993).

There is no single cause of AN. It appears to be a multifactorial disorder, with three broad classes of predisposing risk factors: individual, familial, and cultural. These risk factors combine in a complex and poorly understood way to increase vulnerability to the disorder.

■ *Individual risk factors.* Although several theories have suggested that a disturbance in brain biochemistry or physiology causes AN, scientists have so far been unable to identify any definite biological risk factors for the disease. However, several psychological factors have been implicated. Amidst the turbulence of adolescence, the refusal to eat may represent a subconscious attempt to exert control or maintain order in a life felt to be out of control. Fears of maturity—adulthood, puberty, sexuality, pregnancy—have been proposed as underlying conflicts with which the individual cannot cope.

■ *Familial risk factors.* Young people with AN often come from families that are overprotective, rigid, and demanding. These families often put a very high value on appearance and success. Achievement at school or in sports may be overemphasized at the expense of a developing teenager's feelings of self-worth and identity.

■ *Cultural risk factors.* Almost all cases of AN occur in developed, affluent countries—societies that tend to prize thinness as the ideal body shape, particularly in females. The conflict between the temptations of an abundant food supply and the cultural ideal of a thin, lean body precipitates, in many people, a daily struggle to control their weight, and in vulnerable individuals, a distorted body image.

Anorexia nervosa may be triggered by stresses such as beginning puberty, leaving home, or entering college (Garner 1993). The disorder often begins with normal dieting for weight loss—the diet may be in response to the physical changes of puberty or a comment from a family member or friend that the person is gaining weight. What begins as a simple diet can, in a person predisposed to AN, evolve into long periods of self-induced starvation.

In AN, diet and weight loss become the central focus of life. People with AN do not lose their appetites—thoughts of food and eating constantly intrude—and they must constantly struggle to avoid eating. Yet they deny hunger and claim to be full after eating tiny quantities of low-calorie foods. They may prepare large meals for friends or family, and encourage them to eat, but not eat any of the food themselves.

They often know the exact calorie content of many foods. When eating with others, they cut food up into small pieces and move it around their plates, but eat very little. If they are pressured to eat, they will pretend to eat, hide their food, and later throw it away.

In an effort to lose more weight, many people with AN exercise vigorously (Buckner 1991). The exercise has a driven, relentless, and exhausting quality. Some people with anorexia, when confined to a room, will constantly pace back and forth; asked to stay seated, they may constantly shuffle their feet. Some people with AN purge daily with laxatives and diuretics in an effort to lose more weight.

At the beginning of the illness, people with AN may continue to do well at their jobs or schoolwork, and may be high achievers. But as the disorder progresses, the physical and mental effects of starvation produce weakness, dullness, apathy, and an inability to concentrate, and performance crumbles. Ultimately, the individual becomes withdrawn, hostile, and critical of family and friends. Insomnia and depression are common (Garner 1993). Unless the person receives help, debility and death often result.

Signs and Symptoms of Anorexia Nervosa

In 1689, in England, Dr. Richard Morton wrote: "I do not remember that I did ever in all my practice see one…that was so much wasted with the greatest degree…like a skeleton only clad with skin." This excerpt from what is thought to be the first medical description of anorexia nervosa describes the dramatic emaciation that is characteristic of the disorder (Garner 1993).

The signs and symptoms of AN are essentially the effects of starvation (see Figure 8.14). In AN the entire body slows down in an effort to conserve energy: body temperature and basal metabolic rate fall, body stores of fat and protein are lost, and in young people, growth and development slow.

The brain responds to the severe nutrient deficit by slowing release of the hormones that stimulate ovulation. Menarche is delayed, or if menstrual cycles have begun, amenorrhea may develop (Biller et al. 1989). In young females, these hormonal changes, combined with dietary deficiencies of minerals and protein, result in loss of minerals from the skeleton at a time when reaching maximum bone density is crucial for reducing the risk of osteoporosis. Unlike the other physical signs of AN that predictably correct with proper nutrition, the adverse effects on skeletal development and overall growth (in young adolescents) may be permanent disabilities (Rigotti 1984).

Starvation has profound effects on the digestive tract. The stomach empties poorly, the intestinal lining atrophies, peristalsis slows, and the pancreas fails to produce adequate digestive enzymes. All of these cause malabsorption of the small amounts of food that are rarely eaten.

The skin appears pale and dry and may be covered with fine, pale hair called *lanugo*. The heart shrinks in size as muscle and fat mass is lost, and it pumps poorly—a slow heart rate, low blood pressure, and (rarely) heart

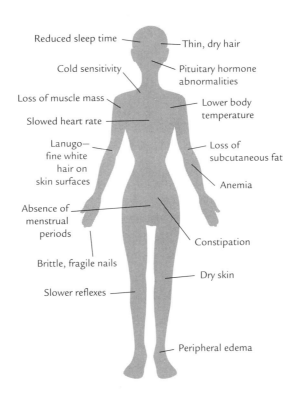

FIGURE 8.14
Physical signs and symptoms of anorexia nervosa

failure can result. Blood potassium falls, which can cause dangerously irregular heart rhythms that can be fatal. Anemia and impaired immune function also develop. People with severe AN may die of overwhelming infection because of their debilitated immune system (Comerci 1990).

Early recognition of developing symptoms and prompt referral for care are crucial. Family, friends, teachers, coaches, and health-care workers should be aware of the frequency of this disorder among young people, and be able to recognize its early symptoms. The longer the duration of disorder, the less the chances are for a complete recovery (Comerci 1991). People with AN will firmly deny their problem and resist what they regard as intrusive efforts by concerned individuals to help them.

Treatment of Anorexia Nervosa

Treatment consists of three major components: correction of medical complications, nutritional rehabilitation, and most importantly, psychotherapy. A variety of drugs have been tried in AN, but none has been found to be effective. Nutritional support should first focus on weight stabilization and improvement of the nutritional state at a low weight. A gradual increase in weight should then occur, through encouraging the patients to choose foods and feed themselves.

Nutritional replenishment must be done with caution—damage to the stomach or kidney, or overload of the heart can occur with too rapid refeeding (Beumont et al. 1993). Often only about 800 kcalories above basal caloric requirements should be consumed as treatment begins.

Starvation adversely affects mental functioning and alters thinking patterns, making it difficult for people with AN to think clearly. Refeeding must correct these cognitive impairments before psychological therapy can begin. Individual and family psychotherapy should attempt to refocus individuals on the underlying emotional and psychological problems, and away from their obsession with weight loss. People in treatment for AN need constant reassurance that treatment will lead to normal weight and appearance, and not obesity. Parents need to be aware that eating disorders are symptoms of underlying issues that often involve the family.

Therapy frequently takes years because people with AN are highly resistant to treatment. AN is often a lifelong struggle—with a high incidence of relapse. Between 40 and 70% recover, and many return to well-adjusted lives. But 9 to 18% of people who develop AN will ultimately die from suicide or complications of the disease (Comerci 1990). Everyone should be aware of the early signs of anorexia nervosa and bulimia listed in Table 8.10. Early recognition and referral can make the difference between life and death.

- Abruptly changing weight goals
- Loss of menstrual periods
- Dieting that leads to increasing criticism of one's body and social isolation
- Hiding food, purging, using diet pills

TABLE 8.10

Early signs of eating disorders

NUTRITION AND PHYSICAL COMPETITIVENESS

Most of us are not competitive athletes; we exercise out of concern for our general good health and view physical activity as part of a lifestyle that promotes cardiovascular fitness and weight control. Proper diet combined with regular exercise during adolescence and adulthood may reduce the risk of many of the chronic diseases of aging. Chapters 9 and 10 contain discussions on the connections between nutrition, exercise, and health. This section focuses on the general role of nutrition in exercise and how diet can contribute to physical competitiveness.

For centuries, athletes and coaches have experimented with food intake, hoping to find a diet that will provide them with a winning edge. Only in the last few decades has research provided scientific information on the link between food choices, diet, and physical performance. Although scientists have made no dramatic discoveries—no special food, diet, or supplement magically boosts athletic performance or compensates for lack of ability or training—athletes do have unique nutritional needs. Optimal nutrition can allow athletes to attain their "personal best."

ENERGY REQUIREMENTS OF ATHLETES

Depending on the intensity and duration of physical training and the size of the individual, energy requirements vary considerably. As one might expect, the more strenuous the exercise and the larger the individual, the higher the energy cost of the physical activity. For example, a person who weighs 60 kg running at a moderate pace (9 miles/hr) will expend about 750 kcal/hr, compared with a 90 kg individual, who would expend over 1200 kcal running for the same time period.

During athletic training, daily caloric needs are often between 3000 and 5000 kcal: a swimmer training four to five hours each day may need up to 7000 kcal/day to maintain his or her weight. Maintenance of a stable body weight is the best criterion of adequate energy intake during athletic training.

Sources of Energy during Physical Exercise

Body stores of fat and carbohydrate serve as the primary energy fuels during exercise (see Table 8.11). Depending on the intensity of the exercise, fat provides 25–90% of the calories. Fat is particularly important as a source of energy during lower-intensity, endurance-type events. Marathon runners typically derive over 75% of their energy needs from metabolism of fat.

As exercise intensity increases, carbohydrate stored as glycogen in muscle and liver cells becomes more important as a fuel source. Storage of carbohydrate as glycogen in muscle and liver is limited and, used alone, can provide energy for only short periods.

Most physical activity is fueled by mixtures of fat and carbohydrate. A working muscle requires both fatty acids (supplied by breakdown of fat stores) and glucose (from glycogen stores in both muscle and liver). Glycogen stores are limited compared to fat stores: during exercise that demands endurance, glycogen stores run out long before fat stores can be depleted. Glycogen depletion causes muscle fatigue and exhaustion. Highly trained athletes are able to protect their glycogen stores by using more energy from fat during exercise (Gollnick 1985).

Diet also has a profound influence on glycogen stores. High-carbohydrate diets stimulate muscles to store more glycogen and can increase endurance. Athletes consuming only 40% of calories from carbohydrates during training are less able to maintain adequate reserves of muscle glycogen, compared with those who consume 70% of calories as carbohydrate. Studies have found that endurance is significantly greater in athletes on diets high in carbohydrate, compared with diets high in fat and protein (Williams 1992). See Table 8.12 and Figure 8.15.

Maximizing Muscle Glycogen

Carbohydrate loading describes several techniques used by endurance athletes to maximize storage of muscle glycogen in preparation for competi-

Body Fuels	Kg	Kcal
Fat cell triglyceride	12.00	110,000
Muscle protein	6.00	24,000
Glycogen from liver	0.07	280
Glycogen from muscle	0.40	1,600
Glucose from body fluids	0.02	80
Free fatty acids from body fluids	0.004	4
Total		135,964

Data from M. N. Goodman and N. B. Ruderman, "Influence of Muscle Use on Amino Acid Metabolism," *Exerc Sports Sci Rev* 10 (1982):1.

TABLE 8.11
Sources of stored energy in a 70 kg male

Diet	Time to Exhaustion (min) while cycling
Mixed	113.6
High fat and protein	56.9
High carbohydrate	166.5

From J. Bergstrom, et al., "Diet, Muscle Glycogen and Physical Performance," *Acta Physiol Scand* 71 (1967):140.

TABLE 8.12

Time to exhaustion in well-trained athletes on three different diets

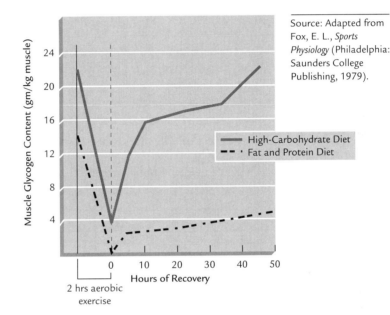

Source: Adapted from Fox, E. L., *Sports Physiology* (Philadelphia: Saunders College Publishing, 1979).

FIGURE 8.15

The effect of different diets on the rate of muscle glycogen replenishment following exercise

tion. In the original regimen, one week before the event, the athlete undergoes exhaustive exercise in an attempt to deplete muscle glycogen stores. This is followed by three days of light-to-moderate activity and a diet very low in carbohydrates. Finally, in the three days prior to the event, the athlete minimizes activity and consumes a diet very high in carbohydrates.

Although glycogen stores in muscle often rebound to three to four times the normal level, this regimen has undesirable side effects, including nausea, dizziness, lethargy, depression, hypoglycemia, and abnormal heartbeats. Several safer variations of the original regimen have been developed.

The best method appears to be a gradual tapering of activity in the week prior to competition, accompanied by a diet very high in carbohydrate (a minimum of 450 g of carbohydrates per day for a person consuming 3000 kcal/day) during the final three days prior to the event. This regimen maximizes glycogen stores and can increase endurance in events requiring prolonged (more than 90 minutes), high-intensity, aerobic exercise (Berning 1993).

PROTEIN REQUIREMENTS OF ATHLETES

Although althletic training increases protein needs, large amounts of extra dietary protein are unnecessary. Exercise scientists have found small increases in muscle protein breakdown during strenuous training, but only modest amounts of extra dietary protein are necessary to restore nitrogen balance. Studies in elite athletes have determined protein requirements of between 1.2 and 1.8 g/kg/day (Tarnopolsky 1988), (Williams 1985), (the U.S. RDA for protein is 0.8 g/kg/day), (NAS 1989). These levels are still below the average amount of protein in the typical protein-rich U.S. diet.

Certain amino acids in muscle (the **branched-chain amino acids**) can be converted to glucose and used as fuel during exercise, and protein supplies about 10–15% of the energy needs of working muscles (Berning 1993). However, muscles burn mostly fatty acids and glucose for energy, not protein. In bodybuilders, increased muscle size is produced by extensive training—not by eating large amounts of protein or amino acid supplements.

In individuals attempting to "bulk up," very little extra dietary protein is required. One pound of new muscle contains only 70–100 g of protein. Adding a half pound of new muscle each week would therefore require only about 5–8 g of additional protein each day (the amount in one egg or a cup of milk). Consuming greater amounts of protein (in food or supplements) will cause the extra amino acids to be converted into fat, while the nitrogen is excreted in the urine. Supplementing athletic diets with single amino acids is of no value, and it may produce detrimental amino acid imbalances.

VITAMIN AND MINERAL REQUIREMENTS OF ATHLETES

branched-chain amino acids: the amino acids leucine, isoleucine, and valine; muscles contain large amounts of these amino acids

Traditionally, many athletes have taken large doses of vitamins and minerals in an attempt to improve performance. However, in the absence of a specific vitamin or mineral deficiency, well-controlled studies have found that supplementation with large amounts of micronutrients does not improve athletic performance (Keith 1989), (Wright et al. 1988). Because certain vitamins (such as thiamin, riboflavin, niacin, and pantothenic acid) and minerals (iron, magnesium) are essential cofactors in energy metabolism, requirements increase during periods of increased energy expenditure. However, athletes who eat balanced, high-calorie diets will consume ample amounts of these micronutrients, and further supplementation is of no value.

Ergogenic Aids

A wide variety of attractively packaged, expensive nutritional supplements are promoted as "ergogenic aids." The term *ergogenic*, meaning capable of producing energy, or work-enhancing, is used often by manufacturers of products that claim to enhance athletic performance. For the great majority of these substances, scientific research does not support these claims. Although most of them are harmless, several have side effects, which may actually impair athletic performance (Contursi et al. 1993).

So-called ergogenic aids of no proven benefit are

Amino acids or protein hydrolysates

Bee pollen

Brewer's yeast

Carnitine

Chromium

DNA and RNA

Gelatin

Ginseng

Octacosanol

Royal jelly

Succinate

Wheat germ oil

Although often promoted as ergogenic aids, supplementation with large doses of vitamins C and E does not improve athletic performance (Gerster 1989).

Iron Requirements

One mineral of concern in athletes is iron. Studies have found that iron deficiency is common among young people who exercise heavily (Ballin et al. 1992), (Buskirk 1990). One study found nearly half of female adolescent runners had low iron stores (Rowland 1987). Iron supplementation in athletes who are iron deficient can improve performance significantly. However, supplements of minerals (such as iron, magnesium, or zinc) in individuals with normal mineral status are of no benefit.

WATER REQUIREMENTS OF ATHLETES

Water is a critical nutrient for the athlete in training and competition. Working muscles produce heat, and water is lost during exercise as the body attempts to dissipate the heat through sweating. (A liter of sweat can dissipate up to 600 kcal of heat energy.) Water losses during heavy exercise are substantial and, if not promptly replaced, can rapidly reduce physical performance. Ninety minutes of strenuous exercise in a 75 kg athlete can produce sweat losses of between 3 and 7 lb of water, depending on air temperature and humidity.

Excessive loss of body water, or *dehydration*, causes a drop in blood volume, and heart rate and blood pressure must increase to compensate for the loss in volume. Dehydration interferes with the ability of the body to circulate oxygen and nutrients and reduces maximal oxygen consumption and performance (Sawka et al. 1984). It also interferes with the body's ability to cool itself and can lead to **hyperthermia** and heat stroke.

Muscular stamina and strength decrease if only 1–2% of body water is lost (Sawka et al. 1984), corresponding to a 2.5 lb loss in a 125 lb individual and 3.5 lb loss in a 175 lb individual. Sweat losses of 5–7% of body weight can cause muscular cramping, heat exhaustion, and collapse. Thirst often lags behind water losses (Fike et al. 1993), so it is important to make a conscious effort to drink fluids before, during, and after activity. In general, about 1–1.5 ml of water is required for each kcal expended in exercise. Drinking ample fluids minimizes dehydration and lessens the rise in internal body temperature during prolonged exercise.

ELECTROLYTE REQUIREMENTS OF ATHLETES

Sweat contains trace amounts of electrolytes, particularly sodium and chloride. Although small amounts of these ions are lost in sweat, because the body loses much more water, concentrations of these ions actually rise slightly in body fluids. Therefore, the need to replace water during and after exercise is much greater than the need to replace lost electrolytes (Fike et al. 1993). Because only very small amounts of electrolytes are lost, heavy sweating for periods of up to two to three hours has insignificant effects on electrolyte concentrations in the body. Only in ultralong endurance events do electrolyte losses in sweat become significant (Noakes et al. 1990). A single postexercise meal amply replaces electrolytes lost in moderate exercise.

THE ATHLETIC DIET

Energy Intakes for Athletes

Maintenance of a desirable body weight should be the goal of energy intake. At least 50%, and preferably 60–70% of total calories should come from carbohydrates (Berning et al. 1993). Daily consumption of 500–600 g of carbohydrates will maintain maximal stores of glycogen during training. Emphasis should be on eating complex carbohydrates because

hyperthermia:
abnormally high body temperature

they contain more of the nutrients needed by athletes (they are richer in B vitamins, minerals, protein, and fiber) compared with simple carbohydrates. Also, because complex carbohydrates are digested more slowly, glucose is absorbed more slowly and evenly into the bloodstream.

For several hours after a strenuous workout, the rate of muscle glycogen synthesis is increased, as muscles replenish depleted stores (Ivy et al. 1988). Recovery of glycogen stores is enhanced if 300–400 calories of carbohydrate are consumed during the immediate postexercise period. Athletes should try to drink three to four cups of fruit juice or eat the equivalent of five to six slices of whole-wheat bread within two hours after exercise.

The Precompetition Meal

The precompetition meal serves several purposes. Most importantly, it replenishes body stores of glycogen and helps support optimal blood glucose levels during exercise. It also keeps the athlete from feeling hungry during competition. In addition, most athletes are accustomed to a favorite pre-event meal, which can give them a psychological lift important for top performance.

Allowing for personal preferences, the pre-event meal should be mainly easily digestible carbohydrates, low in fat and protein, and eaten early enough to allow for digestion. High-fat and protein-rich foods slow stomach emptying and should be avoided just before competition—exercising with a full stomach may cause abdominal pain and nausea.

How much carbohydrate should be eaten? Between 1 and 4 g/kg body weight should be consumed one to four hours prior to exercise; in general, the closer the meal is to the event, the less carbohydrate (1–2 g/kg) should be eaten. Liquid meals (such as blended mixtures of fruit juice, fruit, and low-fat milk) can be eaten closer to competition because they leave the stomach sooner than solid food (Harkins et al. 1993.)

Fat Intake for Athletes

Fat intake should be only 20–30% of total calories. No attempt should be made to consume and store fats for energy, although fats are important fuels during exercise. Body fat stores, even in very lean athletes, contain much more fat than is needed during training or competition (Harkins et al. 1993). Each pound of fat contains about 3500 kcal—a 70 kg athlete with only 10% body fat has over 55,000 kcal stored as fat.

Protein Intake for Athletes

Daily protein requirements are modestly increased in athletes: heavy training increases protein requirements to 1.2–1.8 g/kg/day (Berning et al. 1993), (Williams 1985). However, the typical protein-rich diet in the United States easily provides this much protein each day. Athletes (even those attempting to increase muscle mass) eating balanced, nutritious diets obtain ample dietary protein. Adding extra protein to the diet is of no benefit. Protein should provide about 10–15% of total calories.

Vitamin and Mineral Intake for Athletes

Eating a balanced, nutritious diet should provide vitamins and minerals at levels approximating the RDA for these nutrients. Additional amounts above the RDA appear to be of no benefit to athletes. As stated earlier, athletes should be periodically evaluated for iron status, because of the common occurrence of iron deficiency, particularly in young athletes and long-distance runners (Buskirk 1990). If iron deficiency is present, replacement will usually improve athletic performance.

Fluid and Electrolyte Intake for Athletes

Water is probably the most important nutrient for the athlete in training. If body weight drops more than 2% during exercise, fluid intake has probably been inadequate. In general, at least two cups of cool fluid should be consumed for every pound of body weight lost. Table 8.13 lists guidelines for fluid replacement during exercise.

Cool water is the preferred fluid during exercise lasting less than an hour. Cold fluids are absorbed more rapidly than warmer fluids. Many concentrated fluids (>200 mOsm/L) empty from the stomach more slowly than more dilute fluids. If fruit juices, soft drinks, or concentrated

---

The key to achieving adequate hydration is planned water intake. The following schedule of fluid intake is recommended:

- Two hours before the event, drink 16 oz water or a nonfat, noncarbonated beverage.

- During the half hour before the event, drink 8 to 12 oz of fluid.

- During the event, drink 4 to 8 oz of fluid every 15 to 20 minutes.

- After the event, the athlete should drink additional fluids to replace those lost during the event. The general guide is 16 oz (1 pint) of fluid for every 1 lb loss of body weight.

- Caffeine- and alcohol-containing beverages have a dehydrating effect and should be avoided.

Adapted from M. H. Williams, *Nutrition For Exercise and Sport*, 3d ed. (Dubuque, Iowa: Wm. C. Brown, 1992).

TABLE 8.13
Guidelines for fluid replacement during exercise

sports drinks are used in place of water, they will be more rapidly absorbed if diluted with about two to three parts water.

If exercise exceeds 60 minutes, a beverage containing 6–8% carbohydrate may be beneficial (Harkins et al. 1993). A 6% carbohydrate solution is absorbed as rapidly as water and can provide a source of glucose for energy in muscles running low on glycogen. Glucose consumed during prolonged exercise may increase endurance and prolong time to exhaustion (Coogan 1988). The beverage should contain between 14 and 19 grams of carbohydrate per cup; if it is more concentrated, absorption of the fluid will be slowed.

Although electrolytes are often promoted as important components in sport drinks, they are of little or no value during moderate exercise. Salt tablets should not be consumed during exercise because they can irritate the stomach and cause nausea, and they may increase the body's water needs.

Summary

No single diet best supports physical performance. Optimal athletic performance is sustained by a lifetime of healthful dietary choices, rather than special combinations of foods or food supplements. In general, dietary guidelines for athletes are similar to those for the general population: eat a variety of healthful, nutrient-dense foods, emphasizing carbohydrate and avoiding excess fat, and drink ample fluids. A higher number of servings of fruits, vegetables, and grains is recommended, as shown in Table 8.14. Following these guidelines will not only support athletic performance, but overall good health as well.

NUTRITIONAL GUIDELINES FOR A HEALTHY ADOLESCENCE: REVIEW AND SUMMARY

Many adolescents who feel young, healthy, and invulnerable, often fail to consider how their food choices today may influence long-term health. Providing guidance to teenagers about nutrition is a challenge.

Using the guidelines listed here, adolescents can meet their nutrient needs by eating diets that are varied, balanced, and moderate.

1. To ensure intake of all the essential nutrients, eat a variety of foods, balanced among the five major food groups.

2. Balance energy intake and physical activity to maintain a desirable weight. If overweight, adolescents need to understand the factors involved in weight loss and adopt sensible, healthy dieting practices to attain a desirable weight.

Food Group	Recommended Number of Servings/Day for Teenagers in Training
Milk	3–4
Milk or yogurt, 1 cup	
Cheese, 1 oz	
Cottage cheese, ½ cup	
Ice cream, ½ cup	
Meat	2–3
Cooked lean meat, fish, or poultry, 2–3 oz	
Egg, 1	
Dried beans or peas, ½ cup	
Peanut butter, 2 tbsp	
Cheese, 2 oz	
Fruits and vegetables	8–10
Cooked or juice, ½ cup	
Raw, 1 cup	
Medium piece of fruit 1	
Grains (whole grains, fortified, enriched)	10–12
Bread, 1 slice	
Cereal (ready to eat) 1 oz	
Pasta, ½ cup	

Adapted from *Sports and Nutrition for the 90s,* eds. J. R. Berning and S. N. Steen (Gaithersburg, MD: 1991).

TABLE 8.14
Food choices during training

3. Limit consumption of snacks and foods high in fat, sodium, and simple carbohydrates. Healthful snacks are fresh or dried fruits, carrots, whole-grain or graham crackers, fig bars, low-fat oatmeal cookies with raisins, bran muffins, and yogurt.

4. Dietary fat should contribute 30% or less of total calories, and saturated fat intake should be 10% or less of calories. Saturated fat intake can be decreased by choosing lean meats, chicken without the skin, fish, and low-fat dairy products. To reduce total fat intake, limit consumption of fried foods, mayonnaise, and salad dressings.

5. Most calories should come from complex carbohydrates. Increase intake of starches and complex carbohydrates by eating six or more servings of bread, cereals, rice, pasta, and legumes each day. Emphasize whole-grain products.

6. Eat five or more servings of a variety of vegetables and fruits each day.

Food Group	Recommended Daily Servings

Fruits
  Citrus, melon, berries
  Other fruits
	2–4

Vegetables
  Dark green and deep yellow
  Starchy, including dry beans, peas
  (include dark green and dry beans
  and peas several times per week)
Other vegetables
	3–5

Meat, fish, poultry, eggs (total of 5–7 oz daily from lean choices)	2–3
Milk, yogurt, cheese	3–4

Grains, breads, cereals
  Whole grain and enriched
  (include several servings of whole
  grain products daily)
	6–11

Fats, sweets, alcohol	In moderation; alcohol is not recommended for teenagers

Adapted from U.S. Department of Agriculture, *A Food Guide for Adolescents* (Hyattsville, MD, U.S. Dept. of Agriculture, 1989).

Note: Serving sizes for the different food groups are described in Table 9.28.

TABLE 8.15
USDA food guide for adolescents

7. Diets should emphasize ample sources of calcium and iron. Adequate calcium intake can be achieved by consuming three or four servings of dairy products—milk, yogurt, or cheese—each day, along with other rich sources such as kale, collard and mustard greens, and fortified tofu. Iron-rich foods include fortified cereals, lean meats, and legumes.

8. Avoid vitamin and mineral supplements in excess of the RDAs. Vitamin or mineral supplements at or below the RDAs are safe, but rarely needed.

9. All healthy adolescents should maintain at least a moderate level of physical activity.

The U.S. Department of Agriculture has developed a food guide for adolescents. Its recommendations both fulfill the Dietary Guidelines for Americans and supply the RDAs for the teenage years (see Table 8.15). Achieving good health and good nutrition during adolescence are the goals outlined in the *Year 2000 Objectives for the Nation* published by the

U. S. Department of Health and Human Services (see Table 8.16) (U. S. Public Health Service 1991).

Avoiding the hazards of alcohol, smoking, and illicit drug use will also contribute to nutritional health during adolescence. Optimal nutrition during the teen years ensures normal growth and development and may decrease the risk of future chronic disease.

1. Nutrition

- Prevent overweight, especially in minorities

- Improve knowledge of factors involved in weight loss and sensible dieting practices

- Reduce anemia in low-income pregnant adolescents

- Increase use of Dietary Guidelines for Americans in school meals

- Include nutrition education in school health education

- Increase intakes of calcium and iron

- Reduce dietary fat to no more than 30% of total kilocalories, and saturated fat to no more than 10% of kilocalories

- Increase those in the population who are able to identify diet-related risk factors for heart disease, hypertension, cancer, and osteoporosis

- Increase the proportion of those who can identify the major food sources of fat

- Require nutrition education from preschool to grade 12 as part of comprehensive health education

2. Fitness

- More than 60% should participate in daily physical education classes

- More than 70% should participate in assessments of physical fitness

From Department of Health and Human Services, *Healthy People 2000: National Health Promotion and Disease Prevention Objectives* (Department of Health and Human Services, 1991).

TABLE 8.16

U.S. Department of Health and Human Services goals for adolescent nutrition and fitness

REFERENCES

Aikawa, J. K. *Magnesium: Its Biological Significance* (Boca Raton: CRC Press, 1981).

Allen, K. F., et al., "Teenage Tobacco Use: Data Estimates from the Teenage Attitudes and Practices Survey, United States, 1989," Advance Data from Vital and Health Statistics, 224 (Hyattsville, MD: National Center for Health Statistics, 1992).

American Academy of Pediatrics, *Pediatric Nutrition Handbook* (Elk Grove, IL: AAP, 1993).

American School Health Association, *The National Adolescent School Health Survey: A Report on the Health of America's Youth* (Oakland, CA: Third Party Publishing, 1989).

Bailey, L. and Cerda, J. J., "Iron and Folate Nutriture during the Life Cycle," *World Rev Nutr Diet* 56 (1988):56.

Baker, E. F., "Menstrual Dysfunction and Hormonal Status in Athletic Women," *Fertil Steril* 36 (1981):691–96.

Ballin, A., et al., "Iron State of Female Adolescents," *Am J Dis Child* 146 (1992):803.

Becque, M. D., et al., "Coronary Risk Incidence of Obese Adolescents: Reduction by Exercise Plus Diet Intervention," *Pediatrics* 81 (1988):605–12.

Berning, J., et al., "Fuel Supplies for Exercise," in *Sports Nutrition*, ed. D. Bernardot (Chicago: The American Dietetic Association, 1993).

Best, J. A., et al., "Preventing Cigarette Smoking among School Children," *Annu Rev Publ Health* 9 (1988):161–201.

Beumont, P. J. V., Russell, J. D., and Touyz, S. W., "Treatment of Anorexia Nervosa," *Lancet* 341 (1993):1635–40.

Bigler-Doughten, S. and Jenkins, R. M., "Adolescent Snacks: Nutrient Density and Nutritional Contribution to Total Intake," *J Am Diet Assoc* 87 (1987):1678–79.

Biller, B. M. K., et al., "Mechanisms of Osteoporosis in Adult and Adolescent Women with Anorexia Nervosa," *J Clin Endocrinol Metab* 68 (1989):548–54.

Blair, S. N., et al., "How Much Physical Activity is Good for Health?" *Ann Rev Public Health* 13 (1992):99–126.

Bolton-Smith, C., et al., "Antioxidant Vitamin Intakes Assessed Using a Food Frequency Questionnaire: Correlation with Biochemical Status in Smokers and Nonsmokers," *Br J Nutr* 65 (1991):337–46.

Brown, J. D., and Lawton, M., "Stress and Well-Being in Adolescence: The Moderating Role of Physical Exercise," *J Human Stress* 12 (1986):125–31.

Buckner, E. T., "Do You Have Patients with Anorexia or Bulimia?" *Postgrad Med* 83 (1991):209–15.

Bull, N. L., "Dietary Habits, Food Consumption and Nutrient Intake During Adolescence," *J Adolescent Health* 13 (1992):384-88.

Buskirk, E. R., "Exercise," in *Present Knowledge in Nutrition*, 6th ed., edited by M. L. Brown (Washington D. C.: International Life Sciences Institute Nutrition Foundation, 1990).

Castro, F. G., Newcomb, M. D., and Cadish, K., "Lifestyle Differences between Young Adult Cocaine Users and their Nonuser Peers," *J Drug Educ* 17 (1987):89.

Centers for Disease Control and Prevention, "Childhood Injuries in the U. S.," *Am J Dis Child* 144 (1990):646–727.

Centers for Disease Control and Prevention, "Teenage Pregnancy and Birth Rates," *MMWR* 42 (1993):733–37.

Centers for Disease Control and Prevention, "Body Weight Perceptions and Selected Weight Management Goals and Practices of High-School Students—U. S., 1990," *MMWR* 40 (1991): 741,747–50.

Centers for Disease Control, "Alcohol and Other Drug Use Among High School Students—United States, 1990," *MMWR* 40 (1991):776–77, 783–84.

Centers for Disease Control, "Participation of High School Students in School Physical Education: United States, 1990," *MMWR* 40 (1991):613–15.

Centers for Disease Control, "Recommendations for the Use of Folic Acid to Reduce the Number of Cases of Spina Bifida and Other Neural Tube Defects," *MMWR* 41 (1992).

Centers for Disease Control, "Vigorous Physical Activity among High School Students: United States, 1990," *MMWR* 41 (1992):33–5.

Christiansen, C., Rodbro, P., and Thoger-Nielson, C., "Bone Mineral Content and Estimated Total Body Calcium in Normal Children and Adolescents," *Scand J Clin Lab Invest* 35 (1975):507–10.

Comerci, G. D., "Medical Complications of Anorexia Nervosa and Bulimia Nervosa," *Med Clin NA* 74 (1990):1293–1310.

Comerci, G. D., Kilbourne, K. A., and Harrison, G. G., "Eating Disorders: Obesity, Anorexia Nervosa and Bulimia," *in Adolescent Medicine*, 2d edited by A. D. Hofmann and D. E. Greydanus (Norwalk, CT: Appleton & Lange, 1989).

Contursi, J., Kleiner, S. and Mielcarek, J., "Nutrient and Quasi-Nutrient Supplement Consumption," in *Sports Nutrition*, edited by D. Bernadot (Chicago: American Dietetic Association, 1993).

Coogan, A. R., and Coyle, E. F., "Effect of Carbohydrate Feedings during High Intensity Exercise," *J Appl Physiol* 64 (1988):1480–85.

Craig, W. Y., et al., "Cigarette Smoking-associated Changes in Blood Lipid and Lipoprotein Levels in the 8- to 15-Year-Old Age Group: A Meta-analysis," *Pediatrics* 85 (1990):155–8.

Davies, K. M., et al., "Third Decade Bone Gain in Women," *J Bone Miner Res* 4 (1989):838.

de Zwaan, M., Nutzinger, D. O., and Schoenbeck, G., "Binge Eating in Overweight Women," *Compr Psychiatry* 33 (1992):256–61.

Drinkwater, B. L., et al., "Bone Mineral Content of Amenorrheic and Eumenorrheic Athletes," *N Engl J Med* 311 (1984):277–81.

Driskell, J. A., Clark, A., and Mock, S. W., "Longitudinal Assessment of Vitamin B6 Status in Southern Adolescent Girls," *J Am Diet Assoc* 87 (1987):307.

Elster, A., ed., *American Medical Association Guidelines for Adolescent Preventive Services* (Baltimore: Williams and Wilkins, 1994).

Epstein, L. H., et al., "Ten-Year Follow-Up of Behavioral, Family-based Treatment for Obese Children," *JAMA* 264 (1990):2519–23.

Escobedo, L. G., et al., "Sociodemographic Characteristics of Cigarette Smoking Initiation in the United States," *JAMA* 264 (1990):1550–55.

Fike, S., Kanter, M., and Markley, E., "Fluid and Electrolyte Requirements of Exercise," in *Sports Nutrition*, edited by D. Bernadot (Chicago: American Dietetic Association, 1993).

Forbes, G. B., "Nutritional Requirements in Adolescence," in *Textbook of Pediatric Nutrition*, ed. R. M. Suskind (New York: Raven Press, 1981).

Fraser, G. E., Philips, R. L., and Harris, R., "Physical Fitness and Blood Pressure in Children," *Circulation* 67 (1983):405–11.

Frisch, R. E., and McArthur, J. W., "Menstrual Cycles: Fatness as a Determinant of Minimum Weight for Height Neccessary for Their Maintenance or Onset," *Science* 185 (1974):949–51.

Frisch, R. E., "Body Weight, Body Fat and Ovulation," *Trends Endocrinol Metab* 2 (1991): 191–97.

Garner, D. M., "Pathogenesis of Anorexia Nervosa," *Lancet* 341 (1993):1631–35.

Gerster, H., "The Role of Vitamin C in Atheletic Performance," *J Am Coll Nutr* 8 (1989):636–43.

Gollnick, P. D., "Metabolism of Substrates: Energy Substrate Metabolism during and as Modified by Training," *Fed Proc* 44 (1985):353–7.

Gortmaker, S. L., Dietz, W.H., and Cheung, L. W., "Inactivity, Diet and the Fattening of America," *J Am Diet Assoc* 90 (1990):1247–52.

Gortmaker, S. L., et al., "Social and Economic Consequences of Overweight in Adolescence and Young Adulthood," *N Engl J Med* 329 (1993):1008–12.

Gortmayer, S. L., et al., "Increasing Pediatric Obesity in the United States," *Am J Dis Child* 141 (1987):535–40.

Guenther, P. M., "Beverages in the Diets of American Teenagers," *J Am Diet Assoc* 86 (1986):493.

Hagberg, J. M., et al., "Effect of Exercise Training on the Blood Pressure and Hemodynamic Features of Hypertensive Adolescents," *Am J Cardiol* 52 (1983):763–68.

Harkins, C., et al., "Protocols for Developing Dietary Prescriptions," in *Sports Nutrition*, edited by D. Bernadot (Chicago: American Dietetic Association 1993).

Heimburger, D. C., et al., "Improvement in Bronchial Squamous Metaplasia in Smokers Treated with Folate and B12," *JAMA* 259 (1988):1525–30.

Herzog, D. , et al., "The Course and Outcome of Bulimia Nervosa," *J Clin Psychiatry* 52 (1991):4–8.

Hirschman, R. S., and Leventhal, H., "Preventing Smoking Behavior in School Children: An Initial Test of a Cognitive-Development Program," *J Appl Soc Psychol* 19 (1989):559–83.

Hofstetter, A., et al., "Increased 24-hour Energy Expenditure in Cigarette Smokers," *New Eng J Med* 314 (1986):79–82.

Ivy, J. L., et al., "Muscle Glycogen Synthesis after Exercise: Effect of Time of Carbohydrate Ingestion," *J Appl Physiol* 64 (1988):1480–85.

Joffe, A., "Adolescent Medicine," in *Principles and Practice of Pediatrics*, 2d. ed., F. A. Oski (Philadelphia: J. B. Lippincott, 1994).

Johnston, C. C., et al., "Calcium Supplementation and Increases in Bone Mineral Density in Children," *N Engl J Med* 327(2) (1992):82–7.

Johnston, L. D., O'Malley, P. M., and Bachman, J. G., "Drug Use among American High School Seniors, College Students and Young Adults, 1975-1990. Volume 1: High School Seniors," *DHHS Publ.* no. (ADM) 91-1813 (Rockville, MD: National Institute of Drug Abuse, 1991).

Jonas, J. M., et al., "Eating Disorders and Cocaine Abuse: A Survey of 259 Cocaine Abusers," *J Clin Psychiatry* 48 (1987):47.

Kandel, D. B., and Logan, J. A., "Patterns of Drug Use from Adolescence to Young Adulthood, I. Periods of Risk for Initiation, Continued Use, and Discontinuation," *Am J Public Health* 74 (1984):660–66.

Kaplan, K. M., and Wadden, T. A, "Childhood Obesity and Self-Esteem," *J Pediatr* 109 (1986):367–70.

Keith, R. E., "Vitamins in Sport and Exercise," in *Nutrition in Exercise and Sport*, eds., I. Wolinsky and J. F. Hickson (Boca Raton: CRC Press, 1989).

Kimm, S. Y. S., et al. "Dietary Patterns of U. S. Children: Implications for Disease Prevention," *Prev Med* 19 (1990):432–42.

Kovar, M. G., "Use of Medications and Vitamin-Mineral Supplements by Children and Youths," *Public Health Report* 100 (1985):470-73.

Kushner, R. F., "Body Weight and Mortality," *Nutr Rev* 51 (1993):127–36.

Larkin, F. A., et al., "Dietary Patterns of Women Smokers and Nonsmokers," *J Am Diet Assoc* 90 (1990):230–37.

Leibel, R. L., Edens, N. K., and Fried, S. K., "Physiological Basis for the Control of Body Fat Distribution in Humans," *Ann Rev Nutr* 9 (1989): 417–43.

Levenson, D. I., and Bockman, R. S., "A Review of Calcium Preparations," *Nutr Rev* 52 (1994):221–32.

Life Sciences Research Office, FASEB, "Nutrition Monitoring in the U. S. An Update Report on Nutrition Monitoring," *DHHS Publ.* no. (PHS) 89 1255 (Hyattsville, MD: U. S. Depts HHS, USDA Sept 1989).

Lifshitz, F., and Moses. N., "Growth Failure. A Complication of Dietary Treatment of Hypercholesterolemia," *Am J Dis Child* 143 (1989):537–42.

Lifshitz, F., Tarim, O., and Smith, M. M., "Nutrition in Adolescence," *Endocrinol Metab Clinics NA* 22 (1993):673–83.

Lifshitz, F., and Moses, N., "Nutritional Dwarfing: Growth, Dieting, and Fear of Obesity," *J Am Coll Nutr* 7 (1988):367–76.

Lloyd, T., et al., "Calcium Supplementation and Bone Mineral Density in Adolescent Girls," *JAMA* 270 (1993):841–44.

Loeb, L. A., et al., "Smoking and Lung Cancer: An Overview," *Cancer Res* 44 (1984):5940–58.

Lucas, A. R., and Huse, D. M., "Behavioral Disorders Affecting Food Intake: Anorexia Nervosa and Bulimia Nervosa," in *Modern Nutrition in Health and Disease*, 7th ed., edited by M. E. Shils and V. R. Young (Philadelphia: Lea & Febiger, 1988).

Lucas, A. R., et al., "50-Year Trends in the Incidence of Anorexia Nervosa in Rochester, Minn: A Population-based Study," *Am J Psychiatry* 148 (1991):917–22.

Marcus, M. D., Wing, R. R., and Lamparski, D. M., "Binge Eating and Dietary Restraint in Obese Patients," *Addict Behav* 10 (1985):163–68.

Martin, S., et al., "Self-Esteem of Adolescent Girls as Related to Weight," *Percept Mot Skills* 67 (1988):879–84.

Matkovic, V., et al., "Factors that Influence Peak Bone Mass Formation: A Study of Calcium Balance and the Inheritance of Bone Mass in Adolescent Females," *Am J Clin Nutr* 52 (1990):878–88.

Matkovic, V., "Calcium Intake and Peak Bone Mass," *New Engl J Med* 327(2) (1992):119–20.

McGinnis, J. M., and Lee, P. R., "Healthy People 2000 at Mid Decade," *JAMA* 273 (1995):1123.

McNamara, J. J., et al., "Coronary Artery Disease in Combat Casualties in Vietnam," *JAMA* 216 (1971):1185–87.

Metropolitan Life Insurance Company, "1983 Metropolitan Height and Weight Tables," *Stat Bull Metropol Life Insur Co* 64 (1983).

Mohs, M. E., Watson, R. R., and Leonard-Green, T., "Nutritional Effects of Marijuana, Heroin, Cocaine and Nicotine," *J Am Diet Assoc* 90 (1990):1261–67.

Moses, N., Banilivy, M., and Lifshitz, F., "Fear of Obesity among Adolescent Girls," *Pediatrics* 83 (1989):393–98.

Moss, A. J., et al., "Recent Trends in Adolescent Smoking, Smoking-Uptake Correlates, and Expectations about the Future. *Advance Data from Vital and Health Statistics;* 221 (Hyattsville, MD: National Center for Health Statistics, 1992).

Must, A., et al., "Long-Term Morbidity and Mortality of Overweight Adolescents," *N Eng J Med* 327 (1992):1350–55.

National Academy of Sciences Food and Nutrition Board of the National Research Council, *Recommended Dietary Allowances*, 10th ed. (Washington D. C.: National Academy Press, 1989).

National Center for Health Statistics, "Dietary Intake Source Data," *DHEW Publ.* no. 79-1221 (Hyattsville, MD: NCHS, 1979).

National Cholesterol Education Program (NCEP), "Report of the Expert Panel on Blood Cholesterol in Children and Adolescents," *Pediatrics*, 89 suppl. 89 (1992):525–84.

National Heart, Blood and Lung Institute, "Report of the Second Task Force on Blood Pressure Control in Children—1987," *Pediatrics* 79 (1987):1–25.

Nickerson, H. J., et al., "Causes of Iron Deficiency in Adolescent Athletes," *J Pediatr* 114 (1989):657–63.

Noakes, T. D., et al., "The Incidence of Hyponatremia during Prolonged Ultraendurance Exercise," *Med Sci Sports Exerc* 22 (1990):165–70.

Patton, G. C., "Mortality in Eating Disorders," *Psychol Med* 18(4) (1988):947–52.

Pi-Sunyer, P. X., "Health Implications of Obesity," *Am J Clin Nutr* 53 (1991):1595S-1603S.

Portnoy, B. and Christenson, G. M., "Cancer Knowledge and Related Practices: Results from the National Adolescent Student Health Survey," *J School Health* 59 (1989):218.

Prasad, A. S., "Clinical and Biochemical Spectrum of Zinc Deficiency in Human Subjects," in *Clinical, Biochemical and Nutritional Aspects of Trace Elements. Current Topics in Nutrition and Disease*, Vol. 6, ed. A. S. Prasad (New York: Alan R. Liss, 1982).

Preston, A. M., "Cigarette Smoking: Nutritional Implications," *Prog Food Nutr Sci* 15 (1991):183–217.

Pugliese, M. T., et al., "Fear of Obesity: A Cause of Short Stature and Delayed Puberty," *N Engl J Med* 309 (1983):513–18.

Recker, R. R., et al., "Bone Gain in Young Adult Women," *JAMA* 268(17) 1992:2403-8.

Rigotti, N. A., et al., "Osteoporosis in Women with Anorexia Nervosa," *N Engl J Med* 311 (1984):1601–6.

Rising, R., "The Pathophysiology of Childhood Obesity," in *Childhood Nutrition*, ed. F. Lifshitz (Boca Raton: CRC Press, 1995).

Rochinni, A. P., et al., "Blood Pressure in Obese Adolescents: Effect of Weight Loss," *Pediatrics* 82 (1988):16–23.

Rowland, T. W., Black, S. A., and Kelleher, J. F., "Iron Deficiency in Adolescent Endurance Athletes," *J Adolescent Health Care* 8 (1987):322–26.

Sallis, J. F., et al., "Relation of Cardiovascular Fitness and Physical Activity to Cardiovascular Disease Risk Factors in Children and Adults," *Am J Epidemiol* 127 (1988):933–41.

Sandler, R. V., et al., "Postmenopausal Bone Density and Milk Consumption in Childhood and Adolescence," *Am J Clin Nutr* 42 (1985):270–74.

Savin, J. A., "Skin Disease: A Link with Zinc," *Br Med J* 289 (1983):1477.

Sawka, M. N., et al., "Influence of Hydration Level and Body Fluids on Exercise Performance in the Heat," *JAMA* 252 (1984):1165–69.

Shear, C. L., et al., "Value of Childhood Blood Pressure Measurements and Family History in Predicting Future Blood Pressure Status: Results from Eight Years of Follow-Up in the Bogalusa Heart Study," *Pediatrics* 77 (1986):862-69.

Shimokata, H., Muller, D. C., and Andres, R., "Studies in the Distribution of Body Fat. III. Effects of Cigarette Smoking," *JAMA* 261 (1989):1169–73.

Simons-Morton, B. G., et al., "Health-Related Physical Fitness in Child-hood: Status and Recommendations," *Ann Rev Public Health* 9 (1988):403–25.

Sporn, M. B., Roberts, A. B. and Goodman, D. S., eds, *The Retinoids: Biology, Chemistry and Medicine* (New York:, Raven Press,1994).

Stern, R. S., "The Prevalence of Acne on the Basis of Physical Examina-tion," *J Am Acad Derm* 26 (1992):931–5.

Story, M., et al., "Demographic and Risk Factors Associated with Chronic Dieting in Adolescents," *Am J Dis Child* 145 (1991):994–98.

Subar, A. F., Harlan, L. C., and Mattson, M. E., "Food and Nutrient Intake Differences between Smokers and Nonsmokers in the U. S.," *Am J Publ Hlth* 80 (1990):1223–9.

Sutie, J. W., et al., "Vitamin K Deficiency from Dietary Vitamin K Restriction in Humans," *Am J Clin Nutr* 47 (1988):475–80.

Tager, I. B., et al., "Longitudinal Analysis of the Effects of Cigarette Smoking on Children," *Chest* 85 (1984):8S.

Tanner, J. M., *Fetus into Man: Physical Growth from Conception to Maturity* (Cambridge: Harvard University Press,1990).

Tanner, J. M., "Growing Up," *Sci American*, 1973.

Tarnopolsky, M., "Influence of Protein Intake and Training Status on Nitrogen Balance and Lean Body Mass," *J Appl Physiol* 64 (1988):187–93.

Tolman, E. L., "Acne and Acneiform Dermatoses," in *Dermatology*, 2 ed., edited by S. L. Moschella (Philadelphia: W. B. Saunders, 1985).

U. S. Department of Health and Human Services, "The Health Benefits of Smoking Cessation: A Report of the Surgeon General," *DHHS Publ.* no. (CDC) 90-8416 (Washington, D. C.: Government Printing Office, 1990).

U. S. Department of Agriculture, "Nationwide Food Consumption Survey, Nutrient Intakes: Individuals in 48 States, Year 1977–1978," Report no. I-2 (Hyattsville, MD: USDA, 1984).

U. S. Public Health Service, "Healthy People 2000: National Health Promotion and Disease Prevention Objectives," Publication 91-50212 (Washington D. C.: U. S. Dept. of Health and Human Services, 1991).

U. S. Public Health Service, *Health United States 1989 and Prevention Profile*, (Washington D. C.: U. S. DHHS, 1990).

U. S. Department of Health and Human Services, "Reducing the Health Consequences of Smoking: 25 Years of Progress: A Report of the Surgeon General 1989. *DHHS Publ.* no. (CDC) 89-8411 (Washington D. C.: U. S. Government Printing Office, 1989).

Van Itallie, T. B., "Obesity: Adverse Effects on Health and Longevity," *Am J Clin Nutr* 32 (1979):2723–33.

Wadden, T. A., et al., "Obesity in Black, Adolescent Girls: A Controlled Clinical Trial of Treatment by Diet, Behavioral Modification, and Parental Support," *Pediatrics* 85 (1990):345–52.

Weight, L. M., Myburgh, K. H., and Noakes, T. D., "Vitamin and Mineral Supplementation: Effect on Running Performance of Trained Athletes," *Am J Clin Nutr* 47 (1988):192–95.

Williams, G. H., and Moore, T. J., "Hormonal Aspects of Hypertension," in *Endocrinology*, 3d ed., edited by L. J. DeGroot (Philadelphia: W. B. Saunders, 1995).

Williams, M., *Nutrition for Fitness and Sport*, 3d ed. (Dubuque, Iowa: Wm. C. Brown, 1992).

Williams, M. H., "The Role of Protein in Physical Activity," in *Nutritional Aspects of Human Physical and Athletic Performance*, (Springfield, IL: Charles Thomas, 1985).

Williamson, D. F., et al., "Smoking Cessation and Severity of Weight Gain in a National Cohort," *N Engl J Med* 324 (1991):739–45.

Witischi, J. C., et al., "Sources of Fat, Fatty Acids, and Cholesterol in the Diets of Adolescents," *J Am Diet Assoc* 90 (1990):1429.

World Health Organization, Food and Agricultural Organization "Calcium Requirements. Report of a Joint FAO/WHO Expert Consultation," *FAO Meeting Report* no. 30 (Rome: Food and Agriculture Organization, 1962).

Woynarowska, B., et al., "Blood Pressure Changes During Adolescence and Subsequent Blood Pressure Level," *Hypertension* 7 (1985):695–701.

Young, E.A., et al., "Fast Foods Update: Nutrient Analysis," *Dietetic Currents* 13 no. 6 (1986).

Zack, P. M., et al., "A Longitudinal Study of Body Fatness in Childhood and Adolescence," *J Pediatr* 95 (1979):126–30.

9

adulthood
and nutrition

▼

THE GLOBAL IMPACT OF THE MODERN "AFFLUENT" DIET

For much of human history, the food supply was precarious and starvation frequent. About 10,000 years ago, the agricultural revolution brought profound change—the ability to produce and store food increased dramatically, assuring a more constant supply and wider availability. In the last 200 years, the industrial revolution in the developed world has changed our diet even more radically. New methods of food production, processing, storage, and distribution have expanded the amount and variety of available food enormously.

The global health benefits of the improved food supply have been immediate and far-reaching. In the industrialized countries, the increased quantity and quality of the diet have reduced starvation and nutrient-deficiency diseases. Improvements in nutrition have contributed to a dramatic drop in foodborne and infectious disease, reduced infant mortality rates, and increased life expectancy.

Economic development has expanded wealth, improved material well-being, and allowed for the exercise of dietary preferences. Today, we choose what to eat and drink from an astonishing abundance of foods. Each year, the global food industry introduces over 10,000 new food products and spends over $10 billion to advertise them. All of these changes have led to a dramatic and extremely rapid shift in the nutritional composition of the diet. The amount of animal fats and simple carbohydrates we eat has increased substantially, while the complex carbohydrate and fiber content of the diet has plummeted (Posner et al. 1994). Populations that can afford to eat a lot of fat, sugar, and salt do so because foods rich in these nutrients taste better to most people.

The adverse health effects of this modern "affluent" diet have only appeared in recent decades. The establishment of this diet in a popula-

Source: Adapted from WHO (1990)

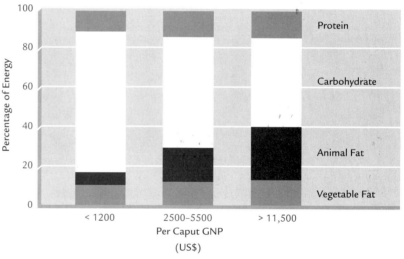

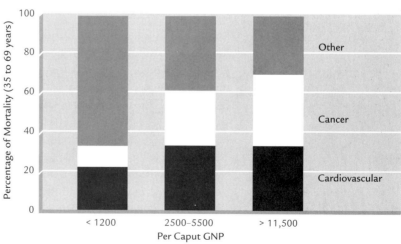

FIGURE 9.1

Diet composition worldwide compared with GNP and causes of mortality

tion, together with a more sedentary lifestyle, leads to the rapid emergence of a range of chronic diseases—including diseases of the heart and blood vessels, cancer, diabetes, obesity, and dental decay (WHO 1990). See Figure 9.1.

The dynamic relationship between changes in a population's diet and changes in its health are magnified in certain countries that have undergone very rapid development in the past 40 to 50 years (Posner et al. 1994). The subtropical island of Mauritius, in the southwestern Indian Ocean, is home to about 1 million people. It is an example of a country where social, economic, and dietary change has occurred very rapidly. In the second half of this century, a manufacturing and tourist boom brought development and prosperity, and many Mauritians began eating foods higher in fat and calories and lower in complex carbohydrates and fiber.

Deaths from foodborne and infectious diseases have dropped, and the infant mortality rate has fallen—but new health problems have emerged. In the 1940s heart disease killed only 1 in 50 people; today, nearly 1 in 2 die of cardiovascular disease. Heart disease, cancer, and diabetes have become the major health problems in Mauritius. Mauritians are living longer, but even when adjusted for age, death rates from heart disease and breast cancer have tripled (Brissonnette and Fareed 1985).

In many parts of the world, hunger and undernutrition remain enormous problems—and for most developing countries the first priority still is finding adequate food. However, like the industrialized world, many developing countries are now also forced to deal with the problem of trying to forestall or arrest a population shift toward a diet high in calories, fat, sugar, and salt (WHO 1990). Table 9.1 shows the causes of death in developing and developed countries.

The rising costs of long-term treatment of cardiovascular disease, cancer, and diabetes threaten the health budgets of many countries. Worldwide, countries as diverse as China, Poland, India, Venezuela, and

	Percentage of Deaths		
Causes of Death	**Developed Countries**	**Developing Countries**	**World Total**
Diseases of the circulatory system	54	19	26
Neoplasms	19	5	8
Infectious and parasitic diseases	8	40	33
Injury and poisoning	6	5	5
Perinatal mortality	2	8	6
All other causes	12	23	21

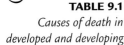

TABLE 9.1

Causes of death in developed and developing countries

Adapted from World Health Organization, "Diet, Nutrition and the Prevention of Chronic Disease: Report of a WHO study group." *Technical Report Series* 797 (Geneva, WHO, 1990).

Mauritius have begun major, populationwide, preventive health campaigns (Posner et al. 1994). These large programs establish nutritional guidelines in an effort to control the explosion of diet-related, chronic diseases. Diet-related chronic disease has reached similar, epidemic proportions in adults in the United States. This chapter describes the unique nutritional needs of adults and shows how dietary choices through adulthood influence long-term health.

ADULTHOOD

The adult stage of the life cycle refers to the years of **maturity** between adolescence and **senescence**. In most people, these are the years 18 through 65, and they will be the focus of this chapter. In the late teenage years, the rapid development that characterizes adolescence slows as maturity approaches. An individual reaches physiological maturity when the body has grown to its full size and strength, and development of the reproductive, musculoskeletal, and other body systems is complete.

Although physiological maturity is reached in the early 30s, psychosocial development continues through adulthood as each person develops along his or her own unique path. Adulthood, for many, means forming relationships, establishing a family, and building a career. Adults who are well nourished can better tolerate the psychological and physiological stresses of family, the workplace, and the many other responsibilities of adult life.

PHYSIOLOGICAL CHANGES AFFECTING ADULT NUTRITION

In contrast to the turbulent growth and development of adolescence, adulthood is characterized by years of functional stability. With the exception of the menopause in women, the adult period is not marked by the timed sequence of physiological changes that take place in earlier development.

Most body systems reach physiological maturity around age 18. An exception is the musculoskeletal system: although maximum height and strength are attained by the early 20s, peak bone mass is not reached until the early 30s (Matkovic et al. 1990). Once the skeleton is complete, the body has reached full maturity. Thereafter, maintenance, equilibrium, and balance are the predominant characteristics of adulthood.

This state of internal balance is referred to as *homeostasis.* Maintenance of homeostasis involves a continual adjustment of the body to environmental factors. For example, if a day is particularly hot, we begin cooling ourselves by sweating and thereby maintain a constant body temperature. If we

maturity: the state of full and complete development or growth

senescence: the process of growing old

skip a meal, we maintain a constant blood glucose level by temporarily increasing the output of glucose from the liver for use as fuel.

This remarkable ability to maintain homeostasis is vital to our health—but there are limits. The adult body can adapt to day-to-day fluctuations in dietary intake—it can conserve nutrients if intake is low, and store many when intake is high. However, as discussed later in this chapter, a lifetime of poor dietary choices can overcome our ability to maintain balance and contribute to poor health in later years.

NUTRITIONAL NEEDS OF ADULTS

RECOMMENDED DAILY ALLOWANCES FOR ADULTS

Table 9.2 lists the RDAs for adults. They are calculated based on the actual median heights and weights of adults in the U.S. population, not on ideal weights. The adult RDAs are divided into recommendations for young (19–24), middle (25–50), and older (51+) adults. An additional subdivision for the elderly (70+) was considered, but the committee that establishes the RDAs concluded that "the data are insufficient to establish separate RDAs" for this group (NAS 1989). (The difficulties of determining nutritional needs in the elderly is examined in the next chapter.) Table 9.3 compares the RDAs to dietary standards published by Canada, the United Kingdom, and the World Health Organization.

Because of gender differences in body size, composition, and physiology, the RDAs for women differ from those for men. The unique nutritional needs of adult women, including the impact of menopause, are discussed later in this chapter.

ENERGY NEEDS

Energy allowances range from 2300 to 2900 kcal/day for men and 1900 to 2200 kcal/day for women (NAS 1989). The energy RDA for men is set higher than that for women because men usually weigh more than women. Also, because women have more body fat and less lean tissue, their resting energy expenditure (REE) per kilogram is about 10% less than that for men. Lean tissue, such as muscle, is more metabolically active than fat and is the main determinant of REE. The energy allowance is reduced in older men and women because aging is associated with lower activity levels and reduced lean body mass.

Fat-Soluble Vitamins

	Age (yrs)	Weight (kg)	(lb)	Height (cm)	(in)	Protein (g)	Vitamin A (µg RE)	Vitamin D (µg)	Vitamin E (mg α-TE)	Vitamin K (µg)
Males	19–24	72	160	177	70	58	1000	10	10	70
	25–50	79	174	176	70	63	1000	5	10	80
	51+	77	170	173	68	63	1000	5	10	80
Females	19–24	58	128	164	65	46	800	10	8	60
	25–50	63	138	163	64	50	800	5	8	65
	51 +	65	143	160	63	50	800	5	8	65

Water-Soluble Vitamins

	Age (yrs)	Vitamin C (mg)	Thiamin (mg)	Riboflavin (mg)	Niacin (mg NE)	Vitamin B6 (mg)	Folate (µg)	Vitamin B_{12} (µg)
Males	19–24	60	1.5	1.7	19	2.0	200	2.0
	25–50	60	1.5	1.7	19	2.0	200	2.0
	51 +	60	1.2	1.4	15	2.0	200	2.0
Females	19–24	60	1.1	1.3	15	1.6	180	2.0
	25–50	60	1.1	1.3	15	1.6	180	2.0
	51 +	60	1.0	1.2	13	1.6	180	2.0

Minerals

	Age (yrs)	Calcium (mg)	Phosphorus (mg)	Magnesium (mg)	Iron (mg)	Zinc (mg)	Iodine (µg)	Selenium (µg)
Males	19–24	1200	1200	350	10	15	150	70
	25–50	800	800	350	10	15	150	70
	51+	800	800	350	10	15	150	70
Females	19–24	1200	1200	280	15	12	150	55
	25–50	800	800	280	15	12	150	55
	51+	800	800	280	10	12	150	55

The allowances, expressed as average daily intakes over time, are intended to provide for individual variations among most normal persons as they live in the United States under usual environmental stresses. Diets should be based on a variety of common foods in order to provide other nutrients for which human requirements have been less well defined.

Estimated Safe and Adequate Daily Dietary Intakes

	Vitamins		Trace Elements				
	Biotin (µg)	Pantothenic Acid (mg)	Copper (mg)	Manganese (mg)	Fluoride (mg)	Chromium (µg)	Molybdenum (µg)
Adults	30-100	4-7	1.5-3.0	2.0-5.0	1.50-4.0	50-200	75-250

Source: Adapted from National Research Council, National Academy of Sciences, *Recommended Dietary Allowances,* 10th ed. (Washington, D. C., National Academy of Sciences, 1989).

TABLE 9.2
Adult RDAs

PROTEIN NEEDS

The previous chapter pointed out that during adolescent growth, protein requirements are high (0.9–1.0 g/kg) to support the synthesis of new cells and tissue, and the body accumulates large amounts of protein. In contrast, in adulthood, protein requirements level off. A balance between new protein synthesis and protein breakdown prevails. Adult protein requirements are 0.8 g/kg/day, or about 50 grams each day for women and 63 grams for men (NAS 1989).

VITAMIN REQUIREMENTS

Fat-Soluble Vitamins

Because vitamin D is readily synthesized in skin exposed to sunlight, most adults who are regularly out in the sun have no dietary requirement for vitamin D (NAS 1989). However, a substantial number of adults in the United States do not get adequate exposure to sunlight, particularly during the winter season (Stryd et al. 1979), (Webb 1988). Therefore, the RDA for vitamin D in adults older than 25 years is set at 5 µg (NAS 1989). This allowance is only half the RDA for adolescence and younger adulthood.

The daily requirement for vitamin K is about 1 µg/kg body weight (Suttie et al. 1988). Based on differences in body weight, the RDA for vitamin K for adults is slightly higher than that during adolescence. The RDAs for vitamins A and E remain unchanged from adolescence (NAS 1989).

TABLE 9.3

Comparison of U.S. (1989), United Kingdom (1985), Canadian (1990), and WHO (1974) dietary standards for the adult male and adult female

Classification	Kcal	Protein (grams)	Calcium (mg)	Iron (mg)	Vitamin A (RE)	Thiamin (mg)	Riboflavin (mg)	Vitamin C (mg)
United States								
Female (63 kg, 1.63 m)	2200	50	800	15	800	1.1	1.3	60
Male (79 kg, 1.76 m)	2900	63	800	10	1000	1.5	1.7	60
United Kingdom								
Female	2150–2500	54-62	500	12	750	0.9-1.0	1.3	30
Male	2150–3350	63-84	500	10	750	1.0-1.3	1.6	30
Canada								
Female (59 kg)	1900	51	700	13	800	0.8	1.0	30
Male (74 kg)	2700	64	800	9	1000	1.1	1.4	40
FAO/WHO								
Female	2200	29	400–500	14–28	750	0.9	1.3	30
Male	3000	37	400–500	5–9	750	1.2	1.8	30

Water-Soluble Vitamins

The RDA for thiamin is set at 0.5 mg/1000 kcal, with a minimum of 1 mg/day even for those adults whose energy intake is less than 2000 kcal/day. Similarly, the RDA for riboflavin is based on energy intake—adults should consume 0.6 mg/1000 kcal, with a minimum intake of 1.2 mg/day. The niacin requirement is also based on energy needs, with 6.6 mg of niacin equivalents/1000 kcal considered adequate (Jacob et al. 1989). One niacin equivalent is equal to 1 mg of niacin or 60 mg of tryptophan (dietary tryptophan can be converted to niacin). Recommended intakes for niacin, riboflavin, and thiamin are approximately 20% less for adults older than 50 years, compared with younger age groups, because they are calculated based on energy requirements, which decline in older adults (NAS 1989).

Because vitamin B_6 functions mainly as a coenzyme in pathways of protein metabolism, requirements are based on a dietary ratio of 0.016 mg vitamin B_6/g protein (Schultz and Leklem 1981). The adult RDA for vitamin B_6 is set in relation to the upper limit for recommended protein intake—that is, twice the RDA for protein, or 126 g/day for men and 100 g/day for women (NAS 1989). This allowance should be sufficient for most adults, as the average protein intakes of adult men and women in the United States are 100 g/day (USDA 1984) and 60 g/day (USDA 1987), respectively.

The adult requirement for dietary folate is approximately 3 µg/kg body weight (NAS 1989). The RDAs are therefore set at 200 µg for adult men and 180 µg for adult women. However, women who are planning a pregnancy may need higher amounts. Increasing evidence suggests that higher folate intake by women during the periconceptional period can reduce the risk of neural tube defects. The U.S. Public Health Service recommends that all women of childbearing age who are capable of becoming pregnant consume at least 400 µg of folic acid per day (Centers for Disease Control 1992). The links between dietary folate and neural tube defects are discussed in Chapter 2.

The adult RDA for vitamin B_{12} is 2 µg (NAS 1989). An adult who maintains this level of intake should accumulate substantial body stores of vitamin B_{12}. Prevalence of vitamin B_{12} **malabsorption** due to atrophic gastritis and **achlorhydria** steadily increases after age 65 (Kassarjian and Russell 1989). For older adults with atrophic gastritis, accumulated stores of vitamin B_{12} can be used to help prevent deficiency in later life. Problems of vitamin B_{12} absorption in older adulthood are discussed in Chapter 10.

The adult RDA for vitamin C is estimated to be 60 mg (NAS 1989). This allowance is set between the amount required to prevent **scurvy** (about 10 mg/day) and the amount beyond which most of the ingested vitamin is excreted in the urine (about 200 mg/day). Intake of 60 mg of

malabsorption: impaired intestinal absorption of nutrients

achlorhydria: absence of hydrochloric acid from the gastric secretions

scurvy: the vitamin C-deficiency disease characterized by bleeding gums and bleeding under the skin, bone and joint abnormalities, and weakness

vitamin C/day will maintain the body pool of vitamin C in the average adult at about 1500 mg (Kallner 1987).

MINERAL REQUIREMENTS

Although full bone length is reached during the teen years, bones continue to increase in density, and **peak bone mass** is not achieved until the early 30s (Matkovic et al. 1990). Therefore, requirements for calcium and phosphorus remain high (1200 mg/day) until age 25. Thereafter, the body no longer needs large amounts of minerals to deposit into bones—it needs only enough for maintenance and repair of the skeleton. The RDAs for calcium and phosphorus are set at 800 mg for middle and later adulthood (NAS 1989). Some scientists think adult women, who are at high risk for osteoporosis, need significantly more calcium than the current RDA (Consensus Development Conference 1993). The debate over calcium requirements is discussed in Chapter 10.

The daily requirement for magnesium during adulthood is estimated to be 4.5 mg/kg body weight—somewhat lower than the 6 mg/kg required to support childhood and adolescent growth (NAS 1989). Therefore, the adult RDA (350 mg for men and 280 mg for women) is lower than that for late adolescence.

Trace Elements

Iron requirements are high during adolescent growth, as large numbers of newly synthesized red blood cells incorporate iron into hemoglobin. In men, iron needs fall in adulthood as red cell synthesis slows. Because adult women have increased iron losses in menstrual blood, their iron needs continue to be high. The RDA for iron in adult men falls to 10 mg from the adolescent RDA of 12 mg; women's needs are 15 mg through adolescence and adulthood. After menopause, women's needs are lowered to 10 mg/day (NAS 1989).

Zinc requirements during adulthood are estimated to be 2.5 mg of absorbed zinc/day. Based on an absorption efficiency of 20% from the average U.S. diet, the zinc requirement for most adults would be about 12.5 mg/day. Taking into account the needs of those adults who regularly consume diets with lower zinc bioavailability (such as vegetarians), the RDA has been established at 15 mg for adult men and 12 mg for adult women (NAS 1989).

The adult RDA for selenium is estimated to be approximately 0.9 μg/kg body weight, or 70 and 55 μg for adult men and women, respectively. The adult RDA for iodine is set at 150 μg for both men and women (NAS 1989). For the remaining trace minerals—copper, manganese, fluoride, chromium, and molybdenum—there is insufficient knowledge to

peak bone mass: the maximal accumulation of bone mineral in the skeleton developed during the first three decades of life, before the age-related loss of bone mineral begins

establish RDAs. Instead, safe and adequate ranges of daily intake has been estimated, as shown in Table 9.2.

ELECTROLYTES AND WATER

Although often overlooked, water and the electrolytes, sodium, potassium, and chloride, are essential nutrients. Their importance in adult diets has recently been underscored by research suggesting that overconsumption of sodium and underconsumption of potassium increase the risk of high blood pressure (Johnson 1992). Although the estimated minimum daily requirement for sodium is set at 0.5 g, the average adult diet in the United States contains five to eight times that amount. Studies have found average salt (sodium chloride) consumption to be between 7 and 11 g/day, or about 2.8 to 4 g sodium/day (Dahl and Love 1993). About 90% of salt intake is from salt added to foods during processing, cooking, and at the table (Sanchez-Castillo et al. 1987). Consumption of large amounts of sodium has no health benefits, and high intakes are clearly detrimental in people susceptible to high blood pressure (Stemler 1993).

The estimated minimum requirement for potassium is 2 g/day (NAS 1989). However, the desirable intake is considerably higher—for adults it is estimated to be approximately 3.5 g/day (NRC 1989). Usual intakes of potassium by adults in the United States vary greatly, depending on diet choices. Average daily intakes range from 1 to 2.5 g/day, but people who consume large amounts of fruits and vegetables obtain up to 11 g/day. Ample intakes of potassium may protect against the development of high blood pressure (Johnson 1992).

Adult requirements for water average about 1.0–1.5 ml/kcal of energy expenditure (NAS 1989). For the average healthy 70 kg adult, water needs are between 2500 and 3000 ml/day. However, high ambient temperatures, dry air, and high altitude all increase water needs, as do fever, diarrhea, or vomiting.

THE ADULT DIET

In most of the developed countries of the world, including the United States and Canada, large food consumption surveys have shown that most adults meet or exceed the RDAs for most nutrients. Only in a minority of the U.S. adult population—the poor, the very sick, the elderly, and heavy alcohol users—is undernutrition still a significant problem (see Table 9.4). In these groups, intakes of protein and certain micronutrients—the minerals calcium, iron, and zinc, and vitamins A, C, and B_6—are often marginal or low (LSRO 1989).

Also, it appears that many healthy adult women are not getting adequate calcium, iron, and folate in their diets (Block and Abrams 1993).

* Poverty

* Chronic Illness

* Older age (> 65 years)

* Alcoholism

TABLE 9.4

*Major risk factors for
undernutrition in adults*

Because of calcium's role in osteoporosis (discussed in Chapter 10), and evidence linking folate deficiency to certain birth defects (discussed in Chapter 2), women should emphasize these nutrients in their diets.

In general, dietary surveys of adults in the United States have pointed to problems of overconsumption, not underconsumption. In the past, our primary concerns were undernutrition—for example, ensuring adequate intake of micronutrients to prevent deficiency diseases, such as pellagra (niacin), scurvy (vitamin C), and rickets (vitamin D). Today, the major dietary concerns in adulthood are how to reduce overconsumption of energy, fat, alcohol, and salt (NRC 1989).

On average, adults in the United States consume 14 to 16% of calories as protein, 42 to 47% as carbohydrate, 34% as fat, and about 10% as alcohol (LSRO 1989), (U.S. Public Health Service 1995). About two thirds of the protein comes from animal sources (meat, eggs, and milk). Contrast this with many other parts of the world, where most protein comes from plant sources (rice, corn, beans, and other plants). Half of the carbohydrate intake in the U.S. diet is from simple sugars. About 60% of the fat comes from animal sources, and 40% comes from plant sources.

The RDAs are valuable in that they focus on specific requirements for protein and the micronutrients and suggest average energy intakes. But they do not provide guidance on dietary fat, types of carbohydrates, or alcohol and salt use. Dietary recommendations in these areas are referred to as *dietary guidelines*. Concerned public health organizations in the United States, including the American Heart Association (AHA 1988), the Surgeon General's Office (USDHHS 1988), the U.S. Department of Agriculture (Dietary Guidelines Advisory Committee 1995), and the National Research Council (NRC 1989), publish sets of dietary guidelines. These guidelines advise consumers how to select a diet from the many types of available foods, not simply to ensure adequate intake but to give the best chances for long-term health. In general, these different sets of dietary guidelines are in close agreement on what constitutes a healthy diet. See Table 9.5.

Most scientists and government officials are convinced that changing the way we eat will have substantial benefits for our health. Reducing the amount of chronic degenerative disease by preventive nutrition has become a significant element in public health policy, not just in the

- Reduce total fat intake to 30% or less of kcalories. Reduce saturated fatty acid intake to less than 10% of kcalories and the intake of cholesterol to less than 300 milligrams daily.

- Increase intake of starches and other complex carbohydrates.

- Maintain protein intake at moderate levels.

- Balance food intake and physical activity to maintain appropriate body weight.

- For those who drink alcoholic beverages, limit consumption to the equivalent of less than 1 oz of pure alcohol in a single day. Pregnant women should avoid alcoholic beverages.

- Limit total daily intake of salt (sodium chloride) to 6 grams or less.

- Maintain adequate calcium intake.

- Avoid taking dietary supplements in excess of the RDA in any one day.

- Maintain an optimal intake of fluoride, particularly during the years of primary and secondary tooth formation.

Adapted from National Research Council, Food and Nutrition Board, *Diet and Health: Implications for Reducing Chronic Disease Risk,* report of the Committee on Diet and Health (Washington, D.C., Academy Press, 1989).

TABLE 9.5
Diet and health recommendations

United States, but worldwide. How does our diet put us at risk? Let's look at the connections between diet and these "diseases of civilization."

NUTRITION AND CHRONIC DISEASES OF ADULTHOOD

Eating and drinking habits in the United States profoundly influence the health of the adult population: diet contributes to four of the ten leading causes of death, including the top three—heart disease, cancer, and stroke (the other is diabetes). Another four of the ten are strongly associated with alcohol use: **cirrhosis** of the liver, accidents, homicides, and suicides. Table 9.6 lists the ten leading causes of death in the United States. Nutritional factors probably contribute to over three quarters of the 2 million deaths in the United States each year (USDHHS 1988). In addition, nutritional risk factors are involved in many common ailments, including osteoporosis, dental decay, dementia, high blood pressure, and obesity.

Unlike the nutritional deficiency diseases, where a cause (a deficiency of vitamin C) and its effect (scurvy) are clearly understood, these chronic diseases do not have clear-cut, single causes. These diseases have compli-

cirrhosis: a disease of the liver, characterized by progressive destruction of the liver cells and their replacement by connective tissue

	Diet	Alcohol	Smoking
1. Heart Disease	√		√
2. Cancer	√	√	√
3. Stroke	√	√	√
4. Accidents		√	
5. Chronic obstructive lung disease			√
6. Pneumonia and influenza			√
7. Diabetes	√		
8. Suicide		√	
9. Cirrhosis of the liver		√	
10. Homicides		√	

TABLE 9.6

Ten leading causes of death in the United States with dietary and lifestyle risk factors

cated origins arising from interactions between genetics, age, gender, lifestyles, environments, and diet.

To define risk factors for these diseases, researchers examine populations that frequently develop a disease, looking for factors that tend to be associated with that disease—such as diet choices, exercise patterns, family history, or smoking. Such associations suggest that when the risk factor is present, the likelihood of developing the disease is increased. It does not necessarily mean that the factor causes the disease, nor does it mean that everyone who has the risk factor will develop the disease.

Research has shown that dietary changes can reduce or modify known risk factors for many of the chronic diseases (USDHHS 1988). For example, elevated blood cholesterol is a strong risk factor for heart attack. Eating less saturated fat lowers blood cholesterol levels and thereby reduces the chances of a heart attack. Because most people will benefit from following dietary recommendations, many governments have decided to confront the problem using a broad, population wide approach. Dietary guidelines, such as those in Table 9.5, are examples of this kind of approach.

Not everybody agrees that this is the best way to approach the problem. Some scientists think that dietary recommendations should be individualized and targeted only at people who are at high risk. However, most experts feel that the population approach will harm very few, benefit many, and is the most feasible approach to nutritional prevention.

Throughout the discussion of these diseases, remember that dietary factors are only part of the story. We've all known people who have followed dietary recommendations and still developed heart disease or cancer, while others ignored these recommendations and lived long lives. People's susceptibility to diseases and their responses to dietary changes are variable. Some of us are very sensitive to levels of cholesterol and salt in the diet, and some of us are not. For most of us, however, adopting healthy dietary habits in early adulthood will significantly influence our health later. As the U.S. Surgeon General's Office has stated: "Besides avoiding cigarettes and excess alcohol, one personal choice seems to influence long-term health prospects more than any other: what we eat" (USDHHS 1988).

NUTRITION AND CARDIOVASCULAR DISEASE

Disease of the heart and blood vessels is the major cause of death in the developed countries of the world. Half of all deaths in North America are due to **cardiovascular disease**, mostly from heart attack and stroke. Nearly 1 million people die of cardiovascular disease in the United States each year—almost three times the death rate from all cancers combined. Close to half of all of these deaths occur in middle-aged adults under 65 (NCHS 1993). What causes this epidemic?

Atherosclerosis and High Blood Pressure

Most cardiovascular disease results from two common disorders that develop slowly and silently over many years: **atherosclerosis** and high blood pressure. Both damage blood vessels and can stop the flow of blood, nutrients, and oxygen to tissues.

Although atherosclerosis causes heart attacks in adults, the disorder begins in childhood (NCEP 1992). Starting early in life, fatty deposits build up on the insides of arterial walls, and as people grow older, these fatty deposits grow larger. Although everyone is affected to a degree, in some people the deposits grow much faster than others. In many adults, the fatty deposits have developed into atherosclerotic plaques.

Plaques are rough patches on the inner walls of arteries composed of smooth muscle and scavenger cells (macrophages), mixed with cholesterol, calcium, and other debris. As atherosclerosis progresses, the plaques form irregular mounds that narrow and stiffen arteries. By age 30, most of us have small atherosclerotic plaques scattered throughout our arteries, plaques that are unstable and changeable—they may grow steadily thicker (often), stay the same size, or slowly shrink (rarely).

Blood pressure is the pressure the blood exerts on the inner walls of arteries as it flows from the heart to the tissues. Blood pressure results from two opposing forces: the push of the heart squeezing new blood into the arteries and the resistance produced as the blood squeezes through small capillaries in tissues. Blood pressure is important because it keeps blood endlessly moving through our bodies. It normally fluctu-

cardiovascular disease: disease of the heart and blood vessels

atherosclerosis: a disorder affecting large and medium-sized arteries in which deposits of plaque accumulate in the arterial wall; in advanced states, the flow of blood through the artery is impeded

ates up and down somewhat throughout the day, in response to signals from the kidney and the nervous system. In many adults, blood pressure rises abnormally and stays elevated for years. This condition is known as high blood pressure, or hypertension. It is very common—approximately one in three adults in the United States suffers from high blood pressure.

Atherosclerosis narrows arteries, increases resistance to blood flow, and raises blood pressure. This is analogous to a garden hose: when the nozzle is open, water flows out easily and there is little pressure on the walls of the hose. If the nozzle is narrowed, the pressure of the water in the hose is much higher. Likewise, the heart must strain and work harder to push blood through narrowed arteries. Elevated blood pressure stresses and injures the cells lining the inside of arteries, which accelerates the development of atherosclerosis.

Hypertension and atherosclerosis damage the cardiovascular system silently and gradually over many years. Often the final event that abruptly closes off a narrowed artery is the formation of a blood clot inside the vessel. Blood clots form when platelets clump together into a sticky mass that includes fibers and proteins from the blood. A blood clot resembles a scab formed on top of a healing break in the skin—blood clots form on top of plaques, as the artery tries to repair damage from the plaques. Clots can grow quickly on a plaque and completely block a narrowed vessel. Or they can break off and move through the bloodstream—lodging suddenly in a smaller artery, shutting off blood flow. Depending on where the blockage occurs, different organs are damaged. If a **coronary artery** is affected, the part of the heart that is supplied by that vessel will die (a heart attack). If a cerebral vessel is blocked, brain tissue will die, producing a stroke.

Lipoproteins and Blood Cholesterol

Cholesterol and other lipids are transported in the blood enclosed in molecular packages called *lipoproteins*. Lipoprotein particles are formed in the liver and in peripheral tissues, and there are several different types, each with a distinct function. Although all are composed of proteins, fats, and cholesterol, the proportions vary, so that the densities of lipoprotein particles vary. Lipoproteins are classified according to their densities (Nicolusi and Schaefer 1992).

Most of the cholesterol in the blood (about 70%) is carried by low-density lipoproteins (LDLs). LDL particles are rich in cholesterol and are responsible for transporting lipids from the liver to other tissues. High levels in the blood strongly increase the risk of cardiovascular disease. In contrast, high-density lipoproteins (HDLs) contain less cholesterol and transport lipids from tissues back to the liver. High levels decrease the

coronary artery: an artery supplying blood to the heart muscle

Source: Adapted from
Goldberg (1992)

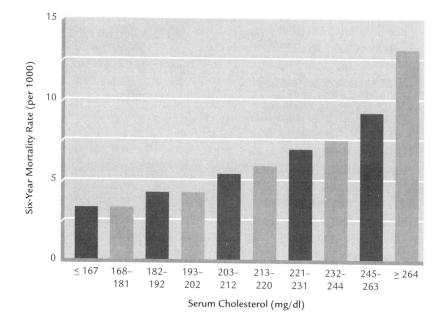

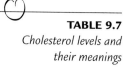

FIGURE 9.2

*Cholesterol levels and
mortality from coronary
heart disease*

	Desirable	**Borderline**	**High Risk**
Total Cholesterol	Below 200	200–239	240 or higher
LDL-Cholesterol	Below 130	130–159	160 or higher
HDL-Cholesterol	Above 45	35–45	Below 35
Ratio*	Below 4.5	4.5–5.5	Above 5.5

*Total cholesterol
   HDLs

TABLE 9.7

*Cholesterol levels and
their meanings*

risk of heart disease (Nicolosi and Schaefer 1992). (An easy way to remember this is to think "healthy" for HDL and "lousy" for LDL.)

Although a high total blood cholesterol increases the risk of coronary heart disease (see Figure 9.2), it is important to look at whether the cholesterol is in the form of LDLs (harmful) or HDLs (protective). People with a normal total cholesterol may still be at high risk if their HDL level is too low; conversely, some people with an elevated total cholesterol (usually women) may not be at high risk if their HDL level is high enough. Therefore, one of the strongest predictors of risk for heart disease is the total cholesterol/HDL ratio: the larger the proportion of total cholesterol to HDL, the greater the risk (Grundy 1990) (see Table 9.7). Diet and lifestyle changes should aim to lower LDL and raise HDL.

Risk Factors for Cardiovascular Disease

There is no single cause of cardiovascular disease, but epidemiological studies have clearly identified a number of dietary and lifestyle factors that are associated with a greater risk of the disease. The links between diet and cardiovascular disease are through these risk factors, shown in Table 9.8.

Changes in diet can reduce the risk of cardiovascular disease by modifying five of these risk factors: hypertension, high total and LDL-cholesterol, low HDL-cholesterol, obesity, and diabetes. Three remaining factors—smoking, stress, and a sedentary lifestyle—are responsive to lifestyle changes. Many adults in the United States have one or more of these risk factors.

Although cardiovascular disease remains the leading cause of mortality in the United States, there is encouraging news. A dramatic decrease in mortality rates from cardiovascular disease during the period 1979–92 has cut death rates from heart attack and stroke by one third (Goldberg 1992). Improved medical care combined with changes in diet and lifestyle have produced this favorable drop—but many experts think we can do even better. They suggest that further dietary and lifestyle modifications could again cut in half rates of coronary heart disease in the United States.

Diet and lifestyle changes can help control atherosclerosis and heart disease, even when the disease is advanced. Researchers studied a group of people with severe atherosclerosis of their heart vessels. For one year, the study participants ate a vegetarian diet extremely low in fat (<10% of calories from fat) and cholesterol (5 mg/day). In addition, they quit

- Hypertension
- High total and LDL-cholesterol
- Low HDL-cholesterol
- Obesity
- Diabetes
- Sedentary lifestyle
- Stress
- Cigarette smoking
- Family history of early heart disease
- Increasing age
- Male gender

TABLE 9.8
Risk factors for cardiovascular disease

smoking, practiced stress management techniques, and participated in a moderate exercise program.

Results were dramatic. On average, total blood cholesterol levels were reduced by nearly 25%, and LDL-cholesterol fell 37%. Not only did the changes prevent further development of atherosclerotic plaques in the coronary arteries, there was also significant regression of the plaque that had been there at the beginning of the study. The participants reported far fewer symptoms of heart disease after following the program for a year (Ornish et al. 1990).

Let's now look at how total and LDL-cholesterol, HDL-cholesterol, obesity, and hypertension—four major risk factors for heart disease—can be strongly modified by nutritional factors. Diabetes, another major risk factor for cardiovascular disease, is considered separately in this chapter.

Total and LDL-Cholesterol

Everyone has heard about the hazards of eating too much fat, and many people know that high-fat diets tend to raise blood cholesterol. However, the kind of fat consumed is as important as how much fat is consumed; some types of fat raise LDL-cholesterol, and some types do not. The amount of saturated fat in the diet strongly influences levels of LDL-

	S	S[1]	M	P	P/S Ratio
Palm kernel	81	72	11	2	0.0
Coconut	87	70	6	2	0.0
Palm	49	45	37	9	0.2
Butter oil	62	39	29	4	0.1
Beef tallow	50	30	42	4	0.1
Cocoa butter	60	26	33	3	0.1
Lard	39	25	45	11	0.3
Chicken fat	30	23	50	21	0.7
Olive	14	11	74	8	0.6
Corn	13	11	24	59	4.6
Soybean	14	10	23	58	4.0
Peanut	17	10	46	32	1.9
Safflower	9	6	12	75	8.2
Canola	7	5	56	33	4.9

S = Total saturated fatty acids

S1 = Cholesterol-raising saturated fatty acid (C12:0, C14:0, C16:0)

M = Monounsaturated fatty acids

P = Polyunsaturated acids

P/S Ratio = Polyunsaturated to saturated fatty acid ratio

TABLE 9.9
Fatty acid composition of selected fats and oils

cholesterol in the blood. High levels of dietary saturated fat increase blood cholesterol (Grundy 1990). Animal fats and a few vegetable oils—palm, coconut, and cocoa butter—are rich in saturated fat.

Scientists have looked closely at the four fatty acids that constitute over 90% of the saturated fat in our diets. Surprisingly, they found differing effects on blood cholesterol. Three of these—lauric acid (C12:0), myristic acid (C14:0), and palmitic acid (C16:0)—raise blood cholesterol; but stearic acid (C18:0) does not raise total or LDL-cholesterol (Grundy and Vega 1988), (Bonanome and Grundy 1988). See Table 9.9.

Saturated fat raises LDL-cholesterol levels by reducing the activity of hepatic LDL receptors, the tiny receptors on the liver cells filtering LDL particles from the blood. Saturated fats slow down the activity of these receptors and decrease clearance of LDLs from the blood—so blood cholesterol levels rise (Nicolosi and Schaefer 1992). On average, an extra 10 g of dietary saturated fat daily will raise LDL-cholesterol about 5 to 8%. Polyunsaturated and monounsaturated fats appear to have little effect on LDL-cholesterol levels—they neither raise nor lower LDLs. However, when substituted for dietary saturated fat, they tend to lower LDL-cholesterol by reducing saturated fat intake (Mensink and Katan 1989).

Dietary cholesterol has been the focus of much attention, but for most people, the amount of cholesterol in the diet has little effect on blood cholesterol levels (Grundy et al. 1988). Why? Our bodies require about 800–1000 mg of new cholesterol each day to replace cholesterol metabolized or excreted. The liver and other tissues synthesize about 800 mg of new cholesterol each day, and the remainder (about 10-20%) comes from dietary sources. The liver is sensitive to changes in dietary cholesterol—if we absorb more cholesterol from dietary sources, the liver will appropriately reduce synthesis of cholesterol, and blood cholesterol will not increase. This is why most of the population is relatively insensitive to changes in dietary cholesterol.

However, about 25 to 30% of people are sensitive to changes in dietary cholesterol (McNamara et al. 1987). In these cholesterol-sensitive people, it appears that the liver does not reduce synthesis of cholesterol when the amount in the diet increases and blood cholesterol rises. These people need to be extra careful about rich sources of dietary cholesterol. For them, reducing dietary intake to less than 300 mg a day may lower blood cholesterol by 5 to 10%. For most of us, saturated fat intake is a much stronger determinant of blood cholesterol than dietary cholesterol. Table 9.10 gives some examples of foods high in saturated fat and cholesterol.

Water-soluble fibers, such as those in fruit, some grains, and legumes, can lower LDL-cholesterol (Anderson et al. 1991), (McIntosh et al. 1991). How fiber works to lower blood cholesterol is only partially understood. The liver synthesizes bile acids from cholesterol and secretes the bile acids into the intestine, where most of the bile acids are reabsorbed and

Saturated Fat	Cholesterol
Hamburger meat	Eggs
Whole milk	Beef steaks, roasts
Cheeses	Hamburger meat
Beef steaks, roasts	Whole milk
Hot dogs, ham, lunch meats	Hot dogs, ham, lunch meat
Doughnuts, cookies, cake	Pork
Eggs	Doughnuts, cookies, cake
Pork	Cheeses
Butter	Liver

TABLE 9.10

Major contributors of saturated fat and cholesterol in the U. S. diet

reused. Water-soluble fibers may bind bile acids in the intestine and carry these compounds out in the feces. The liver must then pull more cholesterol from the blood to synthesize new bile acids, thus lowering blood cholesterol (Eastwood 1992).

In most people, the cholesterol-lowering ability of these dietary fibers is modest—they are most effective when part of a low-fat diet (Truswell and Beynen 1992), (Bell et al. 1990). For example, eating two to three cups of oatmeal a day may lower cholesterol 2 to 5%. However, larger intakes (50–100 g per day of oat bran) in people with elevated blood cholesterol can have a substantial effect, lowering cholesterol levels by up to 25% (Truswell and Beynen 1992). Table 9.11 lists good sources of water-soluble dietary fiber.

Epidemiological studies have been sharply contradictory on the role of coffee in cardiovascular disease (Superko et al. 1991), (Fried et al. 1992), (Bak and Grobbee 1989). Part of the confusion may arise from different preparation methods. Boiling or steeping coffee beans can release a type of fat that can raise blood cholesterol. This fat, however, is captured by paper coffee filters, so filtered coffee does not have this problem. It is unlikely that coffee consumption plays a major role in altering blood cholesterol levels (Anonymous 1992).

Finally, regular aerobic exercise lowers total and LDL-cholesterol levels and raises HDL-cholesterol, thereby lowering the total cholesterol/HDL-cholesterol ratio (Haskell 1986), (Superko and Haskell 1987), (Wood et al. 1988).

In summary, reducing total and saturated fat intake, increasing dietary fiber, and increasing physical activity can lower blood cholesterol. In most people, following dietary and exercise recommendations will lower total cholesterol 10 to 15%. It is estimated that for every 1% difference in cholesterol levels in a population, the risk of heart disease is lowered 2 to 3% (Gordon et al. 1989). Therefore, lowering the average cholesterol level

Apples	Oats
Bananas	Peas
Beans (kidney, lima)	Plums
Broccoli	Potatoes (with skins)
Carrots	Tangerines and oranges

TABLE 9.11

Good sources of water-soluble dietary fiber

10 to 15% in the United States population could prevent around 300,000 of the 1.5 million heart attacks that occur each year.

The National Cholesterol Education Program (NCEP) has published recommended diets (the NCEP Step-One and Step-Two diets shown in Table 9.12) aimed at lowering blood cholesterol and reducing the risk of heart disease (NCEP 1988). Individuals with borderline or high cholesterol levels in their blood are advised to follow the Step-One Diet for three months. If reductions in blood cholesterol are not achieved, the Step-Two Diet is begun under the supervision of a dietitian.

HDL-Cholesterol

Obesity is a common cause of low HDL-cholesterol (Kannel et al. 1979). Overweight people have HDL-cholesterol levels 10 to 20% lower than normal-weight people. In people who are obese, weight loss can significantly raise HDL-cholesterol. The addition of regular aerobic exercise to a weight-reduction program can raise the HDL-cholesterol further than weight loss alone (Wood et al. 1991). Regular exercise can elevate HDLs about 5%, even in normal-weight people.

Unlike LDL-cholesterol, reducing dietary fat has little effect on HDL levels except when fat reduction leads to weight loss. If dietary fat is reduced to very low levels (<20%), HDL levels may fall (as total cholesterol falls). However, LDL levels generally fall to a greater extent than HDL-cholesterol levels, resulting in a more favorable total cholesterol/HDL ratio (Nicolosi and Schaefer 1992).

People who consume one to two alcoholic drinks per day have less heart disease than people who abstain (Stampfer et al. 1988), (Rimm et al. 1991). Moderate alcohol consumption is strongly associated with higher HDL-cholesterol and lower rates of heart attack (Gaziano et al. 1993). Although ingestion of 30 grams of ethanol (two to three drinks) daily can raise HDL-concentrations 5 to 10%, current diet recommendations do *not* advocate alcohol use for the purpose of decreasing the risk of cardiovascular disease. The risks of alcohol use generally outweigh the benefits.

In summary, obesity and a sedentary lifestyle are associated with a low HDL-cholesterol, while regular exercise and moderate alcohol intake may raise HDL-cholesterol.

Step-One Dietary Choices

Food Category	Choose	Decrease
Fish, chicken, turkey, and lean meat	Fish, poultry without skin; lean cuts of pork or veal; shellfish	Fatty cuts of beef, lamb, pork; spare ribs, organ meats; regular cold cuts; sausage; hot dogs, bacon, sardines; roe
Skim and low-fat milk, cheese, yogurt, and dairy substitutes	Skim or 1% fat milk (liquid, powdered, evaporated); buttermilk	Whole milk (4% fat) (regular, evaporated, condensed); cream; half and half; 2% milk; imitation milk products; most nondairy creamers; whipped toppings
	Nonfat or low-fat yogurt	Whole-milk yogurt
	Low-fat cottage cheese (<2% fat)	Whole-milk cottage cheese (4% fat)
	Low-fat, farmer, or pot cheeses (all of these should be labeled no more than 2 to 6 g fat/oz)	All natural cheeses (e.g. blue, roquefort, camembert, cheddar, swiss
	Low-fat or "light" cream cheese, low-fat or "light" sour cream	Cream cheeses, sour cream
	Sherbet or sorbet	Ice cream
Eggs	Egg whites (2 whites = 1 whole egg in recipes), cholesterol-free egg substitutes	Egg yolks
Fruits and vegetables	Fresh, frozen, canned, or dried fruits and vegetables	Vegetables prepared in butter, cream, or other sauces
Breads and cereals	Homemade baked goods using unsaturated oils sparingly, angel food cake, low-fat crackers, low-fat cookies	Commercial baked goods: pies, cakes, doughnuts, croissants, pastries, muffins, biscuits, cookies, high-fat crackers, high-fat cookies
	Rice, pasta	Egg noodles
	Whole-grain breads and cereals (oatmeal, whole-wheat, rye, bran, multigrain, etc.)	Breads in which eggs are major ingredients
Fats and oils	Baking cocoa	Chocolate
	Unsaturated vegetable oils: corn, olive, rapeseed (canola oil). safflower, sesame, soybean, sunflower	Butter, coconut oil, palm oil, palm kernel oil, lard, bacon fat

TABLE 9.12

NCEP Step-One Diet and comparison of the Step-One and Step-Two diets

TABLE 9.12 *Continued*

Food Category	Step-One Dietary Choices	
	Choose	**Decrease**
Fats and oils (continued)	Margarine	Dressing made with egg yolks
	Seeds and nuts	Coconut

Nutrient	Step-One Diet	Step-Two Diet
Total fat	Less than 30% of total calories	
Saturated fatty acids	Less than 10% of total calories	Less than 7% of total calories
Polyunsaturated fatty acids	Up to 10% of total calories	
Monounsaturated fatty acids	10–15% of total calories	
Carbohydrates	50–60% of total calories	
Protein	10–20% of total calories	
Cholesterol	Less than 300 mg/day	Less than 200 mg/day
Total calories	To achieve and maintain desirable weight	

Adapted from the National Cholesterol Education Program, Report of the Expert Panel on Detection, Evaluation, and Treatment of High Blood Cholesterol in Adults. U.S. Department of Health and Human Services, *Public Health Service National Institutes of Health Publication* No. 89-2925, 1989.

Obesity

The most common risk factor for heart disease in the United States is obesity: about a third of all adults in the United States are overweight (McGinnis and Lee 1995). Obesity is a risk factor in both sexes, but particularly in women. Although any level of overweight will increase the risk of heart disease, the more overweight, the higher the risk. Women who gain 10–40 pounds in midlife have a significantly increased risk of heart attack; those who gain over 25 pounds have two to three times the risk (Willet et al. 1995). Moreover, in both women and men, obesity strongly increases the risk of developing hypertension and diabetes (Pi-Sunyer 1991), (Golditz et al. 1995), (Medalie et al. 1975).

Obesity is also associated with high levels of total and LDL-cholesterol. Obese people overproduce LDL-cholesterol: for every 5 kg of excess body weight, daily cholesterol synthesis is increased about 10%. (Kesaniemi and Grundy 1983). At the same time, obese people often have low levels of HDL-cholesterol and an unfavorable LDL/HDL ratio (Kannel et al. 1979).

The distribution of excess body weight appears to be as important as the overall degree of obesity (Bouchard et al. 1990). Fat deposited around the abdomen (android obesity) increases the risk of heart attack and diabetes more than fat deposited equally over the body or fat deposited around the hips and thighs (gynoid obesity) (Leibel et al. 1989). See discussion of obesity and body fat distribution in Chapter 8.

Hypertension

50 to 70 million people in the United States suffer from high blood pressure, or hypertension (Farley 1991). People often have hypertension for years without being aware of it. High blood pressure is a strong risk factor for both heart disease and stroke, and the higher the blood pressure is, the greater the risk (see Figure 9.3).

Blood pressure is measured at two points and is expressed as how high (in mm) the pressure forces a column of mercury (Hg) to rise in a tube. The point of highest pressure, called the systolic blood pressure, occurs as the heart contracts. The point of lowest pressure, measured between con-

Source: Adapted from Kannel, W. B., *Clinical Hypertension* (Baltimore: Williams and Wilkins, 1986).

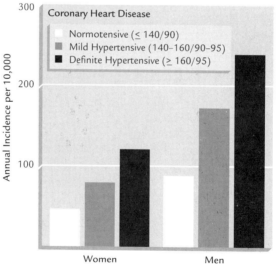

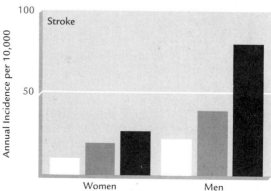

FIGURE 9.3

Hypertension as a risk factor for cardiovascular disease

Degree of Hypertension	Blood Pressure (mm Hg)
Mild	Diastolic 90 to 104 Systolic 140 to 159
Moderate	Diastolic 105 to 114 Systolic >160
Severe	Diastolic >115

FIGURE 9.4

Hypertension standards

tractions, is called the diastolic blood pressure. Normal blood pressure for adults is about 120 mm (systolic) and 80 mm (diastolic). Sustained blood pressure above 140/90 mm Hg is hypertension (see Figure 9.4).

The cause of most high blood pressure is unknown. Atherosclerosis plays a role in the disease by narrowing arteries and reducing blood flow to the kidneys, which disrupts their normal ability to regulate blood pressure. Chronic stress may also contribute to hypertension, as signals from the nervous system during stress can raise blood pressure (Williams and Moore 1995).

Epidemiological studies have identified a number of risk factors associated with hypertension, as listed in Table 9.13. It is much more common with increasing age, often developing in the fifth or sixth decade of life. A family history of high blood pressure triples the risk of developing the disease. Race plays a role—Blacks develop high blood pressure at a rate 50% higher than Whites, and often at earlier ages (Joint National Committee 1988).

Proper nutrition and regular exercise can substantially reduce the risk of developing hypertension. Also, for people who have mild hypertension, the first line of treatment is normally diet and lifestyle changes (Joint National Committee 1988).

Risk Factors for Hypertension: Obesity, Alcohol, and Sodium

Obesity, particularly android obesity, is strongly associated with hypertension. In general, for every 10 kg of excess weight, systolic blood pressure increases by about 3 mm. Weight loss, therefore, can be an effective means of lowering blood pressure (Pi-Sunyer 1991). Weight loss of 10–20 lb can reduce high blood pressure, even in the very obese (Corrigan et al. 1991).

- Obesity
- Alcohol use (more than three drinks per day)
- Increasing age
- Family history of hypertension
- Ethnicity (African Americans are at a higher risk than Whites)

TABLE 9.13

Major risk factors for hypertension

Chronic alcohol use (>3 drinks/day) increases the risk of hypertension and stroke (MacMahon and Norton 1986). Alcohol is a common cause of hypertension in the United States, contributing to 10–20% of all cases in adult men. Many people who drink regularly and have high blood pressure notice a significant drop in blood pressure after just three days of abstention.

Most people can consume salt without it affecting their blood pressure because their kidneys very quickly excrete excess dietary sodium (Dustan and Kirk 1989), (Krakoff 1991).

However, like dietary cholesterol, some people are very sensitive to sodium in the diet. People who have a family history of hypertension, or are older than 55, obese, or African American are most likely to be sodium sensitive (Dustan and Kirk 1989). In these individuals, regular overconsumption of sodium contributes to the development of hypertension. About 25–30% of people with high blood pressure can lower their blood pressure significantly by limiting sodium intake to less than 2 g/day. Futhermore, salt restriction in individuals with high-normal blood pressure can also reduce their blood pressure (The Trials of Hypertension... 1992).

Typical diets of adults in the United States contain between 2.8 and 4.0 g of sodium per day (NAS 1989). Almost all dietary sodium is derived from salt (sodium chloride). Only 10% of dietary sodium originated from natural salt contents of foods, while 15% come from salt added to food at the table, and over 75% come from salt added to foods during processing and manufacturing (Sanchez-Castillo et al. 1987). People who are at risk for hypertension or who suffer from the disorder should avoid processed foods high in sodium, such as canned soups and salty snacks. (Other high-sodium foods are listed in Table 9.14.) They should also limit salt use during cooking and at the table. Recent population studies have shown that cutting sodium intake by about 2.5 g/day produces a modest, but significant reduction in blood pressure (Stamler et al. 1989), (Law et al. 1991).

The Benefits of Exercise, Potassium, and Calcium Although blood pressure temporarily increases during exercise, an aerobic exercise program lowers blood pressure over the long run (Joint National Committee 1988). Regular aerobic exercise reduces the risk of developing high blood pressure, and it is especially beneficial in overweight, hypertensive adults when part of a weight loss program.

There is substantial evidence that increasing dietary intake of potassium may have beneficial effects in reducing hypertension (MacGregor et al. 1982), (Reusser and McCarron 1994). Moreover, potassium intake is inversely correlated with risk of stroke. In a 12-year study of adults in California, women with the lowest intakes of potassium were nearly five times more likely to die from stroke than those with higher intakes (Khawand Barrett-Connor 1987).

Condiments

Pickles, olives, relishes, salted nuts, meat tenderizers, commercial salad dressings, monosodium glutamate, steak sauce, ketchup, soy sauce, Worcestershire sauce, horseradish sauce, chili sauce, commercial mustard, onion salt, garlic salt, celery salt, butter salt, seasoned salt

Breads

Salted crackers

Meat, Fish, Poultry, Cheese, and Substitutes

Cured, smoked, and processed meats such as ham, bacon, corned beef, chipped beef, wieners, luncheon meats, bologna, salt pork, regular canned salmon and tuna; all cheese except low-sodium and cottage cheese; TV dinners, frozen pizza, frozen Italian entrees, imitation sausage and bacon

Beverages

Commercial buttermilk, instant hot cocoa mixes

Soups

Commercial canned and dehydrated soups (except low-sodium soups), bouillon, consommé

Vegetables

Sauerkraut, hominy, pork and beans, canned tomato and vegetable juices

Fats

Gravy, regular peanut butter

Potato or Potato Substitutes

Potato chips, corn chips, salted popcorn, pretzels, frozen potato casseroles, commercially packaged rice and noodle mixes, dehydrated potatoes and potato mixes, bread stuffing

TABLE 9.14
High-sodium foods

In hypertensive adults with low potassium intakes, potassium supplements can modestly reduce blood pressure (MacGregor et al. 1982). People with hypertension (and people at risk for the disease) should consume ample dietary potassium from foods such as potatoes, green vegetables, orange juice, apricots, and bananas.

Poor dietary intake of calcium is associated with a higher risk of hypertension (McCarron et al. 1991), (Witteman et al. 1989). See Figure 9.5. In some people with high blood pressure, increasing intake of calcium to about 2 g/day may modestly reduce blood pressure (McCarron 1992). People with hypertension should make sure to get at least the RDA for calcium (800 mg/day) from rich dietary sources such as low-fat dairy products, leafy greens, and beans.

People with hypertension must be diagnosed early so that dietary and lifestyle changes can be of benefit. Adults should be regularly checked for high blood pressure; many people with hypertension don't know they have it. In the United States over the past 25 years, rates of high blood pressure and stroke have fallen substantially (Goldberg

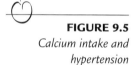

Source: Adapted from
Morris et al. (1987)

FIGURE 9.5

*Calcium intake and
hypertension*

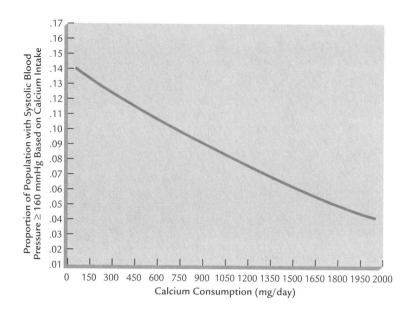

1992). Earlier detection, along with dietary and lifestyle changes, have
contributed to this encouraging trend.

Diet, Antioxidants, and Heart Disease

Scientists have been looking at how elevated LDL-cholesterol levels contribute to atherosclerosis. One intriguing hypothesis that has emerged
and is gaining support states that **oxidative damage** to the LDL particle
may trigger atherosclerosis (Steinberg 1993). Scientists think that
oxidized LDL-cholesterol is much more **atherogenic** than normal LDL-
cholesterol (Luc and Fruchart 1991). It appears that oxidized LDL-
cholesterol is picked up by special scavenger cells in the bloodstream.
These cells become loaded with LDL-cholesterol, migrate into the walls
of arteries, and begin plaque formation.

Antioxidants both within the LDL particle (vitamin E) and circulating
in the blood (beta-carotene and vitamin C) may protect LDL-cholesterol
from oxidation. This hypothesis is supported by experimental data that
antioxidants can retard development of atherosclerosis in animals on
high-cholesterol diets (Steinberg 1993). Also, epidemiological studies have
shown that there is an inverse relationship between antioxidant status and
heart disease (Gay et al. 1991).

A multinational study sponsored by the World Health Organization
revealed that people with higher vitamin E intakes had lower risk of
heart attack than people with lower intakes (Gay et al. 1991). Recently,
over 100,000 men and women participated in a study that showed that
large doses of vitamin E were associated with a lowered risk of heart disease (Rimm et al. 1993), (Stampfer et al. 1993).

oxidative damage: a
process in which a
compound (an oxidant,
such as oxygen) reacts
with biological
macromolecules causing
structural damage and
reduced function; the
reaction involves the
transfer of electrons
from the compound
being oxidized to the
oxidant

atherogenic:
contributing to or
causing atherosclerosis

antioxidant: a com-
pound that protects
other compounds from
oxidation by undergoing
oxidation itself

Dietary antioxidants may also influence risk of stroke (Manson et al. 1993), (Gillman et al. 1995). In a study involving nearly 90,000 women, those who regularly ate foods containing antioxidants (particularly carrots and spinach) had about a 50% lower risk of stroke than those who didn't (Manson et al. 1993). But remember that epidemiological studies like these can only suggest preliminary associations that need to be confirmed by clinical trials. Clinical trials of antioxidants are under way; conclusive evidence that antioxidants can protect against heart disease is still lacking (Rock et al. 1996). Until the results are in, following recommendations to eat generous servings of citrus fruits and green and yellow vegetables will provide ample amounts of dietary antioxidants.

B Vitamins, Homocysteine, and Cardiovascular Disease

Elevated plasma **homocysteine** levels are a risk factor for cardiovascular disease (Boushey et al. 1995). Homocysteine is one of the normal products of methionine metabolism in the body. Although the mechanism is not completely understood, high levels of homocysteine in the blood are thought to damage the endothelium of blood vessels and increase the risk of atherosclerosis.

Because vitamins B_6, B_{12}, and folate play important roles in pathways of homocysteine metabolism, deficiencies of these vitamins may produce increased levels of homocysteine in the blood. In a recent study of over 1000 healthy older adults, poor nutritional status of these vitamins was strongly linked to high levels of circulating homocysteine (Salhub et al. 1993). In another study, increased risk of coronary heart disease was associated with high concentrations of plasma homocysteine and low concentrations of plasma folate (Pancharuniti et al. 1994). An expert panel recently concluded that an elevated level of plasma homocysteine is a strong and independent risk factor for cardiovascular disease (Boushey et al. 1995). In many individuals, the primary cause of elevated homocysteine levels appears to be a dietary inadequacy of folic acid. Although data from clinical trials are not yet available, increasing folic acid intake in individuals with elevated homocysteine levels may help prevent cardiovascular disease (Boushey et al. 1995). Individuals at risk for heart disease should emphasize good sources of folate, B_6 and B_{12} in their diets while awaiting the results of future research.

Fish and Cardiovascular Health

homocysteine: an amino acid that is involved in the metabolism of methionine and is also a precursor in cysteine synthesis

There is evidence that frequent consumption of fish reduces the risk of heart attack (Kromhout et al. 1985), (Norell et al. 1986). Fish are tasty sources of low-fat, high-quality protein and, when substituted for red meat in meals twice a week, significantly reduce dietary saturated fat intake.

Scientists began to look at fish oils when it was noted that Greenland Eskimos, who eat as much fat and cholesterol as the rest of North Americans, had only one-tenth the risk of heart attack (Kronmann and Green 1980). These Eskimos eat mostly fish and other marine animals (most of their daily fat intake comes from fish oils).

Although not all studies agree (Ascherio et al. 1995), most studies indicate that people who eat fish two to three times each week (about 10 oz/week) have significantly lower rates of heart disease than people who rarely eat fish (Kromhout et al. 1985).

What's behind this remarkable connection? Fats from certain fish are rich in a special group of polyunsaturated fats called **omega-3 fatty acids** (see Table 9.15). Other than fish, omega-3 fatty acids are found in some wild animals, but almost nowhere else in human diets. These fats are not essential fatty acids, but we can synthesize only limited amounts of omega-3 fatty acids from alpha-linolenic acid, an essential fatty acid found in a few plant oils (such as soybean oil) (Leaf and Weber 1988).

Our bodies use the omega-3 fatty acids as building blocks for a family of important compounds called *eicosanoids*. Distributed throughout the body, eicosanoids help regulate a remarkable variety of body functions: blood vessel constriction and dilation, platelet activity, and the inflammatory response. Eicosanoids can be built from several different fatty acids, not only the omega-3 fatty acids. But the ones formed from the omega-3 fatty acids are unique in that they appear to have beneficial effects on the three key elements in a healthy cardiovascular system: blood clotting, blood pressure, and blood lipids (Leaf and Weber 1988).

Eicosanoids formed from omega-3 fatty acids slow down blood clotting by reducing the "clumping" activity of platelets—the blood cells responsible for blood clotting. Remember that clot formation on an atherosclerotic plaque is often the final event leading to a heart attack or stroke. This ability to make platelets less "sticky", and less likely to form clots is probably the major benefit of eating fish rich in omega-3 fatty acids two to three times each week.

Scientists have also investigated the effects of the omega-3 fatty acids on blood pressure. Large amounts of fish oil can modestly lower blood pressure (Leaf and Weber 1988), but the amount of omega-3 fatty acids needed to achieve a significant effect in a short time would be nearly impossible to get from dietary sources (about 1 kg of oily fish/day). Omega-3 fatty acids can lower blood triglycerides, particularly in people with very high levels, but they have little effect on cholesterol (Simopoulos 1991).

Although supplements containing fish oils are widely available, experts generally do not recommend their use ("Fish Oil Supplements" 1990). Many questions remain concerning the effectiveness and safety of large, chronic doses of these polyunsaturated fatty acids. Prolonged consumption of fish oil may produce a deficiency of vitamin E. Also, fish oils, par-

omega-3 fatty acid: a polyunsaturated fatty acid in which the first double bond is three carbons from the methyl end of the carbon chain

Fish High In Omega-3 Fatty Acids	Fish Moderately High In Omega-3s
Salmon	Halibut
Mackerel	Bluefish
Herring	Rockfish
Sardines	Rainbow and sea trout
Sablefish	Ocean perch
Lake trout	Bass
Fresh tuna	Hake
Canned albacore tuna	Pollock
Whitefish	Smelt
Anchovies	Mullet

TABLE 9.15
Omega-3 fatty acid content of fish

ticularly if derived from the livers of large fish, may contain pesticides and other contaminants (Leaf and Weber 1988).

The average intake of dietary omega-3 fatty acids in the United States is only 2–3 g/day, and for many, intake is close to zero. In addition to these important fatty acids, fish are also rich in protein, iron, and B vitamins, and low in saturated fat. Following current recommendations to increase our intake of fish to two to three servings a week may pay off in better cardiovascular health.

Hydrogenated Fats and Heart Disease

A good example of the often confusing information on diet and health is the butter versus margarine controversy. Because butter is rich in saturated fat, margarines made from plant oils have been advocated as an alternative. Recently, margarines have been implicated in heart disease because they are hydrogenated—and therefore may be rich in **trans fatty acids** (Mensink and Katan 1990), (Ascherio and Willet 1995).

Hydrogenation is a fat processing technique used by the food industry to modify the characteristics of polyunsaturated fats for use in foods. In the process, hydrogen is added to fats to reduce the number of double (unsaturated) bonds. Changing the chemical structure of the fats makes them more solid and gives them a longer shelf life by making them more resistant to oxidation. Margarine made from corn oil is solid at room temperature because it is partially hydrogenated. While resistance to rancidity and greater plasticity may be desirable characteristics, there appear to be undesirable trade-offs.

One problem with hydrogenation is that it "saturates" double bonds and therefore decreases the polyunsaturated fat content of foods. Because saturated fats are considered more atherogenic than polyunsaturated fats, hydrogenation reduces the potential health benefits from these foods. Also, the process produces fatty acids with double bonds in the *trans* configuration—whereas most natural fats contain only *cis* double bonds. See Figure 9.6.

trans fatty acids: the term *trans* describes the location of the hydrogens surrounding a carbon-carbon double bond; in *trans* fatty acids the hydrogens are located in opposite sides of the bond

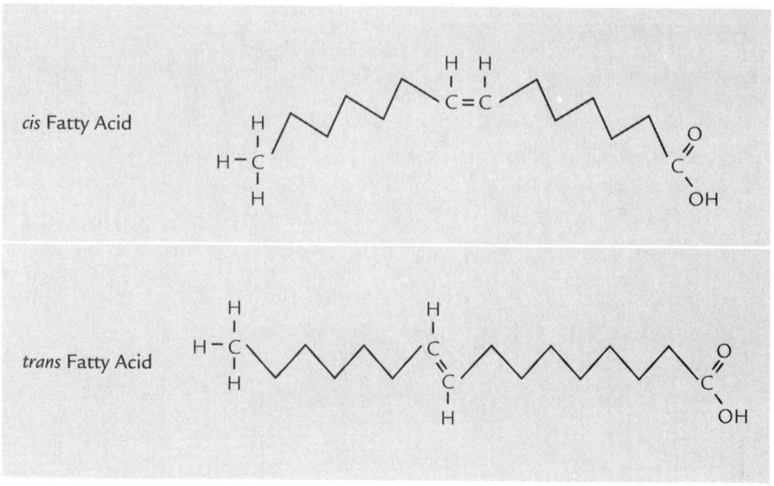

FIGURE 9.6

trans fatty acids

Trans fatty acids are only found in large amounts in products produced by hydrogenation. For example, stick margarine contains about 30% *trans* fatty acids, and certain fast foods such as french fries and fried chicken, as well as processed baked goods such as doughnuts and cookies, contain large amounts. High dietary intake of *trans* fatty acids has been shown to have unfavorable effects on plasma lipids by raising LDL-cholesterol and lowering HDL-cholesterol (Mensink and Katan 1990), (Ascherio and Willett 1995).

A recent study followed over 80,000 women for eight years to look at the impact of dietary *trans* fatty acids on heart disease. After controlling for other risk factors, women who ate the most *trans* fatty acids (from margarine and processed baked goods) had about a 50% greater risk of heart disease, when compared with those who ate the least (Willett et al. 1993).

Because only about 5 to 8% of the energy in our diets comes from *trans* fatty acids (Lichtenstein 1993) (compared with 12% from saturated fat), our main focus should be on reducing saturated fat intake. However, it appears that replacing saturated fat with hydrogenated fats is undesirable (Ascherio and Willett 1995). When necessary, liquid vegetable oils, or "tub" margarines are preferable to stick margarine. Reducing hidden sources of hydrogenated fats found in such foods as crackers, cookies, and other processed baked goods will also contribute to a more healthful diet. The fact remains that most people eat far too much total fat. Debating the benefits of butter versus margarine clouds the real issue: we should be eating less of both. Table 9.16 summarizes lifestyle and nutritional guidelines for lowering risk of cardiovascular disease.

1. Don't smoke cigarettes. If you do smoke, quit. Nonsmokers have a 50–70% lower risk than smokers.

2. Get regular aerobic exercise (3–5 times a week for 20–60 minutes). Adults who participate in regular aerobic exercise have a 45% lower risk than sedentary adults.

3. Attain or maintain your ideal body weight. People who are normal weight have a 35–55% lower risk compared to those >20% above ideal weight. Strategies for weight loss are outlined in the obesity section in Chapter 8.

4. Reduce total blood cholesterol, if possible, into the desirable range shown in Table 9.7.

   Limit saturated fat intake to <10% of total calories.
   Limit total fat intake to <30% of total calories.
   Limit dietary cholesterol to <300 mg each day, particularly if you are
      cholesterol sensitive.
   Limit hydrogenated fat intake.
   Increase intake of dietary fiber, with an emphasis on water-soluble
      fiber, to 25 g/day.

5. Reduce your risk for high blood pressure by maintaining ideal weight, getting regular exercise, limiting alcohol intake, and eating a diet that is high in complex carbohydrates and fiber, low in fat, rich in potassium and calcium, and low in sodium. Pay particular attention to sodium intake if you are likely to be a salt-sensitive person.

6. If you drink, limit yourself to one or two drinks each day.

7. Eat plenty of antioxidant-rich fruits and vegetables; try to get 5–7 servings a day.

8. Reduce stress. Use time-management techniques, listen to music, try regular exercise or meditation—whatever works for you.

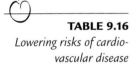

TABLE 9.16
Lowering risks of cardio-vascular disease

NUTRITION AND CANCER

Cancer is responsible for about one quarter of all deaths in the developed countries of the world (WHO 1990). In the United States, cancer is the second leading cause of death after heart disease, causing over half a million deaths each year (NCHS 1994). In grim contrast to the encouraging fall in mortality from heart disease, death rates for cancer during the 1960s, 1970s ,and 1980s in the United States slowly increased, even when considering the changing size, age, and composition of our population (Beardsley 1994).

A recent study by the U.S. National Cancer Institute examined trends in cancer incidence and cancer mortality in the United States from 1975–91. The number of new cases of cancer increased during this period, rising 18% in men and 12% in women. Most of this increase was due to improvements in cancer screening and earlier diagnosis. Overall cancer mortality rates increased 3% in men and 6% in women (Devesa et al. 1995).

Cancer is a disease in which cells begin to divide and multiply out of control, invading the surrounding tissues to form disorganized masses (tumors) that disrupt body functions. The rapidly growing cells can penetrate blood vessels and be carried throughout the body, and they can implant in other tissues and form new tumors.

Cancer develops in three major steps (see Figure 9.7):

1. A cell is exposed to a **carcinogen**.

2. The carcinogen damages the genetic material (the DNA) of the cell—called the initiation step. Chemicals, viruses, radiation from the sun, and many other agents can alter DNA.

3. Cancer development is then "promoted" by factors which push the cell into a cancerous state. Many substances are potential promoters: estrogen, compounds in cigarette smoke, and dietary factors such as alcohol and high levels of dietary fat.

Initiation and promotion of a cancerous cell may take years. During this time, the process can be reversed: the cell may be able to repair its DNA after it has been altered, or the immune system may identify the abnormal cells and destroy them.

Cancer is not a single disease. Cells from any tissue in the body can become cancerous, so there are many different types of cancer. A large number of environmental and genetic factors interact in cancer development. Factors that contribute to colon cancer, for example, may play no role in the development of lung cancer.

carcinogen: a substance that causes cancer

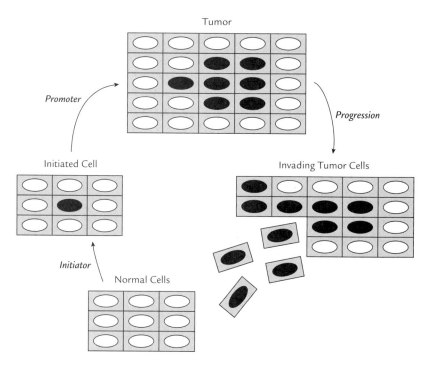

FIGURE 9.7
The process of carcinogenesis

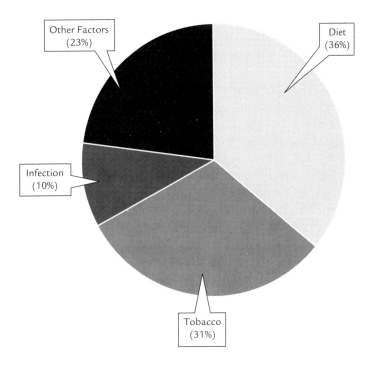

FIGURE 9.8
Estimated importance of factors in the development of cancer in the U.S.

Diet has a substantial impact on cancer rates. Experts estimate that dietary factors contribute to a third of all cancers (WHO 1991), (Doll 1992). See Figure 9.8. However, defining the connections between specific dietary components and different types of cancer has been difficult. Understanding of the risk factors for cancer development is much less complete than that for risk factors involved in diet and the cardiovascular diseases, for example.

Research on Nutrition and Cancer

Most of what is known about diet and cancer comes from epidemiological studies. *Correlation studies* compare cancer rates in different countries with varying diets. An example is a comparison between Japan and the United States (Haenszel 1982). In the 1950s, the average American was five times more likely to die from cancer of the colon as the average Japanese. At that time the American diet contained four times more fat. Conversely, Japanese people had several times the rate of stomach cancer than people in the United States, and Japanese ate much more salt-preserved and smoked foods. Correlation studies like these can only suggest associations, not prove cause and effect: other lifestyle differences besides diet could be causing differences in cancer rates among countries. Figure 9.9 shows other correlation studies of diet and cancer.

Further evidence comes from *migration studies,* which look at cancer rates in immigrants whose diets change as they move to a new country.

Source: Adapted from
Armstrong and Doll (1975)
and Carroll (1975)

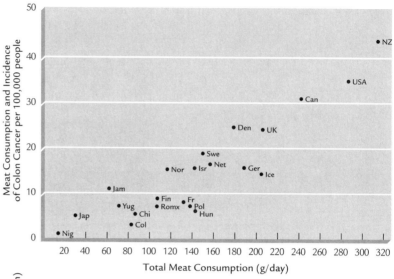

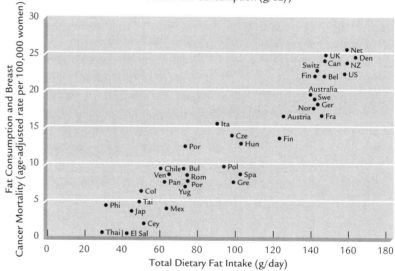

FIGURE 9.9

Diet and cancer

For example, when Japanese immigrants come to the United States, they and their children begin to develop colon and stomach cancer at the same rate as other people in the United States (Haenszel 1982). This is thought to be due primarily to the adoption of a Western diet.

As research defines the links between diet and cancer, results show again and again that what we choose to eat and drink can either increase or reduce our risk of developing cancer (USDHHS 1988). Dietary factors can function as either promoters or antipromoters. Factors that may increase cancer risk include obesity, a high-fat diet, and alcohol use. Fruits and vegetables, and possibly fiber, can act as antipromoters and reduce the risk of cancer. Let's look at each of these as well as other diet and cancer issues that are under study.

Obesity and Cancer

Obesity is a significant risk factor for cancer. Women who are 40% over ideal weight have a 55% greater risk of death from cancer than normal-weight women, and obese men have a 33% greater risk than normal-weight men (Garfinkel 1985). Obesity increases the risk for colon, prostate, breast, and uterine cancer (Pi-Sunyer 1991).

Adipose tissue produces small amounts of estrogens. It appears that, in obese women, increased levels of estrogen in the blood may promote cancer—particularly in the breast and uterus. Also, physical inactivity has been linked to increased cancer risk: men in sedentary jobs have a higher cancer risk than men with more active jobs (Carroll 1986).

Dietary Fat and Cancer

Fat intake has been linked to cancer. High-fat diets can increase risk for colon and prostate cancer—two of the most common types of cancer (Clinton et al. 1992). It appears that some fats, but not all, may act as promoters in cancer development. Evidence points specifically to a possible connection between saturated fat, especially from red meat, and colon cancer. Scientists recently examined the diets of nearly 90,000 women and found that a higher intake (about 5 oz/day) of red meat doubled the risk of developing colon cancer, compared with lower intakes (2 oz/day) (Willett et al. 1990). Fat consumption from chicken, fish, and plant sources was not correlated with increased risk.

One theory of how high amounts of dietary fat may promote colon cancer is that extra fat stimulates excess bile secretion. Colonic bacteria then break down the bile acids into toxic metabolites that damage the cells lining the colon, promoting carcinogenesis in these cells (Reddy 1986). Another potential reason dietary fat is associated with an increased risk for colon cancer is that high-fat diets often are low in dietary fiber, a possible antipromoter.

Fat intake from red meat has also been linked to a different type of cancer—prostate cancer in men. Over 50,000 men were followed for five years, and those who ate the most fat from red meat nearly doubled their risk for advanced prostate cancer (Giovanucci et al. 1993). Fats from dairy products, fish, and vegetable oils did not increase risk.

While evidence supporting a connection between dietary fat and colon and prostate cancer grows, support for the hypothesis that dietary fat increases breast cancer risk has been waning. Recent epidemiological research has produced no convincing evidence to support a breast cancer–dietary fat link (Rogers and Longnecker 1988).

Although saturated fat is most often implicated in human studies of dietary fat and cancer, studies in animals have produced very different results. In animal studies, high amounts of dietary polyunsaturated fats have been repeatedly linked to increased cancer rates. Until these discrepancies are resolved, scientists from the National Cancer Institute recommend cutting back on all fats (National Institute of Health… 1987).

Some scientists looking at the links between fat and cancer question whether it is actually the fat or simply the excess calories from fat that are important . They point to animal studies that have consistently shown that cutting calories from the diet, even if fat intake is not reduced, decreases cancer rates (Pariza 1988), (Anonymous 1992).

Fiber and Cancer

Studies in animals and humans have shown that diets high in fiber are associated with a reduction in the risk of colon cancer (Clinton 1992), (Howe et al. 1992). Fiber absorbs water in the colon and makes stools bulkier. By diluting stools, fiber may decrease contact between carcinogens in the stool and the cells lining the colon. Also, ample dietary fiber decreases the transit time of the stool through the colon, which may limit contact between potential carcinogens in the stool and the colonic cells (Kritchevsky and Klurfeld 1991).

Although most of the evidence suggests that the protective factor in high-fiber diets is the fiber content, other substances typically found in these diets may also be anticarcinogenic. Diets high in fiber often contain large amounts of vegetables and fruits. These foods contain antioxidant nutrients and other nonnutrient substances that may reduce the risk of cancer. Also, high-fiber diets usually are low in dietary fat, a possible promoter.

Many questions remain concerning the relationship between dietary fiber and colon cancer, as there are many different types of fiber and each may affect the risk of cancer differently. For example, several studies have shown that wheat bran appears to be an anticarcinogen while corn bran does not (DeCosse et al. 1989), (Alberts et al. 1990). Until more is known about specific fibers and their role in cancer, consuming very large amounts of fiber (or taking fiber supplements) is not recommended. However, the average fiber intake in the U.S. diet is only about 10–12 g/day. It would be prudent to follow current dietary guidelines and increase dietary fiber to about 25 g/day (NCR 1989), (USDHHS 1988). A diet containing generous amounts of fruits, whole grains, and vegetables will provide ample fiber and may decrease the risk of colon cancer. Table 9.17 gives the fiber content of selected foods.

Food Group	<1 g	1–1.9 g	2–2.9 g	3–3.9 g	4–4.9 g	5–5.9 g	> 6 g
Breads (1 slice)	Bagel White French	Whole wheat	Bran muffin (1)				
Cereals (1 oz)	Rice Krispies Special K Cornflakes	Oatmeal Nutri-Grain Cheerios	Wheaties Shredded Wheat	Honey Bran	Bran Chex 40% Bran Flakes Raisin Bran	Corn Bran	All-Bran 100% Bran
Pasta (1 cup)		Macaroni Spaghetti		Whole-wheat spaghetti			
Rice (½ cup)	White	Brown					
Legumes (½ cup cooked)				Lentils	Lima beans Dried peas		Kidney beans Baked beans Navy Beans
Vegetables (½ cup unless otherwise stated)	Cucumber Lettuce (1 cup) Green pepper	Asparagus Green beans Cabbage Cauliflower Potato, without skin (1) Celery	Broccoli Brussels sprouts Carrots Corn Potato, with skin (1) Spinach	Peas			
Fruits (1 medium unless otherwise stated)	Grapes (20) Watermelon (1 cup)	Apricots (3) Grapefruit (½) Peach, with skin Pineapple (½ cup)	Apple, without skin Banana Orange	Apple, with skin Pear, with skin Raspberries (½ cup)			

Adapted from J. L. Siavin, "Dietary Fiber Classification, Chemical Analyses and Food Sources," *J Am Diet Assoc* 67 (1987): 1164.

TABLE 9.17

Dietary fiber content of selected foods

Fruits and Vegetables and Cancer

Diets high in fruits and vegetables decrease the risk of many different cancers, including the most common types: lung, colon, breast, and prostate (Clinton 1992), (Howe et al. 1992), (Block and Patterson 1992). Yellow and green vegetables—such as broccoli, cabbages, and carrots—appear to be of particular benefit (Ziegler 1991). These foods contain several substances that are potentially anticarcinogenic, but it is uncertain which nutrients in these foods are actually conferring the protective effect.

Yellow and green fruits and vegetables are rich in the antioxidant nutrients beta-carotene, ascorbic acid, and vitamin E. Many scientists think these nutrients are important because oxidative damage to cell membranes and nucleic acids by **free radicals** seems to play an important role in carcinogenesis (Ames et al. 1993).

Beta-carotene, vitamins C and E, and selenium have been shown to be anticarcinogenic in animals (Ames et al. 1993). In several epidemiological studies, beta-carotene and vitamin E have been associated with lower cancer rates (Block et al. 1992), but other human studies have found no associations between antioxidant intake and cancer (Hunter et al. 1993). Researchers are still not sure if these compounds have protective effects in humans (Rock et al. 1996).

A variety of other nonnutrient anticarcinogens have been found in vegetables and fruits (Pierson 1992). The cruciferous vegetables (the cabbage family) contain several compounds, called indoles and isothiocyanates, that appear to be anticarcinogenic. Broccoli, for example, contains sulforaphane, a compound that helps cells produce enzymes against cancer. Certain beans and plant seeds contain protease inhibitors—compounds that may inhibit the spread of cancerous cells in tissues.

Alcohol and Cancer

Alcohol use increases the risk of several types of cancer, such as cancer of the mouth and throat, particularly when combined with cigarette smoking. Alcohol from any source—wine, beer, or spirits—also increases risk of cancer of the liver and breast (Rogers and Conner 1991).

Certain combinations of risk factors for cancer are particularly dangerous. For example, alcohol and smoking independently increase the risk for oral cancer—a moderate drinker has about 60% greater risk and a one-pack/day smoker a 50% greater risk. But smoking and tobacco together increase the risk more than simply adding the risks of the two together. The risk for oral cancer is 400% greater in a person who both smokes and drinks (Rogers and Conner 1991).

Food Pesticides and Cancer

Use of pesticides and other agrichemicals in modern agriculture has greatly increased crop yields and reduced spoilage, and made available a great variety and abundance of fresh foods. However, the residues of these compounds remain on or in many foods we eat, and their potential health hazards are still largely unclear.

The U.S. Environmental Protection Agency (EPA) monitors the food supply and tries to ensure that pesticide residues in foods are below dangerous levels. Synthetic pesticides found to be carcinogenic in animals are

free radicals: compounds that arise during oxidation reactions and contain a single, unpaired electron that is highly unstable; they readily react with nearby molecules

banned, and there is little firm evidence that those currently in use, at the levels found in most foods, have any significant risk. After a substantial review of the evidence concerning pesticides and health, a group of scientists in the United States concluded that while effects of pesticides on human health were difficult to detect, the health risks appeared to be small (Sharp et al. 1986).

However, the EPA is only able to test about 1–2% of the fruits and vegetables (from domestic and imported sources) sold in the United States. As long as the potential toxic effects of pesticides are unknown, it makes sense to minimize exposure to these compounds. Wash fresh fruit and vegetables thoroughly, and trim off and discard the outer leaves of vegetables such as lettuces, cabbages, and celery. Foods such as apples and cucumbers that have a waxy covering should probably be peeled before eating (the wax may seal residues on the fruit). By eating a wide variety of foods, exposure to any one pesticide will be minimized. Some people choose to buy so-called organic foods to limit exposure, but labeling regulations vary considerably, and organically grown doesn't always mean pesticide free.

Cancer and Nonnutrient Compounds in Food

Grilled, charbroiled, and smoke-cured meats contain large amounts of substances called polycyclic aromatic hydrocarbons. These compounds are similar to the tars in cigarette smoke and are carcinogenic in animal studies (Pierson 1992). Cured meats, such as hot dogs and bacon, contain preservatives called nitrites, which inhibit bacterial growth, protect against rancidity, and preserve the pink color of these products. However, nitrates can be converted in our stomachs to nitrosamines, which are carcinogenic. Because of this concern, many food producers have substantially decreased the amount of nitrites used in meat products. Consumption of vitamins C and E with foods containing nitrites may reduce the carcinogenic risk of nitrosamines (Lathia and Blum 1990).

Malnutrition and Cancer

Each year, over a million people in the United States develop cancer, joining many others who have been diagnosed with cancer and are undergoing treatment (Beardsley 1994). Malnutrition is a common problem in people with cancer, because of anorexia, malabsorption of nutrients, or as a side effect of cancer treatment. Encouraging proper nutrition allows people with cancer to tolerate therapy better, and it often improves their quality of life.

Unfortunately, three out of four people with cancer develop severe malnutrition and a rapid, progressive, often fatal wasting condition known as *cachexia* (Norton et al. 1989). Cachexia appears to be caused by the severe stress associated with advancing cancer, and it is often accelerated by infections and chemotherapy. A protein found in the blood of many cancer patients, called cachectin, contributes to the rapid breakdown of tissue. Providing optimal nutrition for people with cancer is a major challenge.

Nutritional Remedies and Cancer

Although some forms of cancer can be effectively treated, particularly if found early, many types of cancer are unremitting, painful, and fatal. This has led to a proliferation of unconventional nutritional therapies that are of unproven value. Food supplements, including megadoses of vitamins, minerals, herbs, and extracts of animal glands are marketed to people with cancer; an assortment of restrictive and imbalanced diets have also been advocated for cancer. Most have little or no scientific support for their use, and some may be harmful, particularly if they delay effective standard treatments. Most of these remedies have been examined by experts who have found no good evidence of benefit (American Cancer Society 1991), (Cassileth et al. 1991).

There are no simple solutions for people with cancer, but optimal nutritional support combined with conventional medical care offers the best hope. The American Cancer Society and the National Cancer Institute have published dietary guidelines to reduce the risk of cancer. (Work Study Group... 1991), (NCS 1987). Their recommendations are summarized in Table 9.18.

NUTRITION AND DIABETES

Diabetes mellitus affects 10–14 million people in the United States alone. It is one of the chief causes of death in the United States—and the number of people with diabetes is steadily increasing (Bennett 1990). Worldwide prevalence of the disease is given in Table 9.19.

In diabetes, the body is unable to control the level of glucose in the blood because of insufficient action of insulin. Insulin is a hormone produced by **beta cells** in the pancreas, and it regulates glucose metabolism throughout the day. Between meals, when no glucose is being absorbed from food, insulin controls the amount of glucose that is released into the blood from stores in the liver. After a meal, insulin signals cells all over the body to take up glucose absorbed from food into the blood for use as fuel and for energy storage. For insulin to do its job, it must be secreted from the pancreas at the proper time in adequate amounts, and

beta cells: the cells of the pancreas that secrete insulin into the bloodstream

1. Avoid obesity; maintain appropriate body weight.

2. Reduce total dietary fat to <30% of total calories.

3. Eat more citrus fruits, and more green, yellow, and cruciferous vegetables.

4. Eat more whole-grain and high-fiber foods. Try to get 20–30 grams of fiber each day.

5. Avoid or limit consumption of salt-cured, pickled, and nitrite-preserved foods.

6. Avoid or limit consumption of smoked foods and foods barbecued or fried at high temperatures.

7. Drink alcohol only sparingly, if at all.

8. Eat a varied and balanced diet.

Source: Adapted from NCI (1987) and ACS (1991)

TABLE 9.18
Dietary recommendations to lower cancer risk

it must send the correct signals to the liver and other tissues. In diabetes, this system breaks down.

Type I and Type II Diabetes

There are two main types of diabetes. About 10% of people with diabetes have *Type I (insulin-dependent) diabetes.* In this type, the pancreas abruptly stops producing insulin because the beta cells die: the body's immune system malfunctions and mistakenly attacks the beta cells and destroys them. What triggers this **autoimmune reaction** is unknown, but beta cells cannot be replaced, and the person will depend on lifelong periodic injections of insulin to control glucose metabolism. People with this type of diabetes are critically dependent on insulin injections and will die within a week if they do not receive insulin. Because Type I diabetes most often begins during childhood or adolescence, it is sometimes called juvenile diabetes.

Ninety percent of people with diabetes have *Type II (noninsulin-dependent) diabetes.* This form of diabetes develops gradually over many

autoimmune reaction: an abnormal immune reaction in which the body's immune system reacts against its own tissues

	1990	2000 (estimated)
Developing countries	60 million	120 million
Developed countries	20 million	40 million
Total	80 million	160 million

Source: From King and Rewers (1993).

TABLE 9.19
Worldwide prevalence of diabetes

years and is usually diagnosed during middle or late adulthood. In contrast to Type I diabetes, the main problem in Type II diabetes is not secretion of insulin (the beta cells continue to produce it), but an inability of tissues throughout the body to recognize and respond to the insulin signal. People with Type II diabetes do not depend on insulin injections to stay alive—their beta cells continue to produce insulin—but the insulin works poorly. Type II diabetes is often referred to as adult-onset diabetes.

In both types of diabetes, glucose levels in the blood are much too high. Insulin no longer restrains the liver's production of glucose, and large amounts of glucose are continually released into the blood. Glucose that is circulating cannot enter muscle and fat cells, so levels climb even higher. Cells become starved for glucose, and in response, large amounts of fats are released into the blood as an alternative fuel.

High levels of glucose in the blood are toxic, causing widespread damage to smaller arteries. Also, the high levels of triglycerides and other lipids released into the blood accelerate the development of atherosclerosis. As a result, diabetics suffer from high rates of heart attack, kidney failure, and blindness—the blood vessels to these tissues are damaged by the disease.

Risk Factors for Diabetes

Poor dietary and lifestyle choices can increase the risk of Type II diabetes (see Table 9.20). Obesity is a primary risk factor for Type II diabetes (Pi-Sunyer 1991), and 80–90% of Type II diabetics are obese. The common misperception that "eating sugar causes diabetes" is only true in that too many calories, from any source, can contribute to diabetes.

Why does diabetes occur so often in overweight people? Many scientists think that year after year of eating too many calories causes cells throughout the body to be exposed to chronic, high levels of insulin in the blood. In genetically susceptible people, cells respond by "turning off"—they become progressively less responsive to the signal from insulin. As a result, glucose levels rise and diabetes develops. People at increased risk of developing diabetes—because of family history, age, or ethnic background—should make every effort to maintain ideal body weight.

Adipose tissue distribution is also an important determinant in risk of diabetes. Central (android) obesity, particularly in women, increases risk for diabetes more than lower body (gynoid) obesity (Kalkhoff et al. 1983). Regular aerobic exercise reduces the risk of developing Type II diabetes. Adults who exercise regularly can reduce their risk for diabetes by about a third (Helmrich et al. 1991).

For people who develop Type II diabetes, weight loss (if overweight) and exercise is the best medicine. Even modest changes in diet and physical activity can significantly reduce high blood glucose and lipid levels. If obese, a modest 5% reduction in weight can improve the body's sensitivity to insulin. Many Type II diabetics avoid using medication and control their disease with diet and lifestyle adjustments. Dietary guidelines for people with Type II diabetes and for people at high risk for diabetes are shown in Table 9.21.

Type II Diabetes in Minorities

Type II diabetes disproportionally affects minority populations in the United States. The prevalence of Type II diabetes in African Americans and Hispanics is higher than in Whites (King 1993). Diabetes is also very common among Native Americans (in some Native-American communities nearly half of the adults have Type II diabetes) (Bennett 1990). Differences in genetic susceptibility, combined with the rising incidence of obesity in minority groups, may be responsible. See Figure 9.10.

Cow's Milk, Infant Feeding, and Type I Diabetes

In some cases of Type I diabetes, the environmental trigger that initiates the autoimmune reaction against the beta cells may come from drinking cow's milk in infancy (Borch-Johnson et al. 1984), (Virtanen et al. 1991). Scientists have recently found a small protein in cow's milk that has a structure similar to a protein on the surface of the beta cells (Karjalainen et al. 1992). In certain genetically susceptible people who have been exposed to cow's milk, antibodies that are produced to attack the cow's milk protein may be confused by the similarity and attack the beta cells. These observations have renewed the debate regarding the suitability of cow's milk in infant diets (AAP 1994).

TABLE 9.20

Risk factors for Type II diabetes

- Obesity

- Sedentary lifestyle

- Family history of Type II diabetes

- Increasing age

- Ethnicity: African Americans, Hispanics, and certain groups of Native Americans are at increased risk.

	Lean Persons	Obese Persons
Energy	Enough to maintain desirable body weight Men and physically active women require 30 kcal/kg desirable body weight Sedentary persons and persons older than 55 years require 28 kcal/kg desirable body weight	Enough to achieve reasonable body weight 20 kcal/kg desirable body weight
Carbohydrate Sucrose	Up to 55 to 60% of total energy Can be included with an individualized diet plan	Same Low nutrient density; limit on low-calorie diets
Fiber	Up to 40 g/day, with emphasis on water-soluble fiber	25 g/1000 kcal
Protein	Recommended dietary allowance is 0.8 g/kg body weight	Minimum of 60 g when restricted to ≤ 1200 kcal diet

Fat
  Polyunsaturated fats
  Saturated fats
  Monounsaturated fats
  Cholesterol
	Ideally <30% of energy
	  Up to 10% of energy
	  < 10% of energy
	  10 to 15% of energy
	  <300 mg/day
		Same

Alternative sweeteners	Use is acceptable	Same
Sodium	Not to exceed 3,000 mg/day	Same
Alcohol	Occasional or no use	Same
Vitamins/Minerals	No evidence that diabetes causes increased need	
Snacks	Individualized on the basis of preferences and glucose patterns; snack should be coordinated with insulin schedule if on insulin	Not necessary; if desired, should be included in total day's meal plan

Adapted from C. A. Beebe, et al., "Nutrition Management for Individuals with Noninsulin-Dependent Diabetes Mellitus in the 1990s," a review by the Diabetes Care and Education Dietetic Practice Group, *J Am Diet Assoc* 91 (1991): 1991.

TABLE 9.21

Nutrition guidelines for people with, or at risk for, Type II diabetes

Source: Adapted from King
and Rewers (1993)

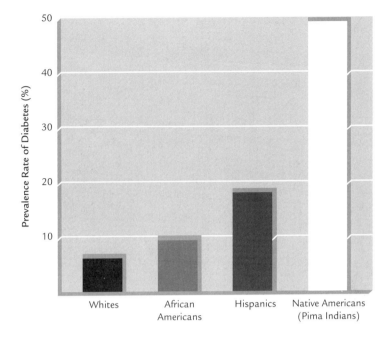

FIGURE 9.10

*Ethnic distribution of
Type II diabetes in the
United States*

NUTRITION AND ALCOHOL

Over 70% of the adult population in the United States consumes alcohol
to some degree (NCR 1989). Taken in moderation, alcohol is widely
accepted as part of our daily lives: it relaxes, reduces inhibitions, and can
be an enjoyable social lubricant. Unfortunately, alcohol is a drug with
enormous potential for abuse and addiction. Alcoholism is common in
all segments of society—young, old, rich, and poor—and causes widespread misery for millions of victims and their families. Alcohol abuse is
particularly common in young and middle-aged adults: 9 to 17% of the
U.S. population between 18 and 45 have problems with alcohol abuse
(Nace 1987). Alcohol is the second leading cause of preventable death in
the United States (after smoking). Figure 9.11 shows the extent of drinking problems in various age groups.

Most alcohols are highly toxic substances, often used as disinfectants
and solvents, and they are poisonous to cells. Ethanol is the alcohol
found in beer, wine, and distilled spirits. Ethanol is less toxic than many
other alcohols, and when diluted and consumed in small amounts, it is a
sedative and produces a mild euphoria.

Heavy alcohol use, however, has significant health risks. In large
amounts, alcohol has toxic effects on virtually every organ in the body.
Alcoholism is a common cause of malnutrition in the United States
(Feinman and Lieber 1994), and regular, heavy alcohol use increases risk
for many chronic diseases, including high blood pressure, stroke, and

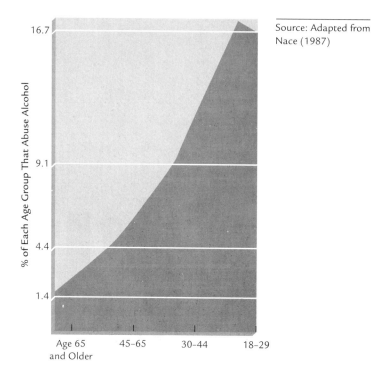

Source: Adapted from Nace (1987)

FIGURE 9.11
Drinking problems in adults in the U.S.

several types of cancer. **Alcoholic liver disease** is a leading cause of death around the world (Arif and Westermeyer 1989). A recent prospective study of nearly 13,000 British men found that, in amounts above two drinks a day (1 oz of ethanol), progressively greater alcohol consumption is associated with progressively higher mortality (Doll et al. 1994).

What Is "Moderate" Alcohol Consumption?

Although tolerances for alcohol differ greatly among people, authorities have defined moderation, for the population, as one to two drinks per day. What is meant by a "drink"? A drink is any dose of alcoholic beverage containing ½ oz of ethanol—about 4 oz of wine, 10 oz of beer, or 1 oz of hard liquor. The concentration of blood alcohol during drinking and its effects on behavior are shown in Table 9.22.

Alcohol and Malnutrition

alcoholic liver disease: damage to the liver caused by alcohol abuse, producing inflammation and accumulation of fat in liver cells and, in later stages, cirrhosis

A substantial number of the total calories in the U.S. diet come from alcoholic beverages. Regular adult drinkers (about two drinks a day) get about 10% of their total daily calories from alcohol, while heavy drinkers may consume between a third and half of their daily calories as alcohol.

Percentage of Blood Alcohol Concentration					
Body Weight (lb)	**Number of Drinks In Two Hours***				
	2	**4**	**6**	**8**	**10**
120	0.06	0.12	0.19	0.25	0.31
140	0.05	0.11	0.16	0.21	0.27
160	0.05	0.09	0.14	0.19	0.23
180	0.04	0.08	0.13	0.17	0.21
200	0.04	0.08	0.11	0.15	0.19

Resulting Condition	
Blood Alcohol Concentration	**Effect**
0.05%	Relaxed state; judgment not as sharp
0.08%	Everyday stress lessened
0.10%**	Movements and speech become clumsy
0.20%	Very drunk; loud and difficult to understand; emotions unstable
0.40%	Difficult to wake up; incapable of voluntary action
0.50%	Coma and/or death

* 1 drink equals 1–1½ ounces of hard liquor, 10–12 ounces of beer, or 4–5 ounces of wine.
** In the U.S., 0.10 is often used as the lowest indicator of driving while intoxicated.

TABLE 9.22
Blood alcohol concentrations and behavior

Alcoholic beverages contain calories (ethanol has 7 kcal/g) but almost no other useful nutrients (Leake and Silverman 1974) (see Table 9.23). Alcohol displaces more nutritious foods from the diet, and as a result, most heavy users have inadequate intakes of carbohydrate, protein, and many vitamins and minerals.

Compounding poor nutrient intake, alcohol is toxic to the cells lining the gastrointestinal tract—damaging the esophagus, stomach, and small intestine (Salaspuro 1993). The injured gastrointestinal cells are less able to absorb nutrients, including thiamin, folate, and vitamin B_{12}. Ethanol has deleterious effects on the storage, metabolism, and excretion of many nutrients. For example, damage to the liver reduces the ability to convert vitamin D to 25-OH vitamin D (Hapner and Roginsky 1975). Metabolism of vitamin B_6 and folate to active forms is impaired, and alcohol increases the excretion of folate and zinc. Because of these changes in nutrient metabolism, chronic, heavy use of alcohol sharply increases the risk of widespread nutritional deficiencies, even in people who eat a reasonable diet (Salaspuro 1993). See Table 9.24.

Beverage	Amt (oz)	Water (%)	Energy (cal)	Protein (g)	CHO (g)	Calcium (mg)	Phos-phorus (mg)	Iron (mg)	Sod-ium (mg)	Potas-sium (mg)	Thia-min (mg)	Ribo-flavin (mg)	Niacin (mg)
Beer	12.0	92.2	151	1.1	13.7	18	108	Trace	25	90	0.01	0.11	2.2
Gin, rum, vodka, whiskey, 80 proof	1.5	66.6	97	—	Trace	—	—	—	Trace	1	—	—	—
Wine													
Dessert	3.5	76.7	141	0.1	7.9	8	—	—	4	77	0.01	0.02	0.2
Table	3.5	85.6	87	0.1	4.3	9	10	0.4	5	94	Trace	0.01	0.1

Adapted from C. Adams, *Nutritive Value of American Foods,* Agriculture Handbook no. 456 (Washington D.C.: U.S. Department of Agriculture, 1975).

TABLE 9.23

Nutrient content of alcohol

Alcohol and Disease

Alcohol is a common cause of high blood pressure in the developed world. Although the link between salt and high blood pressure has been more widely publicized, for most people, alcohol is a greater risk factor for hypertension than dietary sodium. Reducing alcohol intake can significantly lower blood pressure in people with hypertension. In addition, consumption of more than three drinks per day increases risk of stroke independently of its hypertensive effect (MacMahon and Norton 1986).

Alcohol is particularly toxic to the liver—damaging liver cells and disrupting many of its functions (World et al. 1985). Fat metabolism in the liver is impaired: bulging deposits of fat accumulate and clog the cells, and the hepatocytes release large amounts of triglycerides and cholesterol into the bloodstream. There is inadequate synthesis of many important proteins produced in the liver, so that the blood cannot clot well, and the immune system is weakened. Even a single night of heavy drinking damages liver cells and causes fat to accumulate in the liver. Over years, alcohol abuse destroys large sections of the liver and produces cirrhosis, an irreversible and often fatal condition.

Alcohol use has been linked to increased risk of cancer of the mouth, throat, esophagus, liver, breast, and colon (Arif and Westermeyer 1989). Alcohol appears to be a promoter of cancer development at several sites in the body. This effect may be due to direct toxic effects of alcohol on cells, or poor diet quality and malnutrition—or both.

Chronic alcoholism produces brain damage and is a major cause of dementia (Arif and Westermeyer 1989) (discussed in Chapter 10). It takes about ten years of heavy drinking to produce alcoholic dementia, which can be irreversible. Heavy alcohol use also sharply increases the risk of osteoporosis. Figure 9.12 charts the relative risks for certain diseases and other causes of death by number of drinks per day.

Nutrient	Interactions	Consequences
Protein	Decreased dietary intake and impaired hepatic synthesis	Muscle wasting, immune system weakness, reduced blood clotting, reduced ability to transport nutrients (vitamin A, iron) in the blood
Carbohydrate	Damage to intestinal lining causes malabsorption of some sugars; damage to pancreas causes decreased digestive enzyme secretion	Lactose intolerance, diarrhea
Fat	Damage to liver results in decreased bile secretion	Fat and fat-soluble vitamin malabsorption can lead to weight loss and vitamin deficiency
Vitamin A	Damage to liver impairs storage of retinol and increases metabolism and breakdown	Vitamin A deficiency, night blindness
Vitamin D	Decreased dietary intake, impaired hepatic conversion to 25-OH vitamin D	Vitamin D deficiency, decreased bone density, increased osteoporosis and bone fractures
Vitamin K	Decreased dietary intake and malabsorption	Vitamin K deficiency, reduced blood clotting, increased bleeding tendency
Vitamin C	Decreased intake and altered liver metabolism	Vitamin C deficiency
Thiamin	Decreased intake, poor absorption, and impaired conversion to active form (thiamin pyrophosphate)	Thiamin deficiency leading to damage to the brain and heart
Vitamin B_6	Decreased intake, increased breakdown of pyridoxal phosphate	Vitamin B_6 deficiency
Folic acid	Decreased intake, poor absorption, and increased metabolism	Folate deficiency causes anemia, as well as atrophy of the intestinal lining leading to malabsorption
Zinc	Decreased intake, poor absorption, and increased excretion	Zinc deficiency, night blindness, weakened immune system
Iron	Decreased intake, damage to stomach and intestinal lining increases losses from bleeding, impaired iron transport in the blood	Anemia

TABLE 9.24

Effects of alcohol on nutritional status

Source: Adapted from
Boffetta and Garfinkel
(1990)

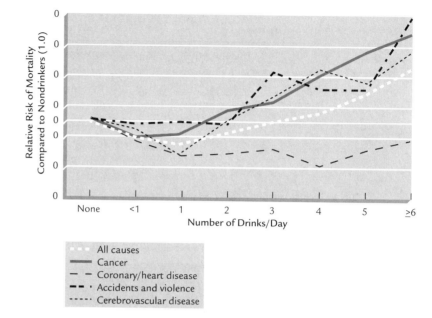

FIGURE 9.12
Mortality and alcohol use

Potential Health Benefits of Alcohol and the French Paradox

What about the potential health benefits of alcohol? Moderate drinking appears to be protective against heart attacks (Friedman and Kimball 1986). Epidemiological studies have shown that heart disease rates are higher in people who don't drink or who drink heavily, when compared with moderate drinkers (one to two drinks per day).

For example, a recent study followed the drinking habits of 88,000 women for eight years: those who consumed three to nine drinks per week had 40% lower risk than those who did not drink (Stampfer et al. 1988). Studies in middle-aged men have shown a more modest, but still significant, reduction in risk (one to two drinks per day lowers risk about 25% compared with nondrinkers).

Alcohol may be protective because it increases the level of high-density lipoproteins (HDL) in the blood, and high HDL levels are associated with decreased risk of heart disease (Handa et al. 1990), (Anonymous 1993).

Worldwide, high amounts of dietary saturated fat in the diet are associated with an increased risk of heart attack. In regions in France, however, people eat diets very high in saturated fat and yet have a low rate of heart disease. This is known as the French paradox. Some scientists think that these findings may be explained by the high consumption of red wine in these areas: per capita, the French consume about ten times more wine than people in the United States.

Red wine is a chemically complex substance—could factors other than ethanol in the wine explain the paradox? Recently, scientists have begun looking at **phenols** in wine. High levels of phenolic substances contribute to the flavor, color, and texture of red wine; and these substances—at least in the test tube—are potent antioxidants (Anonymous 1993). They reduce the oxidation of LDL-cholesterol, which is thought to be an initial step in the development of atherosclerosis. Another component of red wine that may play a role is resveratrol, a compound in grape skins that has an anticoagulant effect. Some scientists feel that the phenols and resveratrol in red wine may explain at least part of the French paradox (Frankel et al. 1993).

Alternatively, other factors in the French diet could explain the reduced rates of heart disease in France. For example, compared with people in the United States, per capita consumption of many fruits and vegetables is substantially higher among the French (Drewnowski et al. 1996). Fruits and vegetables contain many compounds that may reduce risk for heart disease (as discussed earlier in this chapter).

These findings do not imply that nondrinkers should take up drinking to protect their hearts. The high consumption of wine by the French population has detrimental effects: the French have high rates of alcoholic liver disease and are twice as likely as Americans to die of cirrhosis (Arif and Westermeyer 1989). Studies have shown that for every seven people who begin drinking alcohol, one will develop alcohol-related health problems (Nace 1987). Although growing evidence suggests that moderate drinking may be protective against heart disease, most scientists feel that the established health risks of alcohol to society far outweigh the small, and still uncertain, health benefits. Alcohol use is a matter of personal choice, but if you choose to drink, you should be aware of the risks and drink responsibly. Alcohol should certainly be avoided when consumption puts others at risk, such as during pregnancy and before driving.

NUTRITION AND ACQUIRED IMMUNE DEFICIENCY SYNDROME (AIDS)

Acquired immune deficiency syndrome (AIDS) has become a major public health-care problem in the United States and throughout the world. AIDS is caused by infection with the human immunodeficiency virus (HIV), which attacks and disables the immune system. Because the immune system is crippled by HIV, people who are infected are susceptible to **opportunistic** infections and certain cancers. Approximately 1 million individuals—or about 1 in every 250 people in the United States—are infected with HIV.

phenol: a crystalline compound, C_6H_5OH; or a generic term for organic compounds containing one or more hydroxyl groups attached to a carbon ring

opportunistic infection: an infection caused by a microorganism that normally does not cause disease in healthy people but can infect people with weakened immune systems

Wasting and Undernutrition in AIDS

A central feature of AIDS is wasting and undernutrition. Wasting in AIDS is defined by the Centers for Disease Control as involuntary weight loss of >10% of body weight (Centers for Disease Control 1986). Up to 80% of people with AIDS will be affected by wasting during their illness . Weight loss in people with AIDS occurs in a stepwise, episodic pattern related to opportunistic infections (Nerad 1994).

Protein–energy malnutrition (PEM) is the most frequent form of undernutrition seen in people with AIDS, but micronutrient deficiencies are also common (Nerad 1994). Wasting is characterized by a profound loss of fat and muscle tissue, depletion of trace minerals, and further impairment of the immune system.

There are three major factors that contribute to wasting and malnutrition in people with AIDS: inadequate dietary intake, malabsorption of nutrients, and altered metabolism. Poor food intake can result from anorexia, nausea and vomiting, or infections of the mouth and esophagus that cause swallowing to be painful and difficult. Infections of the gastrointestinal tract interfere with nutrient absorption and can produce severe diarrhea. Metabolic alterations in AIDS include fever and increased protein breakdown in response to chronic or repeated infections.

The Role of Nutrition in AIDS

Nutrition plays an important role in supporting people with AIDS. Studies have shown that nutritional counseling combined with food supplements can improve nutrient intake and reduce weight loss (Dowling et al. 1990), (McKinley et al. 1994). Nutritional intervention should begin early in the illness and include dietary counseling on the principles of a balanced, nutrient-dense diet. If an individual cannot consume enough food at meals, high-calorie snacks or supplements can be provided. When wasting is severe, nutritional support by **enteral** or **parenteral** routes may be required.

Although vitamin–mineral supplements at or near RDA levels may be helpful, megadoses of vitamins and minerals should be avoided because they can be toxic. Large doses of certain micronutrients (including vitamins E and C, zinc, and iron) can impair the immune system (Merrill 1995).

Food safety is critical for people with AIDS. Utensils and preparation surfaces should be scrupulously clean. Fresh fruits and vegetables should be washed thoroughly, and raw protein foods (raw eggs, meat, sushi) should be avoided. Foods should be promptly refrigerated, and those suspected of spoilage or contamination should be thrown out.

Proper nutrition for people with AIDS can have many benefits. It can diminish weight loss and improve immune function. It may also help

enteral: within or by way of the intestinal tract

parenteral: the introduction of substances by any other means than through the gastrointestinal tract, such as intravenous injection

people with AIDS maintain independence and enhance their quality of life during the illness (Nerad 1994).

NUTRITIONAL NEEDS OF WOMEN DURING ADULTHOOD

Although the basic nutritional needs of healthy women and men are similar, there are some important differences. Women are generally smaller than men, and their body composition and physiology are unique. Adult women tend to be about 22% fat, while men are only 15% fat. Women also do not have as much lean body tissue as men—they have smaller skeletons and less muscle mass.

Because of these differences, women, on average, require less energy and protein. Also, since micronutrient requirements are often calculated per kilogram of body weight, or are based on energy and protein intake, the RDAs for many nutrients are reduced for women. One exception is iron—women require more iron because of increased losses in menstrual blood. About 1 mg of iron in hemoglobin is lost each day during menstruation.

NUTRITION, THE MENSTRUAL CYCLE, AND FERTILITY

The hallmark of reproductive maturity in women is the female sexual cycle, commonly called the menstrual cycle. Monthly, cyclical changes in the female hormones estrogen and progesterone are associated with changes in the ovaries and uterus. These changes prepare a woman's body for conception and pregnancy. During each cycle, an egg is released by the ovaries (ovulation), and the lining of the uterus grows, thickens, and is then shed (menstruation).

The hormonal changes that occur during the menstrual cycle also affect energy needs and appetite. During the ten days of the cycle just before the onset of menstrual bleeding, basal metabolic rate rises. The increase in energy need stimulates appetite, and caloric intake increases. Studies have found that average daily energy intake during the **luteal phase** is 4–35% greater than intake during the **follicular phase** (Barr et al. 1995). Hormonal changes in the days before menstruation also cause sodium and water retention, and often, modest weight gain.

The menstrual cycle is very sensitive to changes in nutrition. Undernutrition can quickly disrupt the cycle, leading to anovulation and amenorrhea. This can occur in large populations affected by famine, in self-induced starvation in women with anorexia nervosa, and in women on very low calorie diets for weight loss. It is unclear why menstrual cycles become irregular in situations of severe undernutrition; it may be related to low stores of body fat (see the discussion in Chapter 2).

luteal phase: the period of the menstrual cycle after ovulation and before menstruation

follicular phase: the period of the menstrual cycle after menstruation and before ovulation

Missing menstrual periods can be hazardous to bone health. Amenor-rhea causes a fall in estrogen levels (similar to the menopause) and can cause significant bone loss and increase the risk for osteoporosis (see discussions in Chapters 8 and 10).

Oral Contraceptives and Nutritional Health

Oral contraceptives are a popular form of birth control worldwide: in the United States 15 to 19% of women aged 15–45 years use "the pill". Oral contraceptives contain mixtures of estrogen and progesterone that block ovulation, and they are highly effective in preventing pregnancy.

Oral contraceptives used in the 1960s and '70s contained much higher amounts of estrogens than current preparations, and these high-estrogen pills had many undesirable side effects, including cardiovascular complications (stroke and blood clotting abnormalities) and weight gain. They also had a number of nutritional side effects, often causing elevations in blood glucose, triglycerides, and cholesterol; and producing deficiencies of folate and vitamin B_6 (Leklem 1986).

Newer low-dose estrogen–progesterone pills introduced in the late 1970s are much safer than the older forms, although side effects still exist. The newer formulations have negligible effects on carbohydrate and fat metabolism and do not appear to increase the risk of stroke or coronary heart disease in healthy young women (Wenger et al. 1993). However, they should not be used by some women. Smokers over the age of 35, women at high risk for or with a history of stroke or coronary heart disease, and those with impaired liver function have a higher risk of adverse side effects and should avoid oral contraceptives (Drife 1989).

The new lower-dose pills may increase the requirement for vitamin B_6 and folate in some women. An important role of vitamin B_6 in the body is its function as a coenzyme in the synthesis of certain neurotransmitters, including **serotonin**. In a group of women who developed biochemical evidence of vitamin B_6 deficiency and depression while taking the pill, a daily supplement of the vitamin was effective in improving vitamin B_6 status and reducing depression (Adams et al. 1973). Women taking oral contraceptives should be sure to eat foods rich in vitamin B_6 (such as fish, poultry, green leafy vegetables, and fruits) and folate (green vegetables, legumes, fruits, and whole grains).

Nutrition and the Premenstrual Syndrome (PMS)

The premenstrual syndrome is characterized by the cyclic appearance of a variety of symptoms just prior to the menses. Symptoms can be physical, such as breast tenderness, headache, fluid retention, and fatigue; or emotional, including anxiety, depression, and irritability. Appetite

serotonin: a compound (chemical name: 3-(2-aminoethyl)-5 indolol) that functions as a neurotransmitter; it is synthesized from tryptophan

changes and food cravings are described by some women with PMS. Symptoms usually appear in the last seven to ten days of the menstrual cycle. About 2 to 5% of women report that PMS interferes with their usual activities or work (Reid 1991). Many theories have been suggested to explain this disorder—including hormone imbalances, prostaglandin sensitivity, and hypoglycemia—but none has been proven. In some women, vitamin B_6 appears to play a role in the condition.

In a double-blind study, 50 mg of vitamin B_6/day was effective in relieving depression and irritability in women with PMS (Doll et al. 1989). Similar to the effects of vitamin B_6 on depression during use of oral contraceptives, these benefits may be explained by the role vitamin B_6 plays as a coenzyme in the synthesis of serotonin and dopamine, which are neurotransmitters known to affect behavior and mood. However, other studies have found no improvement in PMS with vitamin B_6 supplementation (Kleijnen et al. 1990). Women should be cautious with high doses of vitamin B_6—megadoses (2–6 g/day) can cause severe **neuropathy** (Dalton and Dalton 1987).

NUTRITION AND WOMEN'S HEALTH

Life expectancy is greater for women than for men. In the United States, this "gender gap" in life expectancy is about seven years. Many women are now living into their 80s and 90s, and living a third of their lives after menopause. Consequently, more women are likely to suffer from the debilitating, chronic diseases of old age—including heart disease, osteoporosis (risk for both heart disease and osteoporosis increases sharply after menopause), cancer, and diabetes. Women faced with these challenges have unique nutritional needs.

Cardiovascular Disease in Women

Although breast cancer is the disease most feared by women, heart disease kills nine times more women than breast cancer: in the United States, one in three women will die of heart disease (AHA 1992). Cardiovascular disease is often perceived as a "man's disease," because men more often die of heart attacks in middle adulthood, while the toll in women becomes apparent only after menopause.

Why does heart disease in women occur about 10 to 12 years later than in men? Before menopause, women have high levels of estrogen in their blood, and estrogen has a dramatic protective effect against coronary heart disease (Wenger et al. 1993). Estrogens may produce this effect by raising beneficial HDL-cholesterol levels. When estrogen levels fall at menopause, HDL-cholesterol falls, while total blood cholesterol rises

neuropathy: a general term for functional disturbances and/or pathological changes in the peripheral nervous system

(Kris-Etherton and Krummel 1993). Menopause also influences body fat distribution. In postmenopausal women, more fat accumulates around the abdomen (android obesity), instead of on the buttocks and hips. These two changes—an increasing ratio of total cholesterol/HDL-cholesterol and increasing android obesity—may explain the increase in risk for heart attack in postmenopausal women.

Most of the research looking at risk factors for heart disease have used middle-aged men as subjects. Because of this, some scientists question whether the results of these studies can be applied to women (Wenger 1992). Researchers do know that there are strong similarities in risk factors for both women and men, and that women can lower their risk in many of the same ways. In general, the benefits of lowering blood pressure, avoiding obesity, quitting smoking, and getting more exercise apply to women as well as men (Wenger et al. 1993), (Kris-Etherton and Krummel 1993). For example, Figure 9.13 shows the increased risk of death from cardiovascular disease relative to women's body mass index.

However, there are several gender differences in risk factors for cardiovascular disease. For example, in premenopausal women, a high total cholesterol is not as strong a risk factor for heart disease in women as in men, while a low HDL-cholesterol is an even greater risk factor (Bush et al. 1988). See Figure 9.14. Therefore, women should emphasize dietary and lifestyle changes that raise HDL-cholesterol and put less emphasis on lowering their total cholesterol.

Diabetes is also a particularly dangerous risk factor in women. Diabetes undercuts the protective effects of estrogen, and premenopausal diabetic women develop heart disease as quickly as diabetic men (AHA 1992).

Women's responses to dietary changes may also differ from men's. For example, in men, low-fat, low-cholesterol, high-fiber diets often promptly lower LDL-cholesterol, but in women (particularly post-menopausal women), they may not affect LDL levels as greatly (Jenkins et al. 1993).

Source: Adapted from Goldberg (1992)

FIGURE 9.13

Women, body mass index, and deaths from cardiovascular disease

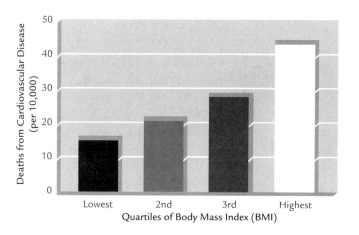

Source: Adapted from
Goldberg (1992)

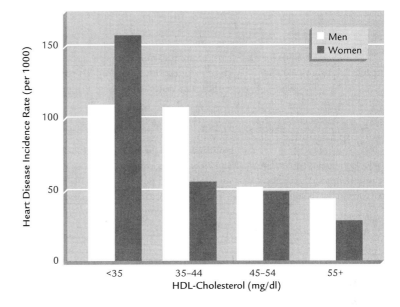

FIGURE 9.14
*Heart disease and
HDL-cholesterol*

Also, women tend to experience greater reductions in HDL-cholesterol on low-fat, lipid-lowering diets (Wood et al. 1991).

As more studies include women as participants, researchers will begin to know more about gender differences in risk factors and responses to diet and lifestyle changes. Until then, women can certainly benefit from following the guidelines for the general population.

Breast Cancer

One out of eight women in the United States will develop breast cancer. For the past 25 years, the incidence of breast cancer has increased steadily, by 1-2% each year (Hankin 1993). Although many more women die of heart disease, stroke, and diabetes each year than from breast cancer, breast cancer is particularly cruel in that it strikes and kills more younger women.

Known risk factors for breast cancer are shown in Table 9.25. However, it is clear that there are other environmental and genetic factors not yet identified, as most women who develop breast cancer have none of these known risk factors (Miller 1992). Dietary factors have been implicated in the development of breast cancer (Hankin 1993), (Howe et al. 1990), (Rohan et al. 1993).

A high-fat diet may increase the risk of breast cancer, but evidence from research has been contradictory. Studies in animals have consistently shown that dietary fat promotes the development of breast cancer (Hankin 1993), but most epidemiological studies in humans have failed to show an association between the two (Willett et al. 1992), (Goodwin and Boyd 1987).

TABLE 9.25

Risk factors for breast cancer

- Older age
- White
- Over 30 years old at birth of first child
- Obesity (postmenopausal women)
- Family history of breast cancer
- Early age menarche
- Late age menopause
- Moderate to heavy alcohol use

Obesity modestly increases risk of breast cancer in postmenopausal women, but not in premenopausal women (Hankin 1993). Women who consume moderate to high amounts of alcohol have an elevated risk of breast cancer (Howe et al. 1991), (Graham 1987) (see Figure 9.15). Low intakes of vitamin A also may increase the risk of the disease (Hunter et al. 1993).

In animal experiments and human studies, diets rich in fiber appear to reduce the risk of breast cancer (Howe et al. 1990). Recently, a large Canadian study followed women for five years to determine whether fiber intake was related to breast cancer risk. The women who ate the most fiber had 30% less breast cancer than the women who ate the least (Rohan et al. 1993). However, diets rich in fiber also tend to be low in fat and high in fruits and vegetables containing antioxidants and many other potential anticarcinogens. Because of this, scientists are unsure whether it is the actual fiber in these diets that is protective, or some other factor.

Further research should answer many of these questions. Until then, a prudent diet for all women, and particularly those at high risk for breast cancer, should emphasize:

1. whole grains, fruits, and vegetables, providing ample dietary fiber (25 g/day);

2. reduction in dietary fat to <30% of total calories;

3. reduction in alcohol use to not more than 1 drink/day

4. weight reduction to ideal body weight, particularly if overweight and postmenopausal.

Osteoporosis

Although many elderly men develop age-related osteoporosis, the disease is five to six times more common in women. Osteoporosis is present in

Source: Adapted from
Longnecker et al. (1988)

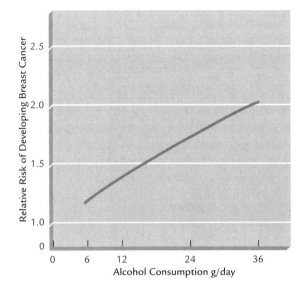

FIGURE 9.15

*Breast cancer and alcohol
consumption*

one in four women over 65 and, by age 85, 40% of women will have suf-fered an osteoporotic bone fracture (Heaney 1993).

Why are women affected more than men? Young men and women have similar bone density, but men have larger bones; so total bone mineral content is about 20% higher in men. As both men and women age, a gradual loss of bone from the skeleton occurs at about the same rates in both sexes. But women start out with less bone and live longer than men—so they have less bone to lose and more time to lose it (Johnston and Slemenda 1992).

The biggest cause of the gender difference in osteoporosis, however, is menopause. At the menopause, the drop in estrogen sharply accelerates bone loss in women, and the loss can be dramatic. Women can lose a third of their bone mass in the decade following menopause (Arnaud and Sanchez 1990).

Dietary and lifestyle factors can strongly influence the risk of osteoporosis. Exercise, a balanced diet, and ample dietary calcium throughout life can help maintain a healthy skeleton (Arnaud and Sanchez 1990). This is discussed in detail in Chapter 10. For women who are lactose intolerant, calcium intake can be increased by following the recommendations in Table 9.26.

Obesity in Women

Obesity is more common in women than men. Gender differences beginning early in life place women on an accelerated course toward increased body fat deposition and weight gain. Twice as many young adult women as men have problems with major weight gain (>20% increase in body

- Try small servings (for example, a small [4 oz] glass) of milk several times daily. Whole milk may be tolerated better than low-fat or skim milk, and taking milk with other foods often helps avoid symptoms.

- Try yogurt containing active, live cultures.

- Try aged hard cheese such as cheddar cheese.

- Try taking lactase tablets or drops when drinking milk. Follow the directions on the package.

- Try lactase-treated milk and milk products.

- Try cultured buttermilk for drinking or baking.

- If calcium intake from foods is inadequate, consider a calcium supplement at the level of the RDA (800 mg).

- Obtain adequate sunlight exposure and vitamin D.

Adapted from Institute of Medicine, Subcommittee for a Clinical Application Guide, *Nutrition during Pregnancy and Lactation: An Implementation Guide* (Washington D.C.: National Academy Press, 1992).

TABLE 9.26

Increasing calcium intake of women who are lactose intolerant

weight). More than a quarter of middle-aged women are overweight, and 10% are severely overweight (St. Jeor 1993).

Obesity greatly increases health risks in women. A woman who is 20 lb overweight doubles her risk for heart disease and triples her risk for diabetes. Obesity also increases the risk of dying from cancer of the uterus, ovaries, and breast (Pi-Sunyer 1991), (St. Jeor 1993), (Garfinkel 1986).

Body fat distribution may be as important as total excess body weight in determining health risks in women. In a study of over 40,000 women, increased accumulation of fat around the abdomen (android obesity) was a stronger predictor of early mortality than overall body weight (Folsom et al. 1993).

Strategies for successful weight loss that are appropriate for both women and men were outlined in the previous chapter in the section on obesity. Women are caught between a physiological tendency toward increased body fat and social pressures that promote thinness as the ideal female form. Women should try to establish realistic and sustainable weight goals because of the substantial health benefits of maintaining ideal or desirable body weight through the adult years (Gillman et al. 1995).

Effects of Alcohol on Women's Health

In general, women are more sensitive to alcohol than men, and they are more likely to develop cirrhosis of the liver and dementia from alcohol abuse. This may be because women have smaller amounts of a stomach enzyme (called alcohol dehydrogenase) that is important in metabolism of alcohol (Frazza et al. 1990).

Alcohol dehydrogenase is found in cells of the stomach and liver, where it breaks down alcohol for energy. Because a woman's stomach has less of the enzyme, it metabolizes incoming ethanol from drinks less quickly, more ethanol reaches the bloodstream, and it circulates longer and at higher levels than in men (Frazza et al. 1990). Earlier in this chapter, the risks and benefits of moderate alcohol intake are discussed. The number of drinks that are considered moderate alcohol use in women (one drink per day) are less than that for men (one to two drinks per day).

THE CHANGING AMERICAN DIET: REVIEW AND SUMMARY

Dietary guidelines began appearing in the early 1970s. Since then we've received a steady stream of recommendations on how to change our diets to lead healthier lives. Are we listening to the experts? Is the American diet changing? Both food consumption surveys and food disappearance data agree—we have changed our diets. Let's look at how we've modified our eating habits since 1970.

We're eating less beef and nearly twice as much chicken and turkey. Egg consumption has fallen, from over 300/person/year in 1970 to 234/person/year in 1989. But we're eating more snack foods, fried foods, salad and cooking oils, and high-fat grain products (biscuits, pancakes, sweet rolls, croissants). Fish consumption continues to be meager—we still eat about six times more beef than fish (USDA 1968-1988).

We have made significant progress toward reducing the fat content of our diets in the past 25 years. The fat content of our diets has fallen from 42 to 34% of total calories, but total fat intake remains higher than recommended levels of <30% of calories (Lewis et al. 1994). And, although consumption of saturated fat has decreased—currently, 12% of total calories come from saturated fat—intake remains about 50% higher than most experts think it should be (NRC 1989). In the period 1988–91 only 21% of the U.S. population met current guidelines for total (<30% of calories) and saturated fat intake (<10% of calories) (Lewis et al. 1994).

We're eating about 20% more complex carbohydrates, mostly from wheat flour and rice. But mean intakes of fiber are estimated to be only 12–16 g/day, which is well below recommended levels of about 25 g/day (NRC 1989).

Fruit and vegetable consumption is rising. We are eating more of some of the healthiest vegetables—twice as much broccoli and cauliflower, and substantially more squash, carrots, and green peppers. However, more than 50% of our intake of fresh vegetables still comes from potatoes and lettuce (USDA 1968-88). Current dietary guidelines recommend consumption of at least five servings of fruits and vegetables each day (NRC 1989). Although current data indicate that the average intake in the U.S. population is four servings of fruits and vegetables per day, less than a third of adults are eating five or more servings per day, and over half of adults do not even obtain one serving of fruit per day (Lewis et al. 1994).

Consumer demand for lower-fat foods is influencing producers. The food industry is introducing many new low-fat products (>1000 new items/year in the 1990s), and they are selling well. Also, consumer demands for leaner meat have led to many farmers raising leaner beef and pork. Overall, it appears we're beginning to change our diets for the better, but we still have a way to go before we reach the levels of consumption recommended by most dietary guidelines.

VEGETARIAN DIETS

The vegetarian approach to eating—a diet high in complex carbohydrates and fiber and low in saturated fat and cholesterol—has substantial health benefits (Dwyer 1988). Studies show that vegetarians have lower risks of heart disease, diabetes, hypertension, and several cancers (particularly cancer of the colon), when compared with nonvegetarians. Also, vegetarians are less likely to be overweight and tend to have lower blood cholesterol levels (Kestin et al. 1989). These health benefits may also be due to lifestyle differences: vegetarians tend to exercise more than nonvegetarians, and many avoid alcohol and smoking.

Vegetarianism is a general term that includes the following:

- *Vegans* are total vegetarians; they avoid all foods of animal origin.

- *Lactovegetarians* include dairy products in their diets.

- *Lacto-ovovegetarians* eat eggs and dairy products but avoid other animal products. Most vegetarians in the United States are lacto-ovovegetarians.

Vegetarian diets, when appropriately planned, are healthful and nutritionally adequate. However, it can be more difficult to get adequate amounts of certain nutrients, listed in Table 9.27. Lacto- and lacto-ovovegetarians choose from most or all of the basic food groups, and their diets are no more likely to be nutritionally deficient than nonvegetarian diets. Vegans must be more careful choosing foods to avoid deficiencies of essential nutrients. In addition to obtaining adequate and complete proteins, vegetarians need to pay particular attention to intakes of five micronutrients: vitamins B_{12} and D and the minerals calcium, iron, and zinc (American Dietetic Assoc. 1988).

Vitamins D and B_{12} are found only in animal products. Adequate exposure to the sun allows most vegetarians to synthesize sufficient vitamin D

Nutrient	Good Vegetarian Sources
Protein	Legumes combined with grains, nuts, or seeds, or any plant food combined with eggs or dairy products
Calcium	Dairy products, dark leafy greens, fortified soy milk, legumes, peanuts, almonds, and seeds
Iron	Legumes, dark leafy greens, torula yeast, dried fruits, whole and enriched grain
Vitamin B_{12}	Dairy products, eggs, nutritional yeast, foods fortified with B_{12}, fermented soy products, supplements
Riboflavin	Dairy products, eggs, whole and enriched grains (if eaten daily), brewer's yeast, dark leafy greens, legume
Vitamin D	Fortified milk, fortified soy milk exposure of skin to sunshine

TABLE 9.27

Nutrients in short supply in vegetarian diets

for their needs. Strict vegans, however, need to consume vitamin B_{12}-fortified foods or a supplement to ensure adequate vitamin B_{12} status.

Rich sources of dietary calcium are hard to find in strict vegan diets. Collard and mustard greens, kale and calcium-fortified tofu are good sources, but the calcium in these foods may not be as efficiently absorbed as the calcium in animal products. Women, because of higher iron requirements, must choose carefully from vegetarian diets to obtain adequate iron. Fortified cereals and breads, legumes, potatoes, and dried fruit are good sources, and vegetarians can enhance iron absorption from these foods by consuming foods rich in vitamin C with meals. Legumes, whole grains, and wheat germ are good sources of zinc for vegetarians. See Table 7.17 for a listing of good vegetarian sources of calcium, iron, and zinc.

Although many plant foods are good sources of protein, the proteins are incomplete. Incomplete proteins contain insufficient amounts of one or more essential amino acids. To obtain all the essential amino acids, vegetarians need to combine complementary plant proteins. When plant foods are properly combined, the amino acids lacking in one food are supplied by the other food, and vice versa. Examples of complementary proteins are provided in the discussion of vegetarian foods in Chapter 2.

A HEALTHFUL DIET FOR ADULTHOOD

Although the nutritional demands of growth and physical development are past, nutrition remains the cornerstone of good health during adulthood. Healthful eating provides nutrients for maintenance and repair and may provide protection from a range of chronic diseases. Reading this chapter, you have probably noticed the similarities among the dietary and lifestyle recommendations for reducing risk from heart disease, cancer, and diabetes. Despite many unanswered questions about the connections between diet and disease, a concensus is emerging on eating habits consistent with good health and long life.

The following recommendations are drawn from dietary guidelines of the American Heart Association (AHA 1988), the National Cancer Institute (NCR 1987), the National Research Council (NRC 1989), and the Surgeon General's Office (USDHHS 1988). Remember that dietary guidelines are aimed at the general population and don't necessarily apply equally to everyone in the population. Variations in age, health, sex, genetics, and environment cause each of us to respond differently to dietary changes and dietary components.

Enjoyment of good food is one of life's great pleasures—and so is enjoyment of good health. The following guidelines allow us to do both.

1. Balance food intake and physical activity to maintain or attain appropriate body weight. Eating a low-fat, high-carbohydrate diet will help, as will exercise. All healthy people should maintain at least a moderate level of physical activity.

2. Eat a variety of foods, balanced among the five major food groups. Consuming a wide variety of foods will ensure intake of all the essential nutrients. It will also limit exposure to pesticides or toxic substances that may be present in one particular food. Specific groups and examples of servings are listed in Table 9.28.

3. Limit total fat intake to 30% of calories or less and saturated fat intake to less than 10% of calories. Intake of saturated fat can be reduced by choosing fish, low-fat or no-fat dairy products, lean meats, and chicken without the skin. Also, to reduce total fat intake, limit consumption of fried foods, mayonnaise, vegetable oils, and salad dressings. Reduce dietary cholesterol to less then 300 mg daily by limiting intake of meats, dairy products, and egg yolks.

4. Increase intake of starches and complex carbohydrates by eating six or more servings of bread, cereals, rice, pasta, and legumes each day. Emphasize whole grains and legumes to obtain adequate dietary fiber (at least 25 g fiber/day). Consume simple sugars in moderation.

5. Eat five or more servings of a variety of vegetables and fruits each day. Emphasize green and yellow vegetables and citrus fruits. These foods are important sources of fiber, micronutrients, and antioxidants. Consumption of ample amounts of these foods may reduce the risk of certain cancers.

6. Maintain protein intake at moderate levels. Protein should compose only about 10–15% of total calories. Try to obtain dietary protein from low-fat sources.

7. Limit daily intake of sodium to less than 2.4 g/day (about the amount in a full teaspoon of table salt). Reduce the amount of salt used in cooking, and avoid heavily salted foods such as canned soups and salted snacks.

8. Limit or avoid alcohol. For those who choose to drink, limit consumption to 1 oz of pure ethanol/day (the amount in two 12 oz beers or two glasses of wine). Pregnant women and women attempting to conceive should avoid alcoholic beverages.

9. Maintain an adequate calcium intake. Ample dietary calcium is important to maintain bone strength and may reduce the risk of hypertension. Try to obtain calcium from low-fat sources, such as low-fat or no-fat dairy products and dark leafy greens.

10. Avoid taking dietary supplements in excess of the RDAs. Vitamin or mineral supplements at or below the RDA are safe but rarely needed.

Food Group	Suggested Servings
Vegetables	3–5 servings
Fruits	2–4 servings
Breads, cereals, rice, pasta	6–11 servings
Milk, yogurt, cheese	2–3 servings
Meats, poultry, fish, dry beans and peas, eggs, nuts	2–3 servings

- One serving of vegetables is 1 cup of raw leafy greens or ½ cup of other vegetables. A ½ cup of cooked dry beans or peas may be counted as either one vegetable or serving or 1 oz of the meat group. Eat dark green leafy and deep yellow vegetables often.

- One serving of fruit is 1 medium piece of fresh fruit, ½ cup of small or diced fruit, or ¾ cup of juice. Have citrus fruits or juices, melons, or berries regularly.

- One serving of grain products is equal to 1 slice of bread; ½ bun, bagel, or English muffin; 1 oz of dry ready-to-eat cereal; ½ cup cooked cereal, rice, or pasta. Emphasize whole grain products daily.

- One serving of milk products is equal to 1 cup of milk or yogurt or 1½ oz of cheese. Choose skim or low-fat milk and fat-free or low-fat yogurt and cheese.

- One serving of meat and other protein products is equivalent to 2 to 3 oz of cooked lean beef, chicken without skin, or fish, or two medium eggs; or 1 cup of cooked dry beans or peas. Trim fat from meat and moderate the use of egg yolks and organ meats.

Modified from U.S. Department of Agriculture, U.S. Department of Health and Human Services, "Preparing Foods and Planning Menus Using the Dietary Guidelines," *Home and Garden Bulletin* 232-8. GPO #125-W. (Hyattsville, MD: U.S. Dept. of Agriculture, 1989).

TABLE 9.28
USDA food guide for adults

REFERENCES

Adams, P. W., et al., "Effect of Pyridoxine Hydrochloride (Vitamin B_6) Upon Depression Associated with Oral Contraception,". *Lancet* 1(1973):897–9.

Alberts, D. S., et al., "Effects of Dietary Wheat Bran Fiber on Rectal Epithelial Cell Proliferation in Patients with Resection for Colorectal Cancers," *J Natl Cancer Inst* 82 (1990):1280.

American Academy of Pediatrics Work Group on Cow's Milk Protein and Diabetes Mellitus, "Infant Feeding Practices and their Possible Relationship to Diabetes Mellitus," *Pediatrics* 94(1994):752–4.

American Cancer Society, *Questionable Nutritional Therapies in the Treatment of Cancer* (Atlanta: American Cancer Society, 1991).

American Dietetic Association, Position of the American Dietetic Association, "Vegetarian Diets," *J Am Diet Assoc* 88(1988):351–55.

American Heart Association, *1993 Heart and Stroke Facts Statistics,* (Dallas: American Heart Association, 1992).

American Heart Association, Nutrition Committee, "Dietary Guidelines for Healthy American Adults. A Statement for Physicians and Health Professionals by the Nutrition Committee, American Heart Association," *Circulation* 77 (1988):721A–24A.

Ames, B. N., Shigenaga, M. K., and Hagan, T. M., "Oxidants, Antioxidants and the Degenerative Diseases of Aging," *Proc Natl Acad Sci* 90 (1993)7915–22.

Anderson, J. W., et al., "Bakery Products Lower Serum Cholesterol Concentrations in Hypercholesterolemic Men," *Am J Clin Nutr,* 54 (1991):836–40.

Anonymous, "Ethanol Stimulates Apo A-1 Secretion in Human Hepatocytes: A Possible Mechanism Underlying the Cardioprotective Effect of Alcohol," *Nutr Rev* 51(1993):151–2.

Anonymous "Health after 50," *Johns Hopkins Medical Letter* (May 1992).

Anonymous, "Regular or Decaf? Coffee Consumption and Serum Lipoproteins," *Nutr Rev* 50 (1992):175–78.

Anonymous,: "Inhibition of LDL Oxidation by Phenolic Substances in Red Wine: A Clue to the French Paradox?" *Nutr Rev* 51(1993):185–87.

Arif, A., and Westermeyer, J., *Manual of Drug and Alcohol Abuse,* (New York: Plenum Publishing, 1989).

Armstrong, R., Doll, R., "Environmental Factors and Cancer Incidence and Mortality in Different Countries, with Special Reference to Dietary Practice," *Int J Cancer* 16 (1975): 617-31.

Arnaud, C. D., and Sanchez, S. D., "The Role of Calcium in Osteoporosis," *Annu Rev Nutr* 10(1990):397–414.

Ascherio, A., and Willett, W., "Metabolic and Atherogenic Effects of *trans* Fatty Acids," *J Intern Med* 238(1995):93–96.

Ascherio, A., et al., "Dietary Intake of Marine n-3 Fatty Acids, Fish Intake, and the Risk of Coronary Disease in Men," *N Engl J Med* 332 (1995):977–82.

Bak, A. A., and Grobbee, D. E., "The Effect on Serum Cholesterol Levels of Coffee Brewed by Filtering or Boiling," *N Engl J Med* 321 (1989):1432–7.

Barr, S. I., Janell, K. C., and Prior, J. C., "Energy Intakes are Higher During the Luteal Phase of Ovulatory Menstrual Cycles," *Am J Clin Nutr* 61(1995):39–43.

Beardsley T., "Trends in Cancer Epidemiology: A War Not Won," *Scientific American* 1994; 270:130-38.

Bell, L. P., et al., "Cholesterol-Lowering Effects of Soluble-Fiber Cereals as Part of a Prudent Diet for Patients with Mild to Moderate Hypercholesterolemia," *Am J Clin Nutr* 52 (1990):1020–26.

Bennett, P. H., "Epidemiology of Diabetes Mellitus," in *Diabetes Mellitus*, 4th ed., edited by H. Rifkin and D. Porte (New York: Elsevier, 1990).

Bjerkedal, T., "Overweight and Hypertension," *Acta Med Scand* 159 (1957):13–26.

Block, G., and Abrams, B., "Vitamin and Mineral Status of Women of Childbearing Potential," *Ann NY Acad Sci* 678 (1993):245–54.

Block, G., Patterson, B., and Subar, A., "Fruit, Vegetables and Cancer Prevention: A Review of the Epidemiological Evidence," *Nutr Cancer* 18 (1992):1–29.

Boffetta, P., Garfinkel, L., "Alcohol Drinking and Mortality Among Men Enrolled in an American Cancer Society Prospective Study," *Epidemiology* 1 (1990) :342-8.

Bonanome, A., and Grundy, S. M., "Effect of Dietary Stearic Acid on Plasma Cholesterol and Lipoprotein Levels," *N Eng J Med* 318 (1988):1244–8.

Borch-Johnsen, K., et al., "Relationship Between Breastfeeding and Incidence Rates of Insulin-Dependent Diabetes Mellitus," *Lancet* ii (1984): 1083–86.

Bouchard, C., Bray, G. A., and Hubbard, V. S., "Basic and Clinical Aspects of Regional Fat Distribution," *Am J Clin Nutr* 52 (1990):946–50.

Boushey, C. J., et al., "A Quantitative Assessment of Plasma Homocysteine Risk Factor for Vascular Disease: Probable Benefits of Increasing Folic Acid Intakes," *JAMA* 274 (1995):1049–57.

Brissonnette, L. G. G. and Fareed, D. S., "Cardiovascular Diseases as a Cause of Death in the Island of Mauritius, 1972–1980," *World Health Stat Q* 38 (1985):163–75.

Bush, T. L., Fried, L. P., and Barrett-Conner, E., "Cholesterol, Lipoproteins, and Coronary Heart Disease in Women," *Clin Chem* 34(1988):B60–70.

Carroll, K. K., "Dietary Fat and Cancer: Specific Action or Caloric Effect?" *Journal of Nutrition* 116(1986):1130.

Carroll, K. K., "Experimental Evidence of Dietary Factors and Hormone Dependent Cancers," *Cancer Res* 35 (1975):3374-83.

Cassileth, B. R., et al., "Survival and Quality of Life among Patients Receiving Unproven as Compared with Conventional Cancer Therapy," *N Engl J Med* 324 (1991):1180–5.

Centers for Disease Control and Prevention, "Classification System for Human T-Lymphotrophic Virus Type III/Lymphadenopathy-Associated Virus Infections," *MMWR* 35(1986):334–39.

Centers for Disease Control, "Recommendations for the Use of Folic Acid to Reduce the Number of Cases of Spina Bifida and Other Neural Tube Defects," *MMWR* 41 (1992).

Clinton, S. K., "Nutrition in the Etiology and Prevention of Cancer," in *Cancer Medicine*, 3d. ed., J. F. Holland, et al. (Philadelphia: Lea & Febiger, 1992).

Colditz, G. A., et al., "Weight Gain as a Risk Factor for Clinical Diabetes in Women," *Ann Intern Med* 122 (1995):481–6.

Consensus Development Conference, "Diagnosis, Prophylaxis and Treatment of Osteoporosis," *Am J Med* 94 (1993):646–50.

Corrigan, S. A., et al., "Weight Reduction in the Prevention and Treatment of Hypertension: A Review of Representative Clinical Trials," *Am J Health Promotion* 5 (1991):208–14.

Dahl, L. K., and Love, R. A., "Etiological Role of Sodium Chloride Intake in Essential Hypertension in Humans," *JAMA* 164(1957):397.

Dalton, K., and Dalton, M. J. T., "Characteristics of Pyrodoxine Overdose Neuropathy Syndrome," *Acta Neurol Scandinav* 76(1987):8–11.

DeCosse, J. J., Miller, H. H., and Lesser, M. L., "Effect of Wheat Fiber and Vitamins C and E on Rectal Polyps in Patients with Familial Adenomatous Polyposis," *J Natl Cancer Inst* 81 (1989):1290.

Devesa, S. S., Donaldson, J., and Fears, T., "Graphical Presentation of Trends in Rates," *Am J Epidemiol* 141 (1995):300–4.

Dietary Guidelines Advisory Committee, "Dietary Guidelines for American, 1995," *Nutr Rev* 53 (1995) : 376-9.

Doll, H., et al., "Pyridoxine (Vitamin B_6) and the Premenstrual Syndrome: A Randomized Cross-Over Trial," *J R Coll Gen Pract* 39(1989):364.

Doll, R., "The Lessons of Life: Keynote Address to the Nutrition and Cancer Conference," *Cancer Res* 52 (1992):2024s–29s.

Doll, R., et al., "Mortality in Relation to Consumption of Alcohol: 13 Years Observations on Male British Doctors," *BMJ* 309(1994):911–18.

Dowling, S., Mulcahy, F., and Gibney, M. J., "Nutrition in the Management of HIV Antibody Positive Patients: A Longitudinal Study of Dietetic Outpatient Advice," *Eur J Clin Nutr* 44(1990):823–29.

Drewnowski, A., et al., "Diet Quality and Dietary Diversity in France: Implications for the French Paradox," *J Am Diet Assoc* 96 (1996):663-9.

Drife, J., "Complications of Oral Contraception," in *Contraception: Science and Practice*, eds. M. Filshie and J. Guillebaud (London: Butterworths, 1989),39–51.

Dustan, H. P., and Kirk, K. A., "Corcoran Lecture: The Case for or Against Salt in Hypertension," *Hypertension* 13 (1989):696–705.

Dwyer, J. T., "Health Aspects of Vegetarian Diets," *Am J Clin Nutr* 48 (1988):712–38.

Eastwood, M. A., "The Physiological Effect of Dietary Fiber," *Annu Rev Nutr* 12 (1992):19–29.

Farley, D., "High Blood Pressure: Controlling the Silent Killer, *FDA Consumer* (December 1991):28–33.

Feinman, L., and Lieber, C. S., "Nutrition and Diet in Alcoholism," in *Modern Nutrition in Health and Disease*, 8th ed., edited by M. E. Shils, J. A. Olson, and M. Shike (Philadelphia: Lea & Febiger, 1994).

"Fish Oil Supplements," *FDA Consumer* (October 1990): 32.

Folsom, A.R., et al., "Body Fat Distribution and 5-year Risk of Death in Older Women," *JAMA* 269(1993):483–87.

Frankel, E. N., et al., "Inhibition of Oxidation of Human LDL by Phenolic Substances in Red Wine," *Lancet* 341(1993):454–7.

Frezza, M., et al., "High Blood Alcohol Levels in Women: The Role of Decreased Gastric Alcohol Dehydrogenase Activity and First-Pass Metabolism," *N Engl J Med* 322(1990):95–99.

Fried, R. E., et al., "The Effect of Filtered-Coffee Consumption on Plasma Lipid Levels. Results of a Randomized Clinical Trial," *JAMA* 267 (1992):811–5.

Friedman, L. A., and Kimball, A. W., "Coronary Heart Disease Mortality and Alcohol Consumption in Framingham," *Am J Epidemiol* 24(1986):481–9.

Garfinkel, L., "Overweight and Cancer," *Ann Intern Med* 103 (1985): 1034–6.

Gaziano, J. M., et al., "Moderate Alcohol Intake, Increased Levels of High-Density Lipoprotein and its Subfractions, and Decreased Risk of Myocardial Infraction," *N Engl J Med* 329 (1993):1829–34.

Gey, K. F., et al., "Inverse Correlation between Plasma Vitamin E and Mortality from Ischemic Heart Disease in Cross-Cultural Epidemiology, *Am J Clin Nutr Suppl* 53 (1991): 326–34.

Gillman, M. W., et al., "Protective Effect of Fruits and Vegetables on Development of Stroke in Men," *JAMA* 273(1995):1113–17.

Giovanucci, E., et al., "A Prospective Study of Dietary Fat and Risk of Prostate Cancer," *JNCI* 85 (1993):1571–9.

Goldberg, R. J., "Coronary Heart Disease: Epidemiology and Risk Factors," in *Prevention of Coronary Heart Disease*, eds. I. S. Ockene, J.K. Ockene, (Boston: Little, Brown, 1992).

Goldberg, R. J., "Temporal Trends and Declining Mortality Rates from Coronary Heart Disease in the United States," in *Prevention of Coronary Heart Disease*, eds. I. S. Ockene, J.K. Ockene, (Boston: Little, Brown, 1992).

Goodwin, P. F., and Boyd, N. F., "Critical Appraisal of the Evidence that Dietary Fat Intake is Related to Breast Cancer Risk in Humans," *J Natl Cancer Inst* 79(1987):473–85.

Gordon, D. J., et al., "High Density Lipoprotein Cholesterol and Cardiovascular Disease. Four Prospective American Studies," *Circulation* 79 (1989):8–15.

Graham, S., "Alcohol and Breast Cancer," *N Engl J Med* 316(1987):1211–13.

Grundy, S. M., "Cholesterol and Coronary Heart Disease: Future Directions," *JAMA* 264 (1990):3053–59.

Grundy, S. M., and Vega, G. L., "Plasma Cholesterol Responsiveness to Saturated Fatty Acids," *Am J Clin Nutr* 47 (1988):822–4.

Grundy, S. M., et al., "Workshop on the Impact of Dietary Cholesterol on Plasma Lipoproteins and Atherogenesis," *Arteriosclerosis* 8 (1988):95–100.

Haenszel, W., "Migrant Studies," in *Cancer Epidemiology and Prevention,* eds. by D. Schottenfeld and J. F. Fraumeni (Philadelphia: W. B. Saunders, 1982), 194.

Handa, K., et al., "Alcohol Consumption, Serum Lipids and Severity of Angiograpically Determined Coronary Artery Disease," *Am J Cardio* 65(1990):287–9.

Hankin, J. H., "Role of Nutrition in Women's Health: Diet and Breast Cancer," *J Am Diet Assoc* 93(1993):994–8.

Haskell, W. L., "The Influence of Exercise Training on Plasma Lipids and Lipoproteins in Health and Disease," *Acta Med Scand* 711, Suppl. (1986):25–37.

Heaney, R. P., "Nutritional Factors in Osteoporosis," *Annu Rev Nutr* 13 (1993):287–316.

Helmrich, S. P., et al., "Physical Activity and Reduced Occurrence of Non-Insulin-Dependent Diabetes Mellitus," *New Engl J Med* 325 (1991):147–52.

Hepner, G. W., and Roginsky, M., "Abnormal Metabolism of Vitamin D in Patients with Cirrhosis," *Clin Res* 23(1975):322.

Howe, G., et al., "The Association Between Alcohol and Breast Cancer Risk: Evidence from the Combined Analysis of Six Dietary Case-Control Studies," *Int J Cancer* 47(1991):707–10.

Howe, G. R., et al., "Dietary Factors and Risk of Breast Cancer: Combined Analysis of 12 Case-Control Studies," *J Natl Cancer Inst* 82 (1990):561–69.

Howe, G. R., et al., "Dietary Intake of Fiber and Decreased Risk of Cancer of the Colon and Rectum: Evidence from the Combined Analysis of 13 Case-Control Studies," *J Natl Cancer Inst* 24 (1992):1887–96.

Hunter, D. J., et al., "A Prospective Study of the Intake of Vitamins C,E, and A and the Risk of Breast Cancer," *N Engl J Med* 329 (1993):234–40.

Jacob, R. A., et al., "Biochemical Markers for Assessment of Niacin Status in Young Men: Urinary and Blood Levels of Niacin Metabolites," *J Nutr* 119 (1989):591–98.

Jenkins, D. J. A., et al., "Effect on Blood Lipids of Very High Intakes of Fiber in Diets Low in Saturated Fat and Cholesterol," *N Engl J Med* 329(1993):21–26.

Johnson, B. F., "Intervention for the Control of Hypertension," in *Prevention of Coronary Heart Disease*, eds. I. S. Ockene and J. K. Ockene (Boston: Little, Brown, 1992).

Johnston, C. C., Slemenda, C. W., "Changes in Skeletal Tissue During the Aging Process," *Nutr Rev* 50(1992):385–87.

Joint National Committee, "The 1988 Report of the Joint National Committee on Detection, Evaluation and Treatment of High Blood Pressure," *Arch Int Med* 148 (1988):1023–38.

Kalkhoff, R. K., et al., "Relationship of Body Fat Distribution to Blood Pressure, Carbohydrate Intolerance, and Plasma Lipids in Healthy Obese Women," *J Lab Clin Med* 102(1983):621–7.

Kallner, A., "Requirement for Vitamin C Based on Metabolic Studies," *Ann N. Y Acad Sci* 498 (1987):418–23.

Kannel, W. B., Gordon, T., and Castelli, W. P., "Obesity, Lipids, and Glucose Intolerance: The Framingham Study," *Am J Clin Nutr* 32 (1979):1238–45.

Karjalainen, J., et al., "A Bovine Albumin Peptide as a Possible Trigger of Insulin-Dependent Diabetes Mellitus," *N Engl J Med* 327(1992):302–7.

Kassarjian, Z., and Russell, R. M., "Hypochlorhydria: A Factor in Nutrition," *Annu Rev Nutr* 9 (1989):271–4.

Kesaniemi, Y.A., and Grundy, S. M., "Increased Low Density Lipoprotein Production Associated with Obesity," *Arteriosclerosis* 3 (1983):170–7.

Kestin, M., et al., "Cardiovascular Disease Risk Factors in Free Living Men: Comparison of Two Prudent Diets, One Based on Lacto-ovovegetarianism and the Other Allowing Meat," *Am J Clin Nutr* 50 (1989):280–87.

Khaw, K. T., and Barrett-Connor, E., "Dietary Potassium and Stroke-Associated Mortality: A 12-year Prospective Population Study," *New Engl J Med* 316 (1987):235–40.

King, M., Rewers, M., "Global Estimates for Prevalence of Diabetes Mellitus and Impaired Glucose Tolerance in Adults," *Diabetes Care* 16 (1993): 157-77.

Kleijnen, J., Ter Riet, G., and Knipschild, P., "Vitamin B_6 in the Treatment of the Premenstrual Syndrome: A Review," *Br J Obstet Gynecol* 97(1990):847–52.

Krakoff, L. R., "Is Reduction of Dietary Salt a Treatment for Hypertension?" *Am J Hyperten* 4 (1991):481–2.

Kris-Etherton, P. M., and Krummel, D., "Role of Nutrition in the Prevention and Treatment of Coronary Heart Disease in Women," *J Am Diet Assoc* 93(1993):987–93.

Kritchevsky, D., and Klurfeld, D. M., "Dietary Fiber and Cancer," in *Human Nutrition: A Comprehensive Treatise, Volume 7, Cancer and Nutrition,* eds. by R. B. Alfin-Slater and D. Kritchevsky (New York: Plenum Press, 1991), 211.

Kromann, N., and Green, A., "Epidemiological Studies in the Upernavik District, Greenland: Incidence of Some Chronic Diseases 1950-1974," *Acta Med Scand* 208 (1980):401–6.

Kromhout, D., Bosschieter, E. B., and de Lezenne Coulander, "The Inverse Relation Between Fish Consumption and 20-year Mortality from Coronary Heart Disease," *N Engl J Med* 312 (1985):1205–9.

Lathia, D., and Blum, A., "Role of Vitamin E as Nitrite Scavenger and N-Nitrosamine Inhibitor: A Review," *Internat J Vit Nutr Res* 59 (1990):430.

Law, M. R., Frost, C. D., and Wald, N. J., "By How Much Does Dietary Salt Reduction Lower Blood Pressure? I: Analysis of Observational Data Among Populations," *Br Med J* 302 (1991):811–5.

Leaf, A., and Weber, P. C., "Cardiovascular Effects of n-3 Fatty Acids," *N Engl J Med* 318 (1988):549–57.

Leake, C. D., and Silverman, M., "The Chemistry of Alcoholic Beverages," in *Biology of Alcoholism,* eds. B. Kissin and H. Begleiter (New York: Plenum Press, 1974).

Leibel, R. L., Edens, N. K., and Fried, S. K., "Physiological Basis for the Control of Body Fat Distribution in Humans," *Ann Rev Nutr* 9 (1989):417–43.

Leklem, J. E., "Vitamin B_6 Requirement and Oral Contraceptive Use: A Concern?" *J Nutr* 116(1986):475–77.

Lerner, D. J., and Kannel, W. B., "Patterns of Coronary Artery Disease Mortality in the Sexes: A 26-Year Follow-Up of the Framingham Population," *Am Heart J* 111(1986):383–90.

Lewis, C. J., et al., "Healthy People 2000: Report on the 1994 Nutrition Progress Review," *Nutr Today* 29(1994):6–14.

Lichtenstein, A., "Trans Fatty Acids, Blood Lipids and Cardiovascular Risk: Where Do We Stand?" *Nutr Rev* 51 (1993):340–3.

Life Sciences Research Office, FASEB, "Nutrition Monitoring in the U. S. An Update Report on Nutrition Monitoring," DHHS publ no. (PHS) 89 1255 (Hyattsville, MD: U. S. Depts. HHS, USDA, Sept 1989).

Longnecker, M. P., et al., "A Meta-analysis of Alcohol Consumption in Relation to Rate of Breast Cancer," *JAMA* 652 (1988): 260.

Luc, G., and Fruchart, J., "Oxidation of Lipoproteins and Atherosclerosis, *Am J Clin Nutr Suppl* 53 (1991):206–09.

MacGregor, G. A., et al., "Moderate Potassium Supplementation in Essential Hypertension," *Lancet* (1982):567–70.

MacMahon, S. W., and Norton, R. N., "Alcohol and Hypertension: Implications for Prevention and Treatment," *Ann Intern Med*; 105 (1986):124–6.

Manson, J. E., et al., "Antioxidant Vitamin Consumption and Incidence of Stroke in Women," *Circulation* 87 (1993):678.

Matkovic, V., et al., "Factors that Influence Peak Bone Mass Formation: A Study of Calcium Balance and the Inheritance of Bone Mass in Adolescent Females," *Am J Clin Nutr* 52 (1990):878–88.

McCarron, D. A., "Calcium Nutrition and Hypertensive Cardiovascular Risk in Humans," *Clin Appl Nutr* 2 (1992):45–66.

McCarron, D. A., et al., "Dietary Calcium and Blood Pressure: Modifying Factors in Specific Populations," *Am J Clin Nutr Suppl* 54 (1991):215–19.

McGinnis, J. M., and Lee, P. R., "Healthy People 2000 at Mid Decade," *JAMA* 273 (1995):1123–29.

McIntosh, G. H., et al., "Barley and Wheat Foods: Influence on Plasma Cholesterol Concentrations in Hypercholesterolemic Men," *Am J Clin Nutr* 53 (1991):1205–09.

McKinley, M. J., et al., "Improved Body Weight Status as a Result of Nutrition Intervention in Adult, HIV-Positive Outpatients," *JADA* 94(1994):1014–17.

McNamara, D. J., "Effects of Fat-Modified Diets on Cholesterol and Lipoprotein Metabolism," *Annu Rev Nutr* 7 (1987):273–90.

McNamara, D. J., et al., "Heterogeneity of Cholesterol Homeostasis in Man: Response to Changes in Dietary Fat Quality and Cholesterol Quantity," *J Clin Invest* 79 (1987):1729–39.

Medalie, J. H., et al., "Major Factors in the Development of Diabetes Mellitus in 10,000 Men," *Arch Intern Med* 135 (1975):811–7.

Mensink, R. P., and Katan, M. B., "Effect of a Diet Enriched with Monounsaturated or Polyunsaturated Fatty Acids on Levels of Low-Density and High-Density Lipoprotein Cholesterol in Healthy Women and Men," *N Engl J Med* 321 (1989):436–41.

Mensink, R. P., and Katan, M. B., "Effect of Dietary Trans Fatty Acids on High-Density and Low-Density Lipoprotein Cholesterol Levels in Healthy Subjects," *N Engl J Med* 323 (1990):439–45.

Merrill, A., "AIDS and Malnutrition: Dual Assaults on the Body," *Home Healthcare Nurse* 13 (1995):56–60.

Miller, B. A., et al., eds. *Cancer Statistics Review: 1973–1989* (Bethesda, MD: National Cancer Institute, 1992).

Morris, C. D., et al. "The Epidemiology of Dietary Calcium and Hypertension," in *Nutrition and Blood Pressure.* Bursztyn, P., ed. (London: John Libbey and Co., 1987).

Nace, E. P., *The Treatment of Alcoholism* (New York: Brunner/Mazel, 1987):19–32.

National Academy of Sciences, Food and Nutrition Board, *Recommended Dietary Allowances*, 10th ed. Washington D. C.: National Academy Press, 1989.

National Center for Health Statistics, "Annual Summary of Births, Deaths, Marriages and Divorces, United States, 1993," *Monthly Vital Statistics Report,* 42 no. 13 (Washington, D. C.: National Center for Health Statistics, October 11, 1994).

National Cholesterol Education Program Expert Panel, "Report on Detection, Evaluation and Treatment of High Blood Cholesterol in Adults," *Arch Int Med* 148 (1988):36–69.

National Cholesterol Education Program NCEP, "Report of the Expert Panel on Blood Cholesterol in Children and Adolescents," *Pediatrics* suppl (1992):525–84.

National Institutes of Health, Public Health Service, U. S. Dept. Health and Human Services, National Cancer Institute, "Diet, Nutrition, and Cancer Prevention: A Guide to Food Choices," NIH Pub. no. 87-28-78 (Washington, D. C.: U. S. Government Printing Office, 1987).

National Research Council, Food and Nutrition Board, *Diet and Health: Implications for Reducing Chronic Disease Risk,* report of the Committee on Diet and Health (Washington, D. C.: National Academy Press, 1989).

Nerad, J. L., "Nutritional Aspects of HIV Infection," *Inf Dis Clin of NA* 8 (1994):499–515.

Nicolosi, R. J., and Schaefer, E. J., "Pathobiology of Hypercholesterolemia and Atherosclerosis: Genetic and Environmental Determinants of Elevated Lipoprotein Levels," in *Prevention of Coronary Heart Disease,* eds. I. S. Ockene, and J. K. Ockene (Boston: Little, Brown 1992).

Norell, S. E., et al., "Fish Consumption and Mortality from Coronary Heart Disease," *Br Med J* 293(1986):426.

Norton, J. A., Peacock, J. L., and Morrison, S. D., "Cancer Cachexia," *CRC Crit Rev Oncol Hematol* 7 (1987):289.

Ornish, D., et al., "Can Lifestyle Changes Reverse Coronary Heart Disease? The Lifestyle Heart Trial," *Lancet* 336 (1990):129–33.

Pancharuniti, N., et al., "Plasma Homocysteine, Folate, and Vitamin B_{12} Concentrations and Risk of Early-Onset Coronary Artery Disease," *Am J Clin Nutr* 59 (1994):940–8.

Pariza, M. W., "Dietary Fat and Cancer Risk: Evidence and Research Needs," *Ann Rev Nutr* 8 (1988):167.

Pi-Sunyer, P. X., "Health Implications of Obesity," *Am J Clin Nutr* 53 (1991):1595S–603S.

Pierson, H. F., "Diet, Chemoprevention and Cancer," in *Cancer Medicine,* 3d. ed., edited by J. F. Holland et al. (Philadelphia: Lea & Febiger, 1992).

Posner, B. M., et al., "Nutrition and the Global Risk for Chronic Disease: The INTERHEALTH Nutrition Initiative," *Nutr Rev* 52 (1994):201–07.

Reddy, B. S., "Diet and Colon Cancer: Evidence from Human and Animal Model Studies," in *Diet, Nutrition, and Cancer: A Critical Evaluation. Vol. 1. Macronutrients and Cancer,* eds. B. S. Reddy and L. A. Cohen (Boca Raton: CRC Press,1986), 47.

Reid, R. L., "Premenstrual Syndrome," *N Engl J Med* 324(1991):1208–10.

Reusser, M. E., and McCarron, D. A., "Micronutrient Effects on Blood Pressure Regulation," *Nutr Rev* 52 (1994):367–75.

Rimm, E. B., et al., "Prospective Study of Alcohol Consumption and Risk of Coronary Disease in Men," *Lancet* 338 (1991):464–8.

Rimm, E. B., et al., "Vitamin E Consumption and the Risk of Coronary Heart Disease in Men," *N Engl J Med* 328 (1993):1450–6.

Rock, C. L., et al., "Update on the Biological Characteristics of the Antioxidant Micronutrients: Vitamin C, Vitamin E and the Carotenoids," *J Am Diet Assoc* 96 (1996):693-702.

Rogers, A. E., and Conner, M. W., "Interrelationships of Alcohol and Cancer," in *Human Nutrition: A Comprehensive Treatise, Volume 7, Cancer and Nutrition,* eds, R. B. Alfin-Slater, D. Kritchevsky (New York: Plenum Press, 1991), 51-68.

Rogers, A. E., and Longnecker, M. P., "Biology of Disease. Dietary and Nutritional Influences on Cancer: A Review of Epidemiological and Experimental Data," *Lab Invest* 59 (1988):729.

Rohan, T. E., et al., "Dietary Fiber, Vitamins A, C, and E, and Risk of Breast Cancer: A Cohort Study," *Cancer Causes Control* 4 (1993): 29–37.

Salaspuro, M., "Nutrient Intake and Nutritional Status in Alcoholics," *Alcohol Alcoholism* 28 (1993):85–8.

Sanchez-Castillo, C. P., Branch, W. J. and James, W. P., "A Test of the Validity of the Lithium-Marker Technique for Monitoring Dietary Sources of Salt in Men," *Clin Sci* 72 (1987):87–94.

Schultz, T. D., and Leklem, J. E., "Urinary 4-Pyridoxic Acid, Urinary Vitamin B-6 and Plasma Pyridoxal Phosphate as Measures of Vitamin B-6 Status and Dietary Intake of Adults," in *Methods in Vitamin B-6 Nutrition: Analysis and Status Assessment,* eds. J. E. Leklem and R. D. Reynolds (New York: Plenum Press, 1981), 297–320.

Selhub, J., et al., "Vitamin Status and Intakes as Primary Determinants of Homocysteinemia in an Elderly Population," *JAMA* 270 (1993):2693–8.

Sharp, D. S., et al., "Delayed Health Hazards of Pesticide Exposure," *Ann Rev Pub Health* 7(1986):441–74.

Simopoulos, A. P., "Omega-3 Fatty Acids in Health and Disease and in Growth and Development," *Am J Clin Nutr* 54 (1991):438–63

St Jeor, S. T., "The Role of Weight Management in the Health of Women," *J Am Diet Assoc* 93(1993):1007–12.

Stamler, J., "Dietary Salt and Blood Pressure," *Ann NY Acad Sci* 676 (1993):122–56.

Stamler, J., et al., "INTER-SALT Study Findings. Public Health and Medical Care Implications," *Hypertension* 14 (1989):570–7.

Stampfer, M. J., et al., "A Prospective Study of Moderate Alcohol Consumption and the Risk of Coronary Disease and Stroke in Women," *N Eng J Med* 319(1988):267–73.

Stampfer, M. J., et al., "Vitamin E Consumption and the Risk of Coronary Disease in Women," *N Engl J Med* 328 (1993):1444–9.

Steinberg, D., "Antioxidant Vitamins and Coronary Heart Disease," *N Engl J Med* 328 (1993):1487–89.

Stryd, R. P., Gilbertson, T. J., and Brunden, M. N., "A Seasonal Variation Study of 25-Hydroxyvitamin D3 Serum Levels in Normal Humans," *J Clin Endocrinol Metab* 48 (1979):771–75.

Superko, H. R., and Haskell, W. L., "The Role of Exercise Training in the Therapy of Hyperlipoproteinemia," *Cardiol Clin* 5 (1987):285–310.

Superko, H. R., et al., "Caffeinated and Decaffeinated Coffee Effects on Plasma Lipoprotein Cholesterol, Apolipoproteins, and Lipase Activity: A Controlled, Randomized Trial," *Am J Clin Nutr* 54 (1991):599–605.

Suttie, J. W., et al., "Vitamin K Deficiency from Dietary Vitamin K Restriction in Humans," *Am J Clin Nutr* 47 (1988):475–80.

The Trials of Hypertension Prevention Collaborative Research Group, "The Effects of Non-Pharmacologic Interventions on Blood Pressure of Persons with High Normal Levels: Results of the Trials of Hypertension Prevention, Phase I," *JAMA* 267 (1992):1213–20.

Truswell, A. S., and Beynen, A. C., "Dietary Fibre and Plasma Lipids: Potential for Prevention and Treatment of Hyperlipids," in *Dietary Fibre: A Component of Food*, eds. T. F. Schweizer, C. A. Edwards (London: Springer Verlag, 1992).

U. S. Department of Agriculture "Nationwide Food Consumption Survey. Continuing Survey of Food Intakes by Individuals: Women 19–50 Years and Their Children 1–5 Years, 4 Days, 1985," *Human Nutrition Information Service Nutrition Monitoring Division* report no. 85–4. (Hyattsville, MD: U. S. Department of Agriculture, 1987).

U. S. Department of Agriculture, Human Nutrition Information Service "Food Disappearance Data," (Hyattsville, MD: USDA, 1968–1988).

U. S. Department of Agriculture, Human Nutrition Information Service, Nutrition Division, "*Nationwide Food Consumption Survey,*" (Hyattsville, MD: U. S. Department of Agriculture, 1984).

U. S. Department of Health and Human Services, *The Surgeon General's Report on Nutrition and Health*, (Washington, D. C.: Government Printing Office, 1988).

Virtanen, S. M. et al., "Childhood Diabetes in Finland Study Group. Infant Feeding in Finnish children < 7 Yr of Age with Newly Diagnosed IDDM," *Diabetes Care* 14(1991):415–17.

Webb, A. R., Kline, L. and Holick, M. F., "Influence of Season and Latitude on the Cutaneous Synthesis of Vitamin D3: Exposure to Winter Sunlight in Boston and Edmonton Will Not Promote Vitamin D3 Synthesis in Human Skin," *J Clin Endocrinol Metab* 67 (1988):373–78.

Wenger, N. K., "Exclusion of the Elderly and Women from Coronary Trials: Is Their Quality of Care Compromised?" *JAMA* 268(1992):1460–1.

Wenger, N. K., Speroff, L., and Packard, B., "Cardiovascular Health and Disease in Women," *N Engl J Med* 329(1993):247–56.

Willett, W. C., et al., "Dietary Fat and Fiber in Relation to Risk of Breast Cancer," *JAMA* 268(1992):2037–44.

Willett, W. C., et al., "Intake of Trans Fatty Acids and Risk of Coronary Heart Disease Among Women," *Lancet* 341 (1993):581–5.

Willett, W. C., et al., "Relation of Meat, Fat, and Fiber Intake to the Risk of Colon Cancer in a Prospective Study Among Women," *N Engl J Med* 323 (1990):1664.

Willett, W. C., et al., "Weight, Weight Change and Coronary Heart Disease in Women. Risk within the 'Normal' Weight Range," *JAMA* 273 (1995):461–5.

Williams, G. H., and Moore, T. J., "Hormonal Aspects of Hypertension," in *Endocrinology*, 3d. ed., edited by L. J. DeGroot (Philadelphia: W. B. Saunders, 1995).

Witteman, J. C. M., et al., "A Prospective Study of Nutritional Factors and Hypertension Among U. S. Women," *Circulation* 80 (1989):1320–27.

Wood, P. D., et al., "Changes in Plasma Lipids and Lipoproteins in Overweight Men During Weight Loss Through Dieting as Compared with Exercise," *N Engl J Med* 319 (1988):1173–9.

Wood, P. D., et al., "The Effects on Plasma Lipoproteins of a Prudent Weight-Reducing Diet, with or without Exercise, in Overweight Men and Women," *New Engl J Med* 325 (1991):461–66.

Work Study Group on Diet, Nutrition, and Cancer, "American Cancer Society Guidelines on Diet, Nutrition, and Cancer," *CA Cancer J Clin* 41 (1991):334–8.

World Health Organization, "Diet, Nutrition and the Prevention of Chronic Diseases: Report of a WHO Study Group," *Technical Report* 797 (Geneva: WHO,1990).

World, M. J., Ryle, P. R., and Thomson, A. D., "Alcoholic Malnutrition and the Small Intestine," *Alcohol* 20(1985):89–124.

Ziegler, R. G., "Vegetables, Fruits, and Carotenoids and the Risk of Cancer, *Am J Clin Nutr Suppl* 53 (1991):251–59.

10

nutrition and
older adulthood

▼

NUTRITION, POPULATION GROWTH, AND LIMITED RESOURCES

demographics: the
statistics and trends
dealing with population
change, including
matters of health,
disease, births, and
mortality

longevity: duration
of life

The major industrialized countries, including the United States, have
entered a period in which their populations are undergoing dramatic
demographic aging. The proportion of older people relative to younger
people is increasing, due to increases in **longevity** and a declining birth
rate. While the total U.S. population is expected to grow by 19% between
1990 and 2050, the number of people over 65 will increase nearly 120%,
and the number over 85 will increase over 400%. By 2050, nearly 1 in 4
people in the United States will be older than 65, and 1 in 20 will be over
85 (see Figure 10.1). Contrast this with earlier in this century, when fewer
than 1 in 25 people were over 65 (U.S. Dept. of Commerce 1990).

This massive demographic shift will create many challenges. Society
will need to define new roles for older adults, as increasing numbers of
vigorous and independent elderly remain productive into their 70s and

Source: Adapted from U. S. Department of Commerce (1990)

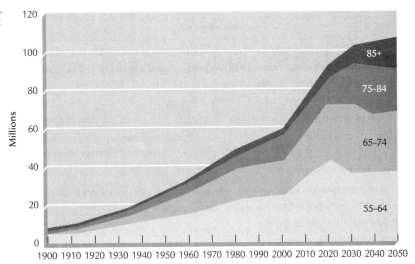

FIGURE 10.1

Growing numbers of Americans over 65

80s. At the same time, because of the rapid growth of those over 85, there will be many more people who are chronically ill and institutionalized. They will strain already overburdened services and greatly increase costs for health care (U.S. Senate 1990). By 2025, some forecasters warn, we could have 10 million people in nursing homes in the United States alone, and taking care of them could cost over $150 billion yearly.

Where will we find the resources to care for the burgeoning social, nutritional, and physical needs of older adults in the twenty-first century? The number of people receiving benefits from age-related entitlement programs such as Social Security and Medicare will grow significantly. In the United States today there are about five people working to support each person over 65; this number will drop to only two people by 2030 (Friedland 1990). Funding for the health care of our aging population will be an enormous challenge, increasing the importance of preventive health measures that will produce a healthy, independent older population.

A major theme of this book is the connection between poor dietary choices and the chronic degenerative diseases such as heart disease, osteoporosis, and diabetes—diseases that predominantly afflict older adults and contribute substantially to health-care costs (see Table 10.1). Optimal nutrition is a cornerstone in the prevention of these diseases, and it is central to the current public policy debate over exploding health-care costs. This chapter examines the nutritional needs of the older adult (those over age 65) and the challenges of providing optimal nutrition to the increasing numbers within this group. It shows that both aging and poor nutrition contribute to the degenerative diseases that affect the elderly and reveals how these diseases grow more significant to the elderly and society with each passing year.

Condition	Prevalence in adults over 65 (%)
Varicose veins	7
Visual impairment	7
Diabetes	9
Chronic sinusitis	14
Skeletal impairment	14
Cataract	16
Heart disease	24
Hearing impairment	25
Hypertension	36
Arthritis	44

Source: U. S. Senate (1990)

TABLE 10.1

Top ten chronic conditions in the U.S.

AGING

LIFESPAN AND LIFE EXPECTANCY

In 1995 a woman in southern France, Jeanne Calment, celebrated her 120th birthday. Mrs. Calment holds the record for documented human longevity and has lived about as long as a human can reasonably hope to live. (The human **lifespan** is estimated to be about 120 years.)

Life expectancy is defined as the average length of life for people in a specific population. In the United States, life expectancy from birth has increased markedly this century. It has risen from 47.3 years in 1900 to 75.5 years in 1990 and is expected to reach 81.2 by 2080 (U.S. Dept. of Commerce 1990). These increases are due to a number of factors, including improvements in living conditions, better health care, decreased mortality from infectious diseases, and improved nutritional health (Kinsella 1992). Table 10.2 shows life expectancies in 15 industrialized countries.

These increases in life expectancy are admirable, but the goal of health education and health care is not just to prolong life: it is to combine good health with a longer life. Increasing disability-free life expectancy is more important than simply increasing life expectancy. Delaying the age at which significant morbidity occurs in older adults assures more healthy, active, and independent years. **Compression of morbidity** is a major goal of preventive health care today, and nutrition plays a central role in achieving that goal. See Figure 10.2.

lifespan: the maximum length of time an organism can be expected to live

compression of morbidity: using good health practices to delay the onset of disability caused by chronic disease

Rank	Nation	Total Population	Men	Women
1	Japan (1990)	78.9	75.9	81.8
2	Iceland (1989–90)	78.0	75.7	80.3
3	Sweden (1990)	77.6	74.8	80.4
4	Switzerland (1989–90)	77.4	74.0	80.8
5	Netherlands (1990)	77.0	73.8	80.1
6	Australia (1990)	77.0	73.9	80.0
7	Norway (1990)	76.6	73.4	79.8
8	Canada (1985–87)	76.4	73.0	79.7
9	France (1987)	76.2	72.0	80.3
10	Germany, F.R. (1986–88)	75.5	72.2	78.7
11	United States (1990)	75.4	72.0	78.8
12	Finland (1989)	75.0	70.9	78.9
13	Denmark (1989–90)	74.9	72.0	77.7
14	United Kingdon (1985–87)	74.8	71.9	77.6
15	New Zealand (1987–89)	74.6	71.6	77.6

From Metropolitan *Life Insurance Company Statistical Bulletin*, July–September 1992.

TABLE 10.2
Comparative life expectancies in industrialized countries

THEORIES OF AGING

Aging begins at birth and refers to changes in body structure and function that occur during growth, maturation, and senescence. It is often used to describe the time-dependent biological and physiological changes that begin in humans around the age of 30 and are degenerative. The sequence and pattern of changes in aging are much the same from person to person, but the rates at which these changes occur can vary greatly among individuals.

As our bodies grow older, our tissues age in several different ways. In some tissues, such as the skin or the intestine, cells are able to divide and replace themselves throughout life—although the rate of division slows and more cells stop dividing and die as the body ages. Other tissues, such as brain tissue, have special cells that do not divide. These long-lived cells become less efficient as they age: they are less able to repair or maintain themselves, less able to synthesize important proteins and enzymes needed for cell metabolism, and are slower to rid themselves of waste products.

Cells in organs throughout the body decline over many years before changes begin to affect the functioning of organs. Many of our organs have substantial reserve capacities. Reserve capacity is the extent to which an organ can maintain its function despite the loss of cells or cell activity, and it is important in maintaining health in the aging body.

FIGURE 10.2

Compression of morbidity: the aim of preventive measures is to shift the proportion of the population without disability toward the survival curve, so that fewer survivors are sick or disabled

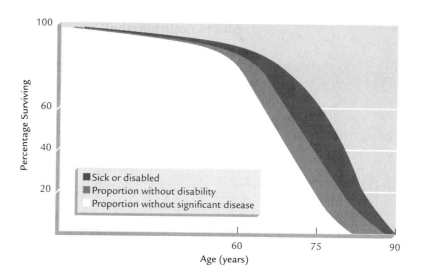

Let's use the kidney as an example. The kidney's ability to filter the bloodstream is greatest between the ages of 5 and 25. After 25, the number of cells in the kidney begins to decline slowly, forcing the remaining cells to work harder. The normal 60-year-old kidney has lost about a third of its original cells, as well as much of its reserve capacity—but it is still able to filter the blood adequately. Only when the kidney has lost over half of its original cells does the extent of cell loss become apparent, as increased levels of waste products appear in the blood (Linderman et al. 1985), (Pfeil et al. 1990).

No one is sure what causes aging. A number of theories have been proposed to explain why our cells slow down and die as we get older. The main theories of aging generally fall into two categories. One group of theorists think that aging occurs through an inevitable, predetermined series of events under genetic control—special genes decide when our "time is up." The second group believes that aging results from wear and tear: an accumulation of environmental stresses damage and eventually overwhelm the cells. Based on observations in laboratory animals and humans, scientists have developed several theories to explain why we age (Miller 1994):

- *Free radical damage.* Highly reactive and destructive charged molecules called free radicals attack membranes, DNA, and proteins and cause accumulating damage to which the cell eventually succumbs.

- *Toxic accumulation of breakdown products.* By-products of lipid and protein metabolism accumulate in cells, interfering with cell function and "choking" the cell. A visible example are the brown age spots found on older skin. These blemishes are collections of fat metabolites that accumulate in the skin.

- *Errors in DNA replication.* The intricate system that the cell uses to make copies of its DNA is not perfect. Gradually, small errors accumulate in the DNA strand; and eventually, the DNA can no longer serve as a template for synthesis of proteins needed for metabolism, and the cell dies.

- *Immune dysfunction.* With age, the immune system gradually deteriorates. The body can no longer efficiently rid itself of destructive viruses and bacteria. Also, the immune system is less able to discriminate between itself and foreign substances and initiates autoimmune reactions that irreversibly damage cells.

- *Programmed division potential.* When cells from different human tissues are grown in cell cultures, they divide only a limited number of times before degenerating and dying. Cells may have a genetically predetermined potential for division.

Most scientists think several of these mechanisms concurrently play a role in aging. In other words, it is likely that aging is part programmed change and part environmental wear and tear.

INFLUENCE OF NUTRITION AND LIFESTYLE ON AGING

Age-associated changes occur simultaneously at molecular, cellular, tissue, and whole organism levels. A fundamental problem in aging research is the difficulty of distinguishing between the so-called programmed changes of normal aging that are predetermined and inevitable, and the changes caused by diet, lifestyle, disuse, and disease (the "environment"). They are so tightly interwoven, it is hard to decide which mechanism is responsible for the changes seen in older tissues.

Through much of this century, scientists attributed the decline in the major body systems in later life to normal aging. However, in the past 25 years, it has become increasingly apparent that many of the changes previously thought to be due to aging are the result of long-term disuse and chronic disease. Gerontologists now view the declines in physiological functions associated with advancing age as a combination of disease, disuse, and normal aging (Sloane 1992). Poor nutrition, a sedentary lifestyle, obesity, stress, smoking, and heavy alcohol use can all contribute to the functional decline of later life. Moreover, all of these factors increase the risk of chronic diseases such as heart disease, stroke, hypertension, and osteoporosis—which also accelerate aging. Therefore, healthy diet and lifestyle choices can strongly influence the aging process.

Again, the aging kidney shows how diet, disease, and normal aging all contribute to age-associated changes in the organ. There is a gradual loss of kidney cells over time due to normal senescence. The rate and extent of loss, however, is profoundly influenced by disease and nutrition.

Chronic consumption of a diet high in protein increases the workload of the kidney and contributes to age-related loss of kidney function (Brenner et al. 1982). Also, poor nutrition increases the risk of atherosclerosis and high blood pressure, conditions which greatly accelerate loss of kidney cells with aging.

AGING AND NUTRITION RESEARCH

The aging of our population has focused increasing attention on the special needs of the elderly. Through the National Institute on Aging, the U.S. government finances and supports many lines of research in **gerontology**. In 1981, the U.S. Department of Agriculture created the USDA Center of Aging at Tufts University, whose goals were to determine the nutritional needs of older adults and to examine the interrelationships between diet and disease in the elderly. These programs, and many others around the world, demonstrate the burgeoning research interest in the relationships between nutrition and aging (Wellman 1994), (DeGroot et al. 1992).

Just as aging within a single organ such as the kidney can be strongly influenced by nutrition and disease, current research is revealing connections between nutrition and longevity in humans and animals. Let's look at several major lines of research.

Animal Research

An intriguing link between diet and longevity comes from studies of caloric restriction in animals. In 1935, McCay, Crowell, and Maynard first reported that restriction of energy intake, with administration of all the essential nutrients, could slow aging and extend the lifespan in rats (McCay et al. 1935). They fed special diets to growing animals sufficient in all nutrients except energy. Young rats raised on diets with only two thirds of the energy they would consume normally lived 40–50% longer than those given unlimited access to food.

These remarkable findings have been confirmed and extended in several animal species, including rats, mice, and fish. Periods of caloric restriction in both growing and adult animals not only prolong life, but also prevent or delay the development of many age-related chronic diseases. Many changes associated with aging, such as reduced kidney and immune system function, occur at much later ages; the animals stay healthy longer (Weindruck 1996).

However, energy restriction in young, growing animals has a number of detrimental effects. Many animals that are restricted during growth and development die when they are young, and although the remainder live a long time, many have a variety of malformations. In contrast, more

gerontology: the scientific study of the problems of aging in all its aspects

moderate energy restriction in adult animals increases life expectancy without apparent ill effects.

No one knows why caloric restriction increases lifespan in animals. Research suggests that changing levels of hormones, slowing of the metabolic rate, or delaying the onset of chronic degenerative diseases may play a role (Weindruck 1996). Scientists are interested in this phenomenon primarily because it may provide valuable insights into the causes of aging.

Human Research

Whether caloric restriction can influence aging or longevity in humans has yet to be determined. Undernutrition in growing children is clearly dangerous, making them more vulnerable to infections and retarding normal developmental processes. However, preliminary research in human adults has provided indirect support for the findings from the animal studies.

Recently, researchers divided 24 normal-weight adult men into two groups (Velthius-te Wierik et al. 1994). One group of men served as a control, while the other men underwent ten weeks of energy restriction (ER)—consuming 80% of their habitual energy intake. Energy restriction had no adverse effects on physical or mental functioning and had several beneficial effects: loss of body fat, lowering of blood pressure, and increased HDL-cholesterol.

Epidemiological studies also lend support to the benefits of moderate caloric restriction. On the Japanese island of Okinawa, the adult intake of dietary energy is 20% less than the national average, and for school children it is about two thirds the national recommendation. At the same time, the diets of the Okinawans are generally nutritious and low in fat and contain ample fish and vegetables. Okinawans live substantially longer than other Japanese (the ratio of centenarians is 2 to 40 times that of other Japanese islands), and death rates from cerebral vascular disease, cancer, and heart disease are nearly half the national average. Although many other factors may also play a role (including lifestyle and genetics) in the Okinawans, moderate dietary restriction is associated with increased longevity and no apparent ill effects (Kagawa 1978).

Further evidence that dietary habits influence aging and longevity in people come from epidemiological studies in the United States. One study examined the lifestyles of 7000 adult men and women, and six health habits were identified as strong modifiers of the aging process (Belloc and Breslow 1972). Three involved nutrition and diet: weight control, eating regular meals including breakfast, and moderate or no use of alcohol. The other three factors were adequate sleep, avoiding smoking, and regular physical activity.

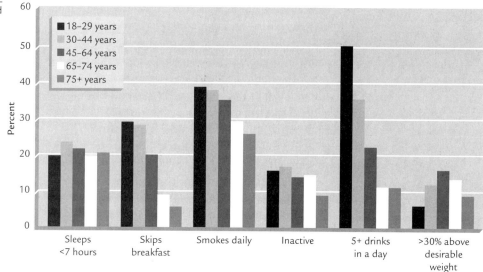

FIGURE 10.3

*Personal habits of adults
over 65 in the United
States compared with those
of younger adults*

functional age:
the relative age of an individual, based not on
the number of years lived,
but on physiological
capacity and level of
functioning

These factors strongly affected longevity. Men who followed at least five of these health practices lived 11 years longer on average than those who followed three or fewer (life expectancy in women increased by 7 years). People who followed these health practices not only lived longer, they were "young for their years": their **functional ages** were often much less than their chronological ages. Figure 10.3 charts the percentages of healthy lifestyle and dietary habits by age group.

Positive health behaviors not only can extend longevity but also reduce the risk of disability and maintain independence in later life. After studying a group of nearly 7000 adults older than 65, researchers reported that four factors—maintaining normal or near-normal body weight, exercising regularly, consuming moderate amounts of alcohol, and avoiding smoking—were associated with increased mobility and independence in later years (Lacroix et al. 1993).

Regular exercise can prolong life. In a study of over 16,000 men between the ages of 35 and 74, those who expended at least 2000 kcal/week exercising had mortality rates a quarter to a third lower than men who were less active (Paffenbarger et al. 1986). Because the researchers controlled for other factors, such as body weight, smoking, and high blood pressure, the difference in mortality rate was attributed to exercise.

From studies like these, it is clear that wear and tear from poor nutri-tion and inactivity can accelerate the changes of aging. This is good news: it means we can influence how we age by eating well, exercising, and maintaining our ideal body weight.

We swallow, digest, and metabolize around 50 tons of food and drink by the time we are 65. A lifetime of poor nutritional choices can have a major impact on health and aging, especially at this stage of life. Proper nutrition can delay or slow disease processes, compress morbidity, and help us reach a maximum life span.

THE NUTRITIONAL HEALTH OF OLDER ADULTS IN THE UNITED STATES

Nutritional surveys in the United States reveal a wide spectrum of dietary intakes among older adults. Although older people are commonly perceived as all "skin and bones," older adults in the United States are generally well nourished, and many are actually overweight. Among older adults (aged 65–74) in the United States, 25% of all men are obese, and 37% of White women and 61% of Black women are obese (USDHHS 1987).

In grim contrast, a significant number of older individuals in the U.S. are undernourished and hungry. It is estimated that 2.5 to 4.9 million older adults suffer from hunger and insecurity about having sufficient food (USDHHS 1990). Certain older adults—particularly the poor, the very old, and the institutionalized—are at high risk of undernutrition (White et al. 1991).

The following sections will look at the various physiological, socio-economic, and behavioral factors that influence nutritional needs in the elderly, and how these changes cause many older people to be at risk for undernutrition.

ENERGY AND PROTEIN INTAKE

Data from several national surveys of adults 65–74 years old in the United States have consistently shown that average energy intake of women (1300–1400 kcal/day) and men (1800–1900 kcal/day) are significantly below current recommendations for this age group (see Figure 10.4). In contrast, protein intake of older adults in the United States consistently meets or exceeds the current RDA (Koehler and Garry 1993).

A significant minority of older adults have very low intakes of energy and protein. Among all **community-dwelling** adults older than 65, 16% consume less than 1000 kcal/day—a level of intake commonly associated with micronutrient deficiencies (Lowenstein 1982). Protein–calorie malnutrition in hospitalized older people is common—surveys have found prevalence rates of 33–68% (Nelson and Franzi 1992). Also, 19–27% of older people living in nursing homes in the U.S. show evidence of protein–calorie malnutrition, probably because of marginal intake and chronic illness (Marley 1993).

community-dwelling: living in the community, that is, not institutionalized or hospitalized

Source: Adapted from
LSRO (1989)

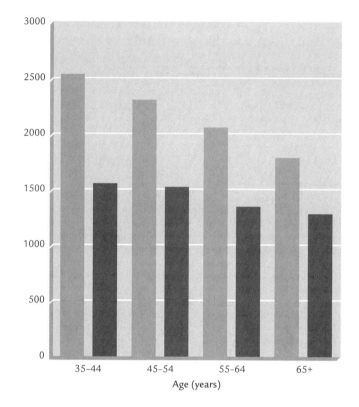

FIGURE 10.4

*Average energy (kcal)
intake by adults in the
United States*

VITAMIN INTAKE

Mean intakes of thiamin, riboflavin, and niacin by individuals over 65 in the United States generally meet or exceed the RDAs. However, a significant minority of older adults obtain inadequate dietary thiamin and riboflavin, usually in association with chronic illness, poverty, or alcoholism. In one national survey, 4–8% of Whites and 17–20% of Blacks over 65 showed signs of riboflavin deficiency (Lowenstein 1982).

Most older people do not obtain adequate vitamin B_6 in their diets, with nearly two thirds of individuals over age 65 consuming less than 70% of the RDA. Poverty, alcoholism, and chronic disease in older people increase the risk of folate and B_{12} deficiency, (Koehler and Garry 1993). Four to 10% of older adults have evidence of impaired folate status (Santi and Pilch 1985). In addition, a recent study of more than 500 community-dwelling adults over 65 found evidence of vitamin B_{12} deficiency in over one third (Lindenbaum et al. 1994). Although mean intake of vitamin C by individuals older than 65 in the United States exceeds the RDA, 22–31% of older adults consume less than 70% of the RDA. In

NHANES II, nearly 8% of older women and 20% of older men had low plasma concentrations of vitamin C.

National surveys have found that mean vitamin A intake by older people in the United States significantly exceeds the RDA . In contrast, many older adults are at risk for poor vitamin D status. In studies in Boston and New Mexico, researchers found that 45 to 74% of older individuals consumed less than 75% of the RDA for vitamin D, and 15% had signs of vitamin D deficiency. (Koehler and Garry 1993).

Overall, surveys in the United States indicate that vitamins B_6, C, and D are nutrients of most concern in older people's diets. A recent survey from the Netherlands reported similar findings. Researchers compared the vitamin status of healthy individuals aged 65–80 years with younger adults (aged 35 years). Mean blood values for folate, 25-OH vitamin D, and pyridoxal phosphate (an active form of vitamin B_6) were significantly lower in the older group (Lowik et al. 1989).

MINERAL INTAKE

Calcium intakes of older women and men in the United States are substantially below recommended levels. Compared with the RDA of 800 mg, the mean calcium intake of men aged 65–74 is only 700 mg/day, and for women it is even lower (about 550 mg/day) (Koehler and Garry 1993). Metabolic studies have indicated that requirements for older women may be significantly higher than the current RDA (Recker and Heaney 1989) (discussed in more detail later in this chapter). Poor calcium intake may jeopardize the bone health of many older adults, and increases the risk of osteoporosis and hip fractures.

In contrast to calcium intake, the mean intake of iron by older adults in the United States meets current recommendations. Moreover, only about 2–3% of men and 3–4% of women older than 65 years in the United States have impaired iron status (USSGO 1988). Compared to younger women, the prevalence of iron-deficiency anemia in postmenopausal women is sharply lower. This is because after menopause women no longer lose iron in the menses, and iron balance typically improves.

Knowledge of dietary intakes and nutritional status of older adults for the remaining minerals is limited. Minerals thought to be lacking in the diets of many older people are zinc, magnesium, potassium, selenium, and chromium. Many older adults have intakes below two thirds of the RDAs for these nutrients (Mertz et al. 1989). Future research should provide more information on the trace mineral nutrition of older adults.

NUTRITIONAL NEEDS OF AGING ADULTS

Knowledge of the nutritional needs of older adults is limited. Most studies examining the nutritional status and requirements of adults have focused on young, healthy adults, and it is not known whether nutritional requirements for young adults are appropriate for those older than 65. Because aging affects digestion and metabolism of nutrients and alters body composition, needs for older adults are likely to be different from those of middle-aged adults (Russell and Suter 1993).

Another barrier to developing recommendations for older age groups is the heterogeneity of older adults. Differences in diet, lifestyle, genetics, and illnesses become more evident as years pass; the physiology, activity patterns, and biological ages of different 75-year-olds can vary greatly.

RECOMMENDED DIETARY ALLOWANCES FOR OLDER ADULTS

The 1989 RDAs divide adults into two groups—those aged 25 to 50 years and those 51 years and above (NAS 1989). The RDAs for older adults are listed in Table 10.3. Although the U.S. Food and Nutrition Board considered a further subdivision of older adults, it concluded, "the data are insufficient to establish separate RDAs for people over 70" (NAS 1989).

As more information on the nutritional requirements of older adults becomes available, specific RDAs for this group may appear. It is very likely that for certain nutrients, the RDAs for a 75-year-old are different from those for a 55-year-old (Wood 1991). For example, the Canadians have established RDAs for adults aged 50–74 and over 75, as shown in Table 10.4. Establishing the unique nutritional needs of older individuals is the focus of much current research.

ENERGY NEEDS

The metabolic needs of older cells do not change: energy requirements per kilogram of fat-free mass are the same in older and younger people. But energy needs decrease with age because older adults tend to be more sedentary, and the age-associated loss of fat-free mass (muscle) lowers the basal energy expenditure (James et al. 1989). In sedentary older adults, the total energy requirement is often close to basal energy expenditure, and needs in many older adults are only 1100–1400 kcal /day.

With such a low daily requirement for energy, careful selection of nutrient-dense foods (and avoidance of nutrient-empty, high-calorie foods such as many baked and processed foods) is essential to maintain adequate daily intake of all nutrients. The incidence of nutrient deficiency is markedly increased in older adults with the lowest energy intakes (White et al. 1991).

| | | Weight[a] | | Height[a] | | | Fat-Soluble Vitamins | | | |
	Age (yrs)	(kg)	(lb)	(cm)	(in)	Protein (g)	Vitamin A (µg RE)	Vitamin D (µg)	Vitamin E (mg α-TE)	Vitamin K (µg)
Males	51+	77	170	173	68	63	1000	5	10	80
Females	51+	65	143	160	63	50	800	5	8	65

Water-Soluble Vitamins

	Age (yrs)	Vitamin C (mg)	Thiamin (mg)	Riboflavin (mg)	Niacin (mg NE)	Vitamin B_6 (mg)	Folate (µg)	Vitamin B_{12} (µg)
Males	51+	60	1.2	1.4	15	2.0	200	2.0
Females	51+	60	1.0	1.2	13	1.6	180	2.0

Minerals

	Age (yrs)	Calcium (mg)	Phosphorus (mg)	Magnesium (mg)	Iron (mg)	Zinc (mg)	Iodine (µg)	Selenium (µg)
Males	51+	800	800	350	10	15	150	70
Females	51+	800	800	280	10	12	150	55

The allowances, expressed as average daily intakes over time, are intended to provide for individual variations among most normal persons as they live in the United States under usual environmental stresses. Diets should be based on a variety of common foods in order to provide other nutrients for which human requirements have been less well defined.

[a] Weights and heights are actual medians for the U.S. population of the designated age. The use of these figures does not imply that the height-to-weight ratios are ideal.

Adapted from National Research Council, National Academy of Sciences, *Recommended Dietary Allowances,* 10th ed. (Washington D. C., National Academy of Sciences, 1989).

TABLE 10.3

The RDAs for older adults

Recommendations for energy sources for older adults do not differ from younger adults. The majority of calories (50–60%) should come from carbohydrates. Total fat intake should be less than 30% of calories, and saturated fat intake should be less than 10% of calories. Most fat calories should come from monounsaturated and polyunsaturated fats (NRC 1989).

PROTEIN NEEDS

The RDA for protein for healthy older adults is identical to that for younger adults (0.8 g protein/kg/day) (NAS 1989). However, recent studies suggest that the protein requirement in older adults is higher—about 1.0 g/kg/day (Campbell et al. 1994). In many older adults, acute and chronic diseases increase protein requirements by causing the breakdown of body protein, and extra dietary protein is necessary to replace losses and support repair of tissues. It may be difficult for sedentary older people to get adequate protein from the diet because energy requirements and food intake are so low.

Age	Sex	Weight (kg)	Protein (g/day) [a]	Fat-Soluble Vitamins			Water-Soluble Vitamins			
				Vitamin A (RE/day) [b]	Vitamin D (µg/day) [c]	Vitamin E (mg/day) [d]	Vitamin C (mg/day) [e]	Folate (µg/day)	Vitamin B$_{12}$ (µg/day)	
50–74	M	73	63	1000	5	7	40	230	1.0	
	F	63	54	800	5	6	30	195	1.0	
75+	M	69	59	1000	5	6	40	215	1.0	
	F	64	55	800	5	5	30	200	1.0	

Age	Sex	Calcium (mg/day)	Phosphorous (mg/day)	Magnesium (mg/day)	Iron (mg/day)	Iodine (µg/day)	Zinc (mg/day)
50–74	M	800	1000	250	9	160	12
	F	800	850	210	8	160	9
75+	M	800	1000	230	9	160	12
	F	800	850	210	8	160	9

Recommended intakes of energy and of certain nutrients are not listed in this table because of the nature of the variables upon which they are based. The figures for energy are estimates of average requirements for expected patterns of activity. For nutrients not shown the following amounts are recommended based on at least 2000 kcalories per day and body weights as given: thiamin 0.4 milligrams per 1000 kcalories: riboflavin 0.5 milligrams per 1000 kcalories; niacin, 7.2 niacin equivalents per 1000 kcalories; vitamin B$_6$ 15 micrograms as pyridoxine per gram of protein. All recommended intakes are designed to cover individual variations in essentially all of a healthy population subsiding upon a variety of common foods available in Canada.

Source: Health and Welfare Canada, *Nutrition Recommendations: The Report of the Scientific Review Committee* (Ottawa: Canadian Government Publishing Centre, 1990) Table 20, p.204.

[a] The primary units are expressed per kilogram of body weight. The figures shown here are examples.

[b] One retinol equivalent (RE) corresponds to the biological activity of 1 microgram of retinol, 6 micrograms of beta-carotene, or 12 micrograms of other carotenes.

[c] Expressed as cholecalciferol or ergocalciferol

[d] Expressed as d-α-tocopherol equivalents, relative to which β- and γ-tocopherol and α-tocotrienol have activities of 0.5, 0.1. and 0.3 respectively.

[e] Cigarette smokers should increase intake by 50 percent.

TABLE 10.4
Canadian recommended nutrient intakes

VITAMIN REQUIREMENTS

For vitamins E and C, and most of the B vitamins, daily requirements for older adults are the same as those for younger adults. However, it appears daily requirements for vitamins A, D, and B$_6$ change with aging (Russell and Suter 1993).

Vitamin B$_6$

In adults over 60 years old, a requirement greater than the current RDA for vitamin B$_6$ may be indicated—especially in women—probably because of less efficient absorption of the vitamin (Russell and Suter 1993), (Ribaya-Mercado et al. 1991). Adequate dietary vitamin B$_6$ for older adults is important because deficiency has been linked to decreased functioning of the immune system (Meydani et al. 1991).

Vitamin D

Although older adults can absorb vitamin D and convert it to 25-OH vitamin D as well as younger people, aging affects many other aspects of vitamin D nutrition. In older people, the kidney is less able to synthesize $1,25\text{-}(OH)_2$ vitamin D in response to a signal from **parathyroid hormone** (Dandona et al. 1990). Also, the aging intestine is less responsive to the signal from $1,25\text{-}(OH)_2$ vitamin D to increase absorption of calcium (Eastell et al. 1991).

Younger people can synthesize significant amounts of vitamin D in sun-exposed skin, but aging skin is less able to synthesize procholecalciferol, which is the compound converted by sunlight to vitamin D. Compounding this, many older adults, particularly those with disabilities, obtain little sunlight exposure. (Russell and Suter 1993).

Many studies have documented deficiencies of vitamin D and its active metabolites in older adults, (Gloth et al. 1995), (Egsmose et al. 1982). Deficiency of the vitamin can have major effects on bone health and increase the risk of osteoporosis. The current RDA for vitamin D may be too low for some older adults, especially those who get little sunlight exposure (Russell and Suter 1993). Diets of older adults at risk for vitamin D deficiency should emphasize fortified milk, dairy products, and fish. Older adults who cannot obtain adequate dietary vitamin D may benefit from vitamin D supplementation.

Vitamin A

Older people should be careful with vitamin A supplements. Compared with a younger adult, a dose of vitamin A given to an older person will result in higher blood levels of the vitamin (Krasinski et al. 1990), (Rasmussen et al. 1991). In older adults, absorption efficiency is increased, and the liver metabolizes circulating vitamin A less efficiently. Because of this, supplements of vitamin A at only two to three times the RDA can cause liver damage in older adults (Krasinski et al. 1989).

Increased absorption and less efficient metabolism of vitamin A may also explain why older adults who do not consume amounts of the vitamin near the RDA do not show biochemical signs of deficiency. Research done in several European cities showed that more than 90% of older adults consumed vitamin A in amounts significantly below wrecommended levels, yet there was little evidence of deficiency (Euront-SENECA "Nutritional…" 1991). These studies and others have persuaded some scientists to recommend lowering the RDA for vitamin A in older individuals (Russell and Suter 1993).

parathyroid hormone (PTH): a hormone produced by the parathyriod gland that raises the calcium level in the blood

MINERAL REQUIREMENTS

Calcium

Intestinal calcium absorption decreases with age (Bullamore et al. 1990). While younger adults respond to low-calcium diets by increasing the efficiency of calcium absorption, older people are less able to do this (Ireland and Fordtran 1973). The diminished response of the intestine to vitamin D with advancing age may be partly responsible for reduced calcium absorption in older adults (Eastell et al. 1991).

There is currently an active debate among scientists regarding calcium requirements for older adults, particularly older women. Some scientists feel that older women, who are at high risk for osteoporosis, need significantly more calcium than the current RDA (up to 1500 mg/day) (Consensus Development Conference 1993). Others feel the evidence is not yet strong enough to recommend higher intakes (NAS 1989).

Zinc and Magnesium

Compared with younger adults, absorption of zinc and magnesium from foods decreases in older adults (Wood 1991), (Turnland et al. 1986). Research is being done to determine whether the diminished absorption of these minerals is due to loss of intestinal function with aging, or whether older people's needs are less, so that they don't need to absorb as much from foods. Marginal zinc deficiency among older adults may be associated with decreased taste acuity and slower wound healing (Keenan and Morris 1993).

NUTRITION AND THE PHYSIOLOGICAL CHANGES OF AGING

Each of us will become a different person as we age. Our ability to taste and digest food, and absorb and metabolize many nutrients changes. There is a gradual loss of muscle mass and bone minerals. Figure 10.5 shows the decline in body levels of calcium and protein as we age. These, and other age-associated changes, affect our nutritional needs as we grow older.

THE MOUTH AND TEETH

Because sensitivity of taste and smell declines with age, many older adults have a diminished appreciation for the taste and aroma of foods (Rolls 1992). As we age, there is a steady decrease in the number of taste buds on the tongue. Elevated thresholds of flavor detection and recognition for

Source: Adapted from
Heymsfield et al. (1989)

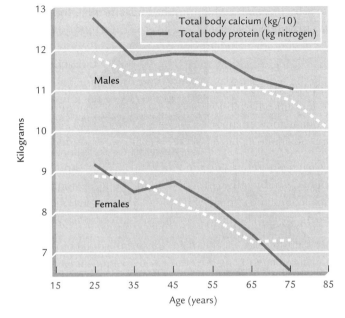

FIGURE 10.5

*Total body calcium and
protein decrease with age*

sweet, bitter, sour, and salty tastes have been reported in older individuals (Schiffman 1994). Losses in taste and smell usually become noticeable
at about 60–70 years of age.

These changes can be compounded by certain drugs and mineral deficiencies (such as zinc deficiency) that further reduce taste sensations.
Sensory changes with advancing age can influence dietary choices (a
preference for salty or spicy foods may develop), limit enjoyment of
food, and reduce appetite and food intake (Rolls 1992).

Teeth are crucial to good nutrition and health, but in the United
States, about 25% of adults over 50 and nearly half of 75-year-old adults
have no teeth at all (USDHHS 1986). Many of the **edentulous** do not
have dentures, or have ill-fitted dentures that interfere with chewing.
Periodontal disease is also common in older adults. It undermines the
bone of the jaw that supports the teeth, loosening remaining teeth and
reducing chewing vigor. In older people, commonly used medications
and dehydration can reduce salivary flow and produce a "dry mouth,"
which influences the taste and palatability of foods.

edentulous: toothless

periodontal disease: a
disorder marked by
destruction of the tissues
and structures
surrounding and
supporting teeth

THE GASTROINTESTINAL TRACT

The stomach has several important functions: it holds swallowed foods,
mixes them thoroughly with digestive juices, and then releases them into
the intestine. All of these functions are diminished in the elderly: stomach motility decreases; foods are mixed less thoroughly with secretions
and released more slowly into the intestine (Russell 1992).

Atrophic gastritis affects one in four adults in their 60s and nearly 40% of 80-year-old adults (Krasinski et al. 1986). This loss of cells lining the stomach creates two significant problems relating to nutritional health. First, the normally low pH of the stomach rises, decreasing the ability to absorb iron, calcium, and the vitamins B_6, B_{12}, and folate (Russell 1992). Reduced acidity may also allow bacterial overgrowth in the small intestine and interfere with the absorption of nutrients (Suter et al. 1991). Second, the decreased secretion of intrinsic factor by stomach cells can further decrease absorption of vitamin B_{12} and lead to anemia.

Liver function declines in older adults, decreasing clearance of many drugs and increasing the potential for side effects from drugs (Zoli et al. 1989). The liver also metabolizes cholesterol and vitamin A less efficiently, contributing to higher serum cholesterol and an increased sensitivity to large doses of vitamin A (Krasinski et al. 1990).

Lactase activity often diminishes with age, and many older people, particularly African Americans and Asian Americans, are lactose intolerant (Flatz 1987). However, some older adults avoid these products unnecessarily because they think milk causes digestive problems. Dairy products are major sources of calcium, vitamin D, and protein for the elderly, and it is important to distinguish lactose intolerance from misperceptions about these foods.

Constipation is a common complaint in older adults. Immobility, dehydration, and foods low in fiber are likely villains in this problem. Increasing physical activity, consuming more dietary fiber, and drinking from six to eight glasses of water per day is often beneficial.

THE MUSCLES

atrophic gastritis: a chronic inflammation of the stomach associated with atrophy of the lining cellls and glands resulting in impaired gastric function

As we age, fat cells and connective tissue gradually replace muscle fibers. This process changes body composition: fat-free mass (mainly muscle) decreases, and body fat increases (Frontera et al. 1991). Fat-free mass decreases about 5% each decade after reaching adulthood. Because fat-free mass is a major determinant of the basal metabolic rate (BMR), the BMR falls about 2% every ten years after age 30 (James et al. 1989). This substantial decrease in BMR reduces resting energy expenditure and decreases daily caloric requirements.

These changes in body composition, although often attributed to normal aging, are more the result of disuse than of age (Evans 1992). Athletes in their 60s who maintain a high level of fitness over their lifetimes have body fat and body compositions similar to active people in their 20s.

Muscle mass	Increased
Bone density	Increased
Body fat	Decreased
Blood pressure	Decreased
Maximum oxygen consumption	Increased
Respiratory function	Increased
Blood LDL-cholesterol	Decreased
Blood triglycerides	Decreased
Blood HDL-cholesterol	Increased

Adapted from D. T. Lowenthal, et al., "Effects of Exercise on Age and Disease," *Southern Medical Journal,* 87, no. 5 (1994):S5–12.

TABLE 10.5

Age-related changes positively influenced by a healthful diet and exercise

"Pumping Up" in the Very Old

Scientists recently studied the effects of exercise in a group of older adults in their 80s and 90s. After eight weeks of strength training, strength improved an average of 160%, and muscle mass increased by 10%. The researchers concluded that even the very old can benefit from short-term exercise programs (Fiatarone et al. 1990).

Regular physical activity has many benefits for older adults. Exercise can preserve muscle mass, increase strength, balance, and mobility, and keep the BMR from falling. Exercise burns calories for energy, increases appetite, and allows older adults to eat more without becoming overweight (Fiatarone et al. 1994). Exercise is also of significant benefit in many diseases of the old—hypertension, heart disease, and diabetes. Table 10.5 lists the age-related changes that are positively influenced by exercise and a healthful diet.

THE BONES AND JOINTS

Bone density decreases as we age, with bone loss beginning in the mid-30s. Both the minerals and the protein matrix that holds the minerals are gradually lost (Johnston and Slemenda 1992). See Figure 10.6.

Osteoporosis

Our bones are metabolically active tissues; they are continually remodeled throughout life. Cells in the bones work constantly, carving out old bone and laying down new. *Osteoclasts* tunnel through old bone,

Source: Adapted from Raisz and Smith (1987)

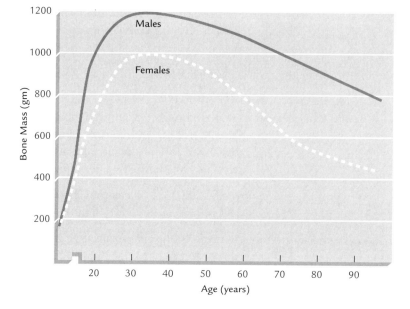

FIGURE 10.6
Gender differences in bone mass

dissolving small chunks of bone into calcium, phosphorus, and small fragments of protein. This process is called bone resorption.

Osteoblasts—bone-forming cells that move in after the osteoclasts are done—replace the bone that has been removed. They secrete a new **protein matrix**, which is slowly filled in by minerals coming from the bloodstream. In adults, about 10% of the skeleton is in the middle of being remodeled at any given time (Kaltenborn 1992). In healthy bone, resorption and formation are carefully balanced: the amount of bone formed is about the same as the amount lost.

Even though about 99% of the calcium in the body is in the bones, the 1% that is dissolved in the bloodstream and in cells all over the body is actually more important. The heart and nervous system are extremely sensitive to changes in the calcium level of the blood, and even slight fluctuations can cause the heart to beat irregularly or produce convulsions.

We lose about 200–400 mg of calcium each day in urine and feces. In order to stay in balance, we need to replace these losses with calcium from the diet. If dietary calcium is unable to cover these losses, the body sets its priorities: it maintains blood levels of calcium at the expense of bone calcium. In this way, the skeleton serves as a calcium "bank." Extra calcium may be deposited in bones if dietary supply is ample and blood levels are normal. If dietary supply is inadequate, calcium will be withdrawn from the bones to maintain blood levels. If net losses continue for long periods, maintenance of blood calcium comes at a high price.

Osteoporosis is a Greek word meaning "a condition of porous bones." In osteoporosis there is a reduction in the amount of bony tissue in the skeleton. Osteoporotic bones generally keep their shape, but they become less dense: the bony tissue that remains is normal, but there is too little of it.

protein matrix: the organic component of bone; a connective tissue that is 90–95% collagen and in which are deposited the calcium-containing crystals

Osteoporosis is a major public health problem affecting older adults, with about 20 million sufferers in the United States (Allen 1993). Chances of developing osteoporosis increase steadily with age. The problem is six times more common in older women than men: it is present in one in four women over the age of 65. The symptoms of this "silent" disease appear only late in its development when the skeleton is already irreversibly weakened. Often the first sign of the disease in the aged is a fracture of one of the bones in the back or the hip from a minor fall. In the United States each year, over 1 million aged people sustain fractures related to osteoporosis (Allen 1993).

There are two main types of osteoporosis. *Postmenopausal osteoporosis* (Type I) affects only women and is caused by the sudden drop in estrogen levels at the menopause. Estrogen normally supports the skeleton by putting a check on bone resorption. When estrogen levels fall, osteoclasts remove bone faster than it can be reformed, leading to rapid bone loss. Up to 5% of the bony mass can be lost each year for as long as eight to ten years after the menopause (Kaltenborn 1992).

Senile osteoporosis, also called age-related or Type II osteoporosis, is due to the cumulative bone loss that both men and women experience with aging (Niewoehner 1993). Bone size and mass increase steadily during childhood and adolescence. Even after we stop growing, bones continue to accumulate minerals and strengthen themselves; bone mass reaches its peak in the early 30s. Around 35, total bone mass begins to slowly decline, and we lose about 3% of our bone each decade (Arnaud and Sanchez 1990). Age-related bone loss is nearly universal.

Osteoporosis is irreversible, and because there is no effective treatment for it, prevention is critical. A healthy diet and lifestyle play major roles in prevention. Risk factors for osteoporosis are shown in Table 10.6. Osteoporosis is more common in people who don't obtain adequate calcium in their diets (Heaney 1993). Calcium from food or supplements can't cure osteoporosis, but it can help prevent the disease by slowing the rate of bone loss. A recent study found that calcium supplementation (1 g/day) protected against bone loss and suppressed bone turnover in postmenopausal women (Reid et al. 1993). See Figure 10.7.

Despite recent recommendations to increase calcium intake, many people consume too little calcium. Women are at higher risk for deficient intakes: 75% of women in the United States consume less than the RDA (800 mg) of calcium each day, and 25% get only 200–300 mg daily (Heaney et al. 1982). Dietary deficiency may be especially detrimental at the beginning and at the end of life. Optimal intake of calcium during growth, adolescence, and early adulthood builds dense, strong bones and decreases the risk of osteoporosis in later years (Matkovic et al. 1990), (Sandler et al. 1985).

Repeated dieting of adolescents and young adults in pursuit of thinness, on the other hand, often comes at the expense of good nutrition.

The following risk factors—most of which are hereditary—cannot
be modified:

- being a Caucasian or Asian female;

- having close relatives with osteoporosis;

- having light skin;

- having a delicate frame;

- experiencing previous bone loss due to illness or medications;

- experiencing early menopause.

Other risk factors—some the result of dietary and lifestyle habits—can be
modified:

- being sedentary;

- inadequate calcium intake;

- smoking cigarettes;

- consuming more than two alcoholic drinks daily;

- being underweight;

- a deficiency of vitamin D;

- undergoing surgical removal of the ovaries before menopause.

Adapted from B. L. Riggs and J. L. Melton, "Involutional Osteoporosis," *N Engl J Med* 314
(1986) 1676–84.

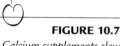

TABLE 10.6

*Risk factors for
osteoporosis*

Source: Adapted from
Reid (1993)

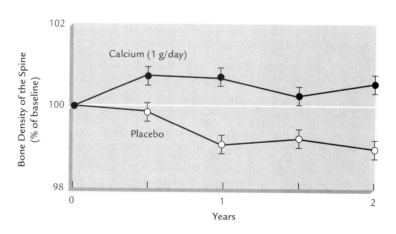

FIGURE 10.7

*Calcium supplements slow
loss of bone density for
postmenopausal women*

Young people who grow up on diet sodas instead of milk lower their calcium intake at a time when they can least afford it; decreased peak bone
mass may contribute to crippling osteoporotic fractures 50 years later.

Unfortunately, the dramatic bone loss of postmenopausal osteoporosis
appears to be unaffected by increasing calcium intake: in older women
who are still within five years of the onset of menopause, high calcium

intakes do not slow down bone loss (Heaney 1993). The great burst of osteoclastic activity caused by the sudden drop of estrogen is unaffected by dietary calcium. After this time, calcium intake once again becomes important in slowing bone loss from senile osteoporosis (Dawson-Hughes et al. 1990).

Bone health can also be adversely affected by vitamin D deficiency, which is common in older people (Gloth et al. 1995). A major function of vitamin D is to increase absorption of dietary calcium. If calcium is unavailable from dietary sources and blood levels of calcium begin to fall, the body reacts by secreting parathyroid hormone (PTH) into the blood. PTH will then pull calcium out of the bones to maintain blood levels of calcium.

Studies in older adults have shown that bone loss increases during the winter, particularly in northern climates (Heaney 1993). The reason seems to be that decreased exposure to sunlight diminishes vitamin D production in the skin. If dietary intake of vitamin D is also low, deficiency will result. Many older people have impaired vitamin D status and high levels of PTH during the winter months; increasing vitamin D intake during the winter can slow PTH-activated bone loss (Dawson-Hughes et al. 1991).

Calcium and vitamin D are close partners in bone health. A recent study showed that supplementation with these nutrients can benefit older adults. In France, 3000 nursing home residents were divided into two groups. One group was given supplements of both vitamin D and calcium for over a year, while the other group received a placebo. Results were dramatic. The group receiving the supplements lost less bone and had far fewer osteoporotic fractures than the group that received the placebo (Chapuy et al. 1992).

Other dietary components may play a role in osteoporosis. Heavy alcohol use is a risk factor for osteoporosis (Bikle et al. 1985). It appears that alcohol impairs osteoblast function and decreases bone formation (Lindholm et al. 1991). Alcohol also interferes with calcium absorption from the intestine.

A recent study in 200 healthy, postmenopausal women found that daily consumption of >450 mg of caffeine (about the amount in three to four cups of brewed coffee) accelerated bone loss in women with calcium intakes below 800 mg/day (Harris and Dawson-Hughes 1994). In contrast, women who had better intakes of calcium did not lose significant amounts of bone when they consumed the same amount of caffeine. Because two recent studies in older women have found a positive association between caffeine intake and hip fracture (Kiel et al. 1990), (Hernandez-Avila et al. 1991), older women, particularly those at high risk of osteoporosis, should minimize consumption of caffeine-containing beverages.

Along with a balanced diet and ample calcium and vitamin D intake, lifestyle choices can influence bone health in later life. Avoiding cigarette smoking is beneficial, as smoking sharply increases the risk of osteoporosis. One of the most effective ways for older adults to slow age-related loss of bone is to participate in regular, **weight-bearing exercise**. Mechanical loading maintains bone health and integrity and retards loss of minerals from the skeleton (Arnaud and Sanchez 1990).

Arthritis

Nearly everyone who lives long enough will experience arthritis. Minor joint pains and morning stiffness are nearly universal with aging. However, arthritis can also be debilitating. Over 40 million people in the United States, most of them older adults, suffer from painful, crippling arthritis—it is the major cause of restricted mobility in the elderly (U. S. Senate 1990).

There are several forms of arthritis. **Osteoarthritis** (also called degenerative joint disease or "wear and tear" arthritis) is the most common form in older adults. Obesity can aggravate osteoarthritis, particularly if the joints affected are weight-bearing joints such as the knees and hips. People with arthritis should balance diet and activity to maintain or attain a desirable weight.

One of the most serious and crippling types of arthritis is **rheumatoid arthritis**. It affects about 7–10 million adults in the United States and can be progressive and debilitating. Research has shown that, for some people with rheumatoid arthritis, foods can aggravate the disorder (Darlington 1991), (Panush 1991). See Table 10.7. A study in Norway divided 50 people with rheumatoid arthritis into two groups. One group was placed on an **elimination diet** followed by a lactovegetarian diet for one year. The control group ate a regular diet. The number of inflamed joints was less, and pain and stiffness improved in the group given the special diet, and improvements persisted for up to one year (Kjeldsen-Kragh et al. 1991).

Some scientists think that small proteins in food can confuse the immune system and cause immune cells to mistakenly attack and damage joints, causing arthritis. The fact that elimination diets work for some people with rheumatoid arthritis doesn't mean they will work for all. Food appears to be only one of the many triggers of rheumatoid arthritis (Darlington 1991).

THE HEART AND BLOOD VESSELS

The heart's effectiveness as a pump declines with age. There is a gradual fall in cardiac output and maximum heart rate (Lakatta 1994). In the average 70-year-old adult, the maximum heart rate is about three quarters that of a 20-year-old. See Table 10.8.

weight-bearing exercise: exercise in which the body weight is supported in the standing position; examples are walking and running

osteoarthritis: a chronic arthritis, usually occurring in older people, characterized by degeneration of the cartilage of the joints

rheumatoid arthritis: a chronic arthritis marked by stiffness and inflammation of the joints, loss of mobility, and joint deformity

elimination diet: one for diagnosing food allergy, based on the sequential omission of foods that might cause symptoms in the individual

- Corn
- Wheat
- Bacon/Pork
- Oranges
- Milk
- Oats
- Rye
- Eggs
- Beef
- Coffee

From L. G. Darlington, "Dietary Therapy for Arthritis," *Rheum Dis Clin of North America* 17, no.2 (1991):273–85.

TABLE 10.7
Foods most likely to cause joint symptoms (pain, swelling) in people with rheumatoid arthritis

Arterial oxygen pressure	90
Total body water	87
Cerebral blood flow	80
Kidney weight	80
Maximum heart rate	76
Resting oxygen consumption	74
Muscle mass	74
Maximum oxygen consumption	65
Resting cardiac output	64
Maximum urinary concentration	64
Kidney blood flow	60
Cardiac reserve	50

Adapted from R. A. Kenney, "Physiology of Aging," *Clinics in Geriatric Medicine* 1 (1985): 37–59.

TABLE 10.8
Anatomic and functional variables in a 70-year-old man expressed as percentages of those of a 20-year-old

Much of the deterioration in the cardiovascular system associated with aging may be the result of poor dietary choices, chronic disease, and a sedentary lifestyle. Hypertension and atherosclerosis damage the heart and blood vessels of many older people and accelerate age-associated functional loss (Bush 1991). Blood pressure tends to rise with age, and nearly half of older adults in the United States have high blood pressure. In many older adults, atherosclerosis affects large and small blood vessels throughout the body. These topics are discussed in Chapter 9.

THE KIDNEY AND WATER BALANCE

In most people, there is a steady decline in kidney function with age. By age 80, the average person can filter the bloodstream at only half to two thirds the rate of a younger adult (Pfeil et al. 1990). However, not all people lose kidney function as they age. Results from a study of 250 healthy older adults in the United States found that nearly one third had no decrease in kidney function with advancing age—even into their 70s and 80s (Lindeman et al. 1985). These findings suggest that declines in kidney function with advancing age, although found in the majority of older adults, are not inevitable.

Diet and disease have a major impact on kidney function (Brenner et al. 1982). Loss of kidney function is sharply accelerated by hypertension and atherosclerosis, and some researchers have linked a diet high in protein with diminished kidney function. As kidney function declines, it takes longer to clear the blood of metabolic wastes, sodium excess, drugs, and megadoses of water-soluble vitamins. Table 10.8 shows the usual changes in renal, cardiovascular, and musculoskeletal function that occur with aging.

Dehydration

With advancing age, ample water intake is particularly important to maintain homeostasis. To function efficiently, our cells and tissues need plentiful water containing optimal concentrations of salts and minerals. Water balance depends on adequate intake from food and drink, balanced by kidney filtration and urinary excretion. Specialized cells, found in the brain, blood vessels, and kidneys, are sensitive to changes in volume and osmolality of body water. They monitor water balance and send constant signals to thirst centers in the brain and cells in the kidney.

In older adults, these cells become less sensitive to changes in body water. As a result, compared with younger adults, many older people feel less thirsty when their body water is low. Also, the kidney is less efficient at concentrating urine in response to a drop in body water volume (Chernoff 1994).

For these reasons, older individuals tend to be much more susceptible to dehydration than younger people, particularly during hot weather or illness. Frail, institutionalized, or disabled older adults who have difficulty getting up and around or who are dependent on others to meet their needs may not receive adequate water. Other factors that increase the risk of dehydration in older adults include medications such as diuretics and laxatives, and chronic illness or infections.

Although dehydration has many deleterious effects, it may be overlooked in older people because it comes on gradually and its presentation is not dramatic (Chernoff 1994). When severe, it can produce serious electrolyte disturbances, decreased blood pressure, and mental confusion. Family, friends, and health-care workers who care for older people should be aware of these factors and make sure older individuals obtain ample water intake. Most older adults need at least six to eight glasses of fluid per day.

THE BRAIN

Although the number of functioning **neurons** steadily decreases as we grow older, the brain has extensive reserve capacity; in most healthy older adults, no decline in cognitive function is apparent before 75. However, like most age-associated changes in function, there is great variability in the rate and degree of mental decline among older people.

The aging brain is extremely vulnerable to changes in its environment. Inadequate nutrition, drugs, diseases in other organs, decreased blood flow and oxygen supply, and many other environmental factors can accelerate functional losses associated with aging. Depending on which part of the brain is affected, neuronal loss can impair balance, memory, cognitive function, speech, or personality. The aging brain is the site of one of the major causes of disability in older adults—the syndrome of dementia.

Dementia

Many older people are disabled by a gradual loss of brain functions—a condition referred to as dementia. Dementia is a Latin word that means "to take away the mind." People with dementia have severely impaired thinking and judgment, personality changes, and may lose the ability to speak or understand people.

Dementia is one of the biggest public health problems in the United States, and its impact will be greater as our population ages. See Figure 10.8. Five to ten percent of people over the age of 65 have dementia, and the incidence increases sharply with age: over 30% of those over 85 are affected (Gray 1989).

Half of all nursing home residents have dementia and require special care, and more than two thirds of the aged with dementia live at home and require constant care by family and friends. The social and medical costs of dementia in the United States are staggering. It is estimated that we spend nearly $40 billion a year caring for people with dementia, and by the year 2040, the estimated cost could reach more than $120 billion (Weiler 1987), (Huang et al. 1988).

Although there are many causes of dementia, it is estimated that dietary factors and alcohol may contribute to the development of about a third of all cases (Gray 1989). Regardless of the etiology of the dementia, proper nutrition is an important part of the care of people with this crippling disorder. Let's look first at Alzheimer's disease, which is the most common cause of dementia in older individuals.

neuron: a nerve cell capable of initiating and conducting nerve impulses, the structural and functional unit of the nervous system

Source: Adapted from U. S.
Senate (1990)

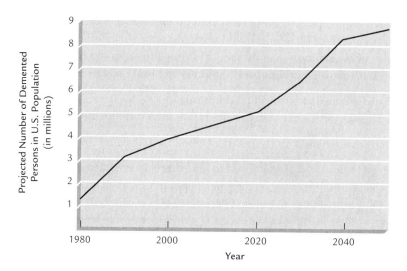

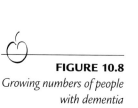

FIGURE 10.8

*Growing numbers of people
with dementia*

In 1907, a German doctor named Alois Alzheimer described in careful detail a puzzling case of severe dementia in one of his older patients. Today, Alzheimer's disease is now recognized as the cause of more than half of all cases of dementia in the United States. Incidence of the disease increases sharply with age: it is rare before age 50 but affects over one in five people over age 85 (Katzman 1986), (Schoenberg 1986).

Alzheimer's disease begins quietly. The first signs are often small lapses of memory such as forgetting appointments or names. People with the disease gradually become more disabled, beginning to lose their way in familiar surroundings and often developing personality changes.

As the disease progresses, people become increasingly confused, forget their own names, cannot recognize family members, and become bedridden. No one knows what causes Alzheimer's disease, but there is a strong hereditary factor: people with a parent with Alzheimer's disease have a much greater chance of developing it than people in the general population (Mohs et al. 1987).

Although there have been many claims that Alzheimer's disease is caused by malnutrition, or a vitamin deficiency, or by a toxin (such as aluminum) in the food or water supply, there is no firm evidence that special diets or megadoses of vitamins and minerals can help people with Alzheimer's disease. The cause and treatment of the disorder remains unknown.

Unlike Alzheimer's disease, other forms of dementia may be preventable or partially reversible through diet and lifestyle, including the second most common type of dementia—*multi-infarct dementia.* This type of dementia is the result of multiple, small cerebrovascular accidents, or strokes, each one damaging a small section of the brain. The strokes occur in an unpredictable, random pattern over months or years, and as more and more brain cells are damaged and lost, the person develops dementia. This usually occurs in people who are in their 60s or 70s and affects men and women equally (Gray 1989).

Because multi-infarct dementia is caused by small strokes, many scientists think that the same dietary changes that are recommended for prevention of stroke (discussed in Chapter 9) may help prevent this type of dementia (Hatchinski 1992). Dietary risk factors for high blood pressure and stroke include avoiding obesity, restricting sodium, increasing potassium and calcium intake, and avoiding excess alcohol. Following these guidelines may decrease the risk of developing multi-infarct dementia and may also slow the progression of the disease in affected individuals by preventing more strokes (Hatchinski 1992).

Not surprisingly, the widespread damage caused by chronic alcohol abuse includes injury to the brain. Chronic abuse of alcohol can cause severe dementia (Lishman 1987). Alcoholic dementia does not refer to the mental impairment of acute intoxication; it is the loss of memory and cognitive function that is present in many alcoholics—even when they are sober. The dementia can be severe, but it may slowly improve over a period of weeks to months if the person continues to abstain (Gray 1989). Scientists are not sure why alcoholism produces dementia. It may be a direct toxic effect of the alcohol on brain cells; or nutritional deficiencies induced by alcohol abuse may also play a role.

Severe deficiencies of niacin or thiamin can cause dementia. Although niacin deficiency is no longer a common cause of dementia in the United States, **pellagra** is still common in some developing countries of Africa and Southeast Asia. The major symptoms of niacin deficiency are the three d's of pellagra: dementia, **dermatitis**, and diarrhea. People with pellagral dementia improve rapidly when given niacin.

Severe thiamin deficiency causes a dementia with profound loss of memory. Although thiamin deficiency is an uncommon cause of dementia in the general population, alcoholics and the aged are at risk for thiamin deficiency (Victor 1983). Only one in four people with dementia caused by severe thiamin deficiency recover completely even when treated with thiamin (Gray 1989).

The most common vitamin-deficiency dementias are those caused by vitamin B_{12} and folate deficiencies (NIH Consensus Conference 1987). Older adults are most often affected by vitamin B_{12} deficiency because of the high incidence of atrophic gastritis in later life. Nearly 40% of people over 80 have atrophic gastritis (Krasinski et al. 1986), and many cannot absorb vitamin B_{12} well. Sometimes dementia can occur unaccompanied by the other usual signs of vitamin B_{12} deficiency, such as anemia and loss of sensation in the fingers and toes (Lindenbaum et al. 1988). Atrophic gastritis can also reduce folate absorption, and severe folate deficiency can cause dementia and memory loss. The dementias caused by folate or vitamin B_{12} deficiency are quickly and completely cured when the appropriate vitamin is given (Gray 1989).

pellagra: the niacin-deficiency disease characterized by skin eruptions, digestive system disorders, and eventual mental deterioration

dermatitis: inflammation of the skin

Nutrition and the Management of Dementia

Dementia has significant nutritional consequences. During the early, mild stages of dementia, memory and judgment are impaired, and the affected person may have difficulty shopping and preparing or cooking food. Programs such as Meals on Wheels (discussed later this chapter), which deliver precooked, nutritious meals can be helpful for people at this stage, and they may allow the person with mild dementia to stay at home (Loney et al. 1987). Food preferences may change as the dementia affects taste and smell, and often a preference for salty or sweet foods develops. If the dementia progresses, the person may lose the ability to use utensils, may no longer recognize food, and refuse to eat.

People with dementia, especially those with Alzheimer's disease, are often agitated and restless. This can increase energy requirements by as much as 1500 kcal/day (Rheaume et al. 1987). At the same time, people with Alzheimer's disease often eat poorly, lose weight, and are undernourished. In people who are institutionalized because of Alzheimer's disease, weight loss often exceeds ten pounds each year, and a third have clear signs of protein–calorie malnutrition (Sandman et al. 1987). Particularly for the aged with mild dementia, correction of malnutrition can maximize remaining functions and improve the quality of life. See Table 10.9.

"Feeding" the Brain

Although there is an inevitable and irreversible loss of neurons during the normal process of aging, researchers have found that many older adults are able to preserve mental functions well into their 80s and 90s. Along with optimal nutrition and a healthful lifestyle, cognitive training and regular mental exercises can help older adults maintain mental skills. Studies show that people in their 70s and 80s can continue to sharpen mental skills, improve spatial orientation and inductive reasoning, and increase their vocabulary.

THE IMMUNE SYSTEM

antibody: an immune protein produced in response to a foreign (or interpreted as foreign) substance or microorganism; they play important roles in immune defense

For many, immunological vigor diminishes with age, as the body's defense systems weaken. Production of **antibodies** is diminished, immune cells react weakly to foreign substances, and white blood cells engulf and destroy bacteria less efficiently (Chandra 1992). These changes make many older people more vulnerable to infection. However, not all older adults have weaker immune systems. Some have immune systems that function as well as those of younger adults, and nutritional status may be partly responsible for these differences.

Behavioral Problem	Interventions
Attention problems	Verbally direct through each step of the eating process Place utensils in hand Make food/fluids available and visible
Combative, throws food	Identify provocative agent, remove Feeder stands/sits on nondominant side Provide nonbreakable dishes with suction holders Give one food at a time Reward appropriate mealtime behavior
Chews constantly	Tell to stop chewing after each bite Serve soft foods to reduce the need to chew Offer small bites
Eats nonedible things	Remove nonedibles from reach Provide finger foods Provide edible centerpiece or table decorations
Eats too fast	Set utensils down between bites Offer food items separately Offer bulky foods that require chewing Use a smaller spoon/cup
Eats too slowly	Monitor eating pace, provide verbal cues: "chew," "take a bite" Serve first to allow more time Use insulated dishes to maintain proper temperatures
Forgetful/disoriented	Simple routines Constant environment Assigned seating Minimize distractions Limit choices
Forgets to swallow	Tell to swallow Feel for swallow before offering next bite Stroke upward on larynx
Plays in food	Serve one food at a time Fill glass/plate half full at refill Give finger foods Cups with covers/spout
Spits	Evaluate chewing/swallowing ability Tell not to spit Place away from others who would be offended Provide mealtime supervision

TABLE 10.9
Nutritional interventions and dementia

Adapted from D. Cohen, "Dementia, Depression and Nutritional Status," *Primary Care* 22, no.1 (1994):107–19.

Nutrition may be a critical determinant of immune competence in old age. We have seen that many older people eat diets that are inadequate in many nutrients. Several of these "problem nutrients" are known to be important in the proper functioning of the immune system. Deficiencies of protein, energy, zinc, and vitamins C, E, and B_6 can weaken the immune system (Chandra 1992), (Meydani 1993).

The link between nutrition and immune function was investigated in a recent study in Canada. One hundred healthy older adults were divided into two groups: one group was given a multivitamin–mineral supplement for one year; the other group received a placebo. The supplement contained 0.5 to 2 times the RDA for most of the essential micronutrients. After a year, the group that took the supplement had better immune function and fewer infections than the group that got the placebo. Many of the participants had micronutrient deficiencies that were corrected by the supplement. Remarkably, improvements occurred even in those who were not deficient in any micronutrients at the beginning of the study (Chandra 1992).

These results were supported by a recent placebo-controlled study in 56 healthy adults aged 60–85 years. Daily supplementation for 12 months with a multivitamin–mineral tablet containing doses of micronutrients at about the RDA level significantly improved the immune response (Bogden et al. 1994).

Supplementation with individual micronutrients has also been shown to benefit older adults. In healthy older adults, supplementation with zinc, vitamin B_6, or vitamin E at moderate doses has produced improvements in immune function (Chandra 1992), (Chandra 1984). However, megadoses of zinc and vitamin E have been shown to weaken the immune system (Prasad 1980).

Many questions remain to be answered. Can micronutrient supplements be of benefit in all older adults? Are they of benefit only in those adults who have dietary deficiencies? Can supplements be of benefit at doses greater than the RDAs? Future research should answer these questions. In the meantime, older adults should eat a healthy, balanced diet containing ample zinc, protein, and vitamins B_6, E, and C.

COORDINATION AND VISION

With age, many people develop a slight tremor, balance and coordination diminish, and reflexes slow. Falls are a major cause of morbidity in older adults, and decreasing muscle strength, arthritis, poor balance, and poor vision all play a role. These disabilities often reduce food choices and variety, and they can have a major impact on nutritional status.

Many older adults, for example, have difficulty driving or boarding public transportation. Carrying shopping bags can be a real challenge, especially if they purchase heavy items such as fresh juice or fruit, milk, or canned goods. Poor vision may make it impossible to read small print on labels or recipes; older adults may avoid preparing nourishing foods that require peeling, cutting, or slicing. Even cooking, with the problems of gas, tests for doneness, or lifting of heavy pots and pans can prove to be a problem.

Cataract and Macular Degeneration

A common cause of visual impairment in older adults is cataract, a leading cause of blindness worldwide. *Cataracts* are opaque regions in the normally transparent lens of the eye. The number of cataracts rises sharply with age: in the United States, while only 5% of people in their 50s have cataracts, nearly half of those over the age of 75 are affected (Kahn et al. 1977). Recently, research has linked cataract to environmental factors—including diet.

Photooxidative damage to proteins in the lens from sunlight exposure may contribute to the development of cataracts. Populations exposed to more sunshine have increased frequency of cataracts (Taylor 1989). Because certain dietary antioxidants can decrease oxidative damage in tissues, researchers are busy investigating whether antioxidant status may play a role in the development of cataracts.

Studies have shown that lower intakes of antioxidant nutrients (carotene, vitamins E and C, riboflavin) increase risk for cataracts (Jacques et al. 1988), (Jacques and Chylack 1991). In a recent prospective study of older women in Boston, those in the highest intakes of carotene, vitamins E, C, and riboflavin had a 40% lower risk of developing cataract, compared to those with low intakes. Risk of cataract was 50% lower in persons who ate spinach or other greens at least five times a week vs. those who ate such foods less than once a month (Hankinson et al. 1992).

Dietary patterns may also reduce the risk of developing *age-related macular degeneration* (AMD)—the other major cause of blindness among people older than 65. In AMD, vision is lost because a small area in the center of the retina, called the macula, deteriorates. Although the cause is unknown and there is no effective treatment, encouraging results from a recent study indicate that food choices can reduce the risk of AMD. In a study of 876 people aged 55 to 80 years, consumption of foods rich in certain carotenoids, particularly those from dark green, leafy vegetables decreased the risk of AMD. Compared to those individuals with lower intakes of dietary carotenoids, people with higher intakes cut their risk by nearly a half (Seddon et al. 1994).

Results from these studies and others suggest that proper food choices can help older adults maintain their visual health. As our population ages, the incidence of cataract and AMD are sharply increasing in the United States. Nutritional strategies to reduce or prevent development of these disorders could have a major impact on visual disability among older adults and reduce health-care costs substantially (Cataract Panel Report 1983).

NUTRITION AND THE SOCIOCULTURAL CONTEXT OF AGING

Aging and chronic disease can jeopardize the nutritional status of older people by changing the absorption, metabolism, or excretion of nutrients. The nutrition of many older people is also impaired by various social, economic, and psychological factors that limit the quantity or quality of food ingested in old age.

Throughout life, food and eating are social activities involving sharing and companionship. Retirement, change of residence, death of a spouse, and loss of friends or relatives isolate many older people (Walker and Beauchene 1991). For many, aging involves disengagement: a gradual withdrawal from social roles and social activity. Many older people, especially women, live alone: a third of women aged 65–74 and over half of women over 75 live alone (Greene et al. 1992). In adults over 60, eating alone is a risk factor for dietary inadequacy (White et al. 1991). Older adults who are alone often skip meals; many have lower intakes of nutrients such as zinc, vitamins C and D, thiamin, and protein (Horwath 1991), (Grotkowski and Sims 1978).

In the United States, the age group over 65 years has high rates of depression (Blazer et al. 1987). Depression can be precipitated by social isolation, retirement, or development of a disability; it can also be a side effect of medications taken by older adults. Depression commonly produces anorexia and loss of interest in food, and is a risk factor for poor nutritional status in older adults (White et al. 1991), (Darton-Hill 1992). Loneliness and depression may also cause some people to focus on food as one of their few remaining pleasures, and may gain excess weight.

Although most older adults live independently and are free of major illness, the risk of functional disability increases sharply with advancing age. Many older people have difficulty shopping because of lack of transportation and disability from illness. Older people often live in poorer sections of cities, and because of their limited mobility and fragility, fear going out into the neighborhood to shop. Many are forced to shop at convenience stores with high prices and limited food selection, as supermarkets move out to the suburbs.

Most older adults in the United States live close to their adult children and maintain close contact with family members. Families provide most of the long-term care needed by older people: 75–80% of disabled older adults living in the community rely solely on informal support by family and friends (Greene et al. 1992). Informal care provided by families—assistance with personal care, transportation, shopping, cooking, and eating—enables many disabled older adults to remain in the community and avoid institutionalization.

However, about 5% of those 65 years or older live in nursing homes (U.S. Senate 1990). Institutionalized people tend to be older, female, White, and have chronic disabilities. Surveys have found that up to a third of older people living in nursing homes in the United States are undernourished due to marginal dietary intake and chronic illness (Morley 1993).

Dietary variety often declines as people get older. Consumption of a varied diet is considered an effective way to ensure adequate nutrient intake (Horwath 1989); variety of diet has been associated with better health in the aged. Infrequent shopping trips can lead to lower consumption of milk, fresh fruit and vegetables, and other perishable products. Convenience foods, often eaten because they require less preparation and cooking, are usually high in salt and fat, and more costly.

Poverty is a major determinant of diet and food choices (Horwath 1991), (Davies 1990). Despite improvements over the past 25 years, older adults in the United States continue to have substantially lower median incomes than the median for the rest of the population. Over 12% of persons 65 and older are living at or below the official poverty line, and 25–37% have annual incomes below $10,000 (U.S. Senate 1990), (Briley 1989). See Figure 10.9.

Retirement often means a loss of buying power, and many adults are living on fixed incomes set when prices were lower. One in every six dollars spent by older adults goes toward food—a higher proportion than in younger groups. Poverty in older age groups leads to consumption of inexpensive, nutrient-empty foods and increases the risk of deficiencies of many nutrients (White et al. 1991), (Horwath 1991). In a large study of adults in the United States over 60, the most important determinant of food habits was income (Grotkowski and Sims 1978). Older adults most likely to be poor are single women, those older then 85, and minorities.

People over 60 generally have the lowest level of nutrition knowledge of any adult age group (Grotkowski and Sims 1978). Nutrition education comes most often from television, newspapers, and magazines. Many older adults think they are eating adequate diets, but are not. They may have eating patterns they have followed for 40–50 years and may not have kept up to date with the changes in food availability, processing, and preservation that have occurred during their lifetimes. Some older people have deeply ingrained ideas about foods, such as the beliefs that fresh fruits and vegetables have too much roughage and are bad for digestion, or that tomatoes are too acidic.

Source: Adapted from U. S.
Senate (1990)

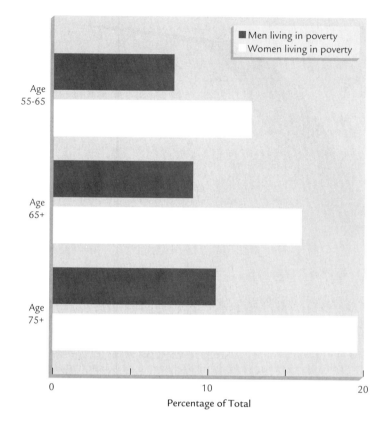

FIGURE 10.9

*Poverty rates among
Americans over 55*

Teaching older adults about proper food selection and cooking techniques can significantly improve their diets (Horwath 1991). Many older people are modifying their diets and eating healthier foods, substituting low-fat for high-fat milk, eating less high-fat meat and more poultry and fish, increasing fruit consumption, and substituting whole-grain for white bread (Popkin et al. 1992), (Koehler 1994). Older adults are also leading consumers of egg substitutes, bran, and decaffeinated coffee.

The major risk factors for malnutrition in older people include the following (White et al. 1991), (Horwath 1991), (Davies 1990):

- *Lifestyle*: lack of sunlight, multiple medications, and high alcohol consumption

- *Eating patterns*: irregularity of meals, lack of fruits and vegetables, lack of stored foods, throwing food away or letting foods spoil, rejection of foods

- *Socioeconomic*: older individuals (over 80 years old), living alone, reduced social contacts, depression, loneliness, or extended bereavement, low income or state of economic hardship, low expenditures for food and poor nutritional knowledge

Significant weight loss over time
5.0% or more of body weight in 1 mo
7.5% or more of body weight in 3 mo
10.0% or more of body weight in 6 mo or involuntary
weight loss of 10 lbs in 6 mo

Significantly low or high weight for height
20% above or below desirable for height at a given age

Significant reduction in serum protein
Serum albumin of less than 3.5 g/dl

Significant change in functional status

Significant and sustained inappropriate intake
Failure to consume the U.S. Dietary Guidelines recommended minimum from
one or more basic food groups, or inappropriate salt, sugar, fat, or alcohol
consumption, for a period of 3 months or more

Significant reduction in mid-arm circumference

Significant increase or decrease in skinfold thickness

Significant obesity
More than 120% desirable weight, or BMI over 27, or triceps skinfold above
95th percentile

Selected nutrition-related disorders
Osteoporosis
Folate deficiency
Vitamin B$_{12}$ deficiency

TABLE 10.10

Indicators of poor nutritional status in older adults

Adapted from *Indicators of Poor Nutritional Status on Older Americans* (Washington, D.C.: Nutrition Screening Initiative, 1991).

- *Physical*: loss of teeth or difficulty chewing, chronic disease, recent surgery on the gastrointestinal tract, immobility or physical disability impairing the ability to shop and prepare foods

Indicators of poor nutritional status in older adults are listed in Table 10.10. Family, friends, and health-care workers who take care of older adults should also be aware of the warning signs of poor nutritional status. A screening tool for assessing nutritional risk in older adults is shown in Table 10.11.

NUTRITIONAL HEALTH AND THE OLDER FEMALE

In general, women live longer than men, with life expectancy at birth for females in the United States 7.5 years longer than for males. Because fewer men than women survive to older ages, women are disproportion-

---

Determine Your Nutritional Health

The warning signs of poor nutritional health are often overlooked. Use this checklist to find out if you or someone you know is at nutritional risk.

Read the statements below. Circle the number in the yes column for those that apply to you or someone you know. For each yes answer, score the number in the box. Total your nutritional score.

	Yes
I have an illness or condition that made me change the kind and/or amount of food I eat.	2
I eat fewer than 2 meals per day.	3
I eat few fruits or vegetables, or milk products.	2
I have 3 or more drinks of beer, liquor, or wine almost every day.	2
I have teeth or mouth problems that make it hard for me to eat.	2
I don't always have enough money to buy the food I need.	4
I eat alone most of the time.	1
I take 3 or more different prescribed or over-the-counter drugs a day.	1
Without wanting to I have lost or gained 10 pounds in the last 6 months.	2
I am not always physically able to shop, cook, and/or feed myself.	2

Total _____

Total Your Nutritional Score. If it's...

0–2 Good! Recheck your nutritional score in 6 months.

3–5 You are at moderate nutritional risk. See what can be done to improve your eating habits and lifestyle. Your office on aging, senior nutrition program, senior citizen center, or health department can help. Recheck your nutritional score in 3 months.

6 or more You are at high nutritional risk. Bring this checklist the next time you see your doctor, dietitian, or other qualified health or social service professional. Talk with them about any problems you may have. Ask for help to improve your nutritional health.

These materials developed and distributed by the Nutrition Screening Initiative, a project of the American Academy of Family Physicians, The American Dietetic Association, and the National Council on the Aging, Inc.

Source: The Nutrition Screening Initiative, Washington, D. C.

TABLE 10.11

Screening for nutritional risk with older adults

ately represented in the older population. In the United States, the ratio of women to men rises steadily from near 1:1 at age 50 to over 2:1 by age 85 (U.S. Dept. of Commerce 1990).

The fact that the majority of people older than 65 years are women has important implications for nutritional health. Older women are at high risk of economic hardship. Fifteen percent of women over age 65 and 22% over 85 live below the poverty level (for men, the poverty rates are 8% and 14% for these age groups) (Greene et al. 1992). Over a third of women older than 65 depend almost entirely on Social Security benefits, and poverty is a major risk factor for undernutrition in older adults .

Older women are much more likely than men to be widowed and living alone. Of older women living in the community, 41% live alone (compared to 15% of men) (Greene et al. 1992). Moreover, older women are more likely to have impairments in functional capacity than men, and they are less likely to be able to afford outside assistance. Women are more likely to live a greater proportion of their lives with a disability and spend longer periods in institutions (Kinsella 1992). Social isolation, disability, and institutionalization strongly increase the likelihood of nutritional deficiencies in older women.

Although energy intake falls with advancing age in both sexes, women reduce their energy intake less than men. Obesity is more common among older women than men; in the United States, 37% of White women aged 65–74 are obese versus 25% of men (USDHHS 1987). At the same time, in the U.S. population older than 65, women are more likely to be undernourished than men. Among older women, being significantly underweight is a greater risk factor for mortality than is being overweight (Tayback et al. 1990).

Certain diseases occur more commonly in older women than men. Because of the marked gender differences in the prevalence of osteoporosis among older adults, hip fractures are twice as common in women than in men (Pierron et al. 1990). The importance of adequate dietary calcium and vitamin D in maintaining bone health is discussed earlier in this chapter. Although the current RDA for calcium for all adult women is set at 800 mg (NAS 1989), metabolic studies have shown that healthy women may need 1000 mg/day to maintain calcium balance before menopause, and up to 1500 mg/day after menopause (Consensus Development Conference 1993). Older women who cannot obtain at least the RDA for calcium from dietary sources should consider taking a calcium supplement at the RDA level.

There are important gender differences in lipid metabolism among older individuals. After the menopause, the risk for cardiovascular disease in women increases sharply. The loss of estrogens at menopause produces a steady fall in HDL-cholesterol (the protective cholesterol fraction in the blood) in women between 50 and 80 years of age, but a similar decrease does not occur in men (Morely 1993). At the same time, levels of cholesterol in the blood associated with the lowest overall mortality in older women are significantly higher than those for older men (Forette et al. 1989). To support HDL-cholesterol levels after the menopause, older women should avoid obesity and, if possible, participate in a regular exercise program.

NUTRITIONAL PROBLEMS OF OLDER MINORITIES IN THE UNITED STATES

As discussed earlier in this chapter, the population over 65 in the United States is expected to double between 1990 and 2050. At the same time, the number of Hispanic older adults will experience a sevenfold increase, and the number of older Blacks will triple. Thus, the proportion of Blacks and Hispanics in older age groups will increase substantially (U.S. Dept. of Commerce 1990).

Compared with the general population, older adults from minority groups in the United States have higher risks of insufficient dietary intake and nutrient deficiencies (Dwyer 1994). They are also more likely to suffer from undernutrition secondary to chronic disorders such as kidney and liver disease.

A number of factors contribute to differences in nutritional health in older minorities. Minority populations bear a disproportionate amount of the physical and social burdens of advancing age. Compared to the general population, they suffer from increased rates of many chronic diseases, such as hypertension and kidney failure in Blacks and diabetes in Hispanics. In addition, poverty strikes older minorities particularly hard. For example, Black women over 80 years of age are among the poorest groups in the United States (Dwyer 1994). In certain ethnic groups, such as Hispanics and Asian Americans, many older individuals migrated to the United States later in life and have retired without significant benefits from Social Security. Disabilities that interfere with daily living (such as walking, shopping, and cooking food) are more common among Hispanics than other older adult groups in the United States (Dwyer 1994).

Like other older adults in the United States, many older minorities are socially isolated and poor, and depend on friends, family, or the community for support. Language barriers and different food preferences and eating customs often limit support available from community nutritional programs. A major challenge for nutritional programs serving older people is to be responsive to cultural differences in diet and accommodate the needs of older minorities.

NUTRITIONAL SUPPLEMENTS

Understandably, many older people fear becoming ill and unable to care for themselves, and they worry about being able to afford the increasing cost of medical care. They are vulnerable to targeted advertising and unscrupulous salespeople who claim their vitamins and assorted supplements will prevent disease and promote longevity. Although many of these products are of doubtful value, the use of nutritional supplements by the elderly is widespread (McIntosh 1990), (Gray et al. 1986).

Most of these supplements are self-prescribed, using information from newspapers, magazines, and television. Few contain the nutrients most often lacking in the diets of elderly people, such as vitamin D, vitamin B6, folate, calcium, and zinc. They often contain micronutrients at levels many times the RDAs.

Although people who promote these products argue that the supplements are, at worst, harmless, they may not be harmless in older adults. Older adults metabolize and excrete many drugs and supplements more slowly, and they are more likely to experience side effects or toxicity. Vitamin A supplements are of particular concern because of the increased susceptibility of the aged to vitamin A toxicity (Russell and Suter 1993).

Vitamin or mineral supplements may, however, be appropriate for some older people. Many elderly people have decreased energy requirements; some require less than 1500 kcal/day. At such low levels of food intake, it is often difficult to obtain sufficient vitamins and minerals. A daily multivitamin–mineral supplement at levels near the RDA can be beneficial, together with menu planning that emphasizes balanced, nutrient-dense meals (Johnson and Kligman 1992).

DRUGS AND NUTRITIONAL HEALTH

Many elderly people suffer from heart disease, arthritis, high blood pressure, and other chronic diseases for which they take drugs daily for many years. Although the elderly make up only 12% of the general population, they consume 25% of all prescription drugs (Roe 1987). Most older people take at least one medication daily, and many take six or more each day.

Most common prescription and over-the-counter drugs have significant nutrient interactions, and the aged are particularly vulnerable to side effects from drugs. The liver and kidneys in older people, for example, are slower to metabolize and excrete drugs than in younger adults. Clinical testing of drugs is often done in younger adults, and the results do not always predict drug action in the elderly. Many of the aged have marginal underlying nutritional status and so are more susceptible to interactions between drugs and nutrients. Some drugs cause dizziness and light-headedness in older people, and this can diminish an older person's ability to shop, prepare, and cook food. Here are a few ways that drugs can affect nutritional status in the aged (Roe 1987):

anti-inflammatory drug: a drug that reduces or prevents inflammation

- Drugs can influence food intake by changing a person's appetite, causing nausea, or altering the way foods taste and smell. Commonly used diuretics can produce anorexia, while aspirin and other **anti-inflammatory drugs** such as ibuprofen can cause gastric irritation and decrease appetite. Digoxin, a drug commonly used for heart disease, can cause nausea and vomiting and decrease the ability to eat. On the other hand, certain antidepressants and antianxiety drugs increase appetite and can contribute to excess weight gain.

■ Drugs can decrease the absorption of nutrients by changing conditions in the gastrointestinal tract, blocking absorption by intestinal cells, or binding to the nutrient and making it impossible to absorb. Antacids and other drugs taken to reduce stomach acidity and treat ulcers raise the pH of the stomach and decrease the absorption of iron, calcium, folate, and vitamin B_{12}. Antiseizure drugs used for epilepsy decrease absorption of calcium by blocking uptake of the mineral by cells lining the intestine. Alcohol interferes with the intestinal cell's ability to bind and absorb thiamin. Many drugs, when taken with meals, bind nutrients in the stomach or intestine and block absorption. Mineral oil (used as a laxative) and some cholesterol-lowering drugs bind fat-soluble vitamins.

■ Drugs can alter nutrient metabolism in the body. They can prevent metabolic conversion of a vitamin to its active form, usually by interfering with a key enzyme. Isoniazid, a drug used to treat tuberculosis, blocks metabolism of vitamin B_6 to its active form. Several commonly-used antibiotics can interfere with folate metabolism.

■ Drugs can increase or decrease excretion of a nutrient. Aspirin can increase losses of iron from the body by damaging the mucosal cells lining the stomach, which causes blood loss leading to iron deficiency and anemia. Alcohol can also cause **gastritis** and increase iron losses. Tetracycline, an antibiotic, can increase excretion of vitamin C by the kidney. Many laxatives, particularly if overused, can cause significant losses of calcium, magnesium, and sodium in the feces. Probably the most common drug–nutrient interaction in the elderly is diuretic-induced mineral depletion. Diuretics used for the treatment of high blood pressure increase urinary excretion of water, sodium, potassium, or calcium.

How can harmful drug–nutrient interactions be prevented? Older adults (and all consumers) should read labels and the package inserts that come with drugs. Manufacturers are required to list any known adverse drug–food interactions. Pharmacists, dietitians, and physicians can also provide information. Eating well and maintaining good nutritional health can minimize problems from medications.

gastritis: inflammation of the lining of the stomach

NUTRITIONAL PROGRAMS FOR AGING ADULTS

Many community agencies and governments have tackled the problem of hunger and social isolation in older adults by developing nutrition assistance programs for the aged. Assistance can range from the family support of a ride to the grocery store, to community-run classes on cooking and menu planning, up to extensive meal programs funded by the government. In the United States, the Nutrition Program for Older Americans was established in the 1970s. It provides nutritious, low-cost meals to the elderly who need assistance. It also brings older people together for group meals that provide opportunities for socializing. There are two basic programs: Meals on Wheels (MOW) and the Congregate Meals Program.

MEALS ON WHEELS (MOW)

Staffed by volunteers, MOW provides home delivery of a hot meal three to seven days each week. A typical meal includes servings of meat or a meat alternative, vegetables or fruit, milk, bread and butter, and dessert. The meals are set up to provide at least one third of the RDA of all nutrients. MOW is designed for the elderly who cannot shop or prepare food, or who have inadequate cooking facilities at home. Many of the participants in MOW are confined to their homes because of a disability, or are convalescing from a recent illness.

The current emphasis on control of health-care costs and short hospital stays has produced a large increase in the demand for home-delivered meals. In the United Kingdom, flexible alternatives to MOW include programs that provide the homebound adult with small refrigerators, microwaves, or steamers. Prepackaged meals are then delivered in batches, and participants can choose and cook the meals according to their own schedules.

THE CONGREGATE MEALS PROGRAM

The Congregate Meals Program is designed to meet both the nutritional and social needs of the aged. Participants are transported to a local center for a nutritious meal in the company of other adults from the community. During the meetings, various social and recreational activities are available. The participants are also offered nutrition education on topics including the nutritional quality of foods, menu planning for single people, and the links between diet and health. In the United States, over 150 million congregate meals are served per year.

Both MOW and the Congregate Meals Program are available to all persons over 60 years old regardless of income. Priority and assistance is given to those who are economically and socially needy. These large programs, along with many smaller, community-based programs, help many older people avoid undernutrition and remain healthy and independent (Ponza and Wray 1990).

NUTRITION AND LIFESTYLE DURING OLDER ADULTHOOD: REVIEW AND SUMMARY

Although most studies linking dietary risk factors to long-term health have focused on younger adults, lifestyle and diet continue to influence quality of life and mortality even at older ages. Reducing risk factors for cardiovascular disease in older adults—for example, by controlling high blood pressure and quitting smoking—can substantially cut the risk of cardiovascular disease after age 70 (Bush 1991).

Dietary recommendations for reducing risks from hypertension and atherosclerosis continue to be appropriate for older adults (La Rosa 1994), (SHEP 1991), (Stone 1994). Similarly, although most of the evidence that regular exercise is beneficial in reducing overall mortality comes from studies in younger and middle-aged adults, recent studies have shown that the benefits of exercise are maintained in later life (Powell et al. 1987). Older men and women who are more active live longer. Even at ages greater than 75 years, regular exercise reduces mortality (Sherman 1994).

A HEALTHFUL EXERCISE PROGRAM FOR OLDER ADULTS

Prolonged inactivity and bed rest can hasten the development of age-related disorders of the cardiovascular, digestive, and musculoskeletal systems. A growing body of research is showing that the most effective way to minimize age-related degenerative changes and maximize functions during later life is to consume a balanced and complete diet and participate in regular aerobic and strength-building exercise (Blumenthal et al. 1991).

Older people should strive to stay active and maintain an ideal body weight. Regular exercise allows older adults to eat more and obtain enough protein and essential micronutrients and at the same time avoid weight gain. Physical activity can slow or reverse the loss of lean tissue and muscle mass typically associated with aging, enhance flexibility and balance, and provide healthful cardiovascular benefits.

It is never too late to begin exercising and increasing physical activity. Even people in their 80s and 90s can strengthen and rebuild muscle. Recent studies indicate that, with proper guidance, many older people can begin exercise programs, and most remain enthusiastic and committed to physical activity (Schmidt et al. 1990).

However, because cardiovascular, neurological, and musculoskeletal problems are more common in older adults, health status must be thoroughly checked by a physician before beginning to exercise. For the oldest old (80+ years) and for those who have exercise limitations because of

Components	Recommendations
Warm-up	Stretching, calisthenics, 10–15 minutes
Frequency	3–5 days per week
Intensity	60% to 90% of maximal heart rate. Low-to-moderate intensity activity of longer duration is recommended for the nonathletic adult.
Duration	20–60 minutes of continuous aerobic activity. Duration depends on intensity of the activity; lower intensity exercise should be done for longer periods.
Types of exercise	Activities that use large muscle groups, can be maintained continuously, and are rhythmic and aerobic, e.g. walking-hiking, running-jogging, bicycling, cross-country skiing, dancing, rope skipping, rowing, stair climbing, swimming, skating
Resistance training	Strength training at moderate intensity, sufficient to develop and maintain fat-free weight, should be an integral part of an adult fitness program. One set of 8–12 repetitions of 8–10 exercises that condition the major muscle groups on at least 2 days per week is the recommended minimum.
Cool-down	10–15 minutes

Adapted from D. T. Lowenthal et al., "Effects of Exercise on Age and Disease," *Southern Medical Journal*, 87, no.5 (1994):S5-12.

TABLE 10.12
Exercise guidelines for older adults

chronic illness, less traditional forms of physical activity, such as gardening and housework, can be helpful.

Aerobic exercise should be performed three to five times per week, with each session including about 10 minutes of warm-up stretches, 20–60 minutes of low to moderate intensity exercise, and 10 minutes of cooling down (American College of Sports Medicine 1991). Strength training for older adults should emphasize low resistance (lower weights) and higher repetitions. A third component of exercise programs for older people is stretching to increase flexibility. See Table 10.12.

A HEALTHFUL DIET FOR OLDER ADULTS

Eating right is a lifelong commitment. A healthful diet can help older people reduce risk of disease and maintain a high level of function. Although older people are often perceived as "set in their ways" and resistant to changing lifelong diet patterns, a recent survey found that many older adults are making healthful changes in their diets—eating less total and saturated fat and substituting low-fat milk products for whole-fat ones (Popkin et al. 1992), (Koehler 1994).

However, common dietary problems in older adults in the United States resemble those found in younger adults: total and saturated fat, refined carbohydrate, and salt intakes are often above recommended levels, while complex carbohydrate and fiber levels are below recommended levels (Horwath 1989). Intake of fruits, vegetables, and dairy products is often inadequate. The recommendations of the Food and Nutrition Board for a balanced diet low in fat and salt and high in grains, fruits, and vegetables are prudent for adults of any age (NRC 1989).

The following guidelines take into account the unique nutritional needs of older adults and can be used as a general guide to healthful eating in later life. Because energy needs in many older adults are only 1100–1400 kcal/day, careful selection of nutrient-dense foods is essential to maintain adequate daily intake of micronutrients. Rich dietary sources of vitamins B_6, C, D, and folate, and the minerals calcium, zinc, and magnesium should be emphasized.

- Older adults should keep a healthy interest in food. They should eat a variety of foods, try new dishes, and when possible, eat with company.

- Older individuals should find out about community resources that can help them eat better, including cooking or menu-planning classes, nutrition education, and meal programs.

- Lean red meat, poultry, fish, or legumes should be eaten regularly. Fat intake during meat preparation can be minimized by trimming visible fat, removing the skin from chicken, and avoiding frying.

- Dairy products should be consumed regularly (about 3–4 servings/ day). For older people who are unable to consume dairy products or are unable to obtain adequate calcium from alternate dietary sources, a calcium supplement should be prescribed by a dietitian or physician. Also, older people who can't drink milk and are out in the sun very little, should consider a vitamin D supplement (Russell and Suter 1993).

- Several servings of fruits and vegetables should be eaten each day, emphasizing those rich in vitamins A and C, beta-carotene, potassium, calcium, and fiber. They should be eaten with the peel when possible for the extra fiber.

- Several servings of grain products should be consumed every day. Whole grains (brown rice and whole-grain bread, cereal, and pasta) are preferable. Ample dietary fiber should be consumed each day to ensure a healthy and regular digestive system.

- Salt intake should be limited to 6 grams a day or less (that is equal to 2.5 grams of sodium) by using less in cooking and none at the table, and avoiding canned and processed foods and salty snacks (NRC 1989).

- Ample fluid intake is important to prevent dehydration. Six to eight glasses of water or juice should be consumed each day; even more if it is a hot or particularly active day.

- Good snacks are fresh or dried fruits, carrots, whole-grain or graham crackers, fig bars, low-fat oatmeal cookies with raisins, bran muffins, and yogurt. Older adults who are occasionally too busy or tired to cook a meal can eat a mixture of these easy-to-prepare snacks instead.

- Older people should generally avoid alcohol and drink moderately if they do drink. Have no more than 1 oz (two glasses of beer or wine, or two bar drinks) each day (NRC 1989).

- Older adults who take medicines regularly (even if they are common, over-the-counter medicines such as aspirin, antacids, or laxatives) should ask their pharmacists or physicians how the medicines interact with foods and nutrients.

- For most healthy older people who eat a regular, balanced diet, a multivitamin or mineral supplement is not necessary. If a supplement is taken, it should include vitamin D, the B vitamins, calcium, iron, magnesium, and zinc at amounts near the RDAs (Johnson and Kligman 1992). Avoid large doses of vitamins and minerals, and ignore the "anti-aging/life extension" supplements—buy some new walking shoes instead.

REFERENCES

Allen, S. H., "Primary Osteoporosis," *Postgrad Med* 93 (1993):43–56.

American College of Sports Medicine, *Guidelines for Exercise Testing and Prescription*, 4th ed. (Philadelphia: Lea and Febiger, 1991).

Arnaud, C. D., and Sanchez, S. D., "The Role of Calcium in Osteoporosis," *Annu Rev Nutr* 10 (1990):397–414.

Belloc, N. B. and Breslow, L., "Relationship of Physical Health Status and Health Practices," *Prev Med* 1 (1972):409–21.

Bikle, D. D., et al., "Bone Disease in Alcohol Abuse," *Ann Intern Med* 103 (1985):42–8.

Blazer, D., et al., "The Epidemiology of Depression in an Elderly Community Population," *Gerontologist* 27(1987):281.

Blumenthal, J. A. et al., "Long-Term Effects of Exercise on Physical Functioning in Older Men and Women," *J Gerontol* 46 (1991):353.

Bogden, J. D., et al., "Daily Micronutrient Supplements Enhance Delayed Hypersensitivity Skin Test Responses in Older People," *Am J Clin Nutr* 60 (1994):437–47.

Brenner, B. M., Meyer, T. W., and Hostetter, T. H., "Dietary Protein Intake and the Progressive Nature of Kidney Disease," *N Engl J Med* 307 (1982):652–59.

Briley, M. E., "Determinants of Food Choices of the Elderly," *J Nutr Elderly* 9 (1989):39–45.

Bullamore, J. R., et al., "Effects of Age on Calcium Absorption," *Lancet* (1990):535–7.

Bush, T. L., "The Epidemiology of Cardiovascular Disease in Older Persons," *Aging* 3 (1991):3–8.

Campbell, W. W., et al., "Increased Protein Requirements in Elderly People: New Data and Retrospective Reassessments," *Am J Clin Nutr* 60 (1994):501–09.

Cataract Panel Report, "Vision Research: A National Plan 1983–1987. Part 3, *NIH publication* no. 83-2473 (Washington D. C.: U. S. Department of Health and Human Services, 1983).

Chandra, R. K., "Effect of Vitamin and Trace-Element Supplementation on Immune Responses and Infectious in Elderly Subjects," *Lancet* 340 (1992):1124–7.

Chandra, R. K., "Nutrition and Immunity in the Elderly," *Nutr Rev* 50 (1992):367–71.

Chandra, R. K., "Nutritional Regulation of Immunity at the Extremes of Life: In Infants and in the Elderly," in *Malnutrition, Determinants and Consequences,* ed. P. White (New York: Alan Liss, 1984):245–8.

Chapuy, M. C., et al., "Vitamin D_3 and Calcium to Prevent Hip Fractures in Elderly Women," *N Engl J Med* 327 (1992):1637–42.

Chernoff, R., "Thirst and Fluid Requirements," *Nutr Rev* 52 (1994):S3–5.

Consensus Development Conference, "Diagnosis, Prophylaxis and Treatment of Osteoporosis," *Am J Med* 94 (1993):646–50.

Dandona, P., et al., "Low 1,25 Dihydroxyvitamin D, Secondary Hyper-parathyroidism, and Normal Osteocalcin in Elderly Subjects," *J Clin Endocrinol Metab* 71 (1990):1288–93.

Darlington, L. G., "Dietary Therapy for Arthritis," *Rheum Dis Clin North America* 17 (1991):273–85.

Darnton-Hill, I., "Psychosocial Aspects of Nutrition and Aging," *Nutr Rev* 50 (1992):476–79.

Davies, L., "Socioeconomic, Psychological and Educational Aspects of Nutrition in Old Age," *Age Aging* 19 (1990):S37–42.

Dawson-Hughes, B., et al., "A Controlled Trial of the Effect of Calcium Supplementation on Bone Density in Postmenopausal Women," *N Engl J Med* 323 (1990):878–83.

Dawson-Hughes, B., et al., "Effect of Vitamin D Supplementation on Wintertime and Overall Bone Loss in Healthy Postmenopausal Women," *Ann Int Med* 115 (1991):505–12.

DeGroot, C. P. G. M., Hautvast, J. G. A. J. and van Staveren, W. A., "Nutri-tion and Health of Elderly People in Europe: The EURONUT-SENECA Study," *Nutr Rev* 50 (1992):185–94.

Dwyer, J., "Nutritional Problems of Elderly Minorities," *Nutr Rev* 52 (1994):S24–27.

Eastell, R., et al., "Interrelationship among Vitamin D Metabolism, True Calcium Absorption, Parathyroid Function and Age in Women: Evidence of an Age-Related Intestinal Resistance to 1-25 Dihydroxyvitamin D Action," *J Bone Miner Res* 6 (1991):125–32.

Egsmose, C., et al., "Low Serum Levels of 25-Hydroxyvitamin D and 1,25-Dihydroxyvitamin D in Institutionalized Old People: Influence of Solar Exposure and Vitamin D Supplementation," *Age Aging* 16 (1987):35–40.

Euronut-SENECA Investigators, "Intake of Vitamins and Minerals," *Eur J Clin Nutr* 45, suppl. no. 3 (1991):121–38.

Euronut-SENECA Investigators, "Nutritional Status: Blood Vitamins A, E, B$_6$, B$_{12}$, Folic Acid and Carotene," *Eur J Clin Nutr* 45, suppl. no. 3 (1991):63–82.

Evans, W. J., "Exercise, Nutrition and Aging," *J Nutr* 122 (1992):796–801.

Fiatarone, M. A., et al., "Exercise Training and Nutritional Supplementation for Physical Frailty in Very Elderly People," *N Engl J Med* 330 (1994):1769–75.

Fiatarone, M. A., et al., "High-Intensity Strength Training in Nonagenarians: Effects on Skeletal Muscle," *J Am Med Assoc* 263 (1990):3029–34.

Flatz, G., "Genetics of Lactose Digestion in Humans," *Adv Hum Genetics* 16 (1987):1–77.

Forette, et al., "Cholesterol as a Risk Factor for Mortality in Aged Women," *Lancet* 1 (1989):868–70.

Friedland, R., *Facing the Costs of Long-Term Health Care* (Washington D. C., Employee Benefit Research Institute, 1990).

Frontera, W., et al., "A Cross-Sectional Study of Muscle Strength and Mass in 45-78 Year Old Men and Women," *J Appl Physiol* 71 (1991):644.

Gloth, F. M., et al., "Vitamin D Deficiency in Homebound Elderly Persons," *JAMA* 274 (1995):1683–86.

Gray, G. E., "Nutrition and Dementia," *J Am Diet Assoc* 89 (1989): 1795–1802.

Gray, G. E., et al., "Vitamin Supplement Use in a Southern California Retirement Community," *J Amer Diet Assoc* 86 (1986):800–02.

Greene, V. L., Monahan, D., and Coleman, P. D., "Demographics," in *Primary Care Geriatrics: A Case-Based Approach*, eds. R. J. Ham and P. D. Sloane (St. Louis: Mosby-Year Book, 1992):20–40.

Grotkowski, M. L., and Sims, L. S., "Nutritional Knowledge, Attitudes and Dietary Practices of the Elderly," *J Am Diet Assoc* 72(1978):499–506.

Hankinson, S. E., et al., "Nutrient Intake and Cataract Extraction in Women: A Prospective Study," *Br Med J* 305 (1992):335–9.

Harris, S. S., and Dawson-Hughes, B., "Caffeine and Bone Loss in Healthy Postmenopausal Women," *Am J Clin Nutr* 60 (1994):573–8.

Hatchinski, V., "Preventable Senility: A Call for Action Against the Vascular Dementias," *Lancet* 340 (1992):645–8.

Heaney, R. P., "Nutritional Factors in Osteoporosis," *Annu Rev Nutr* 13 (1993):287–316.

Heaney, R. P., et al., "Calcium Nutrition and Bone Health in the Elderly," *Am J Clin Nutr* 36 (1982):986–1013.

Hernandez-Avila, M., et al., "Caffeine, Moderate Alcohol Intake, and Risk of Fractures of the Hip and Forearm in Middle-Aged Women," *Am J Clin Nutr* 54 (1991):157–63.

Heymnfield, S. B., et al., "Body Composition in Elderly Subjects: A Critical Appraisal of Clinical Methodology," *Am J Clin Nutr* 50 (1989): 1167.

Horwath, C. C., "Dietary Intake Studies in Elderly People," *World Review of Nutrition and Dietetics* 59 (1989):1–70.

Horwath, C. C., "Nutrition Goals for Older Adults: A Review," *Gerontologist* 31 (1991):811–21.

Horwath, C. C., "Socioeconomic and Behavioral Effects on the Habits of Elderly People," *Int J Biosocial Med Res* 11 (1989):15–30.

Huang, L. F., Cartwright, W. S., and Hu, T. W., "The Economic Cost of Senile Dementia in the United States, 1985," *Public Health Rep* 103 (1988):3.

Ireland, P., and Fordtran, J. S., "Effect of Dietary Calcium and Age on Jejunal Calcium Absorption in Humans Studied by Intestinal Perfusion," *J Clin Invest* 52 (1973):2672–81.

Jacques, P. F., and Chylack, L. T., "Epidemiologic Evidence of a Role for the Antioxidant Vitamins and Carotenoids in Cataract Prevention," *Am J Clin Nutr* 53 (1991):325–55.

Jacques, P. F., et al., "Nutritional Status in Persons with and without Senile Cataract: Blood Vitamin and Mineral Levels," *Am J Clin Nutr* 48 (1988):152–8.

James, W. P. T., Ralph, A., and Ferro-Luzzi, A., "Energy Needs of the Elderly: A New Approach," in *Nutrition, Aging and the Elderly*, ed., Munro, H. H., Danforth, D. E. (New York: Plenum, 1989).

Johnson, K., and Kligman, E. W., "Preventive Nutrition: An Optimal Diet for Older Adults," *Geriatrics* 47 (1992):56–60.

Johnston, C. C., and Slemenda, C. W., "Changes in Skeletal Tissue during the Aging Process," *Nutr Rev* 50 (1992):385–87.

Kagawa, Y., "Impact of Westernization on the Nutrition of the Japanese: Changes in Physique, Cancer, Longevity and Centenarians," *Prev Med* 7 (1978):205–17.

Kahn, H. A., et al., "The Framingham Eye Study: Outline and Major Prevalence Findings. *Am J Epidemiol* 106 (1977):17–326.

Kaltenborn, K. C., "Perspectives on Osteoporosis," *Clin Obstet Gynecol* 35 (1992):901–12.

Katzman, R., "Alzheimer's Disease," *N Engl J Med* 314 (1986):964.

Keenan, J. M., and Morris, D. H., "How to Make Sure Your Older Patients Are Getting Enough Zinc," *Geriatrics* 48 (1993):57–65.

Kiel, D. P., et al., "Caffeine and the Risk of Hip Fracture: The Framingham Study," *Am J Epidemiol* 132 (1990):675–84.

Kinsella, K., "Changes in Life Expectancy 1900-1990," *Am J Clin Nutr* 55 (1992):1196S–202S.

Kjeldsen-Kragh, J., et al., "Controlled Trial of Fasting and One-Year Vegetarian Diet in Rheumatoid Arthritis," *Lancet* 338 (1991):899–902.

Koehler, K. M., "The New Mexico Aging Process Study," *Nutr Rev* 52 (1994):S34–37.

Koehler, K. M., and Garry, P. J., "Nutrition and Aging," in Labbe, R. F., ed., *Laboratory Utilization for Nutritional Support. Clinical Laboratory Medicine,* Vol. 13 (Philadelphia: W. B. Saunders 1993).

Krasinski, S. D., et al., "Fundic Atrophic Gastritis in an Elderly Population," *J Am Geriatr Soc* 34 (1986):800–6.

Krasinski, S. D., et al., "Postprandial Plasma Retinyl Ester Response is Greater in Older Subjects Compared with Younger Subjects," *J Clin Invest* 85 (1990):883–92.

Krasinski, S. D., et al., "Relationship of Vitamin A and Vitamin E Intake to Fasting Plasma Retinol, Retinolbinding Protein, Retinyl Esters, Carotene, α-Tocopherol, and Cholesterol among Elderly People and Young Adults: Increased Plasma Retinyl Esters among Vitamin A-Supplement Users," *Am J Clin Nutr* 49 (1989):112–20.

LaCroix, A. Z., et al., "Maintaining Mobility in Late Life. II. Smoking. Alcohol Consumption, Physical Activity and Body Mass Index," *Am J Epidemiol* 137, no. 8 (1993):858–69.

Lakatta, E. G., "Alterations in Circulatory Function," in *Principles of Geriatric Medicine and Gerontology* 3d ed., edited by W. R. Hazzard et al. (New York: McGraw-Hill, 1994).

LaRosa, J. C., "Dyslipoproteinemia in Women and the Elderly," *Med Clin North America* 78 (1994):163.

LifeSciences Research Office, FASEB, "Nutrition Monitoring in the U. S.: An Update Report on Nutrition Monitoring: DHSS Pub. No 89 1255 (Hyattsville, MD: USDHHS, USDA, Sept. 1989).

Lindeman, R. D., Tobin, J., and Shock, N. W., "Longitudinal Studies on the Rate of Decline in Renal Function with Age," *J Am Geriatr Soc* 33 (1985):278–85.

Lindenbaum, J., et al., "Neuropsychiatric Disorders Caused by Cobalamin Deficiency in the Absence of Anemia or Macrocytosis," *N Engl J Med* 318 (1988):1720.

Lindenbaum, J., et al., "Prevalence of Cobalamin Deficiency in the Framingham Elderly Population," *Am J Clin Nutr* 60 (1994):2–11.

Lindholm, J., et al., "Bone Disorder in Men with Chronic Alcoholism: A Reversible Disease?" *J Clin Endocrinol Metab* 73, no.1 (1991):118–24.

Lishman, W. A., *Organic Psychiatry*, 2d ed. (Boston: Blackwell Scientific, 1987).

Loney, L. E., et al., "Nutritional Concerns for Patients with Alzheimer's Disease," *Texas Med* 8(1987):40.

Lowenstein, F. W., "Nutritional Status of the Elderly in the United States of America," *J Am Coll Nutr* 1 (1982):165.

Lowik, M. R. H., Schrijver, J., and van den Berg, H., "Nutrition and Aging: Nutritional Status of 'Apparently Healthy' Elderly," *J Am Coll Nutr* 9 (1989):18–27.

Matkovic, V., et al., "Factors that Influence Peak Bone Mass Formation: A Study of Calcium Balance and the Inheritance of Bone Mass in Adolescent Females," *Am J Clin Nutr* 52 (1990):878–88.

McCay, C. M., Crowell, M. F., and Maynard, L. A., "The Effect of Retarded Growth upon the Length of Life Span and upon the Ultimate Body Size," *J Nutr* 10 (1935):63–79.

McIntosh, W. A., "The Relationship Between Beliefs about Nutrition and Dietary Practices of the Elderly," *J Amer Diet Assoc* 90 (1990):671–75.

Mertz, et al., "Trace Elements in the Elderly: Metabolism, Requirements and Recommendations for Intakes," in *Nutrition, Aging and the Elderly*, eds., Munro, H. N., Danforth, D. E. (New York; Plenum, 1989).

Meydani, S. M., "Vitamin/Mineral Supplementation, the Aging Immune Response and Risk of Infection," *Nutr Rev* 51 (1993):106–15.

Meydani, S. N., et al., "Vitamin B_6 Deficiency Impairs Interleukin 2 Production and Lymphocyte Proliferation in Elderly Adults," *Am J Clin Nutr* 53 (1991):1275–80.

Miller, R. A. "The Biology of Aging and Longevity," in *Principles of Geriatric Medicine and Gerontology*, W. R. Hazzard et al. (New York: McGraw-Hill, 1994):3–19.

Mohs, R. C., et al., "Alzheimer's Disease: Morbid Risk Among First Degree Relatives Approximates 50% by 90 Years of Age," *Arch Gen Psychiatry* 44 (1987):405.

Morley, J. E., "Nutrition and the Older Female," *J Am Coll Nutr* 12, no. 4 (1993):337–43.

National Academy of Sciences, Food and Nutrition Board of the National Research Council, *Recommended Dietary Allowances*, 10th ed. (Washington, D. C.: National Academy Press, 1989).

National Research Council, Food and Nutrition Board, "Diet and Health: Implications for Reducing Chronic Disease Risk. Report of the Committee on Diet and Health" (Washington D. C.: National Academy Press, 1989).

Nelson, R. C. and Franzi, L. R., "Nutrition," in *Primary Care Geriatrics*, (St. Louis: Mosby-Year Book, 1992):162–93.

Niewoehner, C. B., "Osteoporosis in Men," *Postgrad Med* 93(1993):59–70.

NIH Consensus Conference, "Differential Diagnosis of Dementing Diseases," *JAMA* 258 (1987):3411.

Paffenbarger, R. S., et al., "Physical Activity, All-Cause Mortality, and Longevity of College Alumni," *N Eng J Med* 314 (1986):605–11.

Panush, R. S., "Does Food Cause or Cure Arthritis?" *Rheum Dis Clin North America* 17 (1991):259–72.

Pfeil, L. A., Katz, P. R., and Davis, P. J., "Water Metabolism," in *Geriatric Nutrition* eds. J. E. Morley, Z. Glick, and L. Z. Rubenstein, (New York: Raven Press, 1990):193–202.

Pierron, R. L., et al., "The Aging Hip," *J Am Geriatr Soc* 38 (1990): 1339–52.

Ponza, M. and Wray, L., *Final Results of the Elderly Programs Study. Evaluation of the Food Assistance Needs of the Low-Income Elderly and Their Participation in USDA Programs* (Alexandria, VA: U. S. Department of Agriculture, Food and Nutrition Service, 1990).

Popkin, B. M., Haines, P. S., and Patterson, R. E., "Dietary Changes in Older Americans, 1977-1987," *Am J Clin Nutr* 55 (1992):823–30.

Powell, et al., "Physical Activity and the Incidence of Coronary Heart Disease," *Ann Rev Public Health* 8 (1987):253–87.

Prasad, J. S., "Effect of Vitamin E Supplementation on Leukocyte Function," *Am J Clin Nutr* 33 (1980):606–8.

Raisz, L. G., Smith, J., "Osteoporosis," in *Clinical Endocrinology of Calcium Metabolism,* eds. Martin, T. J., Raisz, L. G. (New York: Marcel Dekker, 1987).

Rasmussen, H. M., et al., "Serum Concentrations of Retinol and Retinyl Esters in Adults in Response to Mixed Vitamin A and Carotenoid Containing Meals," *J Am Coll Nutr* 10(1991):460–5.

Recker, R. R., and Heaney, R. P., "Calcium Nutrition and its Relationship to Bone Health," in *Nutrition, Aging and the Elderly,* eds., H. N. Munro, and D. E. Danford (New York: Plenum, 1989):183–93.

Reid, I. R., et al., "Effect of Calcium Supplementation on Bone Loss in Postmenopausal Women." *N Engl J Med* 328 (1993):460–4.

Rheaume, Y., Riley, M. E., and Volcier, L., "Meeting Nutritional Needs of Alzheimer Patients Who Pace Constantly," *J Nutr Elderly* 7 (1987):43.

Ribaya-Mercado, J. D., et al., "Vitamin B_6 Requirements of Elderly Men and Women," *J Nutr* 121 (1991):1062–74.

Roe, D. A., "Drugs and Nutrition in the Elderly," in *Geriatric Nutrition,* 2d ed. (Englewood Cliffs, New Jersey: Prentice-Hall, 1987).

Roe, D. A., "Nutrition Services," in *Geriatric Nutrition,* 2d ed. (Englewood Cliffs, NJ: Prentice Hall, 1987).

Rolls, B. J., "Aging and Appetite," *Nutr Rev* 50 (1992):422–26.

Russell, R. M. and Suter, P. M., "Vitamin Requirements of Elderly People: An Update," *Am J Clin Nutr* 58 (1993):4–14.

Russell, R. M., "Changes in the Gastrointestinal Tract Attributed to Aging," *Am J Clin Nutr* 55 (1992):1203S–7S.

Sandler, R. B., et al., "Postmenopausal Bone Density and Milk Consumption in Childhood and Adolescence," *Am J Clin Nutr* 40 (1985):270–4.

Sandman, P. O., et al., "Nutritional Status and Dietary Intake in Institutionalized Patients with Alzheimer's Disease and Multi-infarct Dementia," *J Am Geriatr Soc* 35 (1987):31.

Schiffman, S., "Changes in Taste and Smell: Drug Interactions and Food Preferences," *Nutr Rev* 52 (1994):S11–14.

Schmidt, R. M., et al., "Evaluating Outcomes of Healthy Aging Interventions: The Health Watch or Arizona Study in the Sun Cities," *Gerontologist* 30 (1990):8.

Schoenberg, B. S., "Epidemiology of Alzheimer's Disease and Other Dementing Illness," *J Chron Dis* 39 (1986):1095.

Seddon, J. M., et al., "Dietary Carotenoids, Vitamins A,C, and E and Advanced Age-Related Macular Degeneration," *JAMA* 272 (1994):1413–20.

Senti, S. R. and Pilch, S. M, "Analysis of Folate Data from NHANES II," *J Nutr* 115 (1985):1398–402.

SHEP Cooperative Research Group, "Prevention of Stroke by Antihypertensive Drug Treatment in Older Persons with Isolated Systolic Hypertension. Final Results of the Systolic Hypertension in the Elderly Program," *JAMA* 265 (1991):3255–64.

Sherman, S. E., et al., "Does Exercise Reduce Mortality Rates in the Elderly? Experience from the Framingham Heart Study," *Am Heart J* 128 (1994):965–72.

Sloane, P. D., "Normal Aging," in *Primary Care Geriatrics: A Case-Based Approach*, eds. R. J. Ham and P. D. Sloane (St. Louis: Mosby-Year Book, 1992):20–40.

Stone, N. J., "The 75 Year-Old Patient with Hypercholesterolemia: To Treat or Not to Treat?" *Nutr Rev* 52 (1994):S31-–33.

Suter, P. M., et al., "Reversal of Protein-Bound Vitamin B_{12} Malabsorption with Antibiotics in Atrophic Gastritis," *Gastroenterolgy* 101 (1991):1039–45.

Tayback, M., Kumanyika, S., and Chee, E., "Body Weight as a Risk Factor in the Elderly," *Arch Intern Med* 150 (1990):1065–72.

Taylor, A., "Associations Between Nutrition and Cataract," *Nutr Rev* 47 (1989): 225–34.

Turnland, J. R., et al., "Stable Isotope Studies of Zinc Absorption and Retention in Young and Elderly Men," *J Nutr* 116 (1986):1239–47.

U. S Department of Health and Human Services, Healthy People 2000, "National Health Promotion and Disease Prevention Objectives," (Washington D. C.: U. S. Government Printing Office, 1990).

U. S. Department of Commerce, Bureau of the Census, "Projections of the Population of the U. S .by Age, Sex and Race: 1988–2080," *Current Population Reports, Population Estimates and Projections* serial 101-J (Washington D. C.: U. S. Government Printing Office, 1990).

U. S. Department of Health and Human Services, "Anthropometric Reference Data and Prevalence of Overweight: United States 1976–80," *Vital and Health Statistics*, series 11, no. 238. DHHS publication no.

PHS 87-1688. (Washington D. C.: US Government Printing Office, 1987).

U. S. Department of Health and Human Services, Public Health Service, National Center for Health Statistics, "Use of Dental Services and Dental Health," *U. S. Vital and Health Statistics* series 10, no.165, (Washington, D. C.: U. S. Government Printing Office, 1986).

U. S. Senate, "Aging Americans: Trends and Projections," Serial 101-J (Washington D. C.: U. S. Government Printing Office, 1990).

U. S. Surgeon General's Office, *The Surgeon General's Report on Nutrition and Health* (Washington D. C.: U. S. Government Printing Office, 1988):465–89.

Velthius-te Wierik, E. J. M., et al., "Energy Restriction, a Useful Intervention to Retard Human Aging? Results of a Feasibility Study," *Eur J Clin Nutr* 48 (1994):138–48.

Victor, M., "Mental Disorders Due to Alcoholism," in *Handbook of Psychiatry, Vol. 2: Mental Disorders and Somatic Illness,* ed. M. H. Lader (New York: Cambridge University Press, 1983).

Walker, D., and Beauchene, R. E, "The Relationship of Loneliness, Social Isolation and Physical Health to Dietary Adequacy of Independently Living Elderly," *J Am Diet Assoc* 91 (1991):300–04.

Weiler, P. G., "The Public Health Impact of Alzheimer's Disease," *Am J Public Health* 77(1987):1169.

Weindruch, R., "Caloric Restriction and Aging," *Sci American* 274 (1996): 32-8.

Wellman, N. S., "The Nutrition Screening Initiative," *Nutr Rev* 52 (1994):S44–47.

White, J. V., et al., "Consensus of the Nutrition Screening Initiative: Risk Factors and Indicators of Poor Nutritional Status among Older Americans," *J Am Diet Assoc* 91 (1991):783–87.

Wood, R. J., "Mineral Needs of the Elderly: Developing a Research Agenda for the 1990s," *Age* 14 (1991):120–28.

Zoli, M., et al., "Portal Blood Velocity and Flow in Aging Man," *Gerontology* 35 (1989):61–5.

appendix

▼

Recommended Dietary Allowances[a]**. Revised 1989.**

Category	Age (yrs) or Condition	Weight[b] (kg)	(lb)	Height[b] (cm)	(in)	Protein (g)	Fat-Soluble Vitamins Vitamin A (µg RE)	Vitamin D (µg)	Vitamin E (mg α-TE)	Vitamin K (µg)
Infants	0.0–0.5	6	13	60	24	13	375	7.5	3	5
	0.5–1.0	9	20	71	28	14	375	10	4	10
Children	1–3	13	29	90	35	16	400	10	6	15
	4–6	20	44	112	44	24	500	10	7	20
	7–10	28	62	132	52	28	700	10	7	30
Males	11–14	45	99	157	62	45	1000	10	10	45
	15–18	66	145	176	69	59	1000	10	10	65
	19–24	72	160	177	70	58	1000	10	10	70
	25–50	79	174	176	70	63	1000	5	10	80
	51+	77	170	1 73	68	63	1000	5	10	80
Female	11–14	46	101	157	62	46	800	10	8	45
	15–18	55	120	163	64	44	800	10	8	55
	19–24	58	128	164	65	46	800	10	8	60
	25–50	63	138	163	64	50	800	5	8	65
	51+	65	143	160	63	50	800	5	8	65
Pregnant						60	800	10	10	65
Lactating	1st 6 months					65	1300	10	12	65
	2nd 6 months					62	1200	10	11	65

The allowances, expressed as average daily intakes over time, are intended to provide for individual variations among most normal persons as they live in the United States under usual environmental stresses. Diets should be based on a variety of common foods in order to provide other nutrients for which human requirements have been less well defined.

[b] Weights and heights are actual medians for the U.S. population of the designated age. The use of these figures does not imply that the height-to-weight ratios are ideal.

TABLE A.1

Designed for the maintenance of good nutrition of practically all healthy people i the United States.

TABLE A.1 *Continued*

Category	Age (yrs) or Condition	Water-Soluble Vitamins							Minerals						
		Vitamin C (mg)	Thiamin (mg)	Riboflavin (mg)	Niacin (mg NE)	Vitamin B_6 (mg)	Folate (µg)	Vitamin B_{12} (µg)	Calcium (mg)	Phosphorus (mg)	Magnesium (mg)	Iron (mg)	Zinc (mg)	Iodine (µg)	Selenium (µg)
Infants	0.0–0.5	30	0.3	0.4	5	0.3	25	0.3	400	300	40	6	5	40	10
		35	0.4	0.5	6	0.6	35	0.5	600	500	60	10	5	50	15
Children	1–3	40	0.7	0.8	9	0	50	0.7	800	800	80	10	10	70	20
	4–6	45	0.9	1.1	12	1.1	75	1.0	800	800	120	10	10	90	20
	7–10	45	1.0	1.2	13	1.4	100	1.4	800	800	170	10	10	120	30
Males		50	1.3	1.5	17	1.7	150	2.0	1200	1200	270	12	15	150	40
		60	1.5	1.8	20	2.0	200	2.0	200	1200	400	12	15	150	50
		60	1.5	1.7	19	2.0	200	2.0	1200	1200	350	10	15	150	70
		60	1.5	1.7	19	2.0	200	2.0	800	800	350	10	15	150	70
		60	1.2	1.4	15	2.0	200	2.0	800	800	350	10	15	150	70
Females		50	1.1	1.3	15	1.4	150	2.0	1200	1200	280	15	12	150	45
		60	1.1	1.3	15	1.5	180	2.0	1200	1200	300	15	12	150	50
		60	1.1	1.3	15	1.6	180	2.0	1200	1200	280	15	12	150	55
		60	1.1	1.3	15	1.6	180	2.0	800	800	280	15	12	150	55
		60	1.0	1.2	13	1.6	80	2.0	800	800	280	10	12	150	55
Pregnant		70	1.5	1.6	17	2.2	400	2.2	1200	1200	320	30	15	175	65
Lactating															
1st 6 months		95	1.6	1.8	20	2.1	280	2.6	1200	1200	355	15	19	200	75
2nd 6 months		90	1.6	1.7	20	2.1	260	2.6	1200	1200	340	15	16	200	75

'Retinol equivalents.
^{d}As cholecalciferol.
^{e} α-Tocopherol equivalents.
^{f} Niacin equivalents.

Adapted from National Research Council, National Academy of Sciences, *Recommended Dietary Allowances,* 10th ed. (Washington D. C., National Academy of Sciences, 1989).

Estimated Safe and Adequate Daily Dietary Intakes of Selected Vitamins and Minerals

Category	Age (years)	Vitamins	
		Biotin (µg)	Pantothenic Acid (mg)
Infants	0–0.5	10	2
	0.5–1	15	3
Children and adolescents	1–3	20	3
	4–6	25	3–4
	7–10	30	4–5
	11+	30–100	4–7
Adults		30–100	4–7

Category	Age (years)	Trace Elements[b]				
		Copper (mg)	Manganese (mg)	Fluoride (mg)	Chromium (ug)	Molybdenum (ug)
Infants	0–0.5	0.4–0.6	0.3–0.6	0.1–0.5	10–10	15–30
	0.5–1	0.6–0.7	0.6–1.0	0.2–1.0	20–60	20–40
Children and adolescents	1–3	0.7–1.0	1.0–1.5	0.5–1.5	20–80	25–50
	4–6	1.0–1.5	1.5–2.0	1.0–2.5	30–120	30–75
	7–10	1.0–2.0	2.0–3.0	1.5–2.5	50–200	50–150
	11+	1.5–2.5	2.0–5.0	1.5–2.5	50–200	75–250
Adults		1.5–3.0	2.0–5.0	1.5–4.0	50–200	75–250

[a]Because there is less information on which to base allowances, these figures are not given in the main table of RDA and are provided here in the form of ranges of recommended intakes. [b]Since the toxic levels for many trace elements may be only several times usual intakes, the upper levels for the trace elements given in this table should not be habitually exceeded.

Estimated Sodium, Chloride, and Potassium Minimum Requirements for Healthy Persons[a]

Age	Weight (kg)[a]	Sodium (mg)[a, b]	Chloride (mg)[a, b]	Potassium (mg)[c]
Months				
0–5	4.5	120	180	500
6–11	8.9	200	300	700
Years				
1	11.0	225	350	1,000
2–5	16.0	300	500	1,400
6–9	25.0	400	600	1,600
10–18	50.0	500	750	2,000
>18[d]	70.0	500	750	2,000

[a] No allowance has been included for large, prolonged losses from the skin through sweat. [b]There is no evidence tha higher intakes confer any health benefit. [c]Desirable intakes of potassium may considerably exceed these values (–3,500 mg for adults). [d]No allowance included for growth, pregnancy, or lactation.

TABLE A.2

Metric and Common U.S. Equivalents

Metric Unit	U.S. Equivalent
Mass and Weight	
1 microgram (µg)	0.00000004 ounce (oz)
1 milligram (mg)	0.00004 ounce
1 gram (g)	0.04 ounce
28.35 grams	1 ounce
l kilogram (kg)	2.2 pounds (lb)
0.454 kilogram	1 pound

Liquid Capacity	
1 milliliter (ml)	0.035 fluid ounce
29.6 milliliters	1 fluid ounce
1 liter (l)	1.06 quarts (qt)
0.946 liter	1 fluid quart

Heat	
1 kilojoule	0.239 kilocalorie (kcal)
4.18 kilojoules	1 kilocalorie

U.S. Household Measurements and Equivalents

l quart = 4 cups (c)

l cup = 8 fluid ounces = 16 tablespoons (T)

1T = 3 teaspoons (t)

1t = 5 grams dry weight

TABLE A.3
*Measurements Used in
Nutrition*

index
▼